Petrophysical Properties of Crystalline Rocks

Geological Society Special Publications
Society Book Editors

Special Publication reviewing procedures

The Society makes every effort to ensure that the scientific and production quality of its books matches that of its journals. Since 1997, all book proposals have been refereed by specialist reviewers as well as by the Society's Books Editorial Committee. If the referees identify weaknesses in the proposal, these must be addressed before the proposal is accepted.

Once the book is accepted, the Society has a team of Book Editors (listed above) who ensure that the volume editors follow strict guidelines on refereeing and quality control. We insist that individual papers can only be accepted after satisfactory review by two independent referees. The questions on the review forms are similar to those for *Journal of the Geological Society*. The referees' forms and comments must be available to the Society's Book Editors on request.

Although many of the books result from meetings, the editors are expected to commission papers that were not presented at the meeting to ensure that the book provides a balanced coverage of the subject. Being accepted for presentation at the meeting does not guarantee inclusion in the book.

Geological Society Special Publications are included in the ISI Index of Scientific Book Contents, but they do not have an impact factor, the latter being applicable only to journals.

More information about submitting a proposal and producing a Special Publication can be found on the Society's web site: www.geolsoc.org.uk.

GEOLOGICAL SOCIETY SPECIAL PUBLICATION NO. 240

Petrophysical Properties of Crystalline Rocks

EDITED BY

P. K. HARVEY and T. S. BREWER
University of Leicester, UK

P. A. PEZARD
Université de Montpellier II, France

and

V. A. PETROV
IGEM, Russian Academy of Sciences, Russia

2005
Published by
The Geological Society
London

THE GEOLOGICAL SOCIETY

The Geological Society of London (GSL) was founded in 1807. It is the oldest national geological society in the world and the largest in Europe. It was incorporated under Royal Charter in 1825 and is Registered Charity 210161.

The Society is the UK national learned and professional society for geology with a worldwide Fellowship (FGS) of 9000. The Society has the power to confer Chartered status on suitably qualified Fellows, and about 2000 of the Fellowship carry the title (CGeol). Chartered Geologists may also obtain the equivalent European title, European Geologist (EurGeol). One fifth of the Society's fellowship resides outside the UK. To find out more about the Society, log on to www.geolsoc.org.uk.

The Geological Society Publishing House (Bath, UK) produces the Society's international journals and books, and acts as European distributor for selected publications of the American Association of Petroleum Geologists (AAPG), the American Geological Institute (AGI), the Indonesian Petroleum Association (IPA), the Geological Society of America (GSA), the Society for Sedimentary Geology (SEPM) and the Geologists' Association (GA). Joint marketing agreements ensure that GSL Fellows may purchase these societies' publications at a discount. The Society's online bookshop (accessible from www.geolsoc.org.uk) offers secure book purchasing with your credit or debit card.

To find out about joining the Society and benefiting from substantial discounts on publications of GSL and other societies worldwide, consult www.geolsoc.org.uk, or contact the Fellowship Department at: The Geological Society, Burlington House, Piccadilly, London W1J 0BG: Tel. +44 (0)20 7434 9944; Fax +44 (0)20 7439 8975; E-mail: enquiries@geolsoc.org.uk.

For information about the Society's meetings, consult *Events* on www.geolsoc.org.uk. To find out more about the Society's Corporate Affiliates Scheme, write to enquiries@geolsoc.org.uk.

Published by The Geological Society from:
The Geological Society Publishing House
Unit 7, Brassmill Enterprise Centre
Brassmill Lane
Bath BA1 3JN, UK

(*Orders*: Tel. +44 (0)1225 445046
      Fax +44 (0)1225 442836)
Online bookshop: http://geolsoc.org.uk/bookshop

British Library Cataloguing in Publication Data

A catalogue record for this book is available from the British Library.

ISBN 1-86239-173-4

Typeset by Techset Composition, Salisbury, UK
Printed by Cromwell Press, Trowbridge, UK

Distributors

USA
  AAPG Bookstore
  PO Box 979
  Tulsa
  OK 74101-0979
  USA
Orders: Tel. +1 918 584-2555
      Fax +1 918 560-2652
      E-mail bookstore@aapg.org

India
  Affiliated East-West Press PVT Ltd
  G-1/16 Ansari Road, Darya Ganj,
  New Delhi 110 002
  India
Orders: Tel. +91 11 2327-9113/2326-4180
      Fax +91 11 2326-0538
      E-mail affiliat@vsnl.com

Japan
  Kanda Book Trading Company
  Cityhouse Tama 204
  Tsurumaki 1-3-10
  Tama-shi, Tokyo 206-0034
  Japan
Orders: Tel. +81 (0)423 57-7650
      Fax +81 (0)423 57-7651
      E-mail geokanda@ma.kcom.ne.jp

Contents

Preface

Petrophysics is a term synonymous with reservoir engineering in the hydrocarbon industry. However, a significant number of boreholes have been and continue to be drilled into crystalline rocks in order to evaluate the suitability of such rock volumes for a variety of applications, including nuclear waste disposal, urban and industrial waste disposal, geothermal energy, hydrology, sequestration of greenhouse gases and fault analysis.

Crystalline rocks cover a spectrum of igneous, metamorphic rocks and some sedimentary rocks where recrystallization processes have been important in their formation. These occur in a range of continental and oceanic settings. Oceanic crystalline basement has been extensively studied as part of the Deep Sea Drilling Program (1968–1980) and, the Ocean Drilling Program (1980–2003), and will continue as an important area of study. On the continents, crystalline rocks have been drilled as part of a very large number of scientific and environmentally driven programmes.

This volume is the result of the meeting sponsored by the Borehole Research Group of the Geological Society of London. In this volume, a spectrum of activities relating to the petrophysics of crystalline rocks are covered, which fall into the following categories:

(1) Fracturing and deformation of igneous, sedimentary and metamorphic rocks: papers by **Sausse & Genter**, **Giese et al.**, **Zimmermann et al.**, **Ito & Kiguchi**, **Goldberg & Burgdorff**, **Lovell et al.**, **Luthi et al.** and **Petrov et al.**

(2) Oceanic basement: **Haggas et al.**, **Einaudi et al.**, **Iturrino et al.** and **Brewer et al.**

(3) Permeability and hydrological problems: **Bartels et al.** and **Zharikov et al.**

(4) Laboratory-based measurements and the application of petrophysical parameters: **Haimson & Chang**, **Lloyd & Kendall**, **Harvey & Brewer**, **Bartetzko et al.**, **Meju**, **Kulenkampf et al.**, **Přikryl et al.** and **Oila et al.**

The editors are particularly grateful to Janette Thompson, both for organization of the conference and for persistence in coaxing authors, reviewers and editors, and also to Angharad Hills for continuous support in the production of this volume. We also thank all those who undertook the often arduous job of reviewing the manuscripts, and without whose help this volume would have been much poorer.

Peter K. Harvey
Tim S. Brewer
Phillipe A. Pezard
Vladislav A. Petrov

From: HARVEY, P. K., BREWER, T. S., PEZARD, P. A. & PETROV, V. A. (eds) 2005. *Petrophysical Properties of Crystalline Rocks*. Geological Society, London, Special Publications, **240**, vii.
0305-8719/05/$15.00 © The Geological Society of London 2005.

Types of permeable fractures in granite

J. SAUSSE[1] & A. GENTER[2]

[1]UMR 7566, *Géologie et Gestion des Ressources Minérales et Energétiques,*
UHP Nancy 1, BP 239, F-54506 Vandoeuvre Cedex, France
(e-mail: judith.sausse@g2r.uhp-nancy.fr)
[2]*BRGM CDG/ENE, BP 6009, 45060 Orléans Cedex 2, France*

Abstract: This study presents a multidisciplinary approach to understanding and describing types of fracture permeability in the Soultz-sous-Forêts granite, Upper Rhine Graben. At Soultz, during the 1993 stimulation tests in the GPK1 well, it was shown that only a limited number of natural fractures contributed to flow, whereas there are thousands of fractures embedded within the massive granite. In order to understand the flow hierarchy, a detailed comparison between static (fracture apertures based on ARI raw curves) and dynamic data (hydraulic tests) was carried out. We propose that two scales of fracture networks are present: a highly connected network consisting of fractures with small apertures that may represent the far-field reservoir, and another network that contains isolated and wide permeable fractures (that produce an anisotropic permeability in the rock) and allows a hydraulic connection between the injection and production wells.

Quantification and modelling of fluid flow in fractured rocks are extensively studied to solve and predict numerous economic or environmental problems (hydrothermal activity, geothermy, waste storage, etc.). Natural discontinuities such as fractures and cracks are primary potential paths for fluid circulation in crystalline rocks, and thus they have a major impact on the hydraulic properties of rock masses. Percolation in fractured media is a complex phenomenon that depends on the specific geological field context. The main problem in modelling flow in such systems is the frequent and real discrepancy between field observations and models of flow, due to the quality and quantity of the data available.

Permeability calculations deal with a quantitative definition of the fracture apertures. Three main types of aperture are described in the literature: hydraulic, mechanical or geometrical aperture types (Fig. 1).

An ideal fracture is usually defined as two smooth and parallel planes separated by a constant hydraulic aperture (Lamb 1957; Parsons 1966; Snow 1965, 1968a,b, 1969; Louis 1969; Oda 1986). This approach is generally used for regular fracture networks with smooth and widely open fractures. In this case, the calculated fracture aperture is maximal and corresponds to global conductivities controlled by the cubic law. However, this approach cannot take into account the channelling phenomenon described

in natural rough fractures, because fractures have surface asperities and contact points or voids within their walls (Gentier 1986; Gentier *et al.* 1996, 1998; Sausse 2002). Cracks or fractures are heterogeneously percolated by fluids, as is evidenced in Figure 2a, where flow is seen to leave the fracture over short segments of its trace. The main consequence is that the flow field, as well as the resulting fluid–rock interactions and fracture fillings, cannot be realistically predicted without a precise description of the geometry of the fracture walls (Fig. 2a & b).

Natural fractures are complex objects with different surface properties and types of alteration.

These facts strongly influence our conceptual approaches to modelling of fluid flow between fracture walls. Previous work (André *et al.* 2001) shows that low fracture roughness tends to lead to homogeneous flows even at great depth where pre-existing fractures are nearly closed. In the case of a laminar flow, the channelling flow is poorly developed, and the classical models of smooth parallel plates are probably relatively well adapted to determine the real permeability of these fractures. In contrast, fractures embedded in unaltered rocks can have high roughness and very heterogeneous aperture distributions. Their closure results in the formation of well-defined channels which do not cover the whole fracture surface. In this case,

From: HARVEY, P. K., BREWER, T. S., PEZARD, P. A. & PETROV, V. A. (eds) 2005. *Petrophysical Properties of Crystalline Rocks.* Geological Society, London, Special Publications, **240**, 1–14.

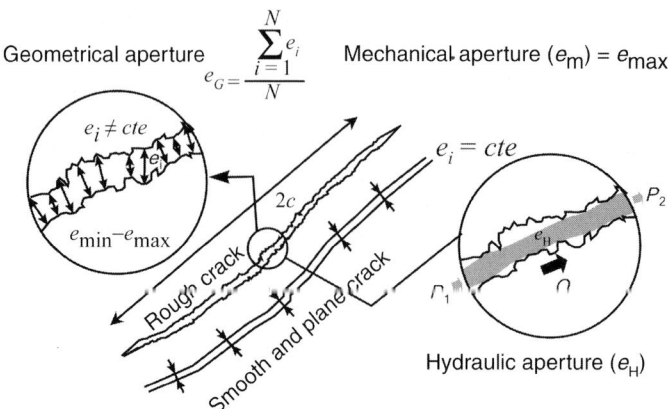

Geometrical aperture $\quad e_G = \dfrac{\sum_{i=1}^{N} e_i}{N} \quad$ Mechanical aperture $(e_m) = e_{max}$

Hydraulic aperture (e_H)

Fig. 1. A smooth and parallel plate model crack, compared to a rough crack. In the case of rough cracks, the local apertures e_i are different. The mean aperture or geometrical aperture is the quadratic or arithmetic mean of the local apertures e_i. The mechanical aperture is usually defined as the maximum value of e_i, and the hydraulic aperture estimation is the result of experimental hydraulic tests. When the differential pressures, the flow and the type of fluid flow (Poiseuille, for example) are defined, the value of e_H is determined. Usually, $e_M < e_G < e_H$ (Gentier 1986).

the cubic law does not adequately describe the hydraulic property of the fracture, and the hydraulic laws have to take roughness into account. Sausse's (2002) results suggest that the alteration phenomena can represent a key factor to characterize the roughness types of fractures. This in turn requires consideration of mineralogical and geochemical factors, in

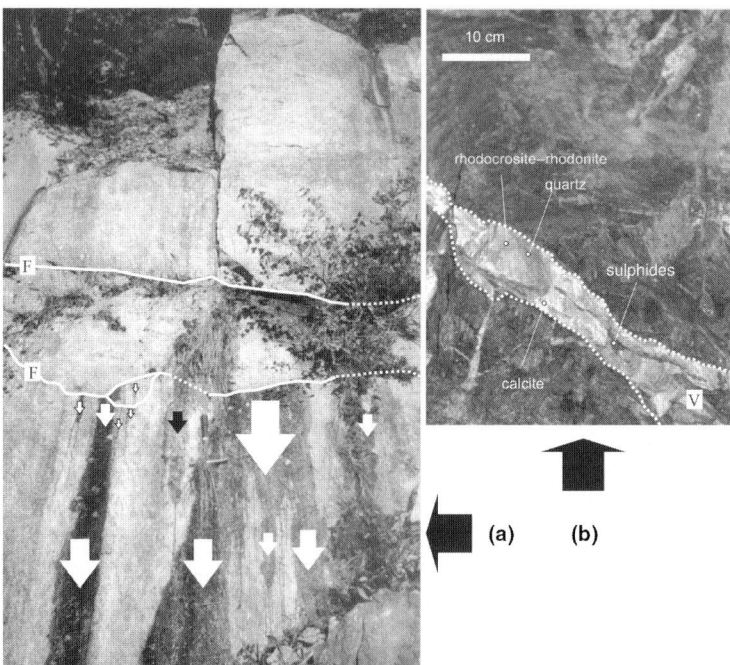

Fig. 2. Two examples showing (**a**) the complexity of fluid flows at the fracture scale, where directional and independent fluxes are going out of the fracture plane (arrows), and (**b**) the related fluid–rock interactions resulting in the multiple precipitation of secondary minerals within cracks.

order to perform more accurate permeability calculations and models of fluid circulation in fracture networks. Thus, different alterations and their intensity may imply different hydraulic laws for fractures.

The aim of this study is to propose a multidisciplinary approach to understand and describe fluid-flow pathways observed in fractures and fracture networks, based on the study of the petrophysical properties of rock and fractures. The rock mass in question is the granite basement of the Rhine Graben near Soultz-sous-Forêts (Bas-Rhin, France) where the 'Enhanced Geothermal System' (EGS) deep geothermal test site is located. This work presents a preliminary interpretation of the complex flow profile of a well, during hydraulic tests conducted during the period 1993–1994, and relates this to the electrical apertures of the fractures from logs, the rock alteration, and the fractures' spatial organization.

Geological context

Soultz-sous-Forêts, located in the Upper Rhine Graben, hosts one of the few deep geothermal 'Enhanced Geothermal Site' test sites in the world. The Palaeozoic granitic basement, is a batholith covered by a thick Tertiary succession (marls and clays) and Triassic sandstones (Fig. 3).

The Soultz granite is a Hercynian monzogranite characterized by phenocrysts of alkali feldspar in a matrix of quartz, plagioclase, biotite and minor amphibole. In its current state of development, the EGS system consists of three boreholes: GPK1 and GPK2, which extend respectively to 3600 m and 5000 m, and a reference hole EPS1 which has been fully cored (Fig. 4). This paper is concerned with observations in GPK1 (an open hole between 2850 and 3600 m) made during and following major hydraulic injections conducted in 1993–1994, before well GPK2 was drilled.

The Soultz boreholes are located inside the graben, 5 km from its western border represented by the main Rhine Fault oriented N030–040°. The geological cross-section, based on old oil-drilling data and seismic work, gives the main relationship between the basement-surface geometry and the normal-fault network (Fig. 4).

A large structural and petrographic database has been collected for GPK1 based on various logging images and cutting analysis between the top of the granite (1400 m) and 3600 m (Genter *et al.* 1997). The extensive logging of GPK1 throughout the open-hole depth range of 2850 and 3610 m gives an opportunity to study the structural organization of the fractures and alteration of the granite.

The granite was strongly altered by successive hydrothermal events (veins and pervasive alterations). As a consequence, the 2998 natural fractures present in the EPS1 well are nearly

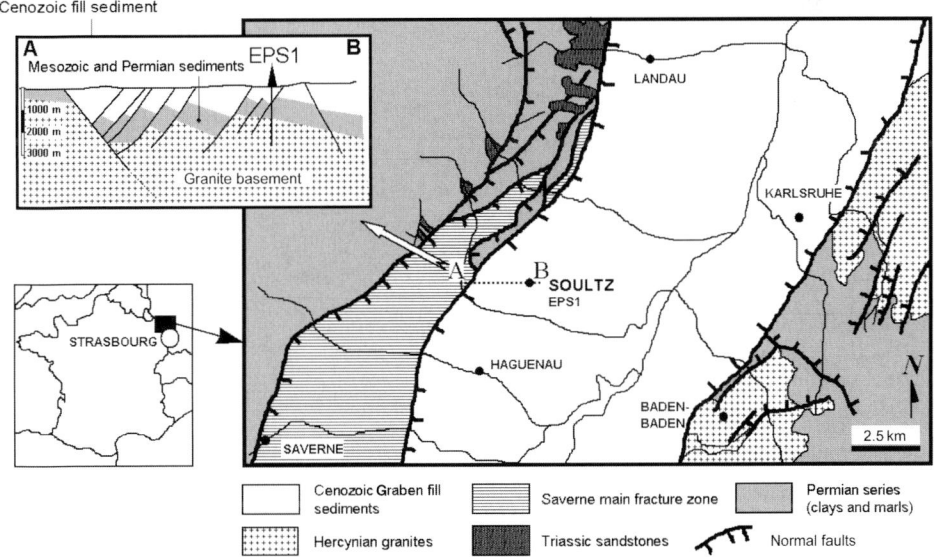

Fig. 3. A schematic geological map of the Rhine Graben and the location of the geothermal drill site of Soultz-sous-Forêts. Vertical section AB: details of cross-section (after Dezayes *et al.* 1995).

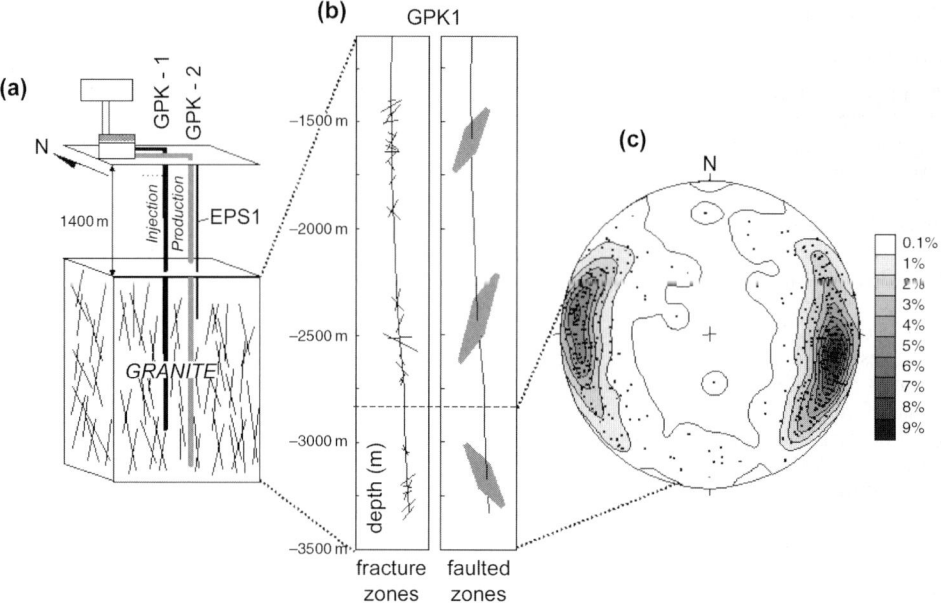

Fig. 4. (**a**) Sketch presentation of the geothermal exchanger at Soultz-sous-Forêts. GPK1 and GPK2 correspond respectively to the two injection and production wells which sample the deep granitic fractured basement down to 5000 m for GPK2, and to 3590 m for GPK1. EPS1 is the cored reference hole. (**b**) and (**c**) Schmidt projection of fracture poles – lower hemisphere. The structural interpretation of GPK1 shows that fractured zones are concentrated in three main intervals (1800, 2800 and 3500 m). The interpretation of UBI logs shows that a nearly vertical conjugated fracture set is oriented NNE–SSW with predominantly westward-dipping fractures (after Genter *et al.* 1995).

systematically sealed by hydrothermal products (29 of them are still opened today). Three distinct alteration types observed on cores were related to the precipitation of the three mineral assemblages of quartz–illite, calcite–chlorite and hematite fill the fracture networks and are related to different palaeo-percolation stages in the granite around EPS1 (Sausse 1998; Sausse *et al.* 1998). Two fractured and altered sections in well GPK1 at depths of 1820 m and 3495 m produced hot salt brines during drilling. This present-day permeability seems to be closely related to open fractures that are partly sealed by late geodic quartz deposits and characterized by extensive wallrock illitization (Genter & Traineau 1992). Anomalies in gases such as methane, helium, radon and carbon dioxide were also recorded during the drilling-mud survey when well GPK1 penetrated fractured and altered granitic sections (Vuataz *et al.* 1990; Aquilina & Brach 1995).

The complex hierarchy and chronology of the fluid palaeo-percolations detected in the Soultz granite could engender a complex hydraulic response during the hydraulic experiments.

The stimulation tests done in GPK1 at the end of 1993 were performed to validate the 'Soultz concept', i.e. to force the water to migrate through a connected fracture system in the basement rock to carry heat for power production. This consists of initially injecting water to great depths under high pressure, in order to establish efficient connections between the deep wells through the natural fracture system embedded within the basement rocks. The pressure is then adjusted in order to force water to migrate between the wells through the natural fracture system (please refer to *http://www.soultz.net* for more details). These experiments were continuously monitored, and different types of data were acquired (microseismicity, flow, spinner and temperature logs, etc.). In this work, the interpretation of fracture permeability during the hydraulic tests is based on the studies of Evans *et al.* (1996) and Evans (2000). These hydraulic data are correlated to the geometry of fractures, and especially to the fracture electrical apertures defined by Henriksen (2000, 2001) on the basis of electrical and acoustic borehole image logs.

Soultz log data

Structural data

The major fracture zones encountered in GPK1 were located through examination of borehole image logs, classical geophysical well-logs, and cutting samples (Genter *et al.* 1995). The schematic vertical west–east cross-section through GPK-1 in Figure 4b, shows that the fractured zones are not randomly distributed with depth, but rather concentrated in three main intervals centred at approximately 1800, 2800 and 3500 m depth. These clusters are interpreted as the traces of megascopic faults, with individual fractured and altered sections representing segments of normal faults. Each one contains at least one permeable section. Their orientation is consistent with normal slip during Oligocene Rhine rifting. The orientation characteristics of all fractures imaged on the UBI logs are shown in Figure 4c. Most of the fractures appear to be members of a nearly vertical conjugated fracture set with a symmetry axis striking NNE–SSW. Structural analysis of EPS1 core shows two types of small-scale fractures filled by hydrothermal products: Mode 1 fractures that show no evidence of shear movement, and Mode 2 fractures which have clearly suffered shearing. The Mode 1 fractures seen in the core are relatively narrow, and thus would be more difficult to detect on borehole image logs than the comparatively wide and sometimes visibly open Mode 2 fractures. Mode 1 fractures are more numerous, in a ratio of 1. Mode 1 fractures are generally related to weak extended fractures with thin apertures, whereas Mode 2 fractures are wide open and therefore easily monitored on electrical images. At Soultz, Mode 2 fractures are clearly Mode 1 fractures which were reactivated by tectonics.

Aperture data

Fracture geometrical properties and their spatial relationships were analysed using direct and indirect data. Fracture aperture data fall into the following categories:

- geometrical, from visual inspection of cores;
- hydraulic, derived from pressure data obtained using flow and temperature logs;
- mechanical, from laboratory tests on cores;
- electrical, from FMI electrical image logs, i.e. Formation MicroScanner (FMS), Fullbore Formation MicroImager (FMI) and Azimuthal Resistivity Imager (ARI).
- acoustic reflectivity, from acoustic image logs, i.e. Ultrasonic Borehole Imager (UBI), BoreHole TeleViewer (BHTV), etc.

Henriksen (2001) analysed a collection of electrical- and acoustic-borehole imaging logs from GPK1 (i.e. FMI, UBI, and ARI) to establish the hierarchy of the near-well fractures in the well between 2850 m and 3505 m depth. On ARI images, the main conductive fractures correspond to large sinusoids traceable across 100% of the image, whereas some fractures are more discontinuous on the trace where only a few per cent of the fracture-plane area produce an electrical response. The qualitative analysis of fractures done for ARI, UBI and FMI images uses the most homogeneous fractures, i.e. fractures where at least 50% of the fracture plane area can be followed continuously on images (Fig. 5b).

In a second step, Henriksen proposed the quantification of the electrical apertures produced by the main ARI fractures. High-resolution imaging tools provide detailed mapping of fractures on the borehole wall. The highly conductive drilling fluid used at Soultz is salty water characterized by a mud weight of $1.070 \, \text{g cm}^{-3}$ and a mud resistivity of 0.106 ohm m measured on 6 December 1992, that filled the open fractures intersected by the well. Moreover, the formation fluid observed in the granite corresponds to brines (Pauwels *et al.* 1993; Dubois *et al.* 1996). Electrical tools measure the contrast between the fluid and the formation resistivity. It is therefore possible to correlate the intensity of the conductive anomaly recorded by the tool as it passes the fracture, with the quantity of fluid within the fracture. Several empirical methods have been developed to estimate the apertures and extension of natural fractures from their conductivity signatures (Sibbit & Faivre 1985; Luthi & Souhaite 1990; Faivre 1993).

Henriksen (2001) estimated the electrical aperture of the fractures in GPK1 using three different methods: ARI conductivity curves (Faivre 1993); LLS and LLD curves of the Dual Latero Log (Sibbit & Faivre 1985); and FMI conductivity curves (Luthi & Souhaite 1990). As an example, ARI analysis only computes the lower limit of the fracture apertures. First, ARI data are reprocessed for aperture calculation. Then, the area of added conductivity (AAC) is computed by restricting the excess conductance between the raw conductivity curve and the background conductivity level to about $0.4 \, \text{mmho m}^{-1}$. Then, the Faivre (1993) formula is used to estimate the fracture aperture based on ARI images:

$$E = a \times \text{AAC}^b \times R_t^c \times R_m^{(1-c)}$$

where E is the fracture aperture, AAC (ohm m) is the area of added conductivity, R_t (ohm m) is the

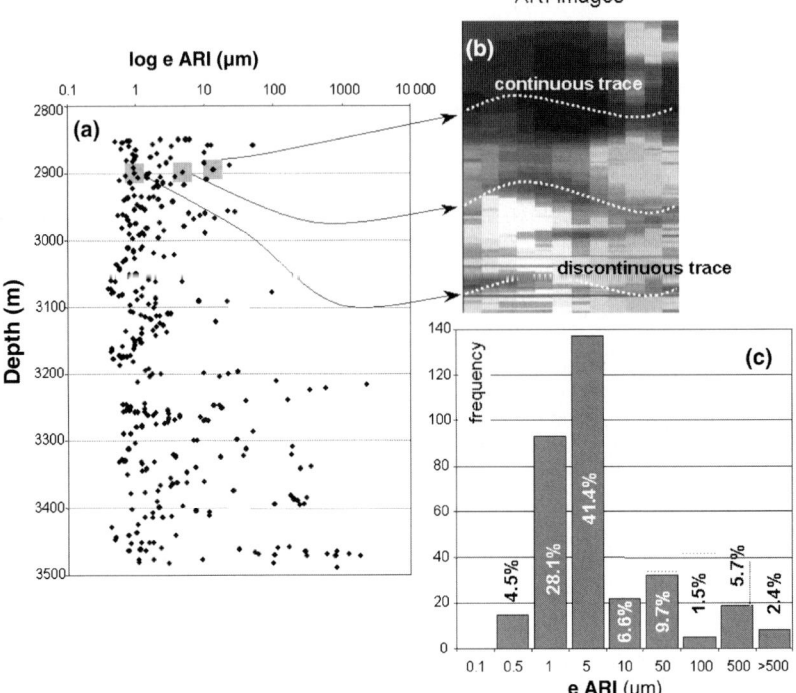

Fig. 5. Log of electrical fracture aperture sizes, labelled e_{ARI} versus depth, (**a**) derived from ARI logs, and (**b**) in well GPK1 (Henriksen 2001). Apertures are calculated based on the interpretation of ARI imageries, and for the example of three main conductive fractures (medium-grey squares). The frequency distribution (**c**) of the electrical apertures is large and shows a modal value of 2.5 μm. Some 80% of the 347 natural fractures are characterized by thin electrical apertures (lower than 10 μm).

matrix resistivity, R_m is the mud resistivity (ohm m) and a (0.9952), b (0.863) and c (0.0048) are all constants. Apertures from the ARI method may reflect the average for a larger penetration depth than with the method using FMI images, and may not be affected by vugs unless the vugs are connected with open fractures forming a conductive network of fluid flow in the reservoir (Henriksen 2000).

The results were compared with the physical apertures measured on EPS1 cores by Genter & Traineau (1996), and with the apertures estimated using the ARI field-print logs (Genter & Genoux-Lubain 1994).

The resulting calculated electrical apertures are shown in Figure 5a, and give a reliable hierarchy between natural fractures detected in GPK1 (Henriksen 2001). The ARI tool was run shortly after the drilling of GPK1. Consequently, the electrical apertures correspond to prestimulation fracture sizes. The fluid conductivity of the borehole mud was 0.1 ohm m.

The highest fracture aperture values are located at 3200 m and 3500 m depths, and correspond to two major permeable zones. One of the major fracture zones at around 3400 m in the well is properly identified by the aperture estimations (Fig. 5b). There is a large distribution of fracture apertures characterized by a modal aperture of 2.5 μm (Fig. 5c). However, 80% of the 347 natural fractures analysed are characterized by thin electrical apertures smaller than 10 μm. These values of electrical apertures do not represent the real geometrical apertures of the fractures, but the assumption was made that there is a correlation between the two types of apertures. Large electrical fractures are probably large opened fractures. The high electrical conductivity anomalies correspond to thick and generally composite fractures, which probably extend some considerable distance from the borehole wall (Henriksen 2001). These natural conductive fractures are compared in this study, with the hydraulic response given by the natural and newly opened fractures during the 1993 hydraulic tests.

Hydraulic data

After the deepening of the GPK1 well in 1992 to 3600 m (with the casing shoe set at 2850 m), large-scale hydraulic tests were carried out in 1993 to first characterize the natural permeability of the rock mass, and then to enhance the permeability of the natural fracture system through massive fluid injections (Jung *et al.* 1995). Supporting activities during the injections included: microseismic monitoring, fluid sampling, and frequent spinner and temperature logs (Baria *et al.* 1993, 1999). The effects of the test were evaluated the following year by conducting relatively low-rate production (June) and injection (July) tests (Baria *et al.* 1999). The profile of flow in the well during the complete test sequence was obtained from analysis of spinner and temperature logs (Evans *et al.* 1996; Evans 2000). Fractures which support flow during the stimulation were identified and precisely located in depth. Each fracture thus identified was assigned by Evans (2000) to one of three categories that broadly reflected the different flow contributions (Fig. 6). These consisted of the major flowing fractures that broadly correspond to important structures that supported more than 5% of the well-head flow; moderately flowing fractures detectable from spinner logs; and minor flowing fractures that produced a temperature disturbance on *T*-logs but are not detectable on spinner logs. Evans (2000) found that, following the injection stimulation, some

20% of the 500 fractures identified by Genter *et al.* (1997) on UBI images supported detectable flow. Prior to the stimulation, less than 1% were recognized as permeable (three fractures at 2815, 3385 and 3492 m depth; Genter *et al.* 1995).

In order to understand this flow hierarchy in terms of fracture aperture, hydraulic response and alteration, a detailed comparison between static (fracture apertures) and dynamic data (hydraulic tests) is carried out.

Comparison between electrical aperture and hydraulic data

Each fracture defined as a flowing structure by Evans (2000) was correlated to the electrical apertures given by Henriksen (2001). Figure 7a shows that the major flowing fractures tend to have broad electrical apertures. For example, three fractures located between 3200 and 3250 m that accepted major flowing features during the stimulation tests, have electrical apertures greater than 100 or 1000 μm. These apertures correspond to wide and extended fractures that were permeable prior to the injection tests (Genter *et al.* 1995). They are pre-existing permeable fractures in the granite. Similar observations relate to the lowest zone of depth in the well around 3500 m, where two large fractures support flow (Fig. 7).

However, a more precise comparison reveals numerous discrepancies between the range of fracture apertures and their hydraulic responses (Fig. 7a). For example, in the lowest part of the well there are two major flowing

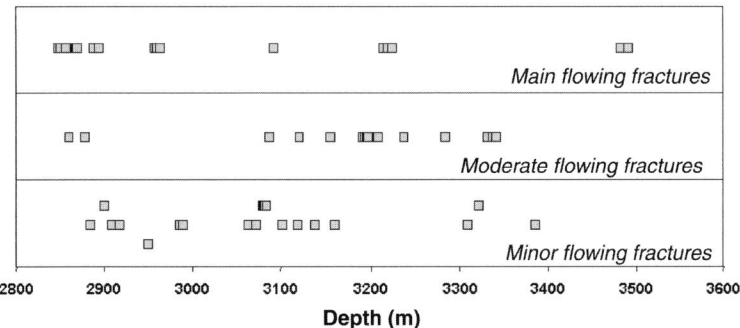

Fig. 6. Distribution of the main permeable fractures in the open-hole section of GPK1 (Evans *et al.* 1996; Evans 2000), based on the analysis of flow profiles, spinner and temperature logs. Major fractures showing flow correspond to wide structures which support more than 5% of the well-head flow. Fractures showing moderate flow are detectable from spinner logs, and fractures showing minor flow produce a temperature disturbance on *T*-logs. This subset, grouping fractures with slight flow, possible flow or no permeability, is derived from flow logs or temperature logs after the stimulation (Evans *et al.* 1996).

fractures – mentioned previously (red dots at 3483 and 3490 m). From a strictly electrical point of view, if the fracture aperture is used in permeability models such as the 'cubic law' types, a first assumption can be that these fractures must be hydraulically equivalent to a group of fractures located just above them (orange dots at 3450–3500 m depth in Fig. 7b). This direct relationship is not systematically observed. Some wide or simply large fractures showing no present-day percolation occur locally in the well, as can be shown in the

lower part of the well around 3500 m, for example (Fig. 7a). Conversely, a lot of thin fractures with electrical apertures lower than 10 μm show evidence of flow on spinner or temperature logs. For example, between 3050 and 3200 m, numerous thin fractures are closely associated and correspond to fractures with minor to moderate flow at the scale defined by Evans (2000) (green dots in Fig. 7a) even though their thin apertures could limit their permeability.

On Figure 7b three types of permeable fractures are summarized that correspond to three

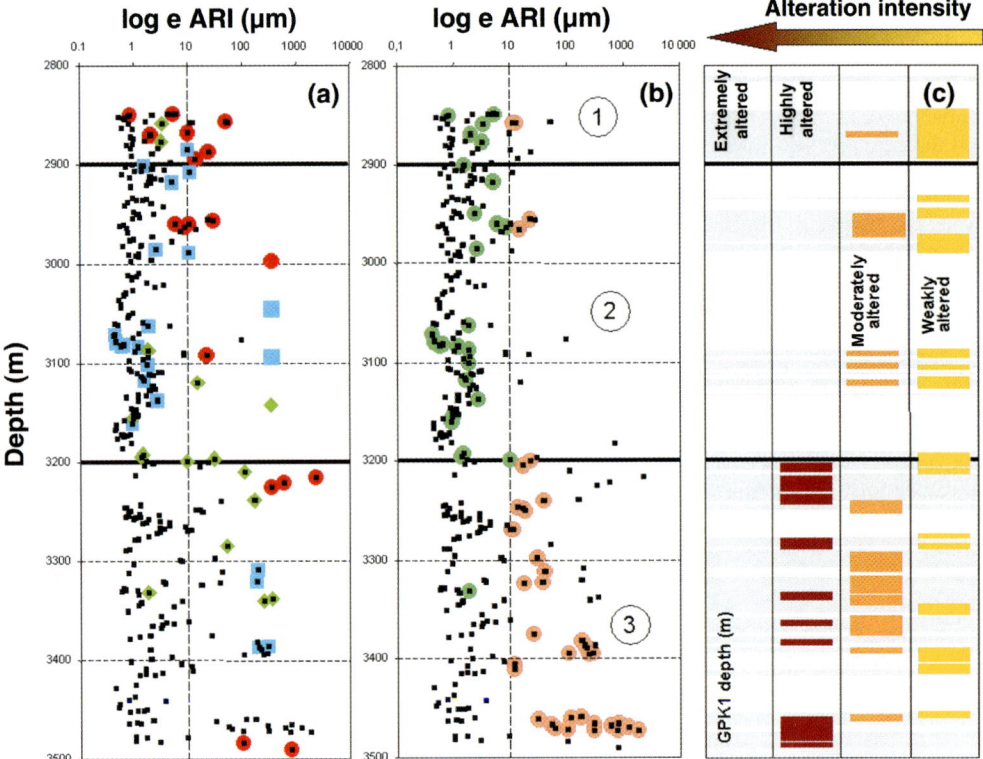

Fig. 7. (a) Comparison between the permeable fractures (Evans 2000), and their electrical apertures derived from ARI logs in well GPK1 (Henriksen 2001). Three categories of permeable fractures are distinguished (as previously defined in Fig. 6): major (red dots), moderately (blue squares) and minor (green diamonds) fractures showing flow (Evans *et al.* 1996). Some 20% of the 500 fractures (Genter *et al.* 1997; Evans 2000) detected on UBI images were flowing during the 1993 stimulation tests. There is a certain spatial relationship between the electrical apertures and the main hydraulic events. A few of the major flowing fractures are characterized by important electrical apertures. (b) Two types of fractures are highlighted. Wide fractures (electrical apertures higher than 10 μm) with no post-stimulation permeability (orange dots) and thin fractures (electrical apertures lower than 10 μm, green dots) that produce flow after the stimulation. An upper depth zone 1 (2850–2900 m) is first defined as a zone of intense damage. Then, numerous thin stimulated fractures are observed between 2900 and 3200 m except for a 2965 m fracture zone permeable prior to the stimulation (depth zone 2). By contrast, wide but non-permeable fractures are broadly located in depth zone 3 (3200–3600 m). (c) A qualitative profile of the granite alteration. Five main categories of granite are distinguished, from weakly to strongly altered facies, based on cuttings and geophysical well-logs. White zones correspond to the presence of unaltered granite.

specific depth zones which are delimited by bold horizontal lines. On this plot, only fractures that do not show a clear positive relation between their electrical aperture and their permeability are circled with orange or green dots. These data correspond respectively to permeable small-scale fractures, or wide fractures without any evidence of flow recorded during the stimulation. Figures 7b & 7a are identical, except that Figure 7b highlights only fractures for which it is difficult to correlate aperture values with the corresponding permeabilities.

The upper part of the open hole section (2850–2975 m), located just below the casing shoe, shows thin fractures with minor till major flowing responses during the stimulation. Fractures in the 2800–2900 m zone are the first fractures that could be reached in depth by the high-pressure injected fluids. They are directly and artificially damaged during the stimulation process. This part of the well is therefore taken as being different from the other zones, in order to avoid some bias in the interpretation of the natural hydraulic behaviour of fractures. This zone is labelled as an intense damage zone, where thin fractures are strongly stimulated (depth zone 1 in Fig. 7b).

At greater depths, two other depth zones are defined (Fig. 7b). Zone 2 (2900–3200 m) is characterized by numerous permeable small-scale fractures, except for two large fractures at 2954 and 2965 m depth that show percolation prior to and after stimulation. This enhancement of the hydraulic properties of small fractures takes into account 105 fractures that present a homogeneous aperture distribution. Some 90% of them show thin apertures lower than 5 μm. Despite these thin electrical apertures, some of them are clearly identified as fractures showing major or moderate flow, by Evans (2000). Between 2900 and 3200 m, these small permeable fractures represent 20% of the whole fractures in this depth zone.

In contrast, depth zone 3 (3200–3500 m) displays wide fractures that do not show any evidence of flow during the stimulation. These fractures are numerous, and 35% have electrical apertures larger than 10 μm. Values of 2.2, 1.2 and 1.8 mm are found at depths of 3125, 3468 and 3472 m respectively. Numerous permeable fractures are clearly identified between 3215 and 3225 m or 3483 and 3490 m on Figure 7a. However, a large majority of them do not show flow, despite their strong resistivity anomalies. These problematic fracture permeabilities correspond to 70% of the fractures characterized by an aperture higher than 10 μm in the 3200–3500 m depth zone.

Figure 7b therefore makes to identify two global depth zones in GPK1 in terms of permeable fracture types during the stimulation tests. The intermediate part of the open well (2975–3200 m) shows evidence of flow, despite the small apertures of the fractures. The fracture permeability seems to be stimulated in this depth zone. On the other hand, the lower part of GPK1, between 3200 and 3500 m, shows a certain inhibition of the hydraulic properties of fractures, with numerous large fractures that are not being percolated.

Alteration of the fractured rock

As was mentioned in previous studies (André et al. 2001; Sausse et al. 1998), fluid percolation at the fracture scale is directly influenced by the type and intensity of alteration. In a first step, a comparison and a correlation between the granite alteration with the hydraulic properties of fractures are carried out. Figure 7c is performed to evaluate the correlation between the previous hydraulic zoning and the location of the main hydrothermal alterations described in the Soultz granite. A qualitative log of the granite alteration (from cuttings analysis) shows the five main categories of granite distinguished: from weakly to strongly altered facies. White zones correspond to the presence of unaltered granite.

Two main types of hydrothermal alteration were seen in the granite core: an early stage of pervasive alteration and subsequent stages of vein alteration (Genter & Traineau 1992). Pervasive alteration affects the granite on a large scale without visible modification of rock texture. Colour variations in the granite, ranging from grey to orange–green, show that low-grade transformation of biotite and plagioclase has occurred. Some of the joints sealed with calcite, chlorite, sulphides and epidote are related to this early stage of alteration. Zones of vein alteration, closely related to fracturing, occur throughout the different wells. They are 1 to 20 m thick, and show strong modification of the petrophysical characteristics of the granite. Water–rock interactions have resulted in the leaching of primary minerals of the granite, and the precipitation of secondary minerals within the fractures and their wallrock (quartz, clays, carbonates, sulphides). Primary biotite and plagioclase are usually transformed into clay minerals. The primary texture of the granite is destroyed in the most altered facies.

The upper part of the open-hole section GPK1 is characterized by unaltered to moderately altered granite. The lower part has very few

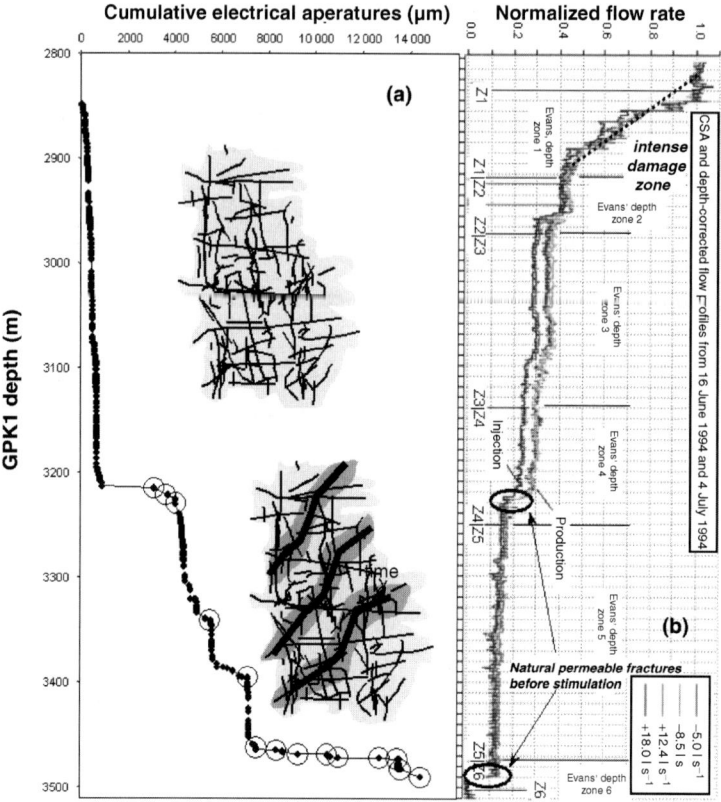

Fig. 8. Theoretical sketch of the types of permeable fractures found in the Soultz granite. (**a**) Cumulative electrical apertures of fractures versus depth, showing a regular and highly connected network of thin fractures (2850–3200 m), or a broken shape corresponding to the superimposition of the previous thin fracture network on the main wide permeable fractures (3200–3600 m). (**b**) Flow profiles of the GPK1 well from spinner logs run during the 1994 post-stimulation characterization tests. The zone boundaries are those defined by Evans (2000).

zones of fresh granite, but is strongly altered (high hydrothermal alteration). The previous zoning described in Figure 7b seems to be correlated with a gradation in the intensity of the granite alteration (Genter *et al.* 2000).

Finally, Figure 7c shows that the GPK1 well cannot be modelled with a homogeneous and single block, but that two zones of depth must be distinguished: a weakly altered zone (2850–3200 m) where thin fractures show percolation, and a strongly altered zone (3200–3600 m) where large structures mainly conduct fluid flows.

Spatial organization of fractures

A second step in the investigation involves studying the spatial organization of fractures in the well. The cumulative electrical apertures of the fractures versus their depth are plotted in

Figure 8a. Once again, a transition to different styles of curve is seen at 3200 m, where a slow, steady increase gives way to a more rugged curve containing large steps (large circles on Fig. 8a at around 3215, 3345, 3387 m, and from 3460 to 3490 m). The upper depth zone is characterized by the presence of thin structures regularly spaced with depth, whereas the deepest zone shows more isolated permeable fractures.

Figure 8b (which is to the same depth scale as Figure 8a), shows a normalized flow log monitored in GPK1 during the 94 relatively low-pressure production and injection tests (Evans 2000). The profile in this log is representative of that which prevailed at the end of September stimulation. Several flow points on the flow log and especially in the depth section below 3200 m, can be seen to correlate with the steps in the cumulative aperture curve. These disturbances correspond to a sudden loss of fluids at precise depths, and are evidence of large permeable structures.

Evans (2000) distinguished six depth zones in the flow log (Fig. 8b), that can be summarized in three global parts:

(1)   The first section between 2850 and 2950 m, located just below the casing shoe, corresponds to an intense damage zone described previously (Evans' depth zones 1 and 2 in Fig. 8b). Some 50 to 60% of the flow enters the rock mass within the series of flowing fractures in this depth interval. However, this zone is not characterized by wide fractures as in the deeper fault zones.

(2)   an intermediate depth section between 2950 and 3230 m is characterized by a hiatus between the injection and production logs (Evans' depth zones 2 and 4 on Fig. 8b). The log deviates strongly at 2960 m, and shows the presence of a large fracture (enabling flow) at this depth. Then, its shape becomes vertical and a hiatus appears between the injection and production logs. This systematic difference between the logs is seen on all logs run during the 1994 and subsequent test series (Evans et al. 1998).

(3)   the lower section of GPK1 between 3230 and 3500 m presents a single injection–production log shape, but is more discontinuous than the previous zone. Two major slope ruptures are present, indicating the presence of large fractures allowing flow (3230 and 3500 m). These permeable fractures are related to the fault zones described in the lower part of GPK1 (Evans' depth zones 5 and 6 in Fig. 8b).

Discussion

This work presents a detailed comparison between fracture electrical aperture and hydraulic data in a deep well penetrating a granitic rock mass. The study makes it possible to point out some different types of fracture organization with depth, in terms of permeable or not permeable fractures. Except for the upper intense damage zone, three main types of permeable fractures can be distinguished in GPK1:

(1)   Between 2975–3200 m, thin fracture apertures match up with permeable fractures. This zone of depth is characterized by a weak pervasive alteration of the granite. Alterations are widely distributed in the rock mass and at the scale of the granitic pluton. However, they correspond to slight modifications of the physical properties of the rock. The granite bulk density or the matrix porosity are not really affected by fluid–rock interaction phenomena. Alteration is produced by the local precipitation of secondary minerals such as calcite, illite and other hydrothermal products which partially fill the porous spaces or microcracks. This pervasive alteration is also associated with the fillings of thin fractures and cooling joints in the granite.

This depth zone shows thin fractures in a slightly altered rock. The fracture distribution versus depth is quite regular, with a mean spacing of 1.6 m along the well, and a coefficient of variation lower than one (e.g. 0.79) corresponding to an anti-clustered organization. The fractures are numerous and regularly spaced, with a mean density of 0.5 fractures per metre. Their thin electrical apertures imply a relatively small extension from the borehole wall. The presence of fractures enabling flow, among them is therefore possible only if they are connected to an extensive, highly connected fracture network. Fluid flow occurs in a thin, regular mesh, where even narrow fractures can produce permeability (Fig. 8a).

(2)   In several depth sections of the GPK1 well, large fracture apertures match up with widely permeable fractures. This broad relationship between fracture apertures and their resulting permeability is not surprising in the case where classic cubic law are envisaged. The granite shows zones where hydrothermal alteration is very important and affects the rock matrix (dissolution–precipitation) and the fractures (precipitation). These phenomena induce strong modifications of the rock's petrophysical properties, with a noticeable increase in the granite's porosity or a decrease in its bulk density. This fractured and altered medium is characterized by fault zones. These wide fractures correspond to normal faults (Mode 2 fractures) and have a different electrical conductivity signature compared to the previous thin fractures of the upper section, which are related to mode 1 fractures (joints). They correspond to major or moderate flowing fractures, which are relatively isolated in GPK1 but are mainly located in the lower part of the well where fracture electrical apertures are largely higher. They control

the fluid flow and limit the role of Mode 1 thin fractures at the same depths. These wide fractures correspond to deterministic flowing fractures. Detailed structural studies based on cores and borehole image logs showed that the wide permeable fracture zones, mainly Mode 2 fractures, have a more complex internal organization than individual thin permeable fractures, mainly Mode 1. The application of the anisotropic present-day stresses would induce mechanical conditions able to enhance voids or channelling and then permeability.

(3) Between 3200 and 3500 m, some large fractures are not permeable. The electrical aperture values are higher than 10 μm and the fluid is potentially available in the formation. Thus, the absence of permeability is surprising. However, the presence of high conductivity during the ARI logging may be related to drilling operations. It is quite usual that hydrothermal products filling the fractures could be washed out during the drilling rotation. It means that thin fracture apertures could be enhanced by the drilling process, introducing a type of size bias during the aperture fracture data analysis. However, this issue does not dominate in hard rocks such as crystalline fractured rocks. Moreover, at Soultz, pre-existing fractures are systematically filled by hydrothermal minerals – even for very thin fractures. However, for geodic deposits or partial fillings, residual free apertures could occur. Then, it is possible that the well had crossed an open fracture at the borehole scale, which is fairly well plugged at a certain distance from the well, inducing non-permeable fracture behaviour.

However, these non-permeable fractures are isolated in the lower part of GPK1. This depth zone shows a lower density of fractures than the 3050–3200 m zone (0.4 fractures per metre), and a mean spacing of 1.45 m. The coefficient of variation is equal to 0.7, and corresponds (as previously) to an anti-clustered distribution of fractures.

The granitic basement is fractured. Two types of fracture organization are superimposed. In the upper part of the well, a wide and regular network of thin fractures is described. In the lower part of GPK1, this thin network is also present, but several wide permeable fractures appear locally. They secondarily affect the granite and control the fluid flows. However, some

of these large fractures are hydraulically inhibited. This behaviour could be explained by the nature of the fillings within the fractures. Fractures that allow flow are generally characterized by geodic quartz deposits, which allow the presence of residual apertures and possibly channel fluid flows. On the opposite, the non-permeable fractures could be partially infilled by other alteration products, such as illite, for example, which can be more obstructive than a geodic quartz growth. These wide fractures are probably locally disconnected from the efficient flow controlled by the faults, e.g. the major flowing fractures.

These different permeability types in granite can be related to the interpretation of flow logs taken during the hydraulic monitoring of GPK1 (Fig. 8b). Apart from the first part of the open-hole section, which is not relevant (2850–2975 m), Evans (2000) distinguishes two major depth zones based on hydraulic flow log data.

1. In the lower part (3200–3500 m) the flow-log responses are equivalent during injection and production tests, which means that the same fracture properties are implicated in the flow. In this case, the main permeable fractures, e.g. deterministic fractures with large apertures play a predominant role. Evans (2000) notes some turbulent-like losses of flow at these depths, which characterize the presence of permeability linked to large-capacity faults (Evans' depth zone 6 in Fig. 8b at 3.5 km).

2. In the intermediate zone between 3000 and 3200 m, the hydraulic response is slightly different between injection and production hydraulic tests (Fig. 8b). Evans (2000) considers that the shift observed between the two logs is due to the presence of a connected network of fractures in the granite which surrounds the well to a depth of 3350 m. This vertical connectivity seems to be expressed on a scale of hundreds of metres.

The zoning proposed by Evans (2000) can be spatially compared to the zoning proposed in this paper. First, there is the presence of an intense damage zone in the upper part of the well (a much greater intensity of the stimulation process at the level of zone 1) (Jones et al. 1995; Evans 2000). Then, at intermediate depths (2930–3230 m), changes in flow profiles occur (hiatus between injection and production curves) implying a diversion of flow within the rock mass (Evans *et al.* 1998; Evans 2000) which is consistent with the present representation of a highly connected network of thin fractures, as illustrated in Figure 8a. Then, there is the superimposition of a thin network of narrow and wide permeable fractures in the lower part of GPK1, which induces large-capacity fluid circulation in the granite.

The Soultz granite therefore displays different types of fracture permeability directly related to the spatial organization of fractures and to their conductivity (electrical apertures). It seems that there are two types fracture networks are present: a small-scale fracture system that may constitute the far-field reservoir, and an isolated but large-scale fault-system which allows the hydraulic connection to the exchanger. These results attempt to demonstrate that a precise description of geological characteristics, such as alteration of the rock or geometric and hydraulic properties concerning the fracture permeability, can give some relevant insights for the better understanding of fluid flows, in order to model fracture permeability.

This work was carried with the financial support of the STREP (Strategic Research Project) 'Pilot Plant' programme – EHDRA (the European Hot Dry Rock Association). Particular thanks are due to K. Evans, for his constructive and helpful comments on the manuscript.

References

ANDRÉ, A. S., SAUSSE, J. & LESPINASSE, M. 2001. Fluid pressure and fracture geometry. Quantification of the paleostresses associated with the sequential sealing of vein systems. *Tectonophysics*, **336**, 215–231.

AQUILINA, L. & BRACH, M. 1995. Characterization of Soultz hydrochemical system: WELCOM (Well Chemical On-line Monitoring) applied to deepening of GPK-1 borehole. *Geothermal Science and Technology*, **4**, 239–251.

BARIA, R., BAUMGÄRTNER, J. & GÉRARD, A. 1993. Heat mining in the Rhinegraben. *Socomine Internal Project Report*, EEIG, Soultz.

BARIA, R., BAUMGÄRTNER, J., GÉRARD, A., JUNG, R. & GARNISH, J. 1999. European HDR research programme at Soultz-sous-Forêts (France) 1987–1996. *Geothermics*, **28** (4–5), 655–669.

DEZAYES, C., VILLEMIN, T., GENTER, A., TRAINEAU, H. & ANGELIER, J. 1995. Analysis of fractures in boreholes of Hot Dry Rock project at Soultz-sous-Forêts (Rhine Graben, France). *Journal of Scientific Drilling*, **5** (1), 31–41.

DUBOIS, M., AYT OUGOUGDAL, M., MEERE, P., ROYER, J. J., BOIRON, M. C. & CATHELINEAU, M. 1996. Temperature of paleo- to modern self-sealing within a continental rift basin: the fluid inclusion data (Soultz-sous-Forêts, Rhine Graben, France). *European Journal of Mineralogy*, **8**, 1065–1080.

EVANS, K. F. 2000. The effect of the stimulations of the well GPK1 at Soultz on the surrounding rock mass: evidence for the existence of a connected network of permeable fractures. *In*: IGLESIAS, E., BLACKWELL, D., HUNT, T., LUND, J., TAMANYU, S. & KIMBARA, K. (eds) *Proceedings, World Geothermal Congress, Kyushu – Tohoku, Japan*, 3695–3700.

EVANS, K. F., KOHL, T., HOPKIRK, J. & RYBACH, L. 1996. *Studies of the Nature of Non-linear Impedance to Flow Within the Fractured Granitic Reservoir at the European Hot Dry Rock Project Site at Soultz-sous-Forêts, France*, Institut für Geophysik, ETH Zurich – Polydynamics Engineering, Zurich.

EVANS, K. F., KOHL, T. & RYBACH, L. 1998. *Analysis of the Hydraulic Behaviour of the 3.5 km Deep Reservoir During the 1995–1997 Test Series, and Other Contributions to the European Hot Dry Rock Project, Soultz-sous-Forêts, France*, Institut für Geophysik, ETH Zurich, Zurich.

FAIVRE, O. 1993. Fracture evaluation from quantitative azimuthal resistivities, *In: Society of Petroleum Engineering, 68th Annual Technical Conference and Exhibition*, Houston, Texas, 179–192.

GENTER, A. & GENOUX-LUBAIN, D. 1994. *Evaluation de la Fracturation dans le Forage GPK1 à Partir de l'Imagerie ARI entre 2870m et 3500m (Soultz-sous-Forêts, France)*, BRGM Report **R 38099**.

GENTER, A. & TRAINEAU, H. 1992. Borehole EPS1, Alsace, France: preliminary geological results from granite core analysis for Hot Dry Rock research. *Scientific Drilling*, **3**, 205–214.

GENTER, A. & TRAINEAU, H. 1996. Analysis of macroscopic fractures in granite in the HDR geothermal well EPS-1, Soultz-sous-Forêts, France. *Journal of Volcanology and Geothermal Research*, **72**, 121–141.

GENTER, A., CASTAING, C., DEZAYES, C., TENZER, H., TRAINEAU, H. & VILLEMIN, T. 1997. Comparative analysis of direct (core) and indirect (borehole imaging tools) collection of fracture data in the Hot Dry Rock Soultz reservoir (France). *Journal of Geophysical Research*, **102** (B7), 15 419–15 431.

GENTER, A., TRAINEAU, H., BOURGINE, B., LEDESERT, B. & GENTIER, S. 2000. Over 10 years of geological investigations within the European Soultz HDR project, France. *In*: IGLESIAS, E., BLACKWELL, D., HUNT, T., LUND, J., TAMANYU, S. & KIMBARA, K. (eds) *Proceedings, World Geothermal Congress, Kyushu – Tohoku, Japan*, 3707–3712.

GENTER, A., TRAINEAU, H., DEZAYES, C., ELSASS, P., LEDÉSERT, B., MEUNIER, A. & VILLEMIN, T. 1995. Fracture analysis and reservoir characterization of the granitic basement in the HDR Soultz project (France). *Geothermal Science and Technology*, **4** (3), 189–214.

GENTIER, S. 1986. *Morphologie et comportement hydromécanique d'une fracture naturelle dans un granite sous contrainte normale. Etude expérimentale et théorique.* PhD thesis, Université d'Orléans, France.

GENTIER, S., BILLAUX, D., HOPKINS, D., DAVIAS, F. & RISS, J. 1996. Images and modeling of the hydromechanical behavior of a fracture. *Microscopy Microanalysis Microstructures: (Les Ulis)*, **7** (5–6), 513–519.

GENTIER, S., RISS, J., HOPKINS, D. & LAMONTAGNE, E. 1998. Hydromechanical behavior of a fracture. How to understand the flow paths. *In*: ROSSMANITH, H. P. (ed.) *Mechanics of Jointed and Faulted Rock*, A. A. Balkema, Vienna, Austria, 583–588.

HENRIKSEN, A. 2000. Near-well fracture hierarchy based on borehole image data from a granite reservoir. *In*: *European Geophysical Society, 25th General Assembly, Nice, Session SE36 'From Fracturing to Faulting: Laboratory, Borehole and Field Studies'*, Geophysical Research Abstracts, Vol. 2, Abstract on CD-ROM.

HENRIKSEN, A. 2001. *Fracture Interpretation Based on Electrical and Acoustic Borehole Image Logs.* BRGM Report BRGM/RP-50835-FR.

JONES, R., BEAUCE, A., JUPE, A., FABRIOL, H. & DYER, B. C. 1995. Imaging induced seismicity during the 1993 injection test at Soultz-sous-Forêts, France. *In*: BARBIER, E., FRYE, G., IGLESIAS, E & PÀLMASON, G. (eds) *Proceedings of the World Geothermal Congress, Florence, Italy*, 2671–2676.

JUNG, R., WILLIS-RICHARD, J., NICHOLLS, J., BERTOZZI, A. & HEINEMANN, B. 1995. Evaluation of hydraulic tests at Soultz-sous-Forêts, European HDR Site. *In*: BARBIER, E. *et al.* (eds) *Proceedings of the World Geothermal Congress, Florence, Italy*, 2665–2669.

LAMB, H. 1957. *Hydrodynamics.* Cambridge University Press, Cambridge (6th Edition).

LOUIS, C. 1969. *A Study of the Groundwater Flow in Jointed Rock and Its Influence on the Stability of the Rock Masses.* Imperial College Rock Mechanics Report No 10, Imperial College, London.

LUTHI, S. M. & SOUHAITE, P. 1990. Fracture apertures from electrical borehole scans. *Geophysics*, **55** (7), 821–833.

ODA, M. 1986. An equivalent continuum model for coupled stress and fluid flow analysis in jointed rock masses. *Water Resources Research*, **22** (13), 1845–1856.

PARSONS, R. W. 1966. Permeability of idealized fractured rock. *Society of Petroleum Engineering Journal*, SPE 1289, 126–136.

PAUWELS, H., FOUILLAC, C. & FOUILLAC, A. M. F. 1993. Chemistry and isotopes of deep geothermal saline fluids in the upper Rhine Graben: origin of compounds and water–rock interactions. *Geochimica et Cosmochimica Acta*, **57**, 2737–2749.

SAUSSE, J. 1998. *Caractérisation et modélisation des écoulements fluides en milieu fissuré. Relation avec les altérations hydrothermales et quantification des paléocontraintes.* PhD thesis, Université Henri Poincaré, Nancy, France.

SAUSSE, J. 2002. Hydromechanical properties and alteration of natural fracture surfaces in the Soultz granite (Bas-Rhin, France). *Tectonophysics*, **348**, 169–185.

SAUSSE, J., GENTER, A., LEROY, J. L. & LESPINASSE, M. 1998. Altération filonienne et pervasive: Quantification des perméabilités fissurales dans le granite de Soultz sous Forêts (Bas-Rhin, France), *Bulletin de la Société Géologique de France*, **169** (5), 655–664.

SIBBIT, A. M. & FAIVRE, O. 1985. The Dual Laterolog response in fractured rocks. *In*: *Society of Professional Well Log Analysts, 26th Annual Logging Symposium*, Dallas, T1–T34.

SNOW, D. T. 1965. *A parallel plate model of fractured permeable media.* PhD thesis, University of California.

SNOW, D. T. 1968*a*. Fracture deformation and change of permeability and storage upon changes of fluid pressure. *Colorado School of Mines Quarterly*, **63**, 201–244

SNOW, D. T. 1968*b*. Rock fractures spacings, openings and porosities. *American Society of Civil Engineering Journal, Soil Mechanics and Foundation Division*, **94** (SM1).

SNOW, D. T. 1969. Anisotropic permeability of fractured media. *Water Resources Research*, **5** (6), 1273–1289.

VUATAZ, F. D., BRACH, M., CRIAUD, A. & FOUILLAC, C. 1990. Geochemical monitoring of drilling fluids: a powerful tool to forecast and detect formation waters. *In*: *SPE Formation Evaluation*, SPE 18734, 177–184.

In situ seismic investigations of fault zones in the Leventina Gneiss Complex of the Swiss Central Alps

R. GIESE, C. KLOSE & G. BORM

*GeoForschungsZentrum Potsdam, Department of GeoEngineering,
Telegrafenberg, D-14473 Potsdam, Germany (e-mail: rudi@gfz-potsdam.de)*

Abstract: Underground seismic tomography investigations have been carried out in the Faido access tunnel of the Gotthard Base Tunnel, Switzerland. Velocity measurements were made over a total length of 2651 m of the adit with the tunnel seismic prediction system, ISIS (Integrated Seismic Imaging System). ISIS provides high-resolution seismic imaging, using an array of rock anchors equipped with 3D-geophones.
The first onsets of the compressional and shear waves were used for tomographic inversion. Two-dimensional seismic-velocity models reveal a disturbed zone between 2 and 3 m inward from the tunnel wall, characterized by strong variations from 3500 to 5800 m s^{-1} in compressional wave velocity V_P, and from 2000 to 3000 m s^{-1} in shear-wave velocity V_S. High-velocity zones correspond to quartz lenses, and low velocities mainly indicate fractured rock. Beyond the excavation disturbance zone, the variations in seismic velocities are generally smaller. The tomographic image of the rock mass also revealed two major fault zones composed of cataclastic shear planes surrounded by wider fracture zones. These structural characteristics are also useful for the prediction of cataclastic zones at other sites.

Since the early 1990s, much effort has been put into the use of seismic methods for characterizing the geotechnical environment in the proximity of tunnels and predicting discontinuities like fault zones ahead of the tunnel face. Acoustic emission and ultrasonic velocity methods have been used to investigate the excavation disturbance zone (EDZ) associated with deep tunnels in hard rock (Falls & Young 1998). The influence of the stress regime and the method of tunnel construction on the EDZ was a main objective of their studies. The width of the EDZ varies according to the type of excavation procedure: between one-tenth of the tunnel radius for a tunnel boring machine (TBM), up to the full radius for conventional tunnelling by drill and blast. The EDZ is characterized by brittle fractures and stress redistribution around the tunnel, induced through excavation work. Fracturing, loosening and weakening of the rock mass lead to a significant decrease in seismic velocities in the immediate neighbourhood of the tunnel wall.

In addition to the detection of changes in rock properties around the tunnel, prediction of discontinuities ahead of the tunnel face is a very important feature. In general, a seismic prediction system is based on two steps. First of all, seismic-wave energy is transmitted: either by firing explosives in drill-holes in the side walls (Dickmann & Sander 1996), or by the use of noise generated by the cutters of the TBM during operation (Neil *et al.* 1999), or by electrodynamic vibrators incorporated in the cutter head of a TBM (Kneib *et al.* 2000). In the second step, the transmitted signals are reflected by geological heterogeneities and recorded by accelerometers or geophones placed in drill-holes along the tunnel or at the head of the TBM. The spatial location of the discontinuities is determined by imaging the reflected seismic energy. The resolution of the latter depends strongly on the degree of heterogeneity of the rock mass. Cataclastic zones may scatter seismic energy because of their frequently irregular branched shapes (Wallace & Morris 1986).

In the following sections, we report on continuous seismic-velocity measurements using the Integrated Seismic Imaging System (ISIS) during tunnel construction in the Leventina Gneiss Complex of the Central Swiss Alps. Measurements of the direct wave field close to the tunnel wall, via tomographic inversion, were used for detection and characterization of fault zones.

ISIS components

The concept of the Integrated Seismic Imaging System (ISIS) was developed by the GFZ

From: HARVEY, P. K., BREWER, T. S., PEZARD, P. A. & PETROV, V. A. (eds) 2005. *Petrophysical Properties of Crystalline Rocks*. Geological Society, London, Special Publications, **240**, 15–24.

Potsdam in co-operation with Amberg Measuring Technique AG, Zurich, Switzerland (Borm *et al.* 1999, 2001). Herein, glass-fibre reinforced polymer resin rock anchors are equipped with 3D-geophones and installed as stabilizing elements (Fig. 1). The geophones are mounted in three orthogonal directions at the tip of the rock anchors. Signals up to 3 kHz and the full seismic vector can be recorded. The receiver anchors are cemented into the drill-holes by a two-component epoxy resin, providing optimum coupling of the geophones to the surrounding rock. Properly oriented, the receiver rods form a radial and axial geophone array close to the tunnel face advance.

A repetitive mechanical hammer is used as the seismic source (Fig. 2). The hammer incorporates a pneumatic cylinder, and the power for impact is supplied by a moving mass of 5 kg. Each impact takes 1 ms and is controlled by a programmable steering unit. Prior to impact, the hammer is prestressed toward the rock with a mass equivalent of 200 kg. This prestress achieves good coupling of hammer and rock. The impact hammer may be used in all directions in combination with a TBM or other machinery. The hammer transmits pulses of frequencies up to 2 kHz, with a repetition rate of five seconds. The maximum error in triggering time is less than 0.1 ms. This small time lag, together with the accurate and reliable repeatability of the transmitted signals at each source point, leads to a significant improvement of the signal-to-noise ratio through vertical stacking. This is a statistical procedure to amplify correlated signals such as reflections from geological discontinuities, and to reduce non-correlated signals such as noise from the TBM. Several thousand of these pulses were fired during the application of ISIS in underground construction work, and seismic reflection energy was recorded at travel distances of up to 250 m.

Field-test case history – the Faido adit of the Gotthard Base Tunnel

The Faido adit is part of the Gotthard Base Tunnel crossing the Central Alps in a north–south direction. After its planned completion in the year 2015, this high-speed railway tunnel with a length of 57 km will be the longest one in the world. The 2651-m long Faido adit is located in the Leventina Gneiss Complex, which is part of the Penninic gneiss zone. Figure 3 shows the geological–geotechnical profile of the Faido adit, excavated during 2000/2001 using a drill and blast technique. The inclination of the tunnel is 12.7%, and the thickness of the overburden is up to 1300 m.

The Leventina gneiss complex consists mainly of granitic gneiss (51% feldspar, 34% quartz, 14% mica and 1% accessory minerals). The gneiss fabric exhibits a wide spectrum of

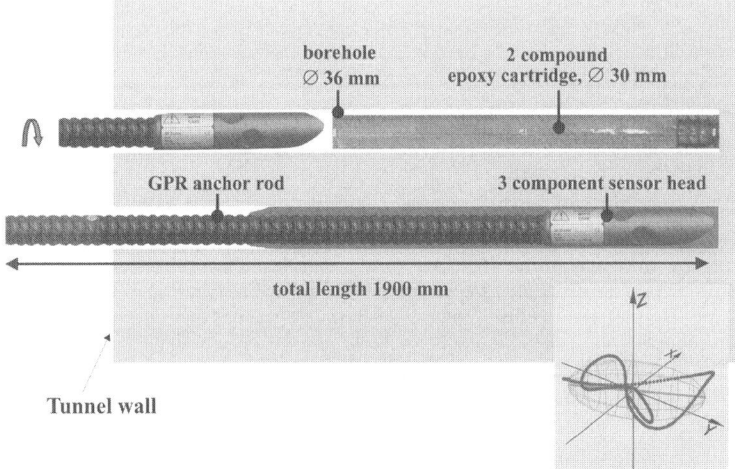

Fig. 1. Schematic diagram of a rock anchor with a three-component geophone. The glass-fibre reinforced polymer resin (GPR) anchor is driven into the drill-hole with high revolution, mixing the two-component epoxy resin and fixing the anchor tightly to the rock mass. The hodograph illustrates the particle motion of the incoming seismic signals and the resulting wave field vector in a given time window. The *y*-axis is parallel and the *x*-axis is orthogonal to the tunnel wall.

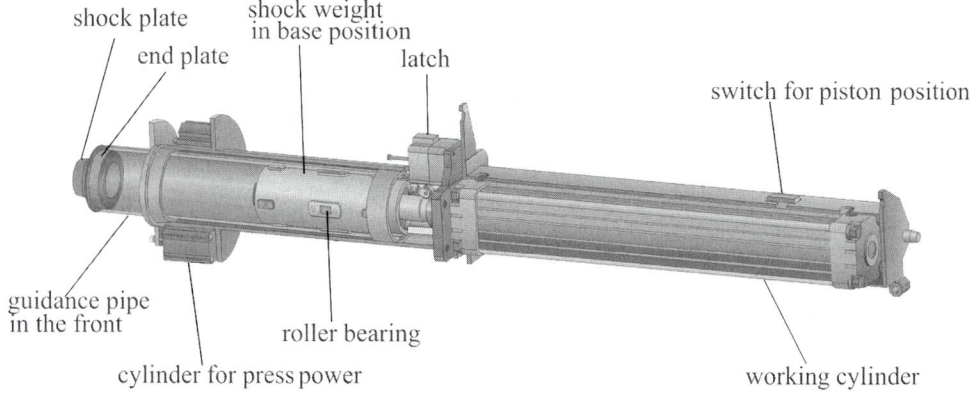

Fig. 2. Mechanical design of the pneumatic impact hammer.

layered, laminated, augen-structured, phacoidal, porphyritic schist and folded varieties at a scale of a few centimetres to metres (Löw & Wyss 1999).

The polyphase metamorphic history of the gneiss has produced dykes of quartz and lenses of biotite, amphibolite and quartz. Ductile deformation folded the Leventina gneiss, and brittle deformation (specifically cataclastic shear) produced various fissured and fractured zones. Five main fracture sets of varying size and shape were observed (Fig. 4). The occurrence frequencies of these are K1a (22%), K1b (30%), K2 (12%), K3 (8%), K4 (8%), and K5 (20%). The preferred orientation of the dominant K1a/b set is parallel to the cataclastic faults. Two cataclastic zones appeared most critical during the tunnel excavation: a 10-m thick fault at tunnel metre Tm 973 (Fig. 3), and an approximately 0.3-m thick fault at Tm 2410.

Layout of the seismic lines

Accompanying the excavation work, seismic measurements were made every 200 m, to gain continuous velocity information along the

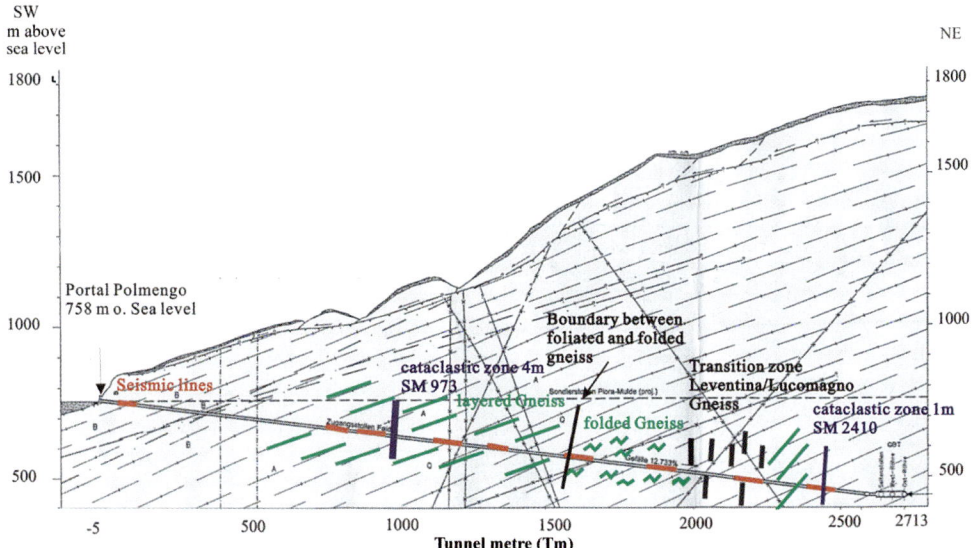

Fig. 3. Geological–geotechnical profile of the Faido adit. Red bars indicate the positions of the seismic measurements, and blue bars mark the location of cataclastic zones at Tm 973 and Tm 2410.

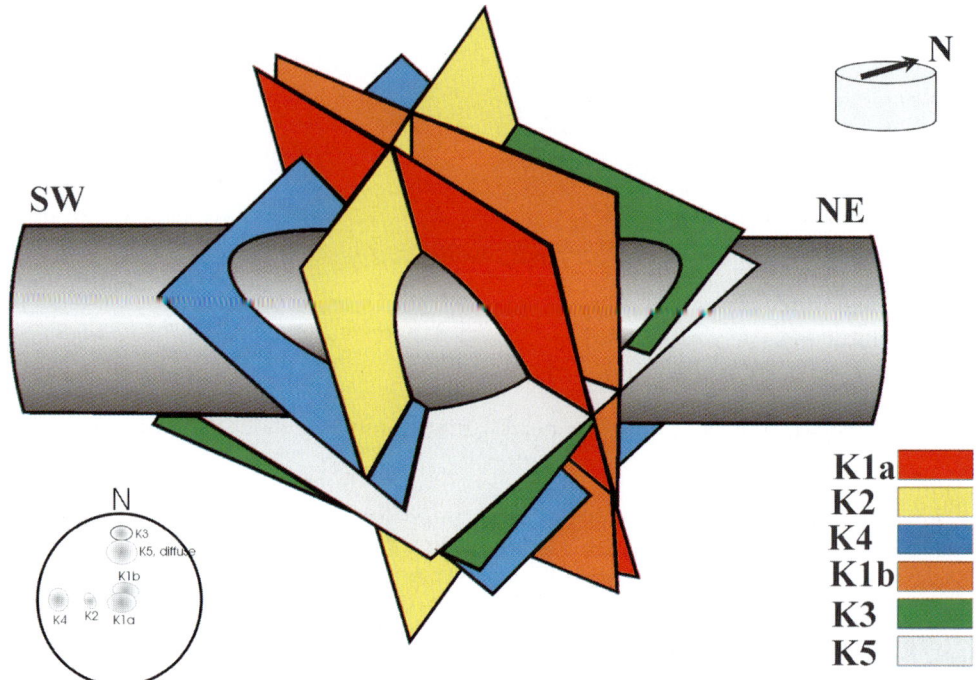

Fig. 4. Orientation of fracture sets with respect to the tunnel being driven in a NE direction. *K*1a/b are the main cataclasites.

complete profile (Fig. 3). For transport and application purposes, the impact hammer was mounted on a motor vehicle. Apart from data acquisition for geological prediction, one of the principal goals of these measurements was to test the performance of the ISIS components under the harsh conditions of underground works. Primary logistical problems involved the choice of an optimum time window for the measurements during low-noise working activities or intermittent pauses. In general, the time breaks during transportation of excavated material were only used when low-frequency background noise occurred (< 100 Hz), which could readily be removed from the data by low-pass filtering.

Table 1. *Information on the seismic measurements in the Faido adit tunnel*

Period	Extension of seismic lines	Number of rods	Number of impacts	Number of source points	V_P of first arrivals	V_S of first arrivals
28–30 March 2000	Tm 86–Tm 137	5(2)	40	5	5300 m s^{-1}	2800 m s^{-1}
18–20 July 2000	Tm 765–Tm 830	8(6)	202	13	5700 m s^{-1}	3100 m s^{-1}
26–28 September 2000	Tm 881–Tm 963	8(8)	411	34	5300 m s^{-1}	2900 m s^{-1}
7–9 November 2000	Tm 1130–Tm 1215	8(8)	498	38	5600 m s^{-1}	3100 m s^{-1}
5–7 December 2000	Tm 1314–Tm 1379	8(8)	536	43	5900 m s^{-1}	3400 m s^{-1}
January 30 to 2 February 2001	Tm 1582–Tm 1663	10(10)	540	38	5900 m s^{-1}	3400 m s^{-1}
13–14 March 2001	Tm 1858–Tm 1942	10(10)	360	27	5800 m s^{-1}	3400 m s^{-1}
April 30 to 2 May 2001	Tm 2128–Tm 2203	10(10)	852	59	5500 m s^{-1}	3200 m s^{-1}
12–14 June 2001	Tm 2360–Tm 2433	10(10)	579	53	5000 m s^{-1}	2700 m s^{-1}
Total	661 m	77(72)	4018	310		

() number of intact anchor rods after setting.

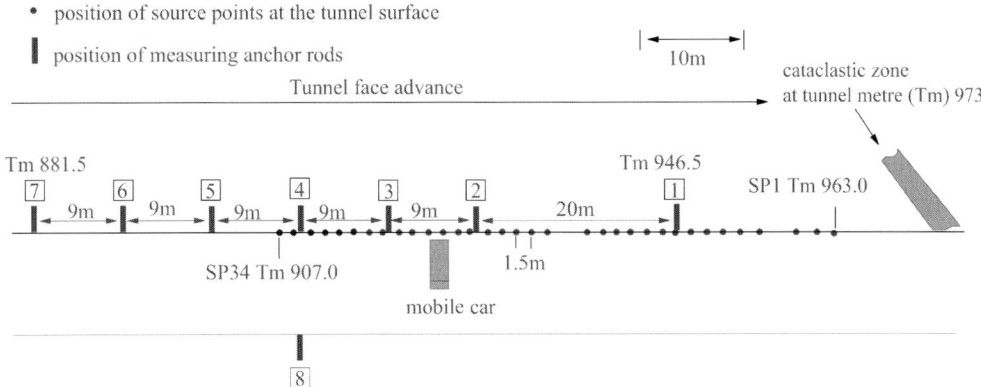

Fig. 5. Horizontal view of the tunnel with source and receiver configuration between Tm 881 and Tm 963.

Layout of the source points and geophone anchors was along the tunnel wall. Usually, an array of eight to 10 geophone anchors of 2 m length was installed on one side of the tunnel wall, and another anchor set on the opposite side to detect guided waves along and around the tunnel surface. Spacing of the geophone anchors was usually 9 m.

Figure 5 shows the horizontal view of a typical measurement configuration. Since the strike directions of the main faults were known from previous geological investigations, the source points were placed on the left-hand side of the tunnel, with a spacing of 1.0 to 1.5 m. The impact hammer was applied at right angles to the tunnel wall. Table 1 summarizes the technical data of the seismic profiles.

In Figure 6, the first 50 ms of the velocity component recorded parallel to the tunnel axis at one of the geophones are shown. Signal phases of the direct P-wave velocity can be seen between 4 and 14 ms at offsets of 16 and 72 m. The first onsets of S-waves follow at 6 ms and 26 ms, respectively. The direct transversal waves are a combination of shear waves and surface waves travelling straight from the source to the receiver or along the tunnel wall.

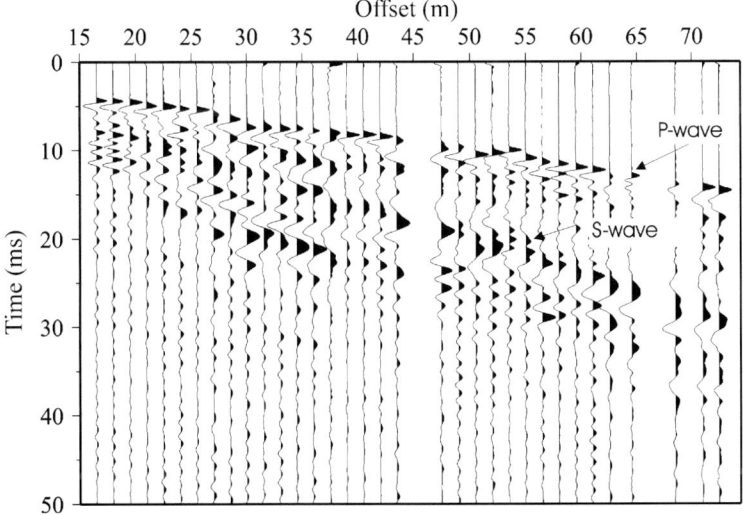

Fig. 6. Seismic data of a horizontal component from the profile between Tm 890 and Tm 955. The first breaks of P- and S-waves in the offset between 16 and 72 m are marked by arrows.

Tomographic travel-time inversion of direct seismic waves

The arrival times of direct P- and S-waves were used for calculation of a 2D tomographic inversion of the velocity field near the tunnel wall. Non-linearity of travel-time inversion requires a starting model and an iterative approach (Zelt & Smith 1992). The starting models used for the P- and S-wave velocity distribution are homogeneous parallel to the tunnel axis, but have radial gradients towards the interior of the rock mass which are characteristic of the EDZ. The P-wave velocity model starts with 4400 m s^{-1} at the tunnel wall, and increases at increasing distances from it, by about 170 m s^{-1} per metre. The velocity of the S-waves at the tunnel surface is taken as 2600 m s^{-1}, and an increase of 100 m s^{-1} per metre is assumed.

The tomographic inversion comprises a forward modelling, the calculation of travel-time differences between the observed and modelled data, and an inversion procedure. A ray tracer is used to simulate the curved ray paths through the rock mass, for the purpose of forward modelling. The theoretical response of the ground modelled is iteratively edged towards the observed data until a model is obtained that sufficiently matches the observed velocity distribution.

The tunnel wall topography was assumed to be smooth, since the undulation amplitude was less than 30 cm over a wavelength of nearly 5 m. As a result of the relatively steep velocity gradient near the tunnel wall, direct waves penetrate the rock mass up to a distance of 10 m from the tunnel wall before reaching the receivers. The tunnel is accessible for geological inspection,

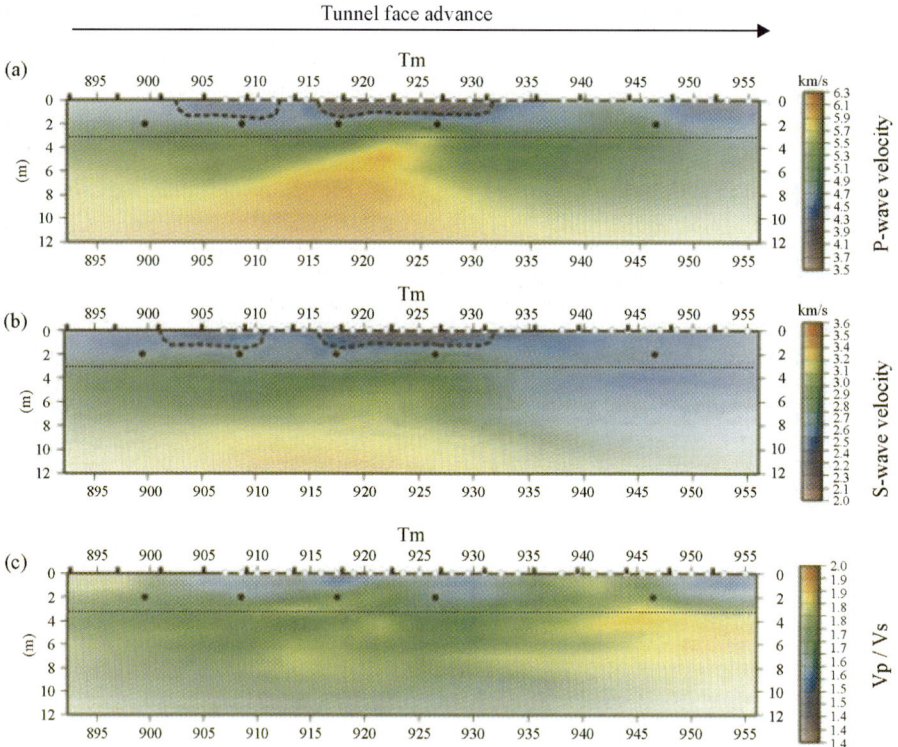

Fig. 7. Tomographic inversion of (**a**) compressional-wave velocity V_P, (**b**) shear-wave velocity V_S, and (**c**) V_P/V_S ratio close to the cataclastic zone at Tm 973. Source and receiver points are symbolized by open and solid circles. Solid bars mark tunnel advances per shift. The velocities are coded with a colour intensity, such that areas with poor ray coverage are light-coloured compared to those with high ray coverage. Dashed lines mark the outcrop of crossing fracture sets between Tm 900 and Tm 912, between Tm 916 and Tm 937, and the position of quartz lenses between Tm 940 and Tm 947. Dotted lines mark the position of the geophones at 3 m distance from the tunnel wall.

meaning that a direct comparison of the geological and the seismic data is possible.

Figure 7a shows the tomographic model for V_P along the left-hand side of the tunnel wall in a horizontal plane through the geophone anchors and the source points. Analysis of the incident angles of the direct waves shows deviation from this plane of less than 5°. The first two to three metres from the tunnel wall mark the radial extent of the EDZ. Here, the V_P-values increase from about 3500 m s^{-1} at the tunnel wall to about 5800 m s^{-1} in the undisturbed interior of the rock mass beyond the EDZ. The steepest gradient in the velocity field is found near the tunnel wall, decreasing as the distance to the tunnel wall increases.

Based on the evidence of geological investigations along the tunnel surface, the low-velocity zones from Tm 900 to Tm 912, and from Tm 916 to Tm 937, coincide with cross-cutting fracture sets. The high-velocity zone between Tm 940 and Tm 947 is caused by quartz lenses striking perpendicularly to the tunnel wall. Starting at Tm 920, the V_P-values between 3 and 8 m radial distance from the excavation wall decrease towards the cataclastic zone at Tm 973 (Fig. 8). This decrease of V_P is continuous and independent of the observed fluctuations in the EDZ.

The tomographic model for V_S is shown in Figure 7b. The distribution of high- and low-velocity zones in the EDZ is in good agreement with that of V_P (cf. Fig. 7a). In the deeper interior of the rock mass, V_S also decreases as the cataclastic zone at Tm 973 is approached, whereby the reduction of V_S is stronger than that of V_P. This is also indicated by the relatively higher V_P/V_S ratio of 1.9 in Figure 7c. On the other hand, a smaller V_P/V_S ratio of 1.5 to 1.6 is seen in the first 2 to 3 m distance from the tunnel

wall. Since the V_P/V_S ratio corresponds to the apparent Poisson ratio of the rock mass, it is a useful indicator of its compressibility – larger in the EDZ than in the undisturbed rock mass.

Figure 8 shows the geological model derived from direct geological inspection of the fault zone crossing the tunnel at Tm 973. A few metres before the cataclastic zone, increased fracturing (K1a/b in Fig. 4) was observed, corresponding to the measured reduction in V_P and V_s. The fracture sets before the cataclastic zone at Tm 973 were water bearing. Since water has a greater impact on the bulk modulus of the rock than on the shear modulus, increasing the water content reduces the V_P-values to a greater extent than the V_S-values.

Travel-time tomographic inversions derived from eight measurements between Tm 765 and Tm 2433 detected general trends of the velocity field related to fault zones close to the tunnel. Figure 9 shows the V_P and V_S values derived from the tomographic inversions along lines of measurement at 3 m distance from the tunnel wall, where maximum coverage of the wave rays is obtained and the influence of the EDZ can be neglected (cf. Fig. 7a & 7b).

A general trend in velocity distribution can be seen with a local minimum of V_P and V_S occurring in the cataclastic fault at Tm 973. From Tm 1200 to Tm 1500, the average velocities increase steadily. The V_P-values decrease again from Tm 1550, reaching a relative minimum at the Tm 2410 cataclastic shear. The V_S-values decrease after Tm 2150.

Figure 10 shows the V_P/V_S ratio taken from the tomographic inversions along the same lines (cf. Fig. 7c). In the fault zones around Tm 973, a local minimum of V_P/V_S is observed, and yet another relative minimum of V_P/V_S is found between Tm 1700 and Tm 2430.

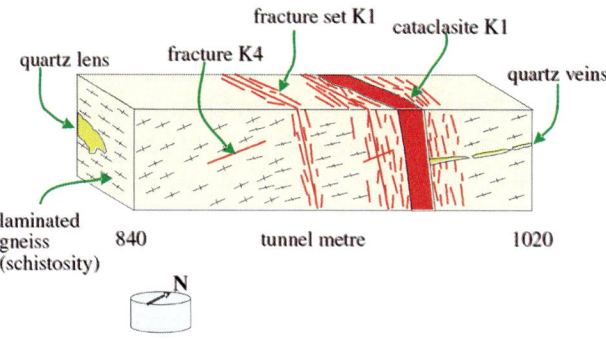

Fig. 8. Geological model of the cataclastic zone at Tm 973, showing the disturbance zone and fracture sets. Vertical scale exaggerated five times.

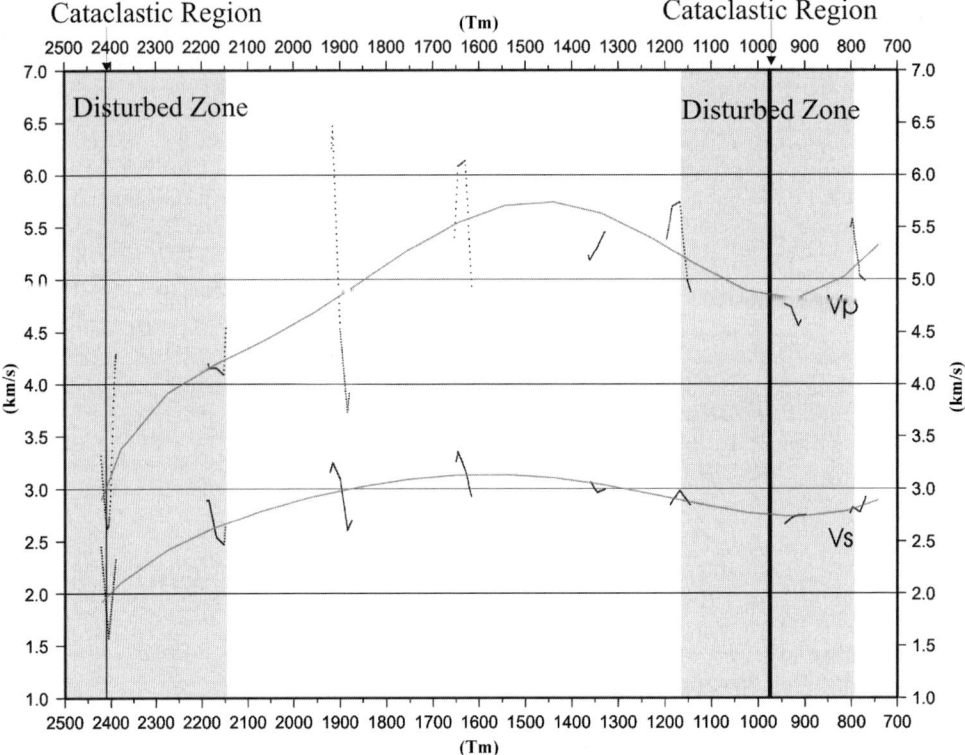

Fig. 9. Seismic velocities V_P and V_S derived from tomographic inversions at 3 m distance from the tunnel wall of the Faido adit (dotted line). The solid lines through the velocity values are polynomial fitting curves. The cataclastic zones at Tm 973 and Tm 2410 are surrounded by wider zones with lower wave velocities, which mark the disturbance zones in the Leventina gneiss (cf. Fig. 7a & b).

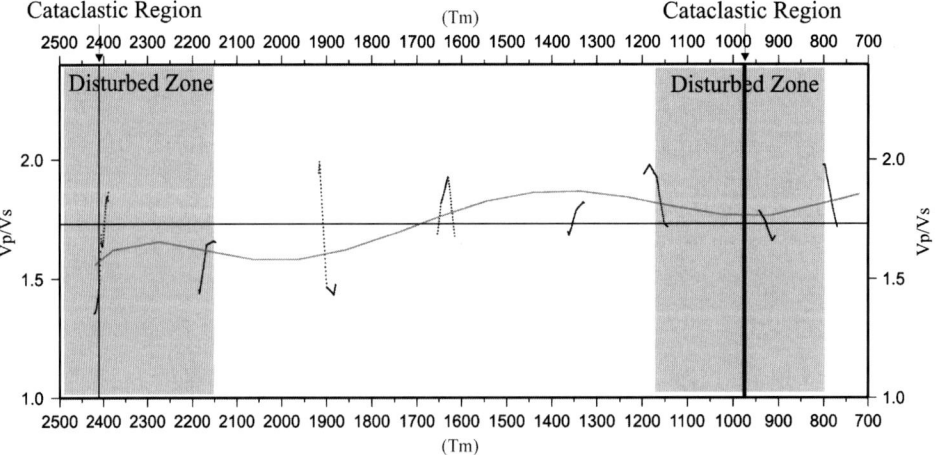

Fig. 10. V_P/V_S ratios derived from tomographic inversions at 3 m distance from the tunnel wall of the Faido adit (dotted line). The solid line connecting the V_P/V_S ratios is a polynomial fitting. A relative decrease of these values indicates a region of reduced strength on approaching a cataclastic fault (cf. Fig. 7c).

Discussion

The general trend in seismic velocity distribution reveals the structure of two major fault zones at Tm 973 and Tm 2410, surrounded by disturbed zones of about 100 m width on each side. The minimum value of V_P in the first fault zone around Tm 973 is 4800 m s^{-1}, whereas a value of 3100 m s^{-1} is reached in the second fault zone at Tm 2410. For the shear wave velocity, a minimum value of 2800 m s^{-1} is found at the first major fault zone, and of 2000 m s^{-1} at the second. The V_P/V_S ratio is 1.7 for the first fault zone, and 1.55 for the second.

One reason for the different behaviour of the velocities in the zone from Tm 1900 to Tm 2150 may be the steeper dip of the gneiss foliation (cf. Fig. 3). If the dip of foliation is almost normal to the tunnel axis, there is an increased tendency to instability at the tunnel wall involved. This may also account for the reduction in the V_P/V_S ratio. On the other hand, the zone of reduced V_P/V_S ratios coincides with the transition zone between the Leventina- and Lucomagno gneiss complexes crossing the Faido adit between Tm 2000 and Tm 2200. The Lucomagno gneiss is rich in mica and is generally softer than the Leventina Gneiss (Schneider 1997). Repeated seismic measurements in the 5500-m long adjacent Piora exploratory gallery revealed V_P/V_S values of 1.7 to 1.9 for the Leventina Gneiss and 1.6 to 1.7 for the Lucomagno gneiss (Dickmann & Sander 1996). Recent seismic tomographic studies made with ISIS in the adjacent Piora exploratory tunnel exhibited V_P/V_S values in the range of 1.5 to 1.6 in the transition zone between these two gneiss types. Thus, the relatively low ratio of $V_P/V_S = 1.6$ at Tm 2000 in Figure 10 may be explained partly by the presence of Lucomagno gneiss.

Conclusions

The seismic tomographic investigations in the Faido adit have shown that relative minima in V_P, V_S and V_P/V_S values measured along the tunnel wall may be used to predict fault zones in the adjacent rock mass. The different absolute values of the velocity, however, require further scrutiny in terms of the analysis of wave attenuation and velocity anisotropy of the rock mass, for example.

The Faido adit seismic case study also indicates the importance of the seismic measurement results' verification by direct inspection of the geological structures and outcropping faults at the tunnel wall. In addition, continuous measurement of geotechnical parameters such as fracture density and spatial orientation of joints would help to improve understanding of the results of seismic tomographic inversion.

The geophone arrays may also be arranged circumferentially around the tunnel wall to allow 3D tomographic inversions and improve our understanding of the impact of stress redistribution on seismic velocities. Influence of the excavation disturbance zone EDZ on the velocity measurements may be reduced by applying a larger base-length of the seismic profiles where the seismic rays can penetrate deeper into the undisturbed rock mass beyond the EDZ.

The authors gratefully acknowledge the support and co-operation of the Amberg Group AG in Zurich, especially F. Amberg and Th. Dickmann. We also thank B. Röthlisberger and F. Walker, from the construction site management of the Faido access tunnel, for their very helpful technical support during the seismic measurements. S. Mielitz, P. Otto and Ch. Selke of the GeoForschungsZentrum Potsdam assisted most efficiently in developing the ISIS hardware and its application to the tunnel seismic investigation reported here.

References

BORM, G., GIESE, R., SCHMIDT-HATTENBERGER, C. & BRIBACH, J. 1999. Verankerungseinrichtung mit seismischem Sensor. (Anchoring system with seismic sensor.) *German Patent Appl. 198 52 455.2; European Patent Appl. 99120626.9-2316; Japanese Patent Appl. HEI 11-322268.*

BORM, G., GIESE, R., OTTO, P., DICKMANN, TH. & AMBERG, F. 2001. Integrated seismic imaging system for geological prediction ahead in underground construction. *Rapid Excavation and Tunneling Conference (RETC), June 11–13, San Diego, USA.*

DICKMANN, T. & SANDER, B. K. 1996. Drivage concurrent tunnel seismic prediction (TSP). *Felsbau,* **14** (6), 406–411.

FALLS, S. D. & YOUNG, R. P. 1998. Acoustic emission and ultrasonic-velocity methods used to characterize the excavation disturbance associated with deep tunnels in hard rock. *Tectonophysics,* **289**, 1–15.

KNEIB, G., KASSEL, A. & LORENZ, K. 2000. Automated seismic prediction ahead of the tunnel boring machine. *First Break,* 295–302.

LÖW, S. & WYSS, R. 1999. *Vorerkundung und Prognose der Basistunnels am Gotthard und Lötschberg. (Prior investigation and predictions at the Gotthard und Loetschberg Base Tunnels.)* A. A. Balkema, Rotterdam, 404 pp.

NEIL, D. M., HARAMY, K. Y., HANSON, D. H. & DESCOUR, J. M. 1999. Tomography to evaluate site conditions during tunneling. *Proceedings of*

Geo-Institute, 99. 3rd National Conference of the GeoInstitute of ASCE Geo-Engineering for Underground Facilities, University of Illinois–Urbana, 13–17 June.

SCHNEIDER, T. R. 1997. *Schlussbericht Sondierstollen Piora-Mulde, Phase 1, Geologie/Geotechnik/Hydrologie/Geothermie. (Final Report on the Priora Mould Testing Gallery, Phase 1, Geology/*

Geotechnique/Hydrology/Geothermics), Reg. No. **144**, 1–30.

WALLACE, R. E. & MORRIS, H. T. 1986. Characteristics of fault and shear zones in deep mines. *Pure and Applied Geophysics*, **124**, 107–125.

ZELT, C. A. & SMITH, R. B. 1992. Seismic traveltime inversion for 2-D crustal velocity structure. *Geophysical Journal International*, **108**, 16–34.

Natural fracturing and petrophysical properties of the Palisades dolerite sill

D. GOLDBERG[1] & K. BURGDORFF[2]

[1]*Lamont–Doherty Earth Observatory, Rte 9W, Palisades,
NY 10964, USA (e-mail: goldberg@ldeo.columbia.edu)*

[2]*GeoMechanics International, Inc., Parmelia House, 191 St George's Terrace, Perth, WA 6000,
Australia (e-mail: burgdorff@geomi.com)*

Abstract: This investigation of naturally occurring fractures in the mafic rocks of the Palisades dolerite sill characterizes the porosity of this crystalline rock sequence, and yields a method of determining the *in situ* porosity when complete down-hole information is not available. Two holes, 229 m and 305 m deep, were drilled 450 m apart through the sill and into the underlying Triassic sediments of the Newark Basin. Both holes were logged with geophysical tools, including the acoustic borehole televiewer (BHTV), to identify intervals of high porosity, fracturing, and potential zones of active fluid flow. Using the BHTV data, 96 and 203 fractures were digitally mapped within the sill in Well 2 and Well 3, respectively. Most fractures dip steeply (76–78°). There is a shift in fracture orientation between Well 2 and Well 3, although the lithology of the sill is continuous. The dolerite penetrated in both holes is fresh and unaltered, and intersects a 7-m thick olivine-rich layer about 15 m above the bottom of the sill. Several fractures identified in the sill have large apparent aperture (>6 cm) that correspond to high-porosity zones (6–14%), measured from both resistivity and neutron logs in Well 2. We use a relationship between porosity and apparent fracture aperture in Well 2 to infer the porosity in Well 3. This correlative method for estimating porosity may be applicable between holes in other crystalline rock environments where down-hole log data are incomplete. Changes in the temperature gradient log also indicate active fluid flow, although flow appears to be most active in fractured and high-porosity zones in the sediments.

The purpose of this study is to investigate the petrophysical properties of naturally fractured rock in a dolerite sill, as well as of the underlying sediments. In this work, we measure the geophysical and mineralogical properties in order to piece together the porosity structure of the sill and sediments in the immediate vicinity of our research site in Palisades, NY, USA. The geology of the sill has been well studied in the past, and it is therefore an ideal location for investigations of the relationship between mineralogy, fracturing, porosity and permeability of crystalline rock. A variety of experiments are ongoing in this area, utilizing two research wells that were drilled on the site. Neither hole was cored, although drilling chips, hand specimens from outcrops, and well-log information were collected in both holes. In this paper, we use these data to evaluate the lithological composition, estimate matrix and fracture porosity of the rocks, and develop a method of determining the *in situ* porosity with incomplete down-hole or core data in one or more drill sites. This approach may be applicable to similar crystalline rock environments where down-hole measurements are limited.

Geological background and site characterization

The Palisades dolerite sill intruded into the Triassic sedimentary rocks of the Newark Basin in the Early Jurassic. It is now found outcropping along the west bank of the Hudson River in New York and New Jersey, opposite New York City, for approximately 80 km (Fig. 1). It is approximately 2.4 km wide in outcrop, and is mainly sheet-like but has some dyke-like features within it (Walker 1969a). The sill is famous for its prominent olivine layer that lies about 15–18 m above the contact between sill and the sediments. This layer most likely resulted from a 'separate late intrusion of olivine-normative magma' and not from gravity settling, as was previously thought (Husch 1990).

From: HARVEY, P. K., BREWER, T. S., PEZARD, P. A. & PETROV, V. A. (eds) 2005. *Petrophysical Properties of Crystalline Rocks.* Geological Society, London, Special Publications, **240**, 25–36.
0305-8719/05/$15.00 © The Geological Society of London 2005.

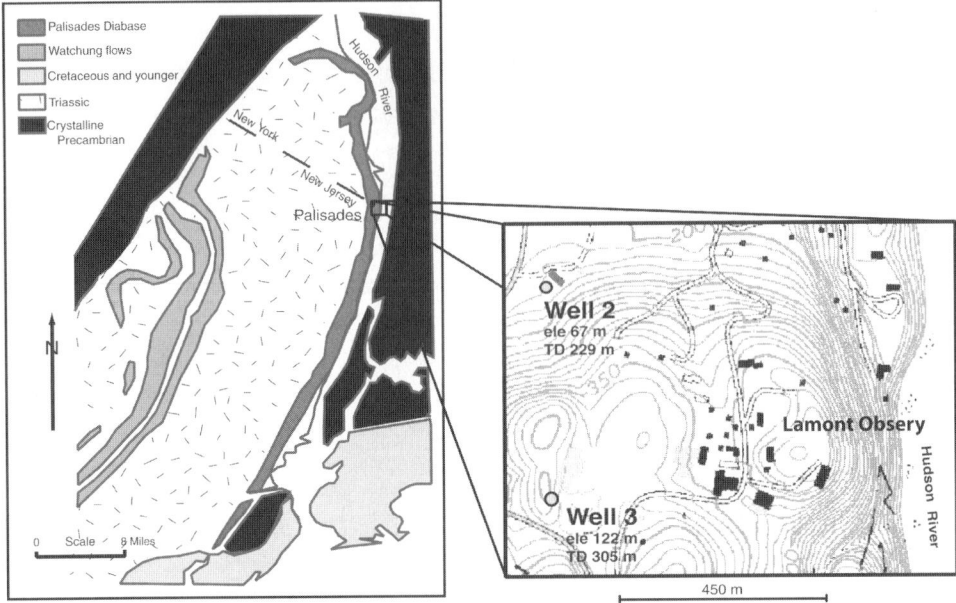

Fig. 1. Geological map of the Palisades Sill and surrounding regions (Walker 1969*b*). USGS topography inset map shows localities of the two drill holes (Wells 2 and 3) on the Lamont–Doherty Earth Observatory campus, with 3.25 m contours. Elevations (ele) and total depths (TD) are marked at each location, and the wells are approximately 450 metres apart.

The contact between the dolerite sill and the underlying sediments north and south of the drill sites is irregular and cuts up- and down-section through both fluvial and lacustrine Newark Basin formations (Olsen 1980). Note the wandering of the sill in the schematic cross-section in Figure 2b. The sill is seen in outcrop to have a general strike of N30°E, dipping 10–15° WNW (Walker 1969*b*).

A hole was drilled through the Palisades Sill (Well 2) on the Lamont–Doherty Earth Observatory campus in 1980, to 229 m depth (Fig. 1). The hole was geophysically surveyed using geophysical and geochemical logging tools (Anderson *et al.* 1990). Goldberg (1997) described the measurements made by these tools in detail. An additional hole (Well 3) was drilled in early 2000 to 305 m, 450 m north of Well 2 and approximately 55 m higher in elevation (Fig. 1). It was surveyed with gamma ray, caliper and temperature logging tools. Both holes were also logged with the borehole televiewer (BHTV). This tool produces an acoustic image record showing the depth and orientation of features intersecting the borehole, such as fractures or bedding planes (e.g. Goldberg 1997). All of the down-hole data were recorded as a function of depth in feet below the surface; the figures present the original units. To convert to SI units, please use the conversion factor of 3.28 ft per metre.

Palisades Sill lithology

Sampling and rock analysis

Well cuttings (drilling chips) provided samples at regular depths in Well 2 and Well 3. We sampled drilling chips every 3 m in the dolerite (0–230 m) and every 1.5 m in the sediments (230–305 m) in Well 3; chip samples were taken every 0.6 m from Well 2. Continuous logs of the drilling chips were created by estimating the percentage of rock types in each sample under a low-power microscope. To examine the mineralogical changes in the dolerite and the sediments, we made 12 thin sections of the drill chips at different depths in Well 3. Six thin sections were made previously in the dolerite and sediments in Well 2. Table 1 lists the depths and rock type of these thin sections.

Seventeen field samples were collected in Palisades State Park, NY, to aid in identifying the different mineralogical compositions in the dolerite and the sediments below. Samples 1 through 12 were taken approximately 7 km south of the drill sites. We took samples above,

(a)

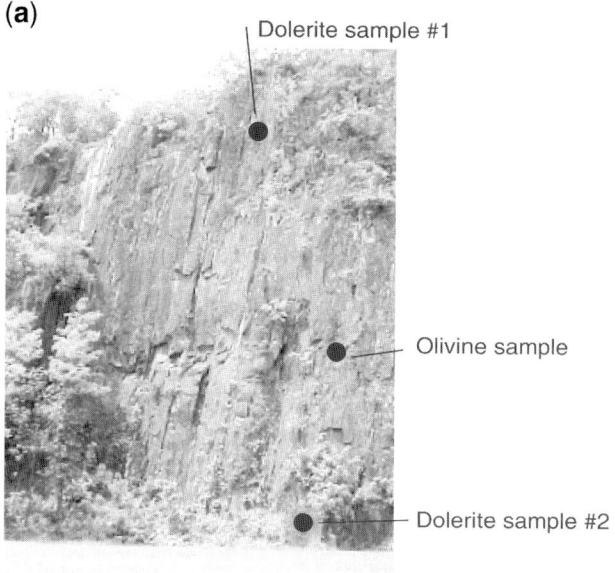

Dolerite sample #1

Olivine sample

Dolerite sample #2

(b)

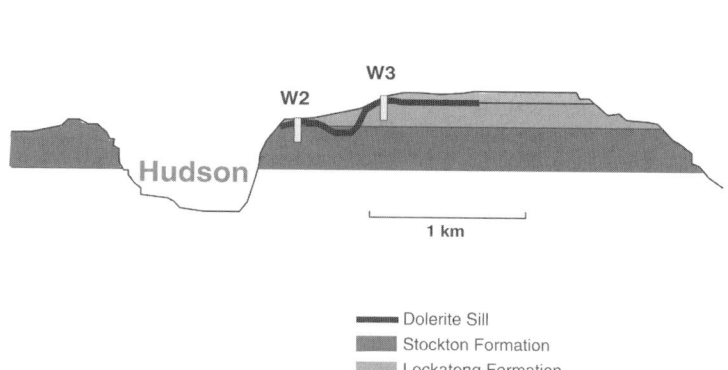

W3

W2

Hudson

1 km

▬ Dolerite Sill
■ Stockton Formation
■ Lockatong Formation

Fig. 2. (**a**). Photograph of Englewood Cliffs at Ross Dock, Palisades State Park, where hand specimens were collected for chemical analysis. (**b**). Schematic cross-section of Hudson River and Newark Basin sediments south of Palisades. The sill (dark) intruded at different stratigraphic levels in the sediments. Wells are shown (with approximate depths) cutting through the sill at different levels of its intrusion into the sediment basin (from Olsen 1980).

below and at the contact between the dolerite and the sediments, to determine the range of sedimentary rock types beneath the sill in the region surrounding the drill sites. Five samples were taken 16 km south of the drill sites, on a large south-facing cliff (Fig. 2a). Fracture zones within the dolerite at the Englewood Cliffs outcrop were noted at about 15 m above the bottom of the dolerite. Hand specimens from this region contained large amounts of olivine.

The photograph in Figure 2a shows the origin of the hand specimen locations.

Dolerite petrography

Examination of the hand specimens and drilling chips in both holes identifies the olivine-rich layer. It is approximately 6 m thick and 15.2 m above the contact between the dolerite and the sediments (Fig. 3). The lithologies identified in

Table 1. *Depths of thin-section samples in Well 2 and Well 3*

Rock type	Well 3		Well 2	
	Metres	Feet	Metres	Feet
Dolerite	12.1	40	180.7	593
	64	210	186	610
	137	**450**		
	158.5	520		
	183	600		
(Olivine-rich layer)	**213.4**	**700**		
	228.6	**750**		
	241	790		
Sediments	234.7	770	187.5	615
	245	805	191	627
	268	880	194.5	638
	303	995	227	745

Depths listed in bold type denote thin sections that are shown and discussed in the text.

the two holes within the dolerite are similar, and correspond well to the previous geological description of the sill (Walker 1969*b*). This is also valid for the mineralogical description of the sill, which is divided into layers that match the description determined by colour and grain size differences (Walker 1969*b*). The sill consists of a thick layer of pigeonite dolerite underlain by hypersthene dolerite (*c*.46 m thick) and bronzite dolerite (*c*.30 m thick). A layer of ferrohypersthene dolerite (21.3 m thick) is present at the top of the sill in Well 3, but not in Well 2. Below the bronzite, there is a 6-m thick layer of olivine-rich dolerite that extends downward to a chilled dolerite interval, which is fine grained and roughly 9 m thick, in contact with the sediments. Stringers of chilled dolerite occur further below the contact and are thinner in Well 2 than in Well 3 (Fig. 3). These do not match up stratigraphically between the holes.

Example images of four thin sections in the dolerite and olivine-dolerite are shown in Figure 4a to 4d. The thin sections show the variability of grain size and and low matrix porosity in the sill rocks (Fig. 4a & b) and illustrate the microstructure and crystal fabric of the mineral components (Fig. 4c & d). Figure 4a shows a section of a chip from 137 m in a typical section of the dolerite. The plagioclase and pyroxene grains are partially intergrown (exhibiting a subophitic texture) that is typical of many dolerites and intrusive dykes (R. Coish, pers. comm. 2001). Figure 4b shows a marked grain-size difference in the fine-grained chilled margin of the dolerite near the contact with the sediments at 228.6 m depth. An example of the olivine-rich dolerite is represented in Figure 4c

& d. This section, taken from a chip at 213.4 m, shows a very large pyroxene phenocryst with plagioclase and olivine grown within, an example of ophitic texture, which seems to be unique to this section. An olivine grain from the same section is identified under cross-polarized light in Figure 4d. Under scanning electron microscope analysis, the composition of this sample is shown to contain forsterite (Fo_{78}), while other olivine grains in the section had forsterite contents ranging from Fo_{66} to Fo_{71}, possibly explaining some of the observed differences.

The lithology of the dolerite sill can be traced continuously between the two drill-holes (Fig. 3). Although there are slight differences in the thicknesses of correlative dolerite sections, individual layers seem to be continuous through this part of the sill. The olivine-rich layer is present in both holes, as shown by examination of the drilling chips and thin sections. The section of ferrohypersthene dolerite from the top layer of Well 3 is not continuous between the wells. This is most likely due to erosion of this part of the sill in Well 2, which is topographically lower than Well 3 and closer to a tributary of the Hudson River (Fig. 1).

Sediment stratigraphy

While the dolerite mineralogy in the two wells is similar, the sediment stratigraphy identified below the dolerite in each hole is very different (Fig. 3). One reason for this may be the different sampling frequency of drilling chips (five times greater in Well 2). However, the absence of red siltstone in Well 3 and their abundance in Well 2, as well as more frequent beds of purple–black shale in Well 3, suggest that the wells penetrate two different sediment sequences (Burgdorff & Goldberg 2001). This may be explained by the variable stratigraphic position of the sill with respect to the underlying Newark Basin sediments. Figure 2 illustrates this schematically in cross-section, based on outcrop evidence south of the drill sites (Olsen 1980). Although the schematic is conjectural, it is likely that the intrusion path of the sill is irregular and undulatory in the area, intersecting different sections of the sedimentary sequence at each drill site.

Analysis of down-hole measurements

Fracture identification from borehole televiewer data

Fracture data from the two wells are compiled from acoustic images taken in the borehole with the borehole televiewer (BHTV). BHTV images

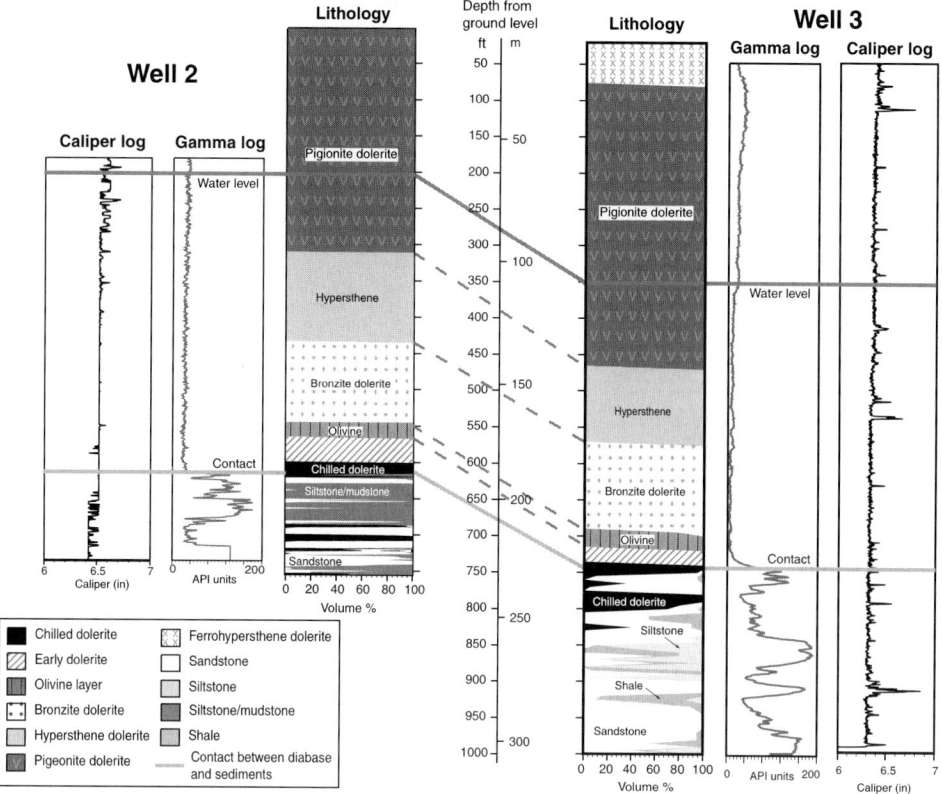

Fig. 3. Volume per cent graphs of the lithology of Well 2 and Well 3 with gamma-ray and caliper logs. Depths are below ground level. The water level and the contact between the dolerite and the sediments are shown between the wells. Dashed lines between the two wells indicate the slight thickness changes of similar mineralogical types within the dolerite. Note the thin olivine layer approximately 15 m above the base of the sill in both wells. High gamma-ray readings (in API units) indicate the occurrence of clay-bearing sediments below the dolerite sill. The caliper logs measure hole diameter.

are created by the acoustic signal emitted by a rotating transducer as the tool is pulled up the hole at a typical logging speed of *c*.1.5 m per minute for optimum resolution (Zemanek *et al.* 1970). The recorded ultrasonic data create a continuous image log of the interior of the borehole (e.g. Keys 1989). BHTV data provide the orientation and depth of features intersecting the borehole wall. Planar features are displayed as sinusoidal banding in the images. Roughness, rock hardness, and sometimes even grain size can be determined, although image quality is usually poor when the conditions of the borehole wall are rough or rock formations are soft. To determine the dip of a feature intersecting the borehole, the height and orientation of the sinusoid are measured from the BHTV image, and the diameter of the borehole is taken from the caliper log (e.g. Goldberg 1997).

Understanding the fracture pattern in the sill is essential for identifying possible active aquifers or potential hydrological flow zones. Using the BHTV images from each hole, we identified and mapped the fractures using digital image analysis software (Fig. 5). The dip direction and plunge of the poles to fracture planes are displayed on a contour stereonet plot to show the dominant orientation of fractures (Fig. 6). In Well 2, the majority of the fractures in the sill generally strike east–west, steeply dipping both to the north and south. Two population centroids of the poles, corrected for the local magnetic declination, are clustered around an orientation of N3°E, plunging 78°NE, and S13°W, plunging 76°SW. The 203 fractures in the Well 3 dolerite are bimodal, with two separate population centroids of poles oriented S49°E and S2°E (again plunging at an average of 78° and 76°SE,

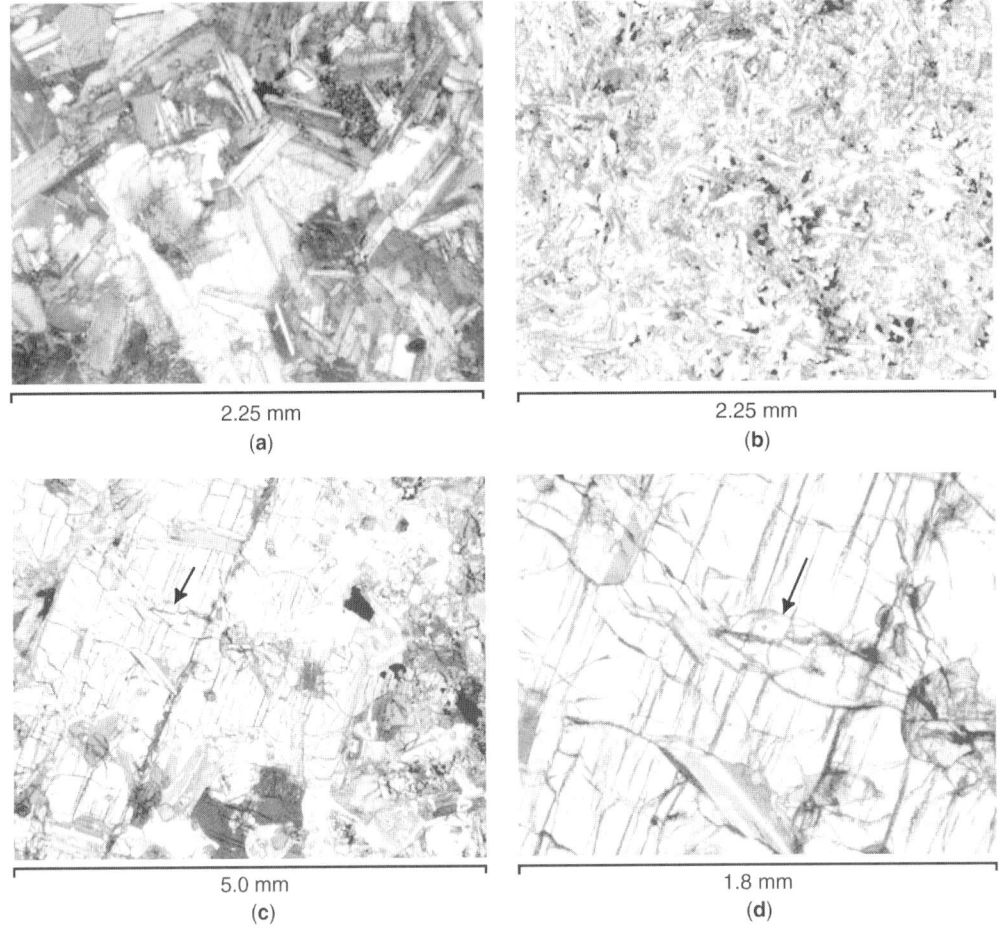

Fig. 4. (**a**) Thin section of chip from 137 m (typical of the massive dolerite) under cross-polarized light. (**b**) Thin section of a fine-grained chip from 228.6 m (at the chilled margin of dolerite above the sediments) under cross-polarized light. (**c**) Thin section of chip from olivine-rich section of the dolerite at 213.4 m under cross-polarized light, with olivine and plagioclase grown within a large pyroxene phenocryst. (**d**) Larger-scale image of Figure 4(c) above, with arrow pointing to an olivine grain in the centre of the field of view.

respectively). The main subvertical fracture planes (given by the normal to the poles) strike predominantly east–west, although a secondary set of subvertical fractures in Well 3 strikes NE–SW. This population of fractures aligns with other regional fractures observed in Newark Basin sediments, and with the inferred maximum compressive stress direction (Goldberg *et al.* 2003). The fracture sets are very similar in both holes.

Analysis of geophysical logs

Natural gamma-ray and caliper logs were recorded in both holes (Fig. 3). Natural gamma logs record the total gamma radiation detected in a borehole, and they are the most widely used nuclear logs for stratigraphic investigations. Sediments are much more radioactive than basalt or other intrusive mafic rocks, and therefore a large increase in gamma-ray activity is observed at the dolerite–sediment contact. Caliper logs measure the diameter of the hole. The two wells appear to be generally smooth, without any large washout zones. Hole diameters in Well 2 and Well 3 range between 16 and 16.5 cm.

A neutron porosity log in Well 2 is shown in Figure 7. With this tool, a radioactive source emits fast neutrons that are then slowed down by collisions with hydrogen nuclei in the rock

Olivine fracture zone in Well-2 (163–173.8 m)

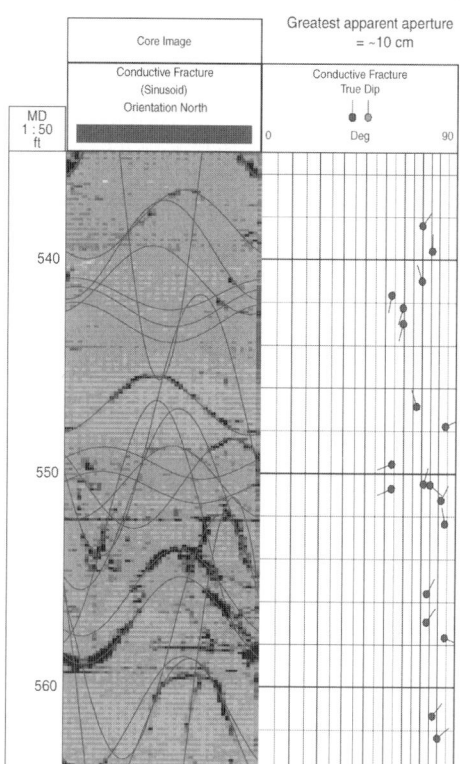

Fracture Zone in Well-2 (97.6–106.7 m)

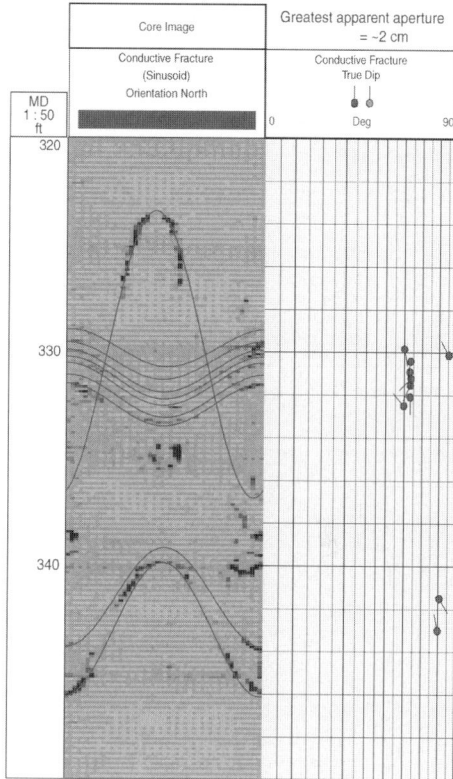

Fig. 5. Two examples of BHTV data from Well 2. Curves were manually picked to illustrate fractures and to digitally record their orientation and dip magnitude. Dip and azimuth of the imaged fractures are shown on a tadpole plot (right). The left-hand image shows a section of Well 2 from 163 to 173.8 m in depth, with a high density of fractures, many with large apparent aperture (>6 cm). The right-hand image is a section from 97.6 to 106.7 m in Well 2, with a high fracture density but smaller apparent aperture (1–2 cm). To convert to SI units, use the conversion factor of 3.28 ft per metre.

(e.g. Broglia & Ellis 1990; Goldberg 1997). These correspond to fluid-filled pore spaces in the dolerite, as well as to fluids in open fractures. Clays and other minerals containing bound hydrogen may affect these measurements, but neutron logs generally provide reliable porosity estimates in unaltered crystalline rocks (Harvey & Brewer 2003). The lithology of the Palisades Sill was shown by geochemical log and sample analysis to contain fresh (unaltered) dolerite and related igneous phases (Anderson *et al.* 1990).

Resistivity logs measure the electrical properties by forcing a current beam into the rock and then receiving it at electrodes located on the logging tool (e.g. Goldberg 1997). The shallow and deep resistivity logs (LLS and LLD) track each other throughout the well, but values diverge in the massive dolerite (uppermost interval), which is highly resistive (Fig. 7). This is

likely due to the tool geometry that emphasizes horizontally oriented features with the shallow resistivity log and deeper and vertically oriented features with the LLD measurement (e.g. Pezard 1990). Therefore, the separation between the two logs indicates that vertical fracturing occurs throughout the sill, but decreases near the bottom where the two logs come together. In the sediments, the LLD and LLS resistivity logs overlie each other and measure much lower resistivity values. In general, deep-reading resistivity logs, like the LLD, more accurately represent undamaged formation properties and are used for porosity and lithology interpretation.

Fractures filled with water are usually more conductive than the surrounding rock. Therefore, zones of high conductivity (low resistivity) in the dolerite may be used to estimate the fracture porosity using the method of Keys (1989) and

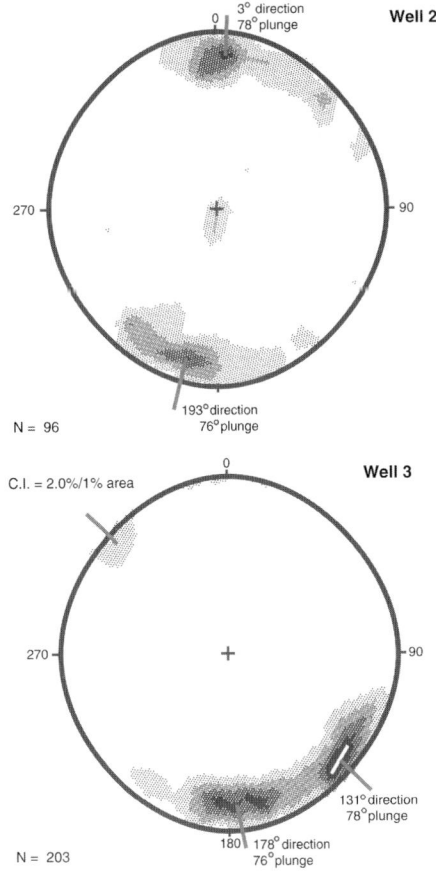

Fig. 6. Contour plots of dip direction and plunge of fracture planes in the dolerite for Well 2 and Well 3. In Well 2, 96 fractures dip steeply (76–78°) with plunge in N3°E and S13°W directions. This depicts fractures that strike approximately east–west with near-vertical dips slightly to either the north or south. Well 3, however, shows a broader distribution of 203 fractures with plunge ranging from S2°E to S49°E directions. Given this shift in orientations, fracture zones in the sill may or may not be continuous over the short lateral distance and north–south offset between the wells.

Archie (1942). For fresh, unaltered rocks, the surface conduction effect of clays and other alteration minerals in the Archie formulation are minor and may be ignored (Revil & Glover 1998). We use the simple relationship between porosity and the LLD resistivity log after Archie (1942):

$$a\phi^m = R_w/R_{LLD}$$

where ϕ = porosity, R_{LLD} is the deep resistivity value, R_w is the resistance of the pore fluid in the formation. To compute porosity with this relationship, constant a and m values are assigned based on rock type and borehole fluid chemistry to fit the average neutron porosity log from Well 2 in low-porosity intervals. We used $a = 10$ and $m = 1.9$ in the dolerite, and $a = 2$ and $m = 1.7$ in the sediments. Published values for m typically fall within the range of 1.0–1.85 for basalt lavas and meta-igneous rocks (Keller *et al.* 1979; Kirkpatrick 1979; Pezard 1990) and near 2.0 for sediments (Archie 1942). We assume a freshwater value of $R_w = 40$ ohm m, roughly two to three times higher than the average of water sample measurements from Well 3 (J. Matter, unpublished data). High R_w values overestimate the a coefficient in this formulation, but do not affect the m value. In these dolerite rocks, accurate values for a probably fall between three and 10. Our resulting estimates of porosity from the resistivity log range from 3% in fresh dolerite to *c*.40% in the underlying sediments (Fig. 7). Divergences of the resistivity-derived and neutron porosity logs in some areas are probably due to the assumption of using constant Archie coefficients when different compositions are present. Furthermore, we assume that neutron absorption is similarly related to porosity in crystalline and sedimentary rocks, but this is not uniformly true (Broglia & Ellis 1990). Nevertheless, the two log estimates of porosity agree extremely well in both the dolerite and sedimentary units, with the constant a and m values estimated above.

Porosity prediction from borehole televiewer imaging

Using the BHTV data over the logged interval in Well 2, we can calculate a fracture density log that indicates the number of fracture occurrences over successive 3.25 m intervals (Fig. 7). Apparent fracture aperture was measured directly from the BHTV images with a measurement error of ±2 cm. There are fracture zones with large apparent apertures throughout the hole, including one that occurs within the olivine layer at approximately 168 m depth. Figure 7 indicates that high-porosity intervals in the dolerite in Well 2 correspond with more intensively fractured intervals with larger apparent apertures.

In order to approximate a relationship between porosity and the fracture aperture in Well 2, we first compute the fracture density over 3.25-m (10-ft) running intervals through the well. We

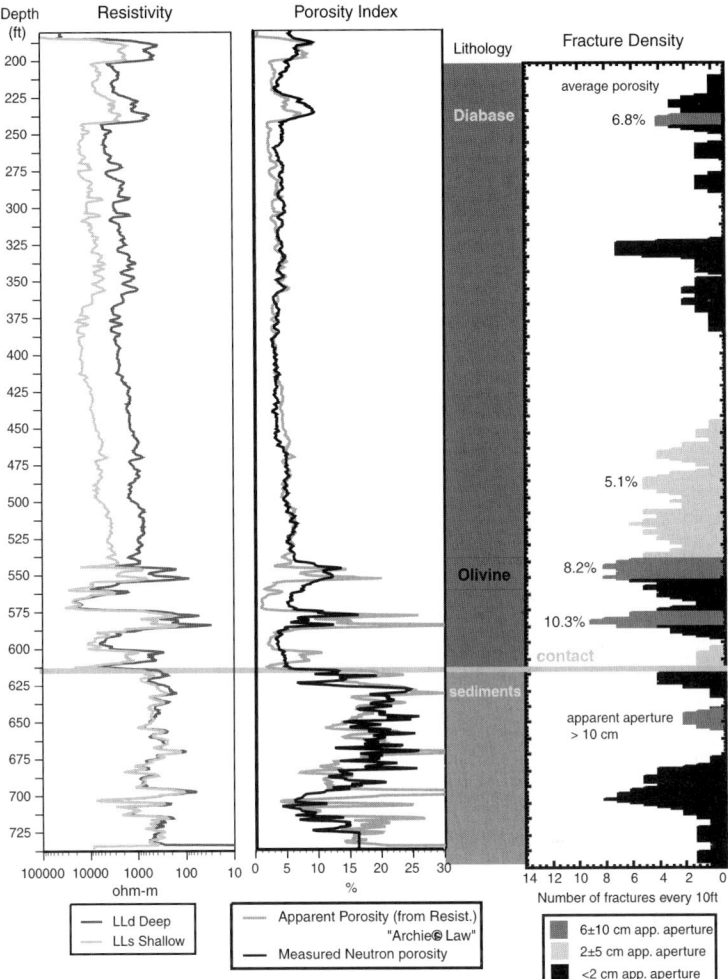

Fig. 7. Deep and shallow resistivity logs (ohm m) on a log scale for Well 2 (left track). Neutron porosity log and apparent resistivity-based porosity (centre track), and fracture density log (right track). Archie's law was used to compute the apparent porosity from the deep resistivity log ($a = 10$, $m = 1.9$ in dolerite; $a = 2$, $m = 1.7$ in sediments). The fracture density log (3-m window) also shows the apparent fracture aperture for increments (6–10 cm), (2–5 cm) and (>2 cm). The average of the porosity logs in high-porosity zones over 3.25-m (10-ft) windows is noted on the fracture density log. High average porosity and high fracture density occur in the olivine layer. To convert to SI units, use the conversion factor of 3.28 ft per metre.

also calculate five different apparent aperture ranges varying from no fractures to apertures of <2 cm, 2–5 cm, 6–10 cm and >10 cm, respectively, within the dolerite sill. The fracture aperture range is superimposed as the three-level grey scale on the fracture density log in Figure 7. Exceptionally large (>10 cm) aperture fracture intervals are noted as well. We then calculate the average porosity using the neutron and resistivity logs over the intervals containing large-aperture fractures in the dolerite sill. The average porosity values

are noted next to the fracture density log, and we use least-squares regression to compute a linear relationship between the average porosity and the aperture range in these zones. Figure 8 shows the resulting relationship, $\phi = 0.7a + 3.8$, where $R^2 = 0.5859$. Although this regression coefficient is somewhat low, we have confidence using it to predict the average porosity from observations of apparent fracture aperture in a similar environment, namely from the BHTV log images in Well 3.

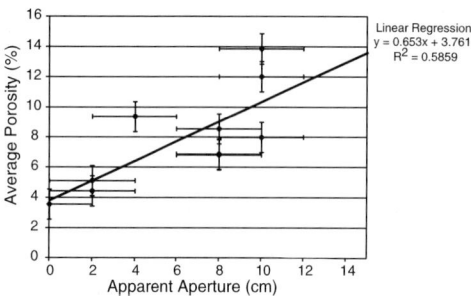

Fig. 8. Relationship between apparent aperture of fractures and average porosity from Well 2. Regression analysis yields: $\phi = 0.653a + 3.76 \ R^2 = 0.59$. Error bars denote ± 2 cm apparent aperture and $\pm 1\%$ average porosity.

In Well 3, the BHTV data were similarly analysed to calculate fracture density and apparent fracture aperture ranges (superimposed grey scale) over successive 3.25 m intervals (Fig. 9). Zones of high fracture density are observed throughout the hole, several of which contain large-aperture fractures. In the absence of resistivity and neutron porosity logs in Well 3, we use the BHTV data to predict the porosity of fractured zones in the dolerite sill by extension of the relationship derived from Figure 8. Values noted on the fracture density log are the aperture-derived porosity values over intervals containing large-aperture fractures. Several of these observed fracture zones have large apparent aperture (and high computed porosity), including one that occurs within the olivine-rich dolerite layer at approximately 216.4 m depth, as well as others in the sill above and in the sediments below. Given that the dolerite rocks in Well 2 and Well 3 are similar in mineralogy, with low matrix porosity, and that fracturing contributes to most of the porosity occurring in the dolerite, the extension of the relationship provides a reliable and effective estimation of porosity from the BHTV images of the apparent fracture apertures.

Discussion

The occurrence of large-aperture, porous fractures in the Palisades dolerite sill, however, does not necessarily imply that these zones are permeable to fluid flow. To address this question, we ran a temperature log in Well 3, and calculated the vertical temperature gradient over successive 4-m depth intervals (Fig. 10). The hole warms by approximately $2\,^{\circ}C$ from top to bottom, with a mean gradient of $0.001\,^{\circ}C/m$. Although the temperature gradient log does not

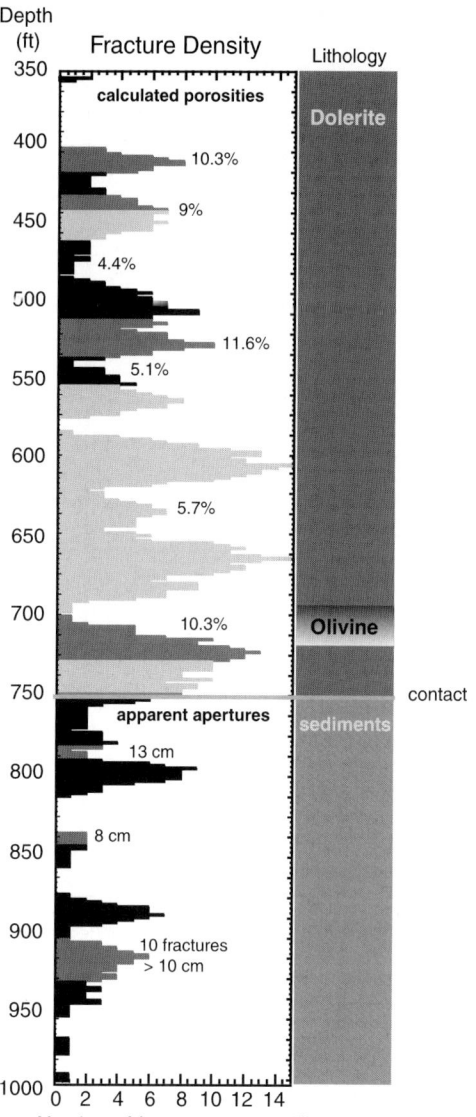

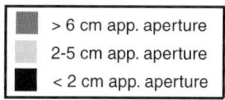

Fig. 9. Fracture density log in Well 3 with large and small apparent aperture ranges shown (same key as Fig. 7). Calculated average porosity values were obtained from the relationship between apparent aperture and porosity in Well 2 (Fig. 8). Note the high-porosity and large apparent aperture zone in the olivine layer. To convert to SI units, use the conversion factor of 3.28 ft per metre.

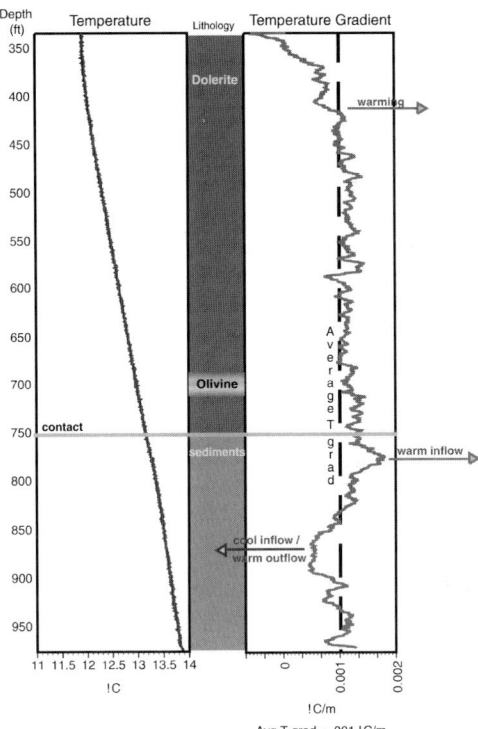

Fig. 10. Temperature log in Well 3 (left) shows warming of approximately 2 °C from 94.5 m (water level) to 304.9 m (bottom of the hole). A temperature gradient log (right) was calculated using a 3.7 m window and a *c*.1.5 m moving average. The average temperature gradient is 0.001 °C m^{-1}. Deviations from this average indicate warming or cooling of the borehole fluid, which is likely due to fluid flow into or out of the well. Changes indicate some flow within the sill, although fluid flow appears to be the most active in the sediments. To convert to SI units, use the conversion factor of 3.28 ft per metre.

substitute for *in situ* permeability measurements, changes in the gradient may be helpful in identifying zones where fluid is actively moving into or out of a borehole. In Figure 10, several active flow zones are indicated in the dolerite, where the gradient deviates from the mean value; the largest changes appear to occur in the sediment layers below the sill, however.

Future work regarding the petrophysical and hydrological characterization of the Palisades Sill will include additional temperature and flowmeter logs and field tests of bulk permeability, to estimate fluid flow potential within isolated intervals (as well as between the two holes), and hydrological modelling of the *in situ* groundwater flow regime. The comparison of these logs and hydrological tests between Well 2 and Well 3 will quantify the permeability in the predicted high-porosity

zones, and enable the lateral continuity of these zones between the holes to be determined.

Conclusions

We use drilling chips and logging data to characterize the geological and geophysical properties of an olivine-rich dolerite containing naturally occurring fractures in the Palisades dolerite sill. We determine the extent of fracturing and the effect of fracturing on porosity in crystalline rocks. A hole-to-hole comparative method is used with BHTV, and geophysical log data where complete information was not available. The methodology may be applied in other low-porosity crystalline rock formations, using image and other geophysical logs that are recorded under similar conditions.

In the Palisades dolerite sill, the igneous petrography formations and fracture orientations map well across the 450-m lateral separation between the two holes considered in this study. We compute apparent porosity from resistivity log data using Archie's law with coefficient values based on these rock types and fresh pore fluid, and match it to the neutron porosity log in low-porosity intervals. The Archie coefficients in the dolerite are $a = 3-10$, depending on the *in situ* fluid resistivity, and $m = 1.9$. From the BHTV images and geophysical logs, we conclude that large apparent aperture fractures correspond with high average porosity values. Without resistivity or porosity logs in both wells, we estimate the average porosity by using the fracture aperture/porosity relationship established in one well and applying it to the apparent aperture measurements in the other. The olivine layer, identified in nearby outcrops and by thin-section analysis of drilling chips, correlates well with a zone of large apparent aperture, high-porosity fractures in both holes. The temperature-gradient log provides an indication of active fluid flow through the sill; however, the most active zones appear to occur in the sediments below. In the future, temperature and flowmeter logs and *in situ* hydrological testing in the dolerite will enable characterization of the *in situ* fluid permeability in this fractured, crystalline rock formation.

We greatly appreciated assistance from G. Guerin, G. Iturrino, W. Masterson, J. Matter, A. Meltser, G. Myers, P. Olsen and M. Reagan at Lamont–Doherty Earth Observatory, E. Scholz from Down-hole Systems, Inc., and R. Coish at Middlebury College for the collection and presentation of these data. We benefited from comments on the manuscript by R. Pechnig and an anonymous reviewer. This project was supported by the Lamont–Doherty

Summer Internship Program, Lamont–Doherty Earth Observatory, and The Earth Institute of Columbia University. Lamont–Doherty Earth Observatory contribution number LDEO 6691.

References

ANDERSON, R. N., DOVE, R. & PRATSON, E. 1990. Geochemical well logs: calibration and lithostratigraphy in basaltic, granitic, and metamorphic rocks. *In*: HURST, A., LOVELL, M. A. & MORTON, A. (eds) *Geological Applications of Wireline Logs*, Geological Society, London, Special Publications, **48**, 177–194.

ARCHIE, G. E. 1942. The electrical resistivity log as an aid in determining some reservoir characteristics. *Transactions AIME*, **146**, 54–62.

BROGLIA, C. & ELLIS, D. 1990. Effect of alteration, formation absorption, and standoff on the response of the thermal neutron porosity log in gabbros and basalts: examples from deep sea drilling project–ocean drilling program sites. *Journal of Geophysical Research*, **95**, 9171–9188.

BURGDORFF, K. & GOLDBERG, D. 2001. Petrophysical characterization and natural fracturing in an olivine-dolerite aquifer. *Electronic Geosciences*, **6** (3), 22 printed pages, 5900 kbytes, *www.springer. de/link/service/journals/10069/tocs.htm*.

GOLDBERG, D. 1997. The role of downhole measurements in marine geology and geophysics. *Reviews of Geophysics*, **35**, 315–342.

GOLDBERG, D., LUPO, T., CAPUTI, M., BARTON, C. & SEEBER L. 2003. In situ stress and fracture evaluation from borehole televiewer and core data in the Newark Rift Basin, *In*: OSLEN, P. E. & LE TOURNEAU, P. M. (eds) *The Great Rift Valleys of Pangea in Eastern North America, Volume 1: Tectonics, Structure, and Volcanism of Supercontinent Breakup*. Columbia University Press, New York, Ch. 7.

HARVEY, P. & BREWER, T. 2005. On the neutron absorption of basic and ultrabasic rocks: the significance of minor and trace elements. *In*: HARVEY, P. & BREWER, T. (eds) *Petrophysics of Crystalline Rocks*, Geological Society, London, Special Publications, **240**, 207–218.

HUSCH, J. M. 1990. Palisades sill: origin of the olivine zone by separate magmatic injection rather than gravity settling. *Geology*, **18**, 699–702.

KELLER, G., GROSE, L., MURRAY, J. & SKOKAN C. 1979. Results of an experimental drill hole at the summit of Kilauea Volcano, Hawaii. *Geophysical Research Letters*, **5** (3), 345–348.

KEYS, W. S. 1989. *Borehole Geophysics Applied to Ground-water Investigations*. National Water Well Association, Dublin, OH.

KIRKPATRICK, R. 1979. The physical state of the oceanic crust: results from downhole geophysical logging in the mid-Atlantic Ridge at 23°N. *Journal of Geophysical Research*, **84**, 178–188.

OLSEN, P. E. 1980. Fossil great lakes of the Newark supergroup in New Jersey. *In*: MANSPEIZER, W. (ed.) *Field Studies of New Jersey Geology and Guide to Field Trips: 52nd Annual Meeting of the New York State Geological Association*. Rutgers University, Newark NJ, 352–398.

PEZARD, P. A. 1990. Electrical properties of mid-ocean ridge basalt and implications for the structure of the upper oceanic crust in hole 504B. *Journal of Geophysical Research*, **95**, 9237–9264.

REVIL, A. & GLOVER, P. 1998. Nature of surface electrical conductivity in natural sands, sandstones and clays. *Geophysical Research Letters*, **25**, 691–694.

WALKER, K. R. 1969a. A mineralogical, petrological, and geochemical investigation of the Palisades sill, New Jersey. *Memoirs, Geological Society of America*, **115**, 175–187.

WALKER, K. R. 1969b.The Palisades sill, New Jersey: a reinvestigation. *Special Papers, Geological Society of America*, **111**, 178 pp.

ZEMANEK, J., GLENN, E., NORTON, L. & CALDWELL, L. 1970. Formation evaluation by inspection with the borehole televiewer. *Geophysics*, **35**, 254–269.

Scale dependence of hydraulic and structural parameters in fractured rock, from borehole data (KTB and HSDP)

G. ZIMMERMANN[1], H. BURKHARDT[2] & L. ENGELHARD[3]

[1]GeoForschungsZentrum Potsdam, Telegrafenberg, D-14473 Potsdam, Germany
(e-mail: zimm@gfz-potsdam.de)
[2]Department of Applied Geophysics, Ackerstr. 76, D-13355 Berlin, Germany
[3]Institute of Geophysics and Meteorology, Mendelssohnstr. 3,
38106 Braunschweig, Germany

Abstract: Fundamental understanding of the origin, geometry, extension and scale dependence of fluid pathways in fractured rock is still incomplete. We analysed fracture networks on different scales, based on data from fluorescent thin sections and borehole televiewer (BHTV) images, to obtain geometrical network parameters and to estimate fracture permeability in the vicinity of a mantle plume (Hawaii Scientific Drilling Project, HSDP).

In the depth interval between 814 and 1088 m below sea-level, we observed microfractures in the fluorescent thin sections, and macroscopic fractures in the corresponding BHTV data from the same depth range. Initial modelling of the microscopic network from the fluorescent thin section taken at 1088 m below sea-level gives a clear indication that in this particular case the preferential hydraulic pathways on the microscopic scale are the microfractures in the olivine crystals. This is the only plausible explanation of high porosity (16.6%, based on core measurement), due to the observed vesicles and the corresponding low permeability of 10 μdarcy. Modelling hydraulic flow and calculation of permeability leads to similar values of permeability of 12.3 μdarcy, assuming a mean fracture aperture of 1 μm and an exponential distribution function of the fractures.

Detected structures from BHTV measurements were used to construct a macroscopic stochastic network to simulate the hydraulic flow. We found 337 fractures in the depth section from 783.5 to 1147.5 m below sea-level, which result in a linear frequency of 0.927 m^{-1}. Assuming horizontal layers and constant fracture apertures of 100 μm for all structures, leads to a first estimate of permeability of 77 mdarcy ($7.7 \times 10^{-14} \text{ m}^2$) in this depth section.

In a recent work, we showed that for data from the Continental Deep Drilling Project (KTB), the fracture density versus fracture length follows a power law. First results from the Hawaiian data suggest a similar relationship, despite all of the differences in the lithological conditions between both sites.

Knowledge of the rock properties controlling fluid movement is a basic prerequisite to understand the dynamical processes, temperature and stress regime of the upper crust. The aim of this work is to transfer methodically a concept concerning hydraulic properties and investigations of their scale effect from the Continental Deep Drilling Program (KTB) (Huenges & Zimmermann 1999; Zimmermann *et al.* 2000*a,b*, 2001, 2003) to the Hawaii Scientific Drilling Project (HSDP). Fracture systems in the different lithological units can be investigated by geometrical networks describing the structure and properties of potential hydraulic pathways.

The aim of this work is the comparison of different hydraulic pathways and the scale dependence of hydraulic parameters to obtain universal connections between these quantities. Key parameters to characterize hydraulic permeability are the aperture, density and connectivity of fractures (as well as additional network parameters from percolation theory at the various scales of investigation). To describe the hydraulic pathways, we use a database of structural borehole measurements (borehole televiewer), core scans and fluorescent thin sections representing different scales of investigation. This leads to the construction of deterministic as well as stochastic networks to estimate the hydraulic flow by finite-element modelling.

Permeability and scale effect

Permeability is one of the essential parameters for the description of hydraulic properties of

From: HARVEY, P. K., BREWER, T. S., PEZARD, P. A. & PETROV, V. A. (eds) 2005. *Petrophysical Properties of Crystalline Rocks*. Geological Society, London, Special Publications, **240**, 37–45.
0305-8719/05/$15.00   © The Geological Society of London 2005.

rocks, and is a prerequisite to describe the hydraulic pathways for fluids. It is a quantity that can take different values depending on the scale of investigation (i.e. it is not scale-invariant) (Brace 1984; Clauser 1992). On the macroscopic scale, it represents an integral value in the vicinity of the borehole, and depends on relatively large, irregular distributed fracture systems; hence it cannot be verified by core measurements. Estimation by hydraulic experiments in drill-holes yield integral values, but are normally limited to a few depth intervals. The core measurements represent a permeability with hydraulic pathways on the microscopic scale and lead to permeability values at simulated *in situ* conditions if taking into account the *in situ* confining pressure. This leads to an estimate of the matrix permeability, but generally cannot be scaled up.

Besides permeability measurements, different kinds of concepts exist to model the hydraulic flow in fractured media. It is a highly sophisticated problem to transfer the complex structures of natural rocks to adequate, equivalent models. One can distinguish between deterministic fracture networks (Kolditz 1992, 1996; David 1993), fractal fracture networks (Kosakowski 1994, 1996; Acuna & Yortsos 1995) and stochastic fracture networks (Wollrath 1990; Zimmermann *et al.* 2000*b*).

The scale dependence of hydraulic properties can be connected to the existence and size of a potential representative elementary volume (REV). So far, this represents an unsolved problem, which has been investigated using various approaches and model concepts. Below the REV, a medium appears as heterogeneous, with high variability of its properties; above the REV it can be considered as a statistically homogeneous and ergodic medium, and hence can be modelled as an 'equivalent homogeneous' medium. An overview concerning the scale effect and the corresponding concepts to model the hydraulic flow was given by Guéguen *et al.* (1996). The description of fracture models and their characteristic parameters is achieved with various theoretical approaches. The connectivity of discrete fracture networks can for instance be described by percolation theory (e.g. Stauffer & Aharony 1992). It describes the fluid flow in a hypothetical medium or network, and was first introduced by Broadbent & Hammersley (1957). Generally, this process can be described as a diffusion of fluids, with corresponding randomness of its pathways in the medium or with the stochastic properties of the medium itself. An introduction to percolation theory and to the fluid transport in fractured media can be found,

for example, in Berkowitz & Balberg (1993), Sahimi (1995), Selyakov & Kadet (1996) and Berkowitz & Ewing (1998).

'Effective medium theory' (EMT) transfers heterogeneous properties into a statistically homogeneous medium (David *et al.* 1990). But this approach is only valid for small heterogeneities and requires a REV (Guéguen et al. 1997). A multi-scale approach was introduced by Gavrilenko & Guéguen (1998). They used a modified 'renormalization-group' (RG) theory in conjunction with a percolation theory approach to describe the scale effect of permeability. Their results are in agreement with the observed measured effective permeability on the regional scale, based on the compilation of permeability measurements as a function of scale by Clauser (1992).

An essential component of percolation theory is the existence of scaling laws near the percolation thresholds. A detailed review of the current state of the art is given by Bonnet *et al.* (2001). For fracture networks close to the percolation threshold, power laws describe hydraulic properties and network parameters in a very general way (Robinson 1983, 1984; Balberg *et al.* 1991; Davy *et al.* 1992; Davy 1993). A frequently analysed quantity, affecting the connectivity of the fracture networks and hence its permeability, is the fracture length l and the length distribution function $n(l)$. This function follows the power law $n(l) \approx l^{-b}$ where b is an exponent varying generally between 1 and 3 (Balberg *et al.* 1991; Bour & Davy 1997, 1998) and hence does not permit an a priori definition of an REV.

Modelling fracture permeability

The aim of this work is to determine the hydraulic transport parameters of fractured rock on different scales and to obtain stochastic properties from these geometrical data to describe the characteristics of fracture networks. To simplify the analysis we assume fractures to be smooth planes without roughness. Generally, individual fracture planes can be described by fracture length and width (surface area), aperture and orientation (dip, azimuth). Projection to a 2D surface reduces the geometry description of the fracture surface area to a fracture length.

We analysed fluorescent thin sections with 2D networks, which represent microfractures with a length scale of up to 100 μm. The macroscopic scale is represented by borehole televiewer (BHTV) measurements, which show images of borehole wall structures on the metre scale. This provides 3D information about the orientation of the fractures, but gives no information about fracture length and width. From these basic geometrical data, discrete deterministic – as well as stochastic – fracture networks can be obtained. Subsequently, fluid flow was simulated with the finite-element program 'ROCKFLOW' (Lege *et al.* 1996) by applying Darcy's law (Guéguen & Palciauskas 1994), which assumes laminar fluid flow.

Calculation of critical parameters

The network parameter describing the transition from a non-connected medium to a connected medium is called the percolation threshold (Stauffer & Aharony 1992). In the case of continuously distributed structures, a quantitative definition of the percolation threshold is given by the percolation parameter. For 2D networks with a constant fracture length l and randomly distributed position and orientation, the percolation parameter can be defined as (Bour & Davy 1997):

$$p = Nl^2/A = \lambda_A l^2 \qquad (1)$$

with $N =$ number of fractures, $l =$ fracture length, $A =$ surface of the network, $\lambda_A = N/A =$ density of fractures (2D). Introducing a fracture length distribution and replacing the constant fracture length by the mean value (expectation value) μ_l leads to a generalization of the formula. According to Priest (1993), a general relationship between the linear frequency of fractures λ_L and the 2D fracture density λ_A is given by:

$$2/\pi \, \lambda_A \mu_l \leq \lambda_L \leq \lambda_A \mu_l \qquad (2)$$

with the lower boundary for spatially random distributed fractures and the upper boundary for parallel fractures. Merging both relations leads to the generalization of the percolation parameter:

$$p = N\mu_l^2/A = \lambda_A \mu_l^2 \geq \lambda_L \mu_l \qquad (3)$$

An alternative to describe quantitatively the connectivity of networks is the mean number of intersections per fracture. Hestir & Long (1990) and Robinson (1984) give critical values of 3.6 and 3.1, assuming randomly distributed fractures in position and orientation.

Results from the Hawaii Scientific Drilling Project (HSDP)

The main objective of the Hawaii Scientific Drilling Project (HSDP) is to obtain stratigraphic sequences of lava flows to provide valuable probes of mantle plume structures and magmatic processes (De Paolo *et al.* 2001). The HSDP programme has been organized in three phases (Stolper *et al.* 1996). The first phase began with a pilot hole in the vicinity of Hilo (Kahi Puka 1, KP1), which reached a final depth of 1056 m. The second phase started with the HSDP2 borehole with a current depth of 3098 metres below sea-level (mbsl).

The main findings concerning hydrology in the pilot hole KP1 (for details, see Paillet & Thomas 1996; Thomas *et al.* 1996) can be outlined as follows: (1) various depth sections exist, with varying mean temperatures and salinities; (2) the deep aquifer (below 710 m in the open borehole section) contains seawater with temperatures and salinities in accordance with those of the open sea at the same depth; the transmissivity of the aquifer is approximately 10^{-3} m^2 s^{-1} (3) a freshwater aquifer exists at the transition from Mauna Loa to Mauna Kea (at 280 m). These findings lead to the construction of a regional hydrological model based on the current data (Thomas *et al.* 1996).

The stratigraphy of the HSDP2 lithological column (detailed description in De Paolo *et al.* 2001) generally can be divided into the subaerial section up to 1079 mbsl and the subjacent submarine section. The subaerial section can be subdivided into the Subaerial Mauna Loa lavas (from the surface to 246 mbsl) and the Subaerial Mauna Kea lavas (to 1079 mbsl). Both sections are characterized by an alternation of aa and pahoehoe lava flows. The submarine section can be subdivided into the upper part with dominantly hyaloclastite debris flows to a depth of 1984 mbsl, and a lower section with pillow lavas and intrusive units down to the final depth of 3098 mbsl.

Hydraulic pathways within the HSDP drillhole differ substantially from the crystalline rock of the KTB, which we have studied in detail. There, the pathways at the various scales can be described by fracture systems with tectonic origin and orientations based on the recent and the palaeostress field. In the environment of the Hawaii drill-hole, fracture-like structures occur between the contact planes of lava flows as a result of cooling processes.

In the subaerial units, the contact planes between the lava flows can be described by geometrical structures, which are similar to those of fracture planes in crystalline rock. Thus, they can be denoted as fractures, despite their differences in origin; hence, the concept can be applied. This reasoning is consistent with the work of Morin & Paillet (1996). They detected fractures in the pilot hole KP-1 from BHTV images, and identified 283 individual fractures and their orientations.

The submarine units consist of pillow lava and massive basalts on one hand, and hyaloclastite debris flows on the other. Hyaloclastites occur predominantly in the upper part of the submarine section. They belong to vesicular volcaniclastic sediments with a pore space similar to a sedimentary rock. Therefore, a concept of fracture analysis cannot be applied to the hyaloclastite units. Pillow basalts show macroscopic fractures from cooling processes, which make a substantial contribution to the fluid transport. Intrusive basalts occur mostly in the deeper part of the borehole, but they are not very abundant, and represent up to 7% of the core in the deepest part below 2.5 km below sea-level.

A review concerning the estimation of permeability in basaltic ocean rocks, which are comparable to the lithological units of the Hawaiian borehole, was published by Fisher (1998). In these rocks, too, the scale effect of permeability can be found; permeability measurements on cores and hydraulic tests differ by about several orders of magnitude.

Data from fluorescent thin sections

On the microscopic scale, fluorescent thin sections (Figs 1 & 2) can be investigated to obtain geometrical parameters like mean fracture lengths, orientation, fracture density and the corresponding distribution functions. The light structures represent the hydraulic pathways, and were rendered visible by a fluorescent epoxy resin, which was forced into the microfractures using a centrifuge. Figures 1 & 2 show different images of thin sections with different resolutions, from cores of the HSDP2 borehole. The thin section in Figure 1 was taken from a core from 814 mbsl, and, according to Büttner & Huenges (2003), was characterized as a moderately olivine-phyric basalt (pahoehoe). They measured core porosity of 30% (with the buoyancy method) and the corresponding gas permeability of 10 μdarcy (10^{-17} m^2). This high porosity and relatively low permeability cannot be explained by the standard Kozeny–Carman approach with the relationship of permeability as a non-linear function of porosity (e.g. Pape *et al.* 1999, 2000). The pore space, which is represented by vesicular structures, is obviously only connected by microfractures. Therefore, we conclude that the microfractures are responsible for the hydraulic flow, whereas the vesicles serve to store fluids or gas. The same finding is obtained from Figure 2, with a measured core porosity (buoy-

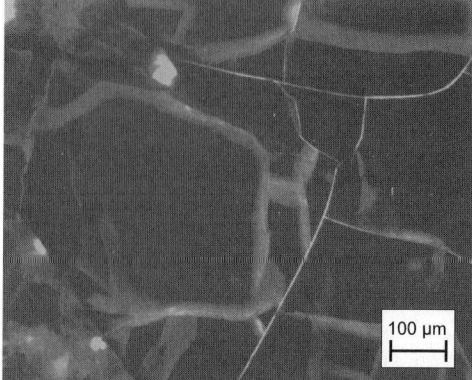

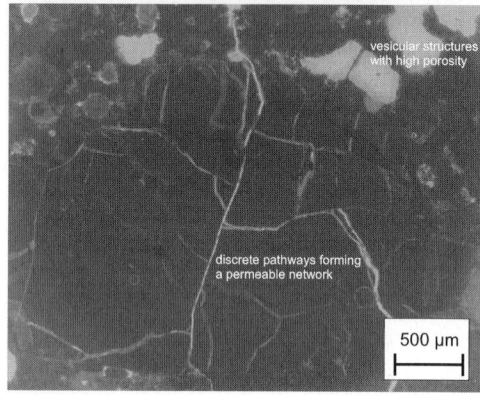

Fig. 1. Fluorescent thin sections from 814 mbsl (moderately olivine-phyric basalt (pahoehoe), measured core porosity is 30%, gas permeability is 10 μdarcy (10^{-17} m^2) (Büttner & Huenges 2003). Porosity is mainly due to the vesicular structures (see upper right-hand side in the bottom image).The pathways of the penetrated fluorescent resin form a network, which can be used to simulate the hydraulic flow on the microscopic scale and to calculate permeability.

ancy method) of 16% and a gas permeability of 10 μdarcy (10^{-17} m^2). Here, we have a thin section of a core of highly olivine-phyric basalt (massive) from 1088 mbsl (Büttner & Huenges 2003). In order to find out if our idea that the microfractures act as the preferential pathways is correct, we have to test whether we can explain this permeability purely with a network of microfractures. Therefore, in Figure 2 the network was superimposed as a finite-element mesh to simulate the hydraulic flow. The flow was calculated from the lower border to the upper border of the olivine crystal. The right and the left sides are so-called 'no-flow-boundaries'. The flow conditions between the nodes were varied stochastically. We applied the exponential distribution function for the fracture aperture between the nodes with a mean of 1 μm, and

Fig. 2. Fluorescent thin section from 1088 mbsl (highly olivine-phyric basalt (massive); the measured core porosity is 16.6% and gas permeability is 10 μdarcy (10^{-17} m^2) (Büttner & Huenges 2003). The pathways of the penetrated fluorescent resin were superimposed by a finite-element mesh to simulate the hydraulic flow.

performed 100 network simulations. The results of permeability calculations are displayed as a histogram in Figure 3. The mean permeability of these 100 simulations was calculated to 12.3 μdarcy (1.23×10^{-17} m^2), according to a log-normal distribution function. Therefore, under the assumption made, we can conclude that the low measured gas permeability can be

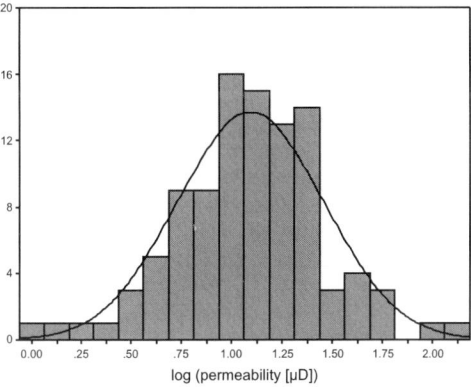

Fig. 3. Permeability calculations for the network from Figure 2, with a stochastic variation of the fracture aperture. It is assumed that the fracture aperture follows an exponential distribution function with a mean value of 1 μm. This leads to a mean permeability of 12.3 μdarcy (1.23×10^{-17} m^2). For comparison, a constant fracture aperture of 1 μm leads to a permeability of 174 μdarcy.

explained by a network of microfractures. For comparison, a network with a constant aperture of 1 μm for all pathways leads to a permeability of 174 μdarcy.

According to equation (3), the percolation parameter as a qualitative measure of connectivity of the network can be calculated. For this calculation, the number of fractures and the mean length of the fractures are needed. Hence the question arises: what is an appropriate definition of a fracture and a fracture length in context with this thin-section data? In accordance with our simplification of fractures as planar-like structures, we assume a fracture to be an individual structure, if it can be described as a straight line in the 2D projection area of the thin-section image. Thereupon, the number of intersections per fracture can be determined.

We determined 109 fractures from Figure 2 with a mean fracture length of 269.2 μm, according to a log-normal distribution function. This leads to a percolation parameter of 1.58, according to equation (3), and a fracture density of 2.18×10^7 m^{-2}. Results from previous analyses of KTB thin-section data showed a critical value of the percolation parameter of 1.3 (Zimmermann *et al.* 2003). Therefore, the network in Figure 2 with a percolation parameter of 1.58 should be well above the percolation threshold. This is evident from visual inspection of the 2D connectivity of the network, too.

The mean number of intersections per fracture was calculated to 3.18. This is in the range of the theoretical values of the critical mean fracture lengths per fracture, according to Hestir & Long (1990) and Robinson (1984), who give critical values of 3.6 and 3.1 intersections per fracture at the percolation threshold.

Data from borehole measurements and core scans

The geometrical parameters of dipping fractures, such as fracture locations (and derived linear fracture frequency) and orientations, can be determined from structural borehole measurements, which create an image of the borehole wall. In the HSDP2 borehole, two different borehole televiewer tools (BHTV) were used: one in a limited depth interval from 795 m to 1160 m (Fig. 4, left-hand side), and the other in the depth interval from 600 m to 1800 m (Fig. 4, middle). For comparison, core scans can be used to identify the structures and support interpretation (Fig. 4, right-hand side).

The images of the borehole televiewer in the depth section from 783.5 to 1147.5 mbsl were

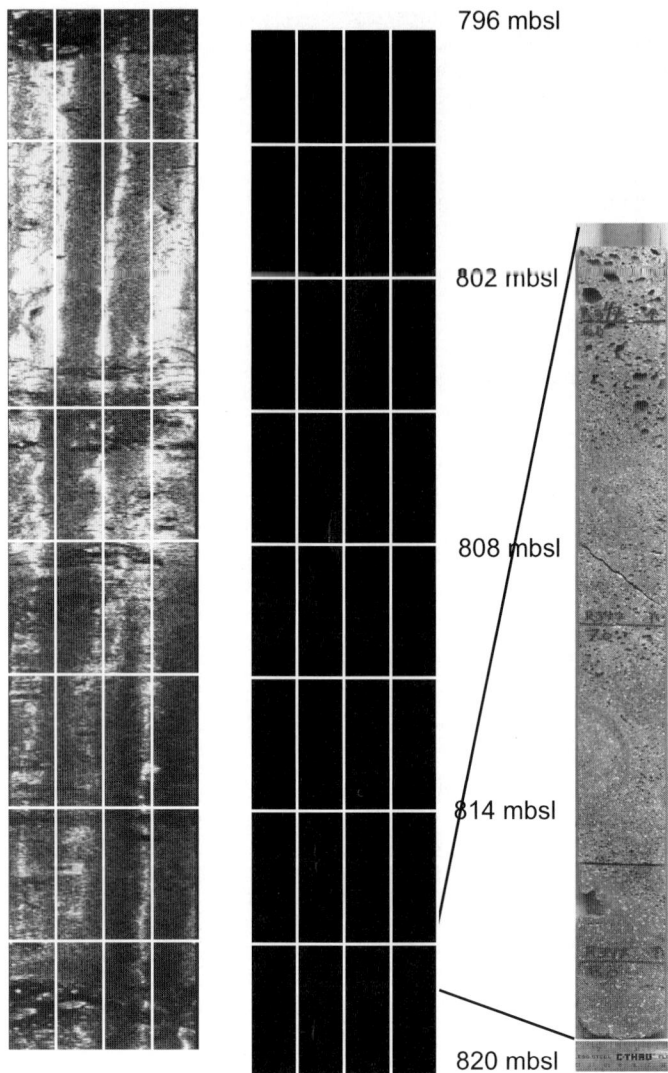

796 mbsl

802 mbsl

808 mbsl

814 mbsl

820 mbsl

Fig. 4. Left-hand side: borehole televiewer image, depth section from 808.3 m to 832.4 m; raw data, left-hand side corresponds to a northerly direction. Middle: second borehole televiewer image; data were corrected by an autocorrelation method, due to mismatch of depth records, and contrasts were enhanced. Right-hand side: core scan of run 347, with indicated depth interval for a comparison of structures with the televiewer data. Detected structures can be used to construct macroscopic stochastic networks to simulate the hydraulic flow.

examined, and 337 fractures were detected from visual inspection. This particular depth section was chosen because data from both televiewers were available in this section. The linear frequency of the fractures λ_L follows as $0.927 \, \text{m}^{-1}$. For a first estimate, we assume horizontal layers and constant fracture apertures of $100 \, \mu\text{m}$ for all structures. This leads to an estimate of permeability of 77 mdarcy $(7.7 \times 10^{-14} \, \text{m}^2)$ according to the following relationship for laminar flow in plane parallel structures:

$$k = \lambda_L \times a^3 / 12 = \lambda_L \times a \times k_{\text{fracture}} \qquad (4)$$

with k = permeability (m^2)
$k_{\text{fracture}} = a^2/12$ = permeability of a single fracture (m^2)
a = fracture aperture (m)
λ_L = linear frequency of fractures (m^{-1})

The linear frequency of the fractures can be applied to calculate the critical fracture length. This is the length, where the corresponding 2D network reaches the percolation threshold. Bour & Davy (1997) give a theoretical value of the critical percolation parameter of $p_{crit} = 5.6$ for infinite large 2D fracture systems with fractures randomly distributed in position and orientation. Assuming these assumptions for the above investigated fractures allows the calculation of the critical length according to:

$$l_{crit} = p_{crit}/\lambda_L = 6.0\,\text{m} \qquad (5)$$

That means, under the assumptions made, a fracture length of at least 6 m is necessary to establish a macroscopic hydraulically permeable fracture network. This initially unknown parameter is substantial for the construction of 2D stochastic networks and simulation of hydraulic flow. With this lower limit, the number of plausible 2D stochastic networks can be restricted.

Comparison of results from KTB and HSDP

In a recent study, Zimmermann et al. (2003) investigated the scale dependence of hydraulic and structural parameters in the crystalline rock of the KTB. We found a power-law relationship between the 2D fracture density and the mean fracture length (Fig. 5), with an exponent of 1.9 ± 0.05. Only those networks which are above the percolation threshold and hence permeable, were included. This results in a straight line in the log–log plot of Figure 5, which can be interpreted as a critical boundary with non-connected networks on the left-hand side and connected networks on the right-hand side of the straight line. We added the results from the Hawaiian data for comparison. The thin-section result is close to the straight line; additional data are required to show whether this can be generalized. The included borehole data are from the BHTV measurements for a depth section from 783.5 mbsl to 1147.5 mbsl. They initially represent 1D information, which was transferred to the 2D density, according to Priest (1993) (see equation (2)). In this case, the fracture length is an unknown parameter, therefore all models are equivalent, according to the initial data. In the case of a 'universal' characteristic of the power law with an attribute as a critical boundary, the networks with a fracture length of 1 m and 3 m (see Fig. 5) should be non-connected. This is in agreement with the result from above, where we estimated a

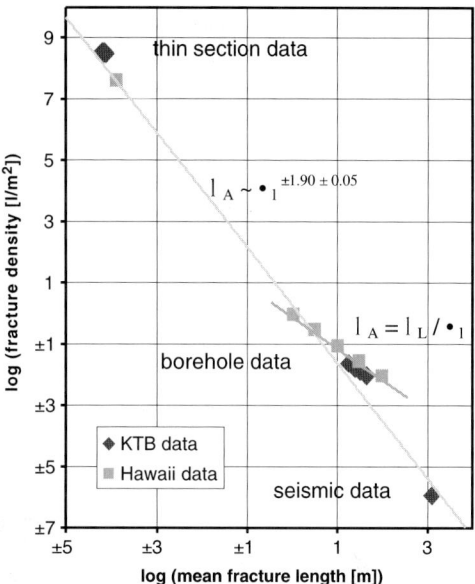

Fig. 5. A log–log presentation of fracture density versus mean fracture length for the connected networks (above the percolation threshold) of the KTB data (dark diamonds) on all investigated scales, and results from the HSDP2 borehole (grey squares) for comparison. Taking into account only the KTB data, the mean fracture length follows a power-law distribution, with an exponent of -1.90 ± 0.05 indicating the fractal nature of fractures in crystalline rock (detailed description in Zimmermann et al. 2003). The calculated fracture densities from the Hawaiian BHTV data (grey squares: depth section from 783.5 to 1147.5 mbsl) were obtained from their linear frequency, and therefore represent equivalent models with different fracture lengths (which are initially unknown without further assumptions and were set to 1 m, 3 m, 10 m, 30 m and 100 m). In the case of a 'universal' characteristic of the power law, all data points on the right side of the straight line should belong to networks above the percolation threshold. Hence, analogously to the KTB data, this results in a lower limit for the fracture lengths of the Hawaiian data. This is also in agreement with the estimation from percolation theory, where a critical fracture length of at least 6 m is necessary for a macroscopic connected network.

critical fracture length of at least 6 m to establish a permeable network. Based on the current data, we can give a lower limit of the fracture length by assuming a macroscopic connectivity, which is evident from inflow zones in this depth interval (Büttner & Huenges 2003).

Conclusions

A concept which had been developed for the fractured rock of the KTB, has been applied successfully to the volcanic rock of the Hawaii Scientific Drilling Project. The results show that this concept can be transferred to the different lithologies. Fluorescent thin-section images and BHTV measurements were analysed, which represent different scales of investigation. Microfractures were observed in these thin sections of olivine-phyric basalts, which might explain the measured low permeability in conjunction with the observed vesicular structures of high porosity. The borehole wall structures from the BHTV measurements represent potentially hydraulically conductive fractures, which are preferential pathways to control the hydraulic flow on the macroscopic scale. In the current situation, the data used for the flow simulation with macroscopic models are very sparse. Therefore, besides the known information about the structures, additional data and scale-dependent relationships, as provided by the concept, are necessary to give a well-constrained estimate of the hydraulic situation in the vicinity of the mantle plume.

We would like to thank G. Büttner from GFZ Potsdam, who kindly made available the thin sections. The borehole televiewer data were provided by R. Wilkens and D. Thomas from the University of Hawaii, and J. Kück from the ICDP Operational Support Group at GFZ Potsdam. The research was financially supported by the Deutsche Forschungsgemeinschaft under grants En391/1 and En391/2.

References

ACUNA, J. & YORTSOS, Y. 1995. Application of fractal geometry to the study of networks of fractures and their pressure transient. *Water Resources Research*, **31** (3), 527–540.

BALBERG, I., BERKOWITZ, B. & DRACHSLER, G. E. 1991. Application of a percolation model to flow in fractured hard rocks. *Journal of Geophysical Research*, **96** (B6), 10 015–10 021.

BERKOWITZ, B. & BALBERG, I. 1993. Percolation theory and its Application to groundwater hydrology. *Water Resources Research*, **29** (4), 775–794.

BERKOWITZ, B. & EWING, R. P. 1998. Percolation theory and network modeling applications in soil physics. *Surveys in Geophysics*, **19**, 23–72.

BONNET, E., BOUR, O., ODLING, N. E., DAVY, P., MAIN, I., COWIE, P. & BERKOWITZ, B. 2001. Scaling of fracture systems in geological media. *Reviews in Geophysics*, **39** (3), 347–383.

BOUR, O. & DAVY, P. 1997. Connectivity of random fault networks following a power law fault length distribution. *Water Resources Research*, **33** (7), 1567–1583.

BOUR, O. & DAVY, P. 1998. On the connectivity of three-dimensional fault networks. *Water Resources Research*, **34** (10), 2611–2622.

BRACE, W. F. 1984. Permeability of crystalline rocks: new in situ measurements. *Journal of Geophysical Research*, **89**, 4327–4330.

BROADBENT, S. R. & HAMMERSLEY, J. M. 1957. Percolation processes. I. Crystals and mazes. *Proceedings of the Cambridge Philosophical Society*, **53**, 629–641.

BÜTTNER, G. & HUENGES, E. 2003. The heat transfer in the region of the Mauna Kea (Hawaii) – constraints from borehole temperature measurements and coupled thermo-hydraulic modeling, *Tectonophysics*, **371**, 23–40.

CLAUSER, C. 1992. Permeability of crystalline rocks. *EOS – Transactions, American Geophysical Union*, **73** (21), 233–237.

DAVID, Ch. 1993. Geometry of flow paths for fluid transport in rocks. *Journal of Geophysical Research*, **98** (B7), 12 267–12 278

DAVID, C., GUÉGUEN, Y. & PAMPOUKIS, G. 1990. Effective medium theory and network theory applied to the transport properties of rock. *Journal of Geophysical Research*, **95** (B5), 6993–7005.

DAVY, P. 1993. On the frequency–length distribution of the San Andreas fault system. *Journal of Geophysical Research*, **98** (B7), 12 141–12 151.

DAVY, P., SORNETTE, A. & SORNETTE, D. 1992. Experimental discovery of scaling laws relating fractal dimensions and the length distribution exponent of fault systems. *Geophysical Research Letters*, **19** (4), 361–363.

DEPAOLO, D. J., STOLPER, E. & THOMAS, D. M. 2001. Deep drilling into a Hawaiian volcano. *EOS Transactions, American Geophysical Union*, **82** (13), 149–155.

FISHER, A.T. 1998. Permeability within basaltic oceanic crust. *Reviews in Geophysics*, **36** (2), 143–182.

GAVRILENKO, P. & GUÉGUEN, Y. 1998. Flow in fractured media: a modified renormalization method, *Water Resources Research*, **34**, 177–191.

GUÉGUEN, Y., GAVRILENKO, P. & Le RAVALEC, M. 1996. Scales of rock permeability. *Surveys in Geophysics*, **17**, 245–263.

GUÉGUEN, Y., CHELIDZE, T. & LE RAVALEC, M. 1997. Microstructures, percolation thresholds, and rock physical properties. *Tectonophysics*, **279**, 23–35.

GUÉGUEN, Y. & PALCIAUSKAS, V. 1994. *Introduction to the Physics of Rocks*. Princeton University Press, Princeton, New Jersey.

HESTIR, K. & LONG, J. C. S. 1990. Analytical expressions for the permeability of random two-dimensional Poisson fracture networks based on regular lattice percolation and equivalent media theories. *Journal of Geophysical Research*, **93**, (B13), 565–581.

HUENGES, E. & ZIMMERMANN, G. 1999. Rock permeability and fluid pressure at the KTB: implications from laboratory – and drill hole – measurements. *Oil & Gas Science and Technology, Review IFP*, **54** (6), 689–694.

KOLDITZ, O. 1992. Kontinuum–Konzept für das geklüftete Medium. *NLfB-Bericht* No. **110 327**, Hannover.

KOLDITZ, O. 1996. *Physikalisch–Mathematische Modellbildung, Lehrgang: Prozeßsimulation im Kluftgestein – Aquifere und Geologische Barriere, 24.06.–26.06.1996 Hannover.*

KOSAKOWSKI, G. 1994. Simulation von Strömung und Wärmetransport im variszischen Grundgebirge: vom natürlichen Kluftsystem zum numerischen Gitternetzwerk. *NLfB-Bericht*, No. **112 046**, Hannover.

KOSAKOWSKI, G. 1996. Modellierung von Strömungs- und Transportprozessen in geklüfteten Medien: vom natürlichen Kluftsystem zum numerischen Gitternetzwerk. *VDI Verlag*, **7**, 304.

LEGE, T., KOLDITZ, O. & ZIELKE, W. 1996. Strömungs- und Transportmodellierung. *Handbuch zur Erkundung des Untergrundes von Deponien und Altlasten*, **2**, Springer-Verlag, Berlin.

MORIN, R. H. & PAILLET, F. L. 1996. Analysis of fracture intersecting Kahi Puka Well 1 and its relation to the growth of the island of Hawaii. *Journal of Geophysical Research*, **101** (B5), 11 695–11 699.

PAILLET, F. L. & THOMAS, D. M. 1996. Hydrogeology of the Hawaii Scientific Drilling Project borehole KP-1 1. Hydraulic conditions adjacent to the well bore. *Journal of Geophysical Research*, **101** (B5), 11 675–11 682.

PAPE, H., CLAUSER, C. & IFFLAND, J. 1999. Permeability prediction based on fractal pore-space geometry. *Geophysics*, **64** (5), 1447–1460.

PAPE, H., CLAUSER, C. & IFFLAND, J. 2000. Variation of permeability with porosity in sandstone diagenesis interpreted with a fractal pore space model. *Pure and Applied Geophysics*, **157**, 603–619.

PRIEST, S. D. 1993. *Discontinuity Analysis for Rock Engineering.* Chapman and Hall, London.

ROBINSON, P. C. 1983. Connectivity of fracture systems – a percolation theory approach. *Journal of Physics A: Mathematical and General*, **16**, 605–614.

ROBINSON, P. C. 1984. Numerical calculations of critical densities for lines and planes. *Journal of Physics A: Mathematical and General*, **17**, 2823–2830.

SAHIMI, M. 1995. *Flow and Transport in Porous Media and Fractured Rock – From Classical Methods to Modern Approaches.* VCH, Weinheim.

SELYAKOV, V. I. & KADET, V. V. 1996. *Percolation Models for Transport in Porous Media – With Applications to Reservoir Engineering.* Kluwer Academic Publishers, Dordrecht.

STAUFFER, D. & AHARONY, A. 1992. *Introduction to Percolation Theory*, Taylor and Francis, London.

STOLPER, E. M., DEPAOLO, D. J. & THOMAS, D.M. 1996. Introduction to special section: Hawaii Scientific Drilling Project, *Journal of Geophysical Research*, **101** (B5), 11 593–11 598.

THOMAS, D. M., PAILLET, F. L. & CONRAD, M. E. 1996. Hydrogeology of the Hawaii Scientific Drilling Project borehole KP-1 2. Groundwater geochemistry and regional flow patterns. *Journal of Geophysical Research*, **101** (B5), 11 683–11 694

WOLLRATH, J. 1990. *Ein Strömungs- und Transportmodell für klüftiges Gestein und Untersuchungen zu homogenen Ersatzsystemen.* PhD thesis, Universität Hannover.

ZIMMERMANN, G. BURKHARDT, H. & ENGELHARD, L. 2000a. Percolation thresholds of fracture networks on different scales in crystalline rock. *EOS Transactions, American Geophysical Union*, **81** (48), Fall Meeting Supplement, Abstract H72B–29.

ZIMMERMANN, G., BURKHARDT, H. & ENGELHARD, L. 2001. Scale dependence of hydraulic and structural parameters in fractured rock. *Geophysical Research Abstracts*, **3**, 26th General Assembly of the European Geophysical Society, Nice, France.

ZIMMERMANN, G., BURKHARDT, H. & ENGELHARD, L. 2003. Scale dependence of hydraulic parameters in the crystalline rock of the KTB. *Pure and Applied Geophysics*, **160**, 1067–1085.

ZIMMERMANN, G., KÖRNER, A. & BURKHARDT, H. 2000b. Hydraulic pathways in the crystalline rock of the KTB. *Geophysical Journal International*, **142** (1), 4–14.

Brittle fracture in two crystalline rocks under true triaxial compressive stresses

B. HAIMSON[1] & C. CHANG[1,2]

[1]Department of Materials Science and Engineering and Geological Engineering Program University of Wisconsin, 1509 University Avenue, Madison, Wisconsin 53706–1595, USA (e-mail: bhaimson@wisc.edu)

[2]Department of Geology, Chungnam National University, Daejeon, South Korea

Abstract: We employed our new polyaxial cell to carry out true triaxial compression tests on dry (jacketed) rectangular prisms of two crystalline rocks, in which different magnitudes of the least and intermediate principal stresses σ_3 and σ_2 were maintained constant, and the maximum stress σ_1 was increased to its peak level in strain control. Both Westerly granite (Rhode Island, USA) and KTB amphibolite (Bohemian Massif, Germany) revealed similar mechanical behaviour, much of which is missed in conventional triaxial tests in which $\sigma_2 = \sigma_3$. Compressive failure in both took the form of a main shear fracture, or fault, steeply dipping in the σ_3 direction. Compressive strength rose significantly with the magnitude of σ_2, suggesting that the commonly used Mohr-type strength criteria, which ignore the σ_2 effect, predict only the lower limit of rock strength. The true triaxial strength criterion for each of the crystalline rocks can be expressed as the octahedral shear stress at failure as a function of the mean normal stress acting on the fault plane. We found that the onset of dilatancy increases considerably for higher σ_2. Thus, σ_2 extends the elastic range for a given σ_3 and, hence, retards the onset of the failure process. The main fracture dip angle was found to increase as σ_2 rises, providing additional confirmation of the strengthening effect of σ_2. SEM inspection of the micromechanics leading to specimen failure showed a multitude of stress-induced microcracks localized on both sides of the through-going fault. Here too the effect of σ_2 is noted, in that microcracks gradually align themselves with the $\sigma_1-\sigma_2$ plane as the magnitude of σ_2 is raised.

Brittle fracture and compressive strength of crystalline rocks have been studied for centuries because of the need to design safe excavations for extracting minerals from the Earth, and for the purpose of using hard rock as a building material (Scholz 1990). But it was only in the last century that rigorous formulations of rock strength were made, based on the physics of the phenomenon and supported by experimental results. *Rock strength* is the maximum stress that the rock can support under specified conditions, and is commonly represented by strength criteria. The most widely used criteria are based on the hypothesis that the intermediate principal stress σ_2 has no role in the process of brittle fracture leading to failure. For example, the Mohr criterion and a number of other criteria, such as the linearized and commonly used Mohr–Coulomb criterion (Jaeger & Cook 1979, p. 96), as well as those due to Griffith (1924), Hoek & Brown (1980), Ashby & Sammis (1990), and others, all define failure in compression in one of the following general relationships:

$$\sigma_1 = A + B\sigma_3$$
$$\tau = C + D\sigma \tag{1}$$

where σ_1 and σ_3 are the minimum and the maximum principal stresses, respectively; τ and σ are two-dimensional shear stress and normal stress, respectively, and both are functions of only σ_1 and σ_3; and A, B, C and D are interrelated constants of the material.

The common way of obtaining such criteria is to conduct a series of conventional triaxial tests in a pressure vessel where cylindrical specimens are subjected to constant confining pressures ($\sigma_2 = \sigma_3$) and increasing axial stresses (σ_1), until brittle fracture occurs and compressive strength is reached. However, a growing number of *in situ* measurements at shallow to intermediate crustal depths have shown that the principal stresses in the Earth's crust are almost always anisotropic, i.e. $\sigma_1 \neq \sigma_2 \neq \sigma_3$ (Fig. 1).

From: HARVEY, P. K., BREWER, T. S., PEZARD, P. A. & PETROV, V. A. (eds) 2005. *Petrophysical Properties of Crystalline Rocks*. Geological Society, London, Special Publications, **240**, 47–59. 0305-8719/05/$15.00 © The Geological Society of London 2005.

(a)

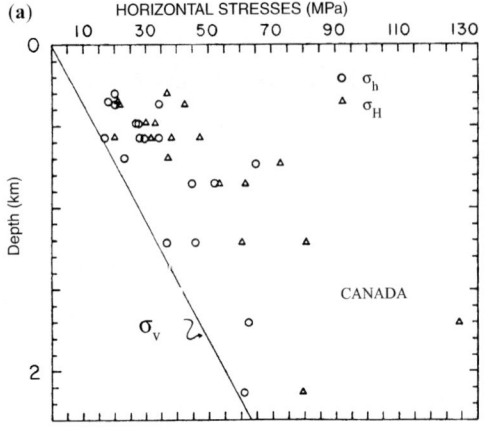

(b)

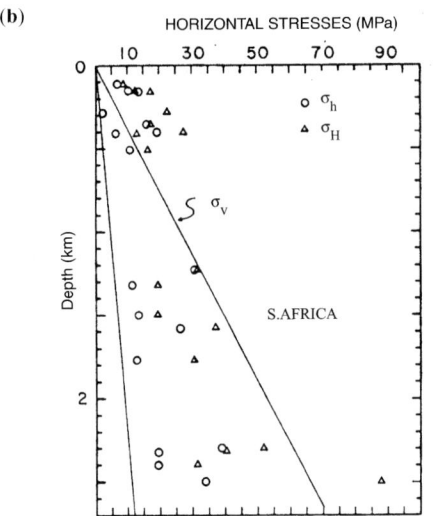

Fig. 1. Collection of measured *in situ* principal stresses in (**a**) Canada and (**b**) South Africa in which principal stress magnitudes σ_v, σ_h and σ_H (vertical, least horizontal and largest horizontal, respectively) are clearly unequal (after McGarr & Gay 1978).

Moreover, criteria neglecting the σ_2 effect have been found to be lacking in a number of important cases. For example, Vernik & Zoback (1992) found that use of the Mohr–Coulomb criterion in relating borehole breakout dimensions to the prevailing *in situ* stress conditions in crystalline rocks did not yield realistic results. They suggested the use of a more general criterion that accounts for the effect on strength of the intermediate principal stress. Recently, Ewy (1998) reported that, for the purpose of calculating the critical mud weight necessary to maintain well-bore stability, using the Mohr–Coulomb

criterion leads to over-conservative predictions, because it neglects the perceived strengthening effect of the intermediate principal stress.

Early attempts to examine the intermediate principal stress effect on rock strength were made by Murrell (1963), who compared the results from two different series of triaxial tests conducted in Carrara marble: triaxial compression tests ($\sigma_1 > \sigma_2 = \sigma_3$) and triaxial extension tests ($\sigma_1 = \sigma_2 > \sigma_3$). He noted that the rock strength for any given σ_3 was larger in triaxial extension than in triaxial compression, implying that strength is not σ_2-independent (Fig. 2). Similarly, Handin *et al.* (1967), reporting on triaxial compression and triaxial extension tests they carried out in Solenhofen limestone, Blair dolomite and Pyrex glass, showed that rock strength was higher when σ_2 was equal to σ_1. Concurrently, Wiebols and Cook (1968) derived an energy-based theoretical strength criterion and found that under true triaxial compressive stress conditions ($\sigma_1 \neq \sigma_2 \neq \sigma_3$) the intermediate principal stress has a pronounced effect, quantitatively predictable if the coefficient of sliding friction between crack surfaces is known. In particular, Wiebols and Cook determined from their model that if σ_3 is held constant and σ_2 is increased from $\sigma_2 = \sigma_3$ to $\sigma_2 = \sigma_1$, then the peak σ_1 first increases, reaches a maximum at some value of σ_2 and then gradually decreases to a level higher than that obtained in a triaxial test, i.e. when $\sigma_2 = \sigma_3$.

Since then, there have been several important experimental studies on the effect of σ_2. Mogi (1971) designed the first apparatus capable of applying three independent and mutually perpendicular loads to the faces of a rectangular

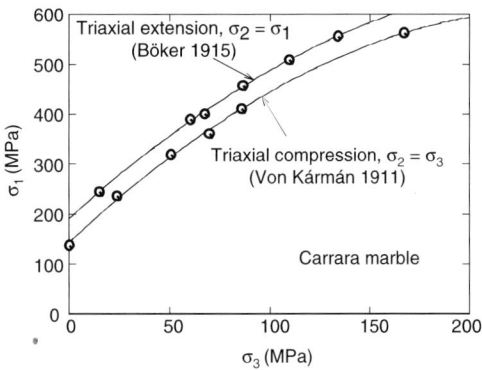

Fig. 2. Carrara marble triaxial compressive strength under two different intermediate principal stress σ_2 settings. In one $\sigma_2 = \sigma_1$ (triaxial extension), and in the other $\sigma_2 = \sigma_3$ (triaxial compression) (after Murrell 1963).

prismatic specimen, with minimum friction. He demonstrated experimentally that the strength of Dunham dolomite, and other rocks, is a function of the intermediate principal stress in a manner similar to that predicted theoretically by Wiebols and Cook. Other experimental investigations (Michelis 1985; Takahashi & Koide 1989; Crawford *et al.* 1995; Wawersik *et al.* 1997) yielded similar results in other rock types. However, the true triaxial testing devices used had relatively low loading capacities and were not suitable for testing compressive failure of hard rocks under high σ_2 and σ_3.

At the University of Wisconsin we have recently designed and fabricated a new true triaxial loading system capable of determining the compressive strength of hard rocks subjected to the most general stress conditions ($\sigma_1 \geq \sigma_2 \geq \sigma_3$). Two types of crystalline rocks were selected for the initial rock strength tests: a granite and an amphibolite, both of which are representative crustal constituents (Chang & Haimson 2000; Haimson & Chang 2000). This paper reviews the results reported separately in these two papers, compares them and attempts to draw general conclusions regarding true triaxial mechanical behaviour of crystalline rocks.

True triaxial loading apparatus

The new apparatus used for the application of three mutually perpendicular high compressive loads on to rectangular prismatic specimens consists of two main parts: a pressure vessel and a biaxial cell (Fig. 3). The rock specimen ($19 \times 19 \times 38$ mm) is inserted inside the pressure vessel and is subjected to a longitudinal load, σ_1, and one lateral load, σ_2 (capacity of both 1.6 GPa), applied through high-strength metal anvils, as well as to a second lateral load applied directly by the confining fluid pressure, σ_3 (capacity 0.4 GPa). The pressure vessel fits inside the biaxial cell, which applies hydraulically activated loads to the two pairs of anvils. All three loads are servo controlled. The entire true triaxial loading system is compact and easily transportable, and does not require the use of a loading frame (Haimson & Chang 2000).

Friction between the anvils and rock specimen is minimized by inserting between them a thin copper shim, over which stearic acid has been smeared as a lubricant (Labuz & Bridell 1993). Tests are conducted by controlling the strain in the least principal stress direction (i.e. in the direction of σ_3, and the strain is typically maintained at 5×10^{-6} s^{-1}. To accommodate strain measurements and control in the σ_3 direction, we designed a thin beryllium–copper beam, the

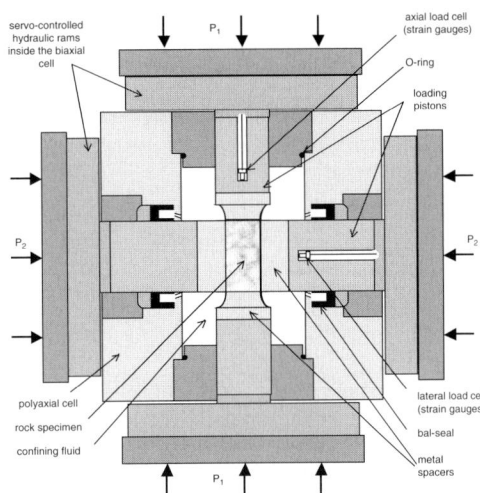

Fig. 3. Cross-section of the true triaxial cell, showing load transmission from the biaxial apparatus (P_1 and P_2) to rock specimen. The load in the third direction is applied directly by the confining fluid pressure.

centre of which is forced to make contact with a pin bonded to the specimen's σ_3 face, while its ends are rigidly fixed (Fig. 4a). As the rock specimen expands in the σ_3 direction during compressive longitudinal loading, the beam bends, allowing the strain gauges mounted near its ends to monitor the least principal strain. Strains in the other two directions are monitored through strain gauges mounted directly to the same pair of specimen faces subjected to σ_3 loading (Fig. 4b). The exposed σ_3 surfaces are then coated with a thin layer of polyurethane, in order to prevent confining fluid infiltration. Careful testing and calibration of the equipment have revealed near-perfect uniformity of loading in all three directions, and load readings correct within an error of $\pm 1.5\%$.

Rocks tested

We used the new apparatus to carry out two exhaustive series of true triaxial compressive strength tests, one in Westerly granite and the other in KTB amphibolite. Westerly granite is a fine- to medium-grained Late Pennsylvanian to Permian crystalline rock found mainly in the southeast area of Rhode Island (Quinn 1954). Its mineral composition is: 28% quartz, 36% microcline, 31% plagioclase and 5% biotite (Wawersik & Brace 1971). Its physical properties are characterized by high strength, very low porosity, almost complete isotropy, elastic linearity, and homogeneity (Krech *et al.* 1974).

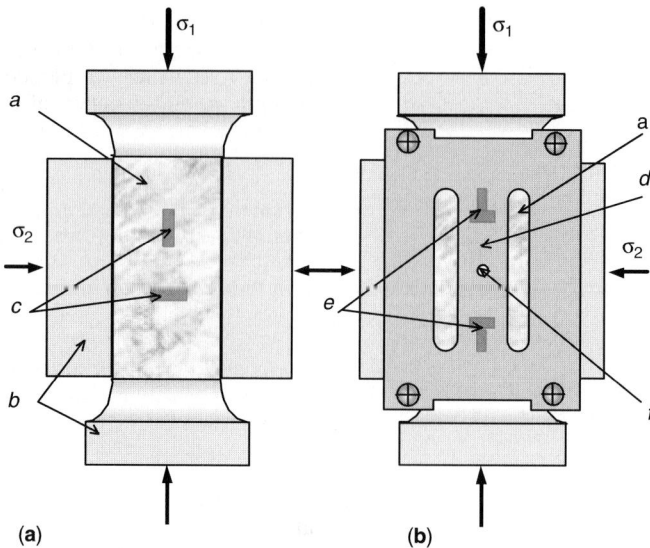

Fig. 4. (a) Rock specimen *a* with metal spacers *b* attached, showing position of strain gauges *c* on the specimen faces subjected to confining fluid (measuring strain in the σ_1 and σ_2 directions). (b) Strain-gauged beam *e* mounted on specimen to determine strain in the σ_3 direction. The beam flexes when the fixed pin *f* displaces as the specimen deforms during loading.

The mechanical properties of Westerly granite have been tested by a large number of experimenters (Brace 1964; Brace *et al.* 1966; Wawersik & Brace 1971; Wong 1982; Lockner *et al.* 1991). However, we know of no attempt being made to investigate rock strength and deformability of Westerly granite under true triaxial stress conditions.

The KTB amphibolite was extracted from the superdeep scientific borehole drilled by the German Continental Deep Drilling Programme (KTB) in Bavaria, Germany. Amphibolite is the dominant rock in the hole between 3 and 7 km depth, and the samples used in our study come from a depth of 6.4 km. The KTB amphibolite is a fine-grained massive metamorphic rock. The mineral composition is: 58% amphibole, 25% plagioclase, 5% garnet, 2% biotite and 7% minor opaque minerals. It is characterized by high strength, some inhomogeneity, and very low porosity (Rauen & Winter 1995). Basic mechanical properties determined by uniaxial and triaxial tests demonstrate near isotropy and linear elasticity (Vernik *et al.* 1992; Röckel & Natau 1995).

Experimental program

We conducted several sets of true triaxial tests in each of the two rocks. Tests were run on dry specimens at ambient temperature. Each set comprised of five to ten tests in which the least principal stress was kept constant throughout and the intermediate principal stress σ_2 was varied from test to test. The range of σ_3 was from 0 to 100 MPa in the Westerly granite and to 150 MPa in KTB amphibolite, and that of σ_2 was from $\sigma_2 = \sigma_3$ to $\sigma_2 > 5\sigma_3$. Each instrumented specimen, prepared as described above, was inserted in the polyaxial cell and subjected first to linearly increasing hydrostatic loading. As σ_3 and then σ_2 reached their preset magnitudes, they were kept constant for the remainder of the test. Thereafter, the maximum stress σ_1 was raised to failure and beyond by controlling the least principal strain at a constant rate of $5 \times 10^{-6}\,\mathrm{s}^{-1}$ (Fig. 5). Selected specimens were used for thin-section analysis and scanning electron microscope (SEM) observation of the micromechanics of compressive failure under true triaxial stress conditions.

True triaxial compressive strength of crystalline rocks

A common visual observation in our true triaxial tests is that, in crystalline rocks (as represented by Westerly granite and KTB amphibolite) brittle fracture takes the final form of a shear band that appears as a steeply inclined fault plane, striking parallel to σ_2 direction and dipping in

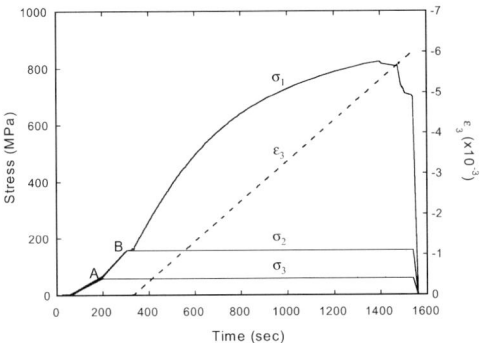

Fig. 5. Typical loading path during true triaxial testing. All three principal stresses are applied concurrently first, until the preset magnitude of σ_3 is reached (A). Then, only the other two principal stresses are increased simultaneously until σ_2 attains its planned value (B). Thereafter, σ_1 is increased by controlling the strain in the σ_3 direction (ε_3).

the σ_3 direction, regardless of the magnitudes of the applied stresses. This failure type is generally similar to that observed in conventional triaxial tests, except that true triaxial loading confirms the expectation that the strike direction of the fault plane is along σ_2 (Fig. 6).

Test results showing the state of stress at failure in the two crystalline rocks are shown in Figure 7, in which we plotted the true triaxial strength σ_1 as a function of the applied σ_2 for different constant values of σ_3. Experiments in which the least and intermediate principal stresses were equal (conventional triaxial tests) reveal a typical monotonic increase in the strength as $\sigma_2 = \sigma_3$ was raised from 0 to 150 MPa (see solid lines in Fig. 7). Plotted separately as σ_1 v. $\sigma_2 = \sigma_3$, the curve best fitting the experimental points represents the Mohr strength criterion for each rock.

Figure 7 demonstrates the similarity between the two crystalline rocks with respect to the dependence of σ_1 at failure as a function of the intermediate principal stress σ_2. As the magnitude of the latter increased beyond that of σ_3, the compressive strength in both rocks was always higher than that when $\sigma_2 = \sigma_3$. Despite the similarity of behaviour between the two crystalline rocks tested, it is evident from Figure 7 that the KTB amphibolite experimental points are considerably more scattered than those in the Westerly granite. This is probably the consequence of the amphibolite's greater inhomogeneity and limited anisotropy (Vernik *et al.* 1992).

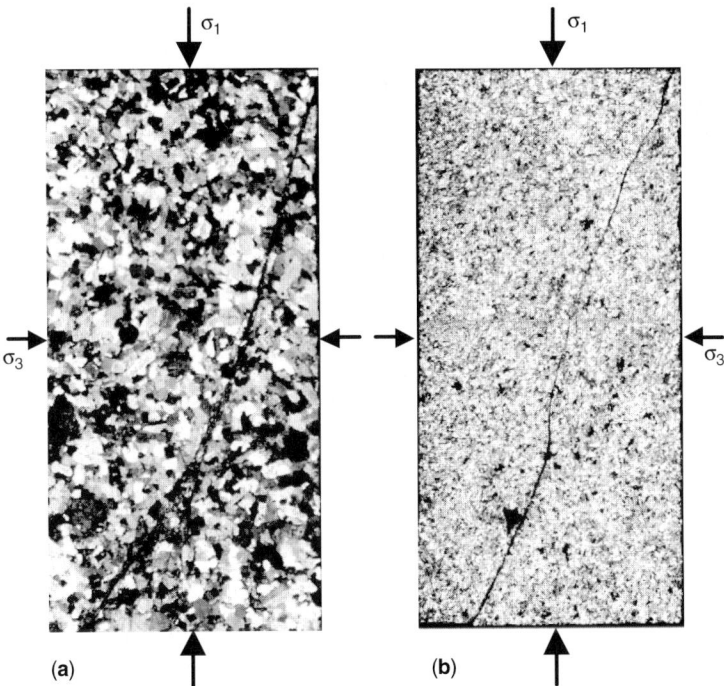

(a)   (b)

Fig. 6. Profiles of failed specimens along a plane normal to σ_2, showing steeply inclined faults dipping in the σ_3 direction: (**a**) Westerly granite; (**b**) KTB amphibolite (after Chang & Haimson 2000).

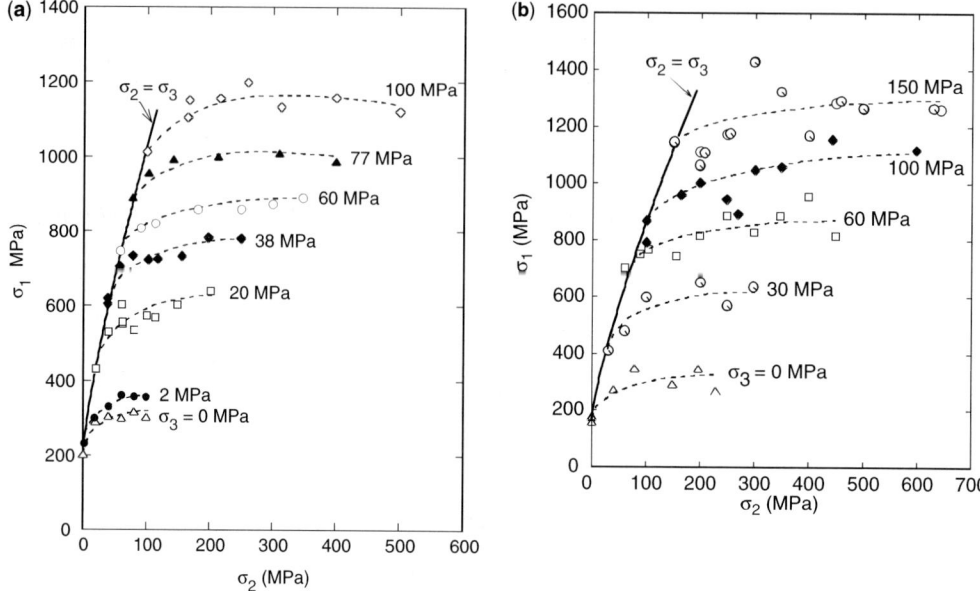

Fig. 7. Relationship between peak σ_1 and σ_2 for different constant magnitudes of σ_3 in (**a**) Westerly granite, and (**b**) KTB amphibolite. The solid line is the best-fit curve to experimental points when $\sigma_2 = \sigma_3$, i.e. the Mohr criterion. Dashed lines show the trend of the true triaxial strength as σ_2 increases.

Nonetheless, the increase in strength as a function of σ_2 for constant σ_3 is quite obvious.

The dashed lines in Figure 7 represent trends in the variation of the true triaxial strength σ_1 with the magnitude of σ_2, for each family of tests for which σ_3 was held constant. They reveal a rise in σ_1 with the magnitude of σ_2, until a plateau is reached where strength appears to level off. This behaviour supports Wiebols & Cook's (1968) theory, based on the strain energy stored in a rock body due to sliding along microcracks, which predicts that when σ_3 is held constant and σ_2 is increased from $\sigma_2 = \sigma_3$ to $\sigma_2 = \sigma_1$, the compressive strength first increases, reaches a maximum at some value of σ_2 and then decreases, but always to a level higher than that obtained in a conventional triaxial test. However, for the ranges of σ_2 used in our tests, there is no clear trend towards a decrease in strength in either of the rocks. The results reported here are also in support of previous observations in Dunham dolomite by Mogi (1971).

Figure 7 reveals that, for true triaxial compressive strength, σ_1 increases by as much as 50% or more over the commonly used conventional triaxial strength (Mohr-based strength criterion), depending on the σ_2 magnitude. The increase in strength is less pronounced at higher magnitudes of σ_3, but, even for the highest least principal stress, the true triaxial strength is larger than

that inferred from the Mohr criterion by nearly 20%. Test results shown in Figure 7 strongly suggest that in crystalline rocks Mohr-based criteria (such as Mohr, Mohr–Coulomb, Hoek and Brown) are only correct for the special case of *in situ* stress conditions in which $\sigma_2 = \sigma_3$, but are not representative of strength under more general, and more realistic, *in situ* stress conditions in which all three principal *in situ* stresses differ in magnitude. The most that can be said is that the almost universally used Mohr-based strength criteria represent the lower limit of compressive strength in crystalline rocks. Since *in situ* measurements rarely indicate that two of the *in situ* principal stresses are equal (McGarr & Gay 1978), Mohr-based criteria appear to yield overly conservative strength estimates.

True triaxial compressive strength criterion

We attempted to establish a unifying strength criterion for each tested rock that would represent the experimental points shown in Figure 7. We considered several brittle-failure theoretical criteria, which incorporate all three principal stresses (Nadai 1950; Freudenthal 1951; Drucker & Prager 1952). These criteria are of the form:

$$\tau_{\text{oct}} = f(\sigma_{\text{oct}}) \qquad (2)$$

where τ_{oct} is the octahedral shear stress, equal to $[(\sigma_1 - \sigma_2)^2 + (\sigma_2 - \sigma_3)^2 + (\sigma_3 - \sigma_1)^2]^{0.5}/3$, σ_{oct} is the octahedral normal stress, equal to $(\sigma_1 + \sigma_2 + \sigma_3)/3$, and f is a monotonically increasing function. None of these criteria seemed to correctly fit our results.

Mogi (1971) noted that the above criteria, for which the independent variable is the three-dimensional mean normal stress σ_{oct}, were more appropriate for characterizing yielding that occurs over the entire volume of rock. The brittle failure of the crystalline rock that we tested was, however, in the form of parting and shearing along a single plane striking in the σ_2 direction. For that failure mechanism, Mogi proposed that a more appropriate independent variable in the above criterion is obtained by downgrading σ_{oct} to its two-dimensional form $(\sigma_1 + \sigma_3)/2$, which represents the mean stress acting on the failure plane:

$$\tau_{oct} = f[(\sigma_1 + \sigma_3)/2]. \qquad (3)$$

Our test data in both rocks are strikingly well represented by Mogi's (1971) brittle failure criterion in the form of monotonically increasing functions, as demonstrated by Figure 8. The true triaxial strength criterion given by equation 3 is clearly more general than the Mohr-based criteria (equation 1), and includes conventional triaxial test results as just a special case.

Fault plane dip and the intermediate principal stress

At the conclusion of each test, we recorded the strike and dip of the induced fault plane. Despite its roughness and local inclination variations, an approximate fault-plane attitude could be determined within an error of just few degrees. The strike was consistently (within $\pm 20°$) aligned with the direction of σ_2. The dip was always steeper than 45° and oriented towards σ_3. Here, another major effect of the intermediate principal stress was observed as the fracture angle for the same σ_3 generally increased as the level of σ_2 was raised (Fig. 9). The trend is similar in both rocks, despite some data scatter.

Within the σ_2 ranges tested, fracture dip angles increased by substantial amounts (up to 20°) from their base values when $\sigma_2 = \sigma_3$. Similar results were obtained by Mogi (1971) in Dunham dolomite. The mechanism responsible for this mechanical behaviour is not clearly understood, but suggests that fracture plane steepening is related to the strengthening of the rock

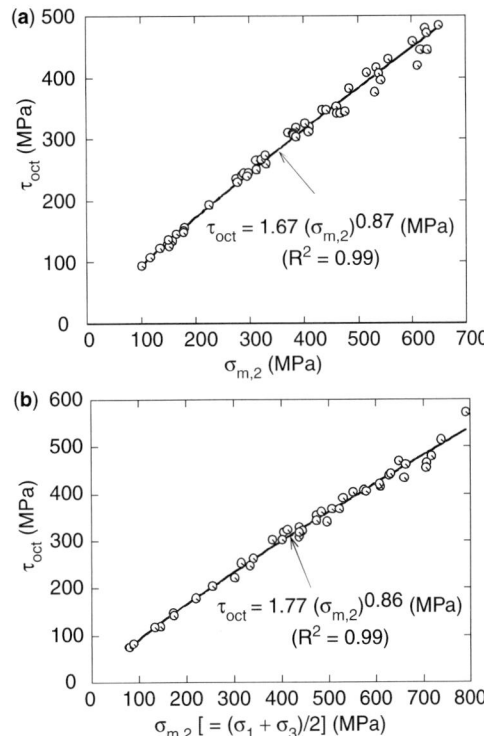

Fig. 8. True triaxial strength criterion best-fitting our experimental results, in the form of the octahedral shear stress τ_{oct} as a function of the mean normal stress acting on the fault plane $\sigma_{m,2}$: (**a**) Westerly granite; (**b**) KTB amphibolite.

with increasing intermediate principal stress magnitude applied.

True triaxial stress–strain relationship

In selected true triaxial tests we monitored the strain in all three principal directions using strain gauges bonded to the σ_3 faces for the measurement of strains in the σ_1 and σ_2 directions, and employing the strain-gauged beam for both strain control of the tests and for recording strain in the σ_3 direction (Fig. 4). Figure 10 shows typical stress–strain records in both rocks. The lower segment of the curves (between points A and B in Fig. 5) represents the stage in which both σ_1 and σ_2 were raised while σ_3 was kept constant. The remainder of each record gives the stress–strain relationships between $(\sigma_1 - \sigma_3)$ and each of the measured strains for constant σ_2 and σ_3, as the strain rate in the σ_3 direction is raised at a constant rate. We note that in both the Westerly granite and the KTB amphibolite, all three stress–strain curves are quasi-linear

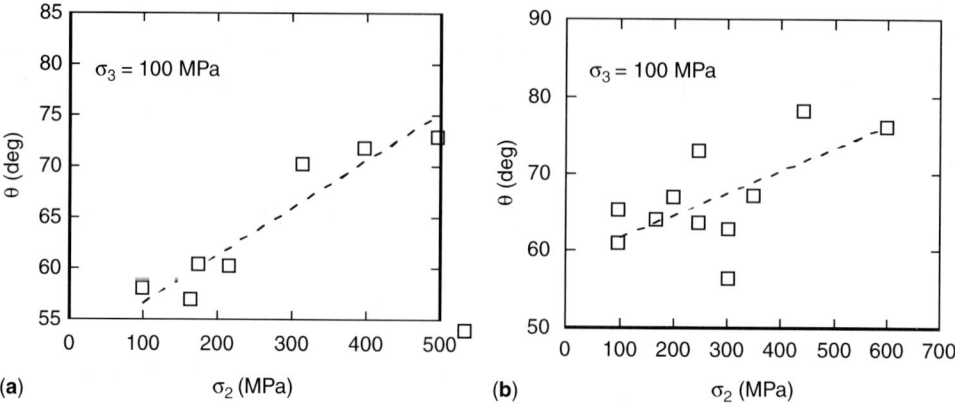

Fig. 9. Typical variation of fault dip θ with the magnitude of the intermediate principal stress σ_2 for a constant least principal stress σ_3 in (**a**) Westerly granite and (**b**) KTB amphibolite.

for the first half to two-thirds of the stress range in this segment. At higher stress levels these stress–strain relationships show increasing softening of the rocks. However, the stress–strain behaviour in the σ_3 direction is significantly more non-linear, suggesting that most of the stress-induced microcracks, leading to eventual failure, are sub-parallel to the σ_1–σ_2 plane and open up primarily in the σ_3 direction.

Another mechanical property common to both rocks is the post-peak (also called post-failure) stress–strain behaviour, which can be character-ized as class II, following Wawersik & Brace (1971). In these crystalline rocks the maximum principal stress σ_1 and the strains in the σ_1 and σ_2 directions are reduced after the peak stress is reached, while the strain in the σ_3 direction

keeps growing, indicating increased cracking or widening in the σ_3 direction of previously devel-oped cracks (Fig. 10).

Figure 10 also depicts the relationship between stress $(\sigma_1-\sigma_3)$ and volumetric strain $\Delta V/V$ (the summation of all three principal strains). The plots show the extent of linear elasticity, characterized by a nearly straight line in the first portion of the loading segment after σ_3 and σ_2 have reached their preset values (see Fig. 10). At some point, the curve becomes non-linear, with the volumetric strain decreasing at a diminishing rate, due to initiation of microcracks or reopening of old ones. This phenomenon, known as dilatancy, has been correlated to internal microcracking responsible for expanding the volume and for

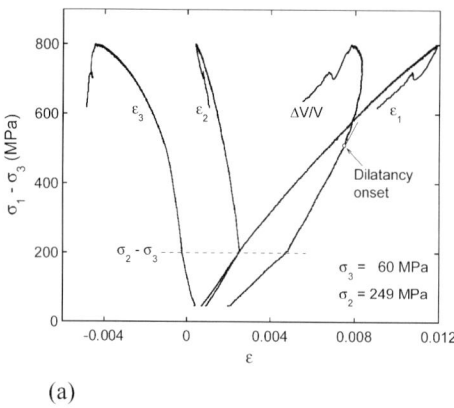

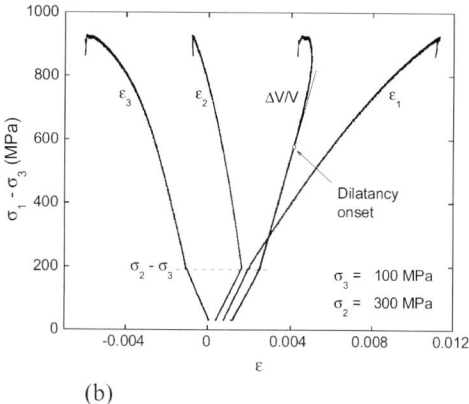

Fig. 10. Stress–strain curves under true triaxial loading in (**a**) Westerly granite and (**b**) KTB amphibolite. ε_1, ε_2, and ε_3 are strains measured in the σ_1, σ_2, and σ_3 directions, respectively; $\Delta V/V$ is the volumetric strain. (Dilatancy onset is the point where $\Delta V/V$ as a function of $(\sigma_1-\sigma_3)$ departs from linearity).

leading eventually to the creation of the fracture plane (Brace *et al.* 1966). The point of departure from linearity is defined as the onset of dilatancy. We plotted the volumetric stress–strain curves and marked the points of dilatancy onset in several tests for which σ_3 was maintained constant, specifically: $\sigma_3 = 60$ MPa in both Westerly granite and KTB amphibolite (Fig. 11). To overcome the ambiguity in determining dilatancy onset directly from the stress–strain curves, we singled it out from a continuous plot of the derivative of $(\sigma_1 - \sigma_3)$ with respect to volumetric strain as a function of $(\sigma_1 - \sigma_3)$. Such a plot yields unambiguously the stress level at which the curve departs from a straight horizontal line, and that value was marked in Figure 11 as the point of dilatancy onset. We estimate the margin of error in the selection of the dilatancy onset by this method to be ± 10 MPa.

The common phenomenon observed in both rocks is the gradual rise in dilatancy onset with the increase in the magnitude of σ_2. We suggest that the apparent increase in the linearity of the stress volumetric strain curve for higher applied intermediate principal stresses, as exhibited by the shift upwards of the dilatancy onset, indicates a retardation of microcrack initiation, which is perhaps also responsible in part for the higher peak σ_3 at failure (Fig. 11).

Micromechanics of brittle failure under true triaxial stress state

We studied aspects of the brittle failure process in the two crystalline rocks by inspecting sections of failed specimens under a scanning electron microscope (SEM model JEOL JSM-6100). Sections were cut along one of two planes, orthogonal to the σ_2 direction, exposing the

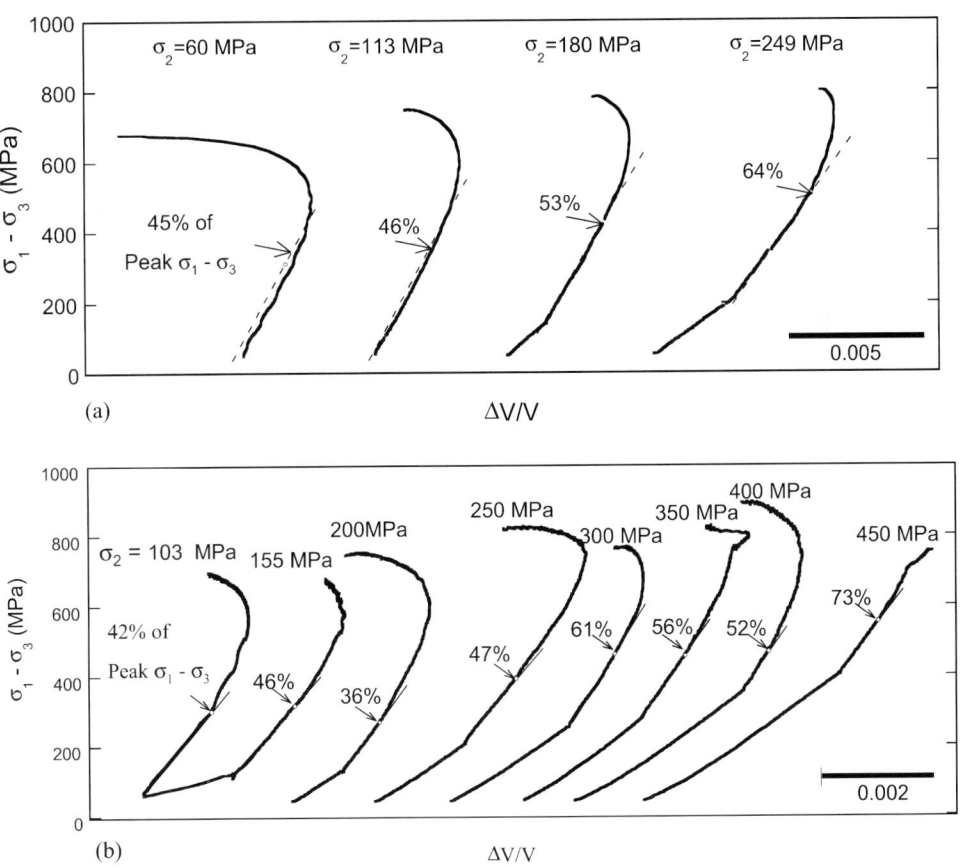

Fig. 11. Stress $(\sigma_1 - \sigma_3)$ variation with volumetric strain $\Delta V / V$ and dilatancy onset (indicated by arrows) for different σ_2 magnitudes and constant σ_3 (60 MPa) in (**a**) Westerly granite and (**b**) KTB amphibolite.

entire profile of the created through-going fault, or orthogonal to the σ_1 direction, revealing the cross-section of the fault. The sections were ground flat and polished down to 0.05 μm. The surface was then sputter-coated with a 0.06-μm-thick carbon layer. Photomicrographs shown here of sample profiles have an identical orientation, with σ_1 acting vertically and σ_3 laterally; those of sample cross-sections are oriented with σ_3 in the vertical, and σ_2 in the lateral directions.

Figure 12 juxtaposes SEM images of segments of the main fault developed during brittle fracture in the two rocks. Common to both crystalline rocks is the development of a multitude of microcracks that are subparallel to the major principal stress σ_1, and localize along a band that upon total failure becomes a steeply inclined fault dipping in the σ_3 direction. Evidence of the shear displacement of the fault is provided by both samples. In Figure 12a, a quartz grain has been split by the fault developed in Westerly granite, with the left half displaced downward (see white arrows). In Figure 12b, debris of apparent fracture asperities that have been sheared off by the fault movement is visible in the KTB amphibolite. Notable in both rocks is the concentration of microcracks in the zone of localization, and their diminishing frequency away from the fault. Two microcracks in a microcline grain left of the fault zone, indicated by black arrows in Figure 12a, demonstrate two typical ways by which they interact. Crack 1, nearest the left edge, reveals *en passant* interaction; crack 2 is developed through en echelon linking (Kranz 1979). The biotite crystal adjacent to the fault is almost free of microcracks, which bypass it and advance intergranularly around it.

We discovered that the main fault does not always extend in a clean planar fashion. Obstacles posed by some crystals or pre-existing structures can cause the fault to follow a tortuous path. One typical phenomenon is that of en echelon arrangement, which can be observed in both rocks (Fig. 13). Biotite grains appear to be the main cause for the offset in the fracture tip. In both samples, microcracks have coalesced and formed a network, similar to observations by Wong (1982).

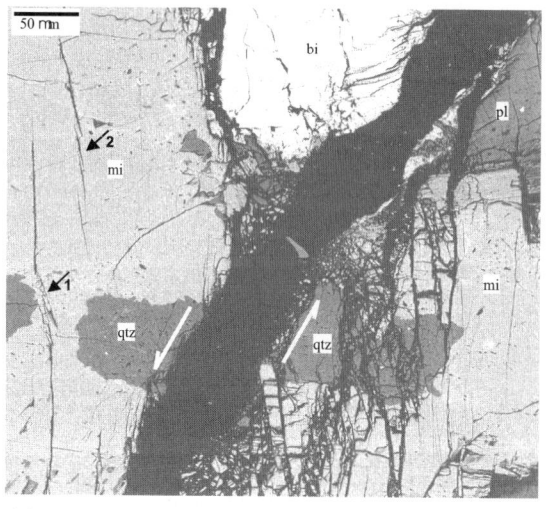 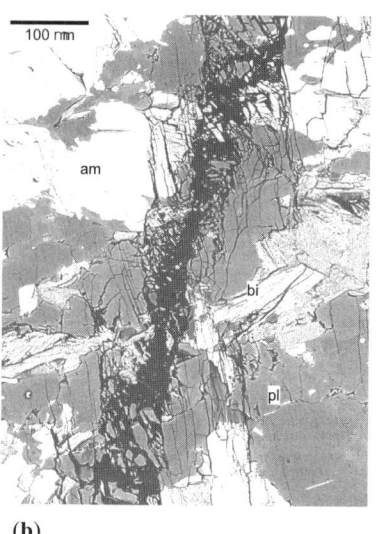

(a)   (b)

Fig. 12. SEM micrographs of fault-plane profile induced during true triaxial testing in (**a**) Westerly granite and (**b**) KTB amphibolite. Microcrack localization is evident on both sides of the fracture surfaces in both rocks. The σ_3 direction is lateral and the σ_1 direction is vertical. White arrows in (**a**) show a quartz (qtz) grain that was split and sheared by the fault. Black arrows in (**a**) indicate two different ways in which microcracks interact in microcline (mi): *en passant* (1) and en echelon (2). Note also in (**a**) that a biotite (bi) grain near the top of the picture is almost free of cracks, which bypass it intergranularly. The original subvertical microcracks (subparallel to σ_3 faces) in both rocks appear to have created a network that includes short transverse cracks. Both pictures, and particularly (**b**), clearly reveal the diminishing of crack concentration away from the fault. Minerals like biotite (bi) and amphibole (am) in (**b**) are almost free of cracks, which tend to concentrate in plagioclase (pl). Inside the fault in (**b**) one can see debris of grains that apparently have been crushed during shear movement.

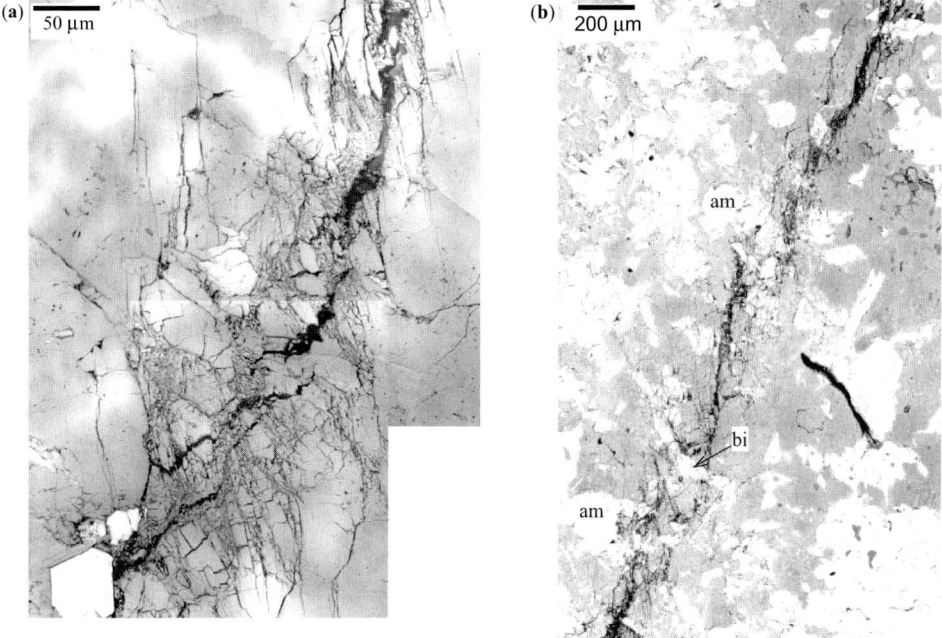

Fig. 13. SEM micrographs, showing the tortuous path taken by the main fracture or fault during true triaxial testing. The σ_3 direction is lateral and the σ_1 direction is vertical. In these examples, both (**a**) Westerly granite, and (**b**) KTB amphibolite, show en echelon interaction (after Chang & Haimson 2000). In (**b**) it is apparent that the en echelon arrangement is caused by amphibole (am) and biotite (bi) grains offsetting the fracture tip. The microcrack network is evident in (**a**).

A different perspective of the development of the fault is obtained from viewing cross-sections of specimens in which failure has occurred. Figure 14 depicts four such views of the fault in Westerly granite under the same least horizontal stress $\sigma_3 = 60$ MPa, and σ_2 varying from 60 to 113 to 180 to 249 MPa. In all cases the fault plane is subparallel to σ_2. We note that when $\sigma_2 = \sigma_3$, the zone immediately next to the fault is fragmented into blocks created by randomly oriented cracks, as should be expected when these two principal stresses are equal. As the magnitude of σ_2 is increased from one test to the next, we note an increased alignment of the localized cracks with the direction of the fault. A very similar observation was made in the KTB amphibolite (Chang & Haimson 2000).

Conclusions

Using a newly designed polyaxial loading apparatus, we completed two extensive series of true triaxial compression tests – one in Westerly granite and the other in the KTB amphibolite. The two crystalline rocks exhibited substantial similarity in their mechanical behaviour. The first important result emerging from the tests is that the effect of the intermediate principal stress σ_2 on the compressive strength cannot be ignored in crystalline rocks, as is tacitly implied in Mohr-type strength criteria. On the contrary, depending on the level of the least and intermediate principal stresses, rock strength may increase by as much as 50% or more, as compared with the strength under the condition used to obtain Mohr-like criteria ($\sigma_2 = \sigma_3$). Generally, the reported tests strongly suggest that use of the conventional Mohr-type criteria in crystalline rocks may lead to overly conservative predictions of strength. Instead, a true triaxial strength criterion that accommodates both tested crystalline rocks, can be expressed as the octahedral shear stress at failure as a function of the mean normal stress acting on the developed fault plane. True-triaxial-strength experimental points for both the granite and the amphibolite are surprisingly well fitted by a power function in this domain.

Brittle failure under the true triaxial stress condition ultimately takes the form of a steeply dipping fault striking in the direction of σ_2. For the same σ_3, fault dip angle increases significantly

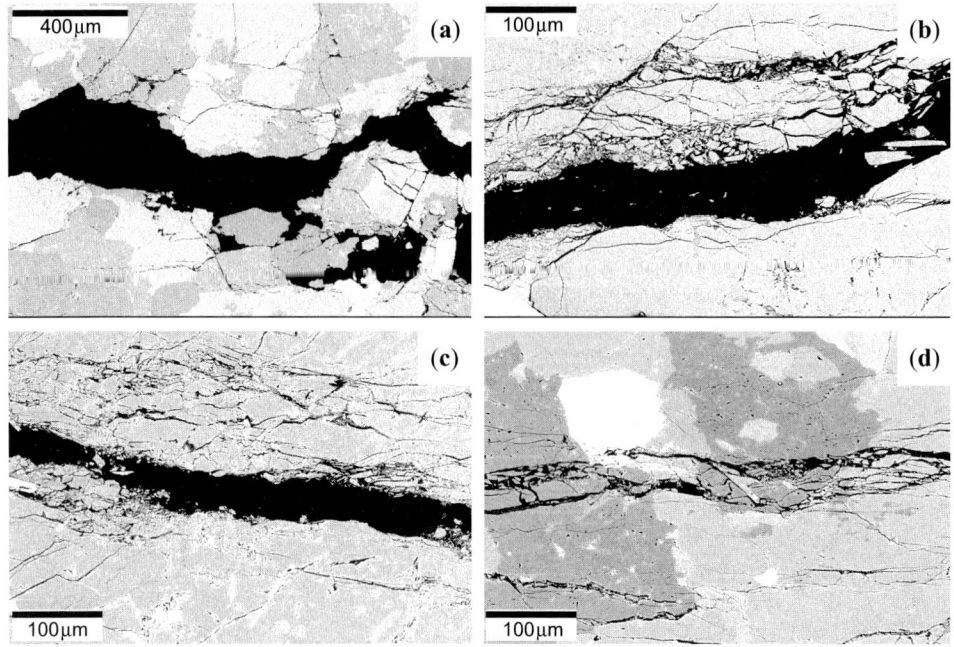

Fig. 14. SEM micrographs of cross-sections of four specimens tested under the same σ_3 (= 60 MPa), but different σ_2 magnitudes (σ_2 = 60, 113, 180, and 249 MPa in (**a**), (**b**), (**c**), and (**d**), respectively). The σ_2 direction is lateral and the σ_3 direction is vertical. The main fracture in all cases is subparallel to σ_2. Localized microcracks align themselves gradually with the σ_2 direction as the stress difference ($\sigma_2 - \sigma_3$) increases.

with the magnitude of the intermediate principal stress. Micromechanically, the fault evolves from the development of microcracks aligned with the $\sigma_1 - \sigma_2$ plane as σ_1 increases relative to the other two principal stresses. Microcracks localize along a steeply inclined band, coalesce, and give rise to the through-going shear fracture or fault. Microcracks link in several ways, such as in en echelon or *en passant* fashion. The fault plane itself develops en echelon linkages when encountering some biotite and amphibole grains that may impede its intragranular growth.

Measurements of strain show accelerated lengthening in the σ_3 direction with increase in σ_1. This lengthening is significantly larger than in the σ_2 direction, suggesting that microcracks subparallel to σ_1, open mainly in the σ_3 direction. This inference is supported by SEM micrographs of specimen cross-sections in the $\sigma_2 - \sigma_3$ plane. Measured volumetric strain variation with the magnitude of σ_1 indicates that, for the same σ_3, dilatancy onset generally increases with the rise in the intermediate principal stress, reflecting a widening of the elastic range as the intermediate principal stress is increased for the same least stress. This suggests a retardation of the onset of the failure process, which

may in part be responsible for the σ_2-related increase in strength in both tested crystalline rocks.

This research was supported by the US National Science Foundation, grant no. EAR-9418738. Some figures in this paper are copyrighted with the American Geophysical Union, and are reproduced or modified by permission.

References

ASHBY, M. F. & SAMMIS, C. G. 1990. The damage mechanics of brittle solids in compression. *Pure and Applied Geophysics*, **133**, 489–521.

BÖKER, R. 1915. Die mechanik der bleibenden formanderung in kristallinish aufgebauten Körpern. *Verhandl. Deut. Ingr. Mitt. Forsch.*, **175**, 1–51.

BRACE, W. F. 1964. Brittle fracture of rocks. *In*: JUDD, W. R. (ed.) *State of Stress in the Earth's Crust.* Elsevier, New York, 111–174.

BRACE, W. F., PAULING, B. W. & SCHOLZ, C. H. 1966. Dilatancy in the fracture of crystalline rocks. *Journal of Geophysical Research*, **71**, 3939–3953.

CHANG, C. & HAIMSON, B. 2000. True triaxial strength and deformability of the German Continental Deep Drilling Program (KTB) deep hole amphibolite. *Journal of Geophysical Research*, **105**, 18 999–19 013.

CRAWFORD, B. R., SMART, B. G. D., MAIN, I. G. & LIAKOPOULOU-MORRIS, F. 1995. Strength characteristics and shear acoustic anisotropy of rock core subjected to true triaxial compression. *International Journal of Rock Mechanics and Mining Science & Geomechanics Abstracts*, **32**, 189–200.

DRUCKER, D. C. & PRAGER, W. 1952. Soil mechanics and plastic analysis or limit design. *Quarterly of Applied Mathematics*, **10**, 157–165.

EWY, R. T. 1998. Wellbore stability predictions using a modified Lade criterion. *In: Proceedings of Eurock 98: SPE/ISRM Rock Mechanics in Petroleum Engineering*, Society of Petroleum Engineers, **1**, SPE/ISRM Paper no. 47251, 247–254.

FREUDENTHAL, A. 1951. The inelastic behavior and failure of concrete. *In: Proceedings of First US National Congress of Applied Mechanics, American Society of Mechanical Engineers*, New York, 641–646.

GRIFFITH, A. A. 1924. The theory of rupture. *In:* BIEZENO, C. B. & BURGERS. J. M. (eds) *Proceedings of First International Congress on Applied Mechanics*. Tech. Boekhandel en Drukkerij, Delft, Netherlands, 55–63.

HAIMSON, B. C. & CHANG, C. 2000. A new true triaxial cell for testing mechanical properties of rock, and its use to determine rock strength and deformability of Westerly granite. *International Journal of Rock Mechanics and Mining Sciences*, **37**, 285–296.

HANDIN, J., HEARD, H. C. & MAGOUIRK, J. N. 1967. Effect of the intermediate principal stress on the failure of limestone, dolomite, and glass at different temperature and strain rate. *Journal of Geophysical Research*, **72**, 611–640.

HOEK, E. & BROWN, E. T. 1980. Empirical strength criterion for rock masses. *Journal of Geotechnical Engineering*. ASCE, **106**, 1013–1035.

JAEGER, J. C. & COOK, N. G. W. 1979. *Fundamentals of Rock Mechanics*, 3rd edn. Chapman and Hall, New York.

KRANZ, R. L. 1979. Crack–crack and crack–pore interactions in stressed granite. *International Journal of Rock Mechanics and Mining Science & Geomechanics Abstracts*, **16**, 37–47.

KRECH, W. W., HENDERSON, F. A. & HJELMSTAD, K. E. 1974. A standard rock suite for rapid excavation research. *Bureau of Mines Report of Investigations*, **7865**.

LABUZ, J. F. & BRIDELL, J. M. 1993. Reducing frictional constraints in compression testing through lubrication. *International Journal of Rock Mechanics and Mining Science & Geomechanics Abstracts*, **30**, 451–455.

LOCKNER, D. A., BYERLEE, J. D., KUKSENKO, V., PONOMAREV, A. & SIDORIN, A. 1991. Quasi-static fault growth and shear fracture energy in granite. *Nature*, **350**, 39–42.

MCGARR, A. & GAY, N. C. 1978. State of stress in the earth's crust. *Annual Review of Earth and Planetary Sciences*, **6**, 405–436.

MICHELIS, P. 1985. A true triaxial cell for low and high-pressure experiments. *International Journal of Rock Mechanics and Mining Science & Geomechanics Abstracts*, **22**, 183–188.

MOGI, K. 1971. Fracture and flow of rocks under high triaxial compression. *Journal of Geophysical Research*, **76**, 1255–1269.

MURRELL, S. A. F. 1963. A criterion for brittle fracture of rocks and concrete under triaxial stress, and the effect of pore pressure on the criterion. *In: Proceedings, Fifth US Rock Mechanics* Symposium, Pergamon Press, 563–577.

NADAI, A. 1950. *Theory of Flow and Fracture of Solids*, **1**, McGraw-Hill, New York.

QUINN, A. W. 1954. Bedrock geology of Rhode Island. *Transactions of the New York Academy of Sciences*, 264–269.

RAUEN, A. & WINTER, H. 1995. Petrophysical properties, *KTB Report 95-2, Niedersächsisches Landesamt für Bodenforschung, Hannover, Germany*, D1–D45.

RÖCKEL, T. & NATAU, O. 1995. Rock mechanics. *KTB Report 95-2, Niedersächsisches Landesamt für Bodenforschung, Hannover, Germany*, E1–E9.

SCHOLZ, C. H. 1990. *The Mechanics of Earthquakes and Faulting*. Cambridge University Press, New York.

TAKAHASHI, M. & KOIDE, H. 1989. Effect of the intermediate principal stress on strength and deformation behavior of sedimentary rocks at the depth shallower than 2000 m. *In:* MAURY, V. & FOURMAINTRAUX, D. (eds) *Rock at Great Depth*, A. A. Balkema, Netherlands, 19–26.

VERNIK, L. & ZOBACK, M. D. 1992. Estimation of maximum horizontal principal stress magnitude from stress-induced well bore breakouts in the Cajon Pass scientific research borehole. *Journal of Geophysical Research*, **97**, 5109–5119.

VERNIK, L., LOCKNER, D. & ZOBACK, M. D. 1992. Anisotropic strength of some typical metamorphic rocks from the KTB pilot hole, Germany. *Scientific Drilling*, **3**, 151–160.

VON KÁRMÁN, Th. 1911. Festigkeitsversuche unter allseitigem Druck. *Verein Deutscher Ingenieure Zeitschrift*, **55**, 1749–1759.

WAWERSIK, W. R. & BRACE, W. F. 1971. Post-failure behavior of a granite and diabase. *Rock Mechanics*, **3**, 61–85.

WAWERSIK, W. R., CARLSON, L. W., HOLCOMB, D. J. & WILLIAMS, R. J. 1997. New method for true-triaxial rock testing. *International Journal of Rock Mechanics and Mining Sciences*, **34**, Paper no. 330.

WIEBOLS, G. A. & COOK, N. G. W. 1968. An energy criterion for the strength of rock in polyaxial compression. *International Journal of Rock Mechanics and Mining Science & Geomechanics Abstracts*, **5**, 529–549.

WONG, T. F. 1982. Micromechanics of faulting in Westerly granite. *International Journal of Rock Mechanics and Mining Science & Geomechanics Abstracts*, **19**, 49–64.

Distribution and properties of fractures in and around the Nojima Fault in the Hirabayashi GSJ borehole

H. ITO & T. KIGUCHI

Geological Survey of Japan/AIST, 1-1-1 AIST Tsukuba Central 7, Tsukuba, Ibaraki 305-8567, Japan (e-mail: hisao.itou@aist.go.jp)

Abstract: A borehole penetrating the Nojima Fault was drilled in 1996, one year after the magnitude 7.2 1995 Kobe earthquake. The major fracture distribution detected by a borehole imaging tool (FMI) in the borehole was compared with the S-anisotropy from aftershocks. The fractures far from the Nojima Fault (more than 50 m) are aligned in an east–west direction, that is, parallel to the tectonic stress direction, whereas the fractures close to the Nojima Fault are normal to the fault. In the fault zone, a fault-parallel fault system was observed over short depth intervals, otherwise the fracture system in the fault zone is random. The fault-normal fracture suggests the complete reduction of the shear stress of the Nojima Fault.

On 17 January 1995, the 1995 Kōbe (Nanbu, Hyōgo-ken) earthquake (M 7.2) occurred beneath the Akashi Strait, and a surface trace about 10 km long appeared on Awaji Island (Fig. 1). The Geological Survey of Japan drilled a borehole penetrating the Nojima Fault one year after the 1995 Kobe earthquake (Ito *et al.* 1996).

The drilling site was 74.6 m from the surface trace of the Nojima Fault (Fig. 1). Cores were recovered for almost entire depth interval from 152 to 746 m. The borehole was drilled through

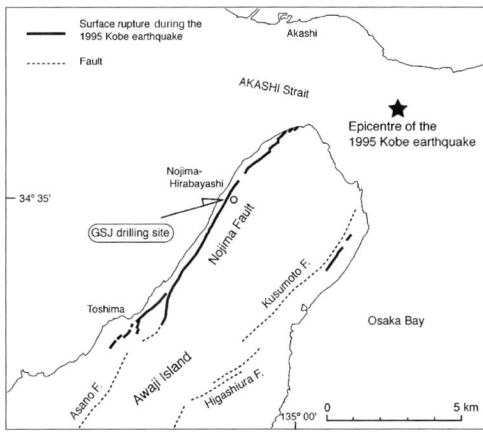

Fig. 1. The location of the Geological Survey of Japan (GSJ) drilling site (modified from Awata *et al.* 1996). The drilling site is located at Nojima–Hirabayashi, along the northern part of the Nojima Fault.

300 m of granodiorite rock mass from the surface. The borehole then penetrated layers of protolith granodiorite and porphyry, and the fault zone from 426 m to the bottom of the hole. The fault was characterized by deformed and altered rocks (Ito *et al.* 1996; Ohtani *et al.* 2000; Ohtani *et al.* 2001; Tanaka *et al.* 2001). The fault core of the Nojima Fault consists of a fault gouge, and was found at 623.1–625.3 m. In the fault zone (426–746 m), fault gouges and cataclasite were detected. Figure 2 shows the geological structure of the borehole.

We conducted conventional borehole logging (Ito *et al.* 1996). In addition to conventional logging, FMI* (Fullbore Formation MicroImager) and DSI* (Dipole Shear Sonic Imager) were conducted (Ito *et al.* 1996). The fault zone assigned by core inspections is characterized by low electrical resistivity, low density and high neutron porosity. Resistivity and density gradually decrease with depth within the fault core to a depth of 623.1 m. Between 623.1 m and 625.3 m depth, the fault gouge has extremely low electrical resistivity of several tens of ohm m. P- and S-wave velocities were determined by the DSI tool, and, as in the case of electrical resistivity and density, both gradually decrease within the fault core to 623.1 m. The P-wave velocity drops from 4 km s^{-1} (just above the fault gouge at 623.1 m) to 2.6 km s^{-1} (at the fault core of 623.1 m to 625.3 m).

Subsurface fracture systems have an important influence on rock strength, seismic velocities, seismic anisotropy and permeability. It is very

From: HARVEY, P. K., BREWER, T. S., PEZARD, P. A. & PETROV, V. A. (eds) 2005. *Petrophysical Properties of Crystalline Rocks*. Geological Society, London, Special Publications, **240**, 61–74.
0305-8719/05/$15.00 © The Geological Society of London 2005.

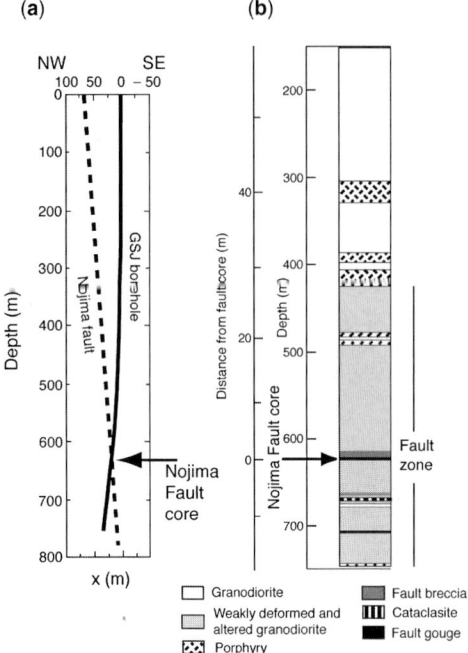

Fig. 2. (a) Trajectory of the GSJ drilling borehole and the Nojima Fault. (b) The geological structure of the GSJ borehole.

important to characterize the orientation, frequency and anisotropy of fractures in the fault zone in order to understand the earthquake fault mechanics. The GSJ Hirabayashi borehole has the advantages that it penetrates the fault zone that activated a recent large earthquake; the geology is simple; and the effect of faulting on fracture distribution is straightforward, so a comparison between the major fractures detected by borehole logging and remote observations (for example, S-wave anisotropy from aftershock observation) is possible.

Imaging data in the GSJ Hirabayashi borehole

FMI images provided a detailed picture of fracture distribution. In this study, we analysed the FMI fracture data from within and outside of the fault zone, in order to characterize the fracture system in the Nojima Fault. The FMI tool provides micro-resistivity images of the borehole wall (Ekstrum *et al.* 1987). We were able to detect 2683 fractures at depths from 152 m to 746 m. Figure 3 shows an example of an FMI image and the fractures detected. A correction for fracture dip for fracture population is neces-

sary in order to analyse fracture data statistically (Hudson & Priest 1983; Barton & Zoback 1992). The probability of intersecting a horizontal fracture in a vertical borehole is one, whereas the probability of intersecting a vertical fracture is zero. We applied the statistical correction for fracture dip to fracture populations. The corrected population predicts the frequency of fractures of a given orientation that could be intersected if the borehole were drilled normal, to each fracture plane. The correction applied to generate the corrected population is $1/\cos\theta$ where θ is the fracture dip.

Structural characterization from microstructural observation

Host rocks are present at depths shallower than 426 m, and consist of granodiorite and porphyry. The fault zone was characterized by deformed and altered rocks (Ohtani *et al.* 2000; Ohtani *et al.* 2001; Tanaka *et al.* 2001). At the mesoscopic scale of observation, the fault zone encompasses the depths of 426 m to 746 m, i.e. the area characterized by alteration and deformation.

We recognized four distinct zones in the fault zone from the results of microstructural observations and chemical analysis of the recovered core samples (Ohtani *et al.* 2000; Tanaka *et al.* 2001). One is the upper fault core (611–624 m), which is located just above the fault core, and is characterized by volume loss and mechanical wearing due to coseimic faulting. The second is the lower fault core (624–641 m), which is located just beneath the fault core and is also regarded as a coseismic fault zone, based on the texture of brecciation. These two fault cores are surrounded by a damaged zone. The third and fourth zones are the upper damaged zone (426–611 m) and lower damaged zone (641–746 m), respectively. The upper and lower fault zones are regarded as a dilatant, coseismic zone at the marginal part of the main shear surface. The fault core-damaged zone model is quite reasonable for understanding the fault-zone mechanics.

We were able to detect resistive fractures as resistive sinusoids observed on the FMI image. The materials infilling the fractures are considered to cause resistivity on the FMI image. Figure 4 shows the distribution of the resistive fractures. The number of resistive fractures is markedly less than that of the conductive fractures. It is noteworthy that most resistive fractures are distributed in the zone shallower than the fault core, at 623.1–625.3 m, and there is only one

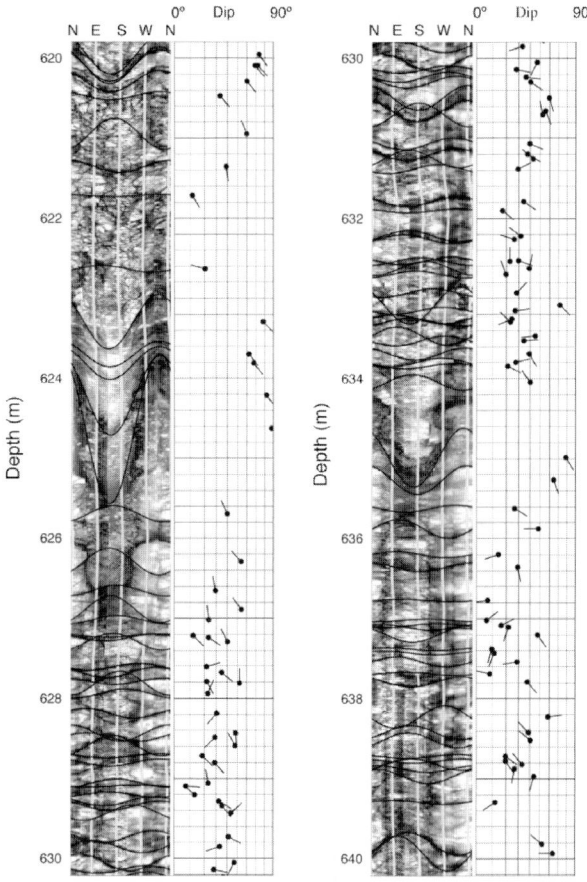

Fig. 3. FMI image at the fault core (620–640 m) of the Nojima Fault. Sinusoidal curves on the middle track indicate interpreted fractures. Dip angle and dip direction of the fractures are shown in the right track. Note that the fractures at the fault core (623.1–625.3 m) have a high dip angle, whereas low dip angle fractures (20–30°) are dominant below the fault core.

resistive fracture at the depth range from 600 m to the bottom of the hole. The trend of the strike of the resistive fractures is NE–SW.

Fracture distribution

Both the conductive and the resistive fracture frequency per 5 m are plotted in Figure 5, over the interval 152–746 m detected from FMI images. The P-wave velocity determined by monopole sonic waveforms is also shown in Figure 5. Dense fractures both outside and within the fault zone were detected as shown in Figure 5. The fracture frequency per 1 m outside the fault zone (152–426 m) is 4.20 (fracture m^{-1}) and inside (426–746 m) it is 4.80. Outside the fault zone there is no systematic change in the fracture population with increasing depth, although frac-

ture frequencies outside the fault zone increase in several short intervals that correspond to low-velocity zones.

Within the fault zone, the fracture frequency is different between the hanging wall and footwall of the Nojima Fault. The fracture frequencies in the upper damaged zone and upper fault core are almost the same, 3.99 and 4.08, respectively. Each of these fracture frequencies is less than outside the fault zone, where the frequency is 4.19.

The footwall of the Nojima Fault has a larger fracture frequency than the hanging wall. The fracture frequency at the lower fault core is 5.24, which is 1.28 times larger than the upper fault core, and in the lower damaged zone it is 6.23, which is 1.56 times larger than the upper damaged zone. Particularly notable is the

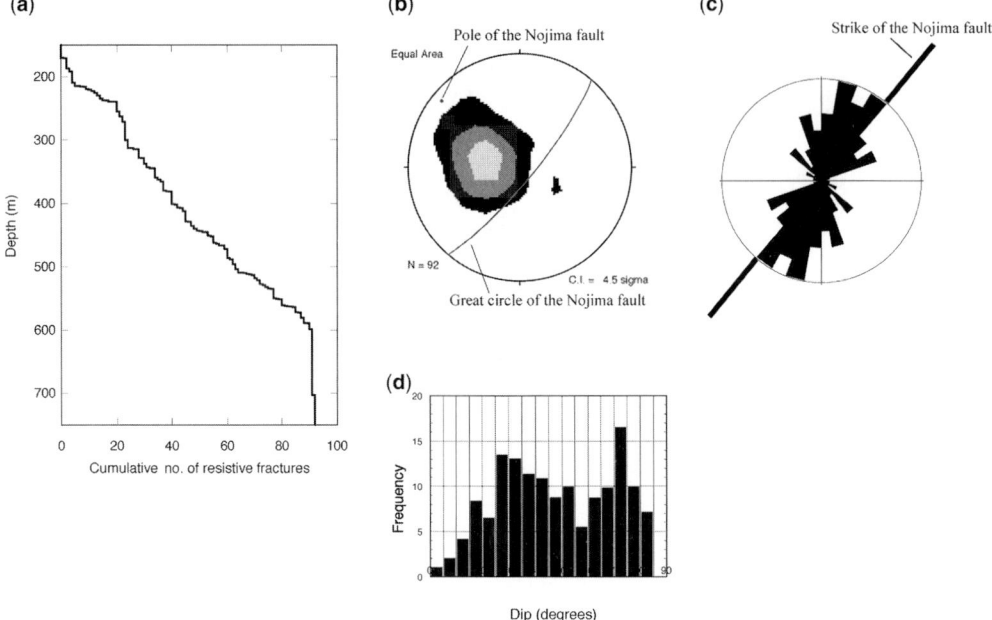

Fig. 4. (a) Distribution of the frequency of the resistive fractures per 5 m. (b) Lower-hemisphere equal-area projection for the resistive fractures (*N* = 92). The distribution of strike (c) and dip (d) of the resistive fractures.

frequency of the fractures with a dip angle lower than 20 degrees: in the footwall it is 1.2, which is 3.4 times larger than in the hanging wall.

The P-wave velocity gradually decreases within the fault core down to 623.1 m. The decrease in P-wave velocity at the fault core is very sharp. Then, at the footwall of the fault zone, the P-wave velocity gradually increases with depth. Thus, there is no simple trend in velocity within the fault zone, and there is no clear correlation between P-wave velocity and fracture frequency.

Fracture orientation

The fracture orientation trend was categorized as four shear zones and the area outside the fault zone. The orientation of all fractures inside the fault zone is shown in Figure 6a, in a lower-hemisphere equal-area projection. The shading pattern gives the standard deviation from a random distribution. Figures 6b and c show the distribution of strike and dip respectively of all fractures inside the fault zone. Figure 7 shows the same plot as Figure 6, but for the fractures outside the fault zone. As shown in Figure 6b, the strike of the fractures inside the fault zone is rather dispersed. On the other hand, the fractures outside the fault zone concentrate mostly around N60°W (Fig. 7). This trend for fractures

outside the fault zone is almost perpendicular to the surface trace of the Nojima Fault. Figures 7b–e shows that fracture orientations of the fault core and damaged zone inside the fault zone have different characteristics. In both the upper and lower parts of the fault core, the fractures strike strongly N45°E, which is almost parallel to the strike of the Nojima Fault (Fig. 7c,d). The trend of fracture orientation in the upper damaged zone is clearly N45°W, which is perpendicular to that of upper and lower parts of the fault core (Fig. 7b). The strike of the fractures in the lower damaged zone is rather dispersed and has a weak north–south orientation (Fig. 7e).

Seismic anisotropy and fracture trend

The seismic anisotropy along the borehole was analysed using DSI logging data. The azimuths of the fast shear-wave polarizations estimated from the DSI data are shown in Figure 8. Outside the fault zone, the fast shear azimuth is localized in a N60°W direction. This trend implies that seismic anisotropy outside the fault zone is caused by the aligned fractures.

Within the fault zone, the fast shear azimuths are rather dispersed, in the same manner as the fracture trend. However, the shallower part of the upper damaged zone and deeper part of the

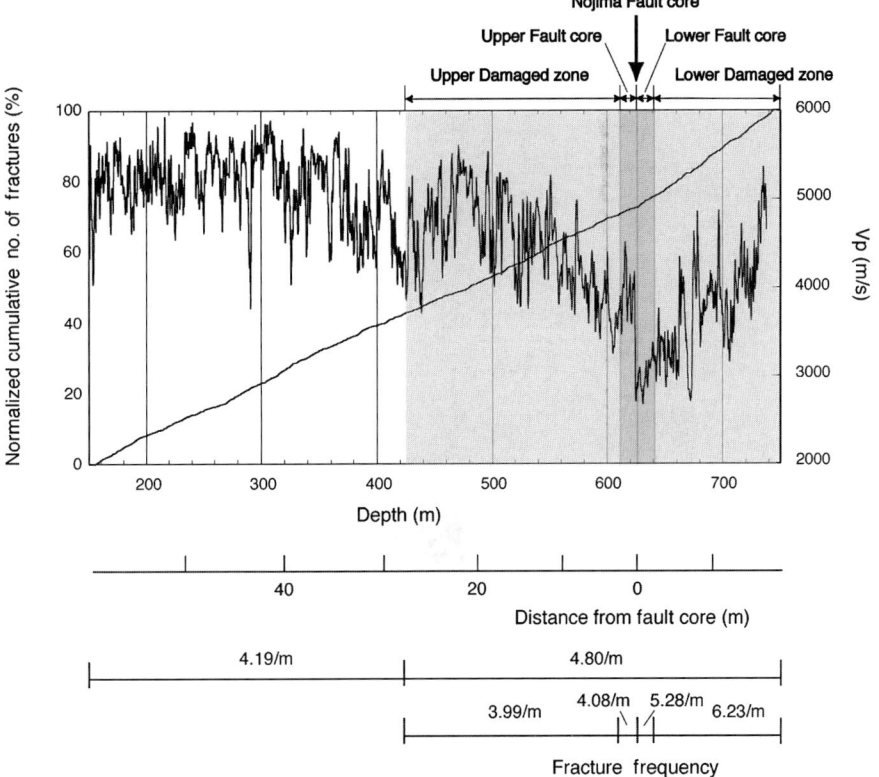

Fig. 5. Distribution of the fracture frequency per 5 m and P-wave velocity along the borehole. The shaded area indicates the fault zone.

lower damaged zone has a N60–70°W trend of fast shear azimuth. This azimuth of anisotropy corresponds almost exactly to the strike of fractures in the upper damaged zone. An estima-tion of the azimuth of the fast shear wave was not possible at several intervals in both the upper and lower parts of the fault core, because of weak seismic anisotropy. The azimuth of the

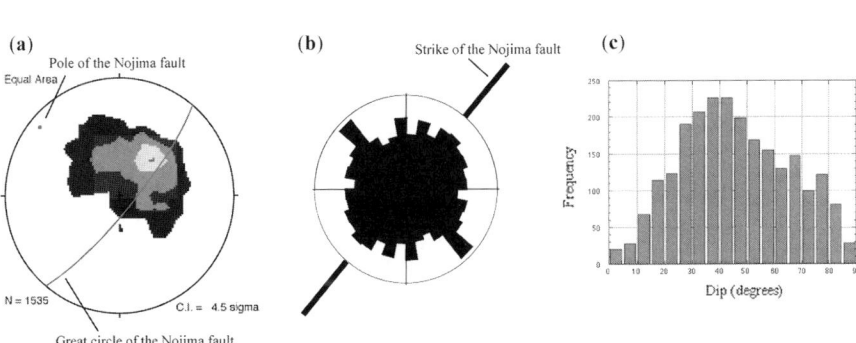

Fig. 6. (a) Lower-hemisphere equal-area projection for the fractures of whole fault zone (426–745 m). The contour interval is 4.5 σ. The distribution of strike (**b**) and dip (**c**) of the fractures of whole fault zone. The surface strike of the Nojima Fault is shown in the figure (**b**).

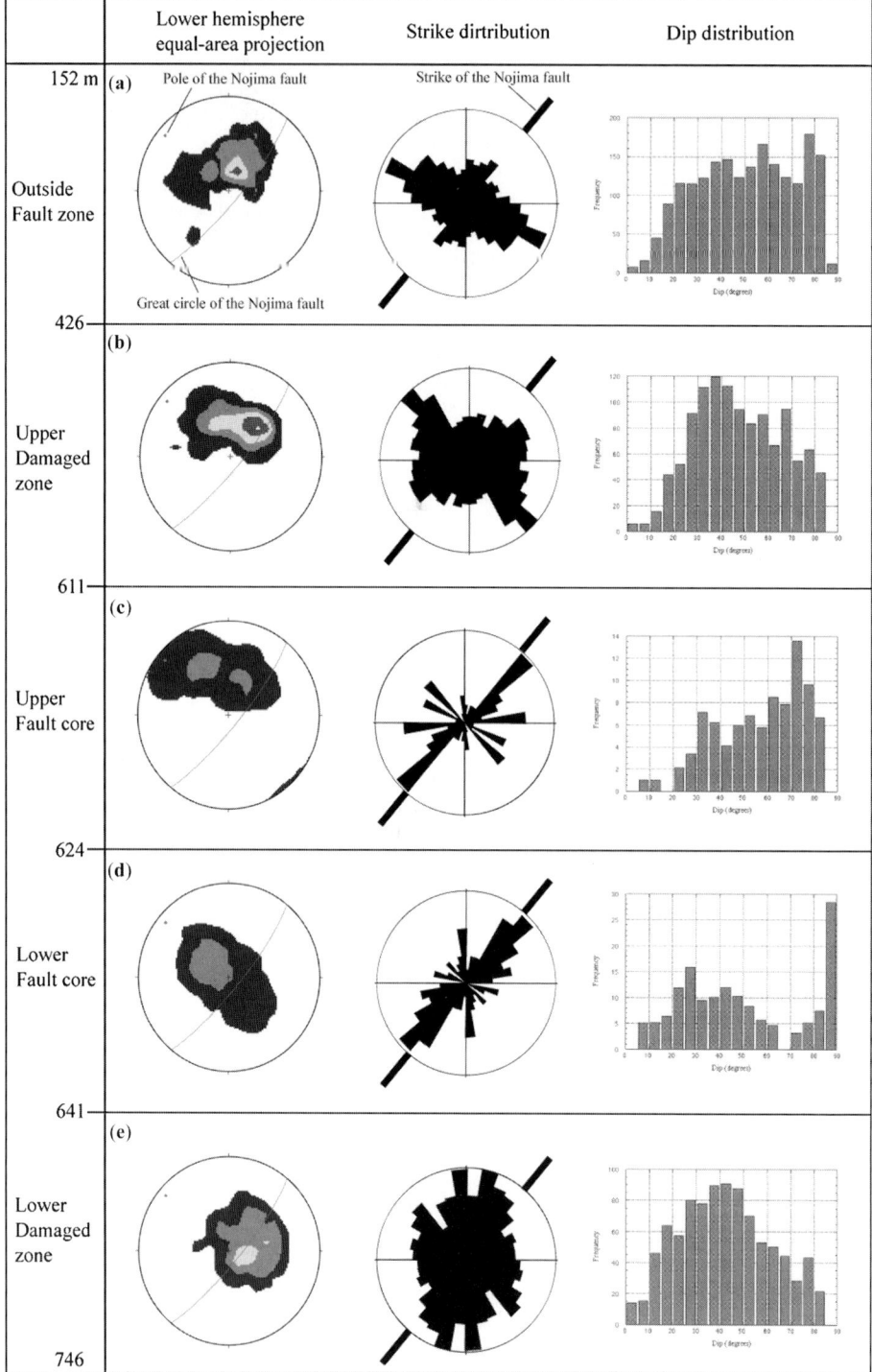

Fig. 7. Lower-hemisphere equal-area projection for the fractures (**a**) outside the fault zone (152–426 m); (**b**) of the upper damaged zone (426–611 m); (**c**) of the upper fault core (611–624 m); (**d**) of the lower fault core (624–641 m); (**e**) of the lower damaged zone (641–746 m).

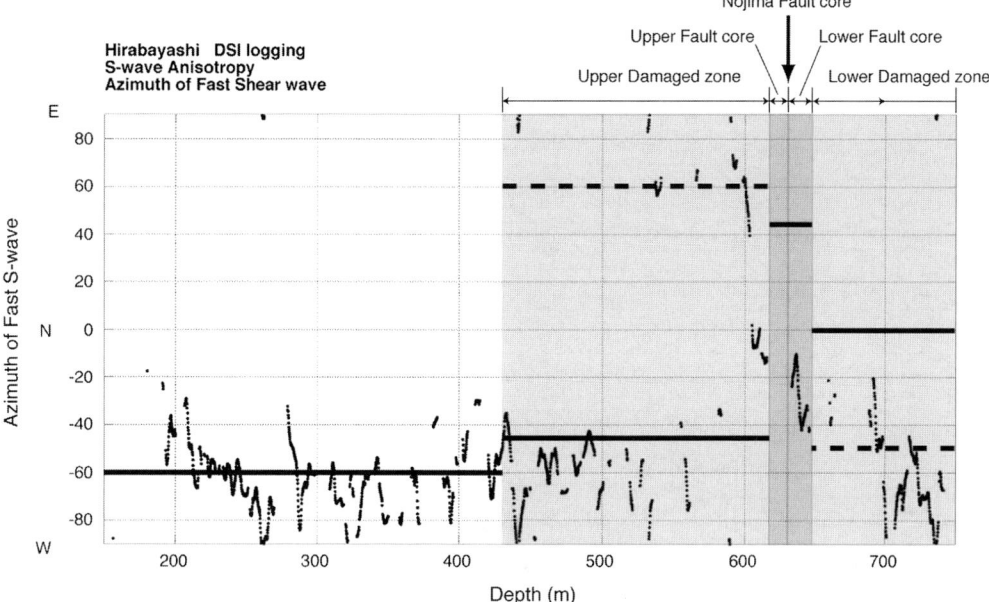

Fig. 8. Azimuth of the fast shear wave estimated from the DSI logging data.

fast shear wave, which was partly estimated in the lower fault core, is N 30°W. This azimuth of seismic anisotropy does not correspond to the dominant strike of the fractures in the fault core.

Fracture dip

As mentioned above, a correction of fracture population for fracture dip was made in order to analyse fracture data statistically. Fracture dip distributions outside the fault zone, in the whole fault zone, and in four shear zones are shown in Figures 6c and 7. The average and variance of fracture distribution were calculated for each zone, in order to estimate normal distributions. Figure 9 parts (a)–(f) show the comparison between fracture dip distribution and normal distribution estimated for each zone. The correlation coefficient is shown in Figure 9. Outside the fault zone, the dip distribution does not fit the normal distribution, because the dominant dip angle is 75–80° (Fig. 9b). The dominant dip angle of the whole fault zone is 35–40°, and the fracture dip distribution corresponds to a normal distribution (Fig. 9a). However, the trends in the dip distribution in the damaged zone and the fault core are quite different. The dominant fracture dip in the upper and lower damaged zone is 35–40° and 40–45°, respectively, and the dip distribution of each damaged zone fits a normal distribution

(Fig. 9c & f). On the other hand, the fracture dip distribution in the upper and lower fault core does not fit a normal distribution. Furthermore, the trend in dip distribution between the upper and lower fault core is different. The upper fault core has higher number of high dip angle fractures than the lower fault core (Fig. 9d & e).

The fracture dip distribution of the lower fault core has a trend fitting a bimodal normal distribution. Figure 10 shows that the dip distribution of this zone, at dips ranging from 0 to 70 degrees, fits the normal distribution.

Fracture spacing

The distribution of fracture spacing for each zone was estimated as shown in Figure 11a–f. According to Priest & Hudson (1976, 1981) and Hudson & Priest (1979), a homogeneous rock would produce a random distribution of fracture spacings and lead to a negative exponential distribution. The negative exponential distribution is mathematically expressed as:

$$f(x) = \lambda \times \exp(-\lambda \times x)$$

where $f(x)$ is the frequency of a fracture spacing x, and λ is the average fracture frequency per metre.

In contrast to the random distribution, the clustered fracture distribution will produce a power-law distribution of fracture spacing (Barton &

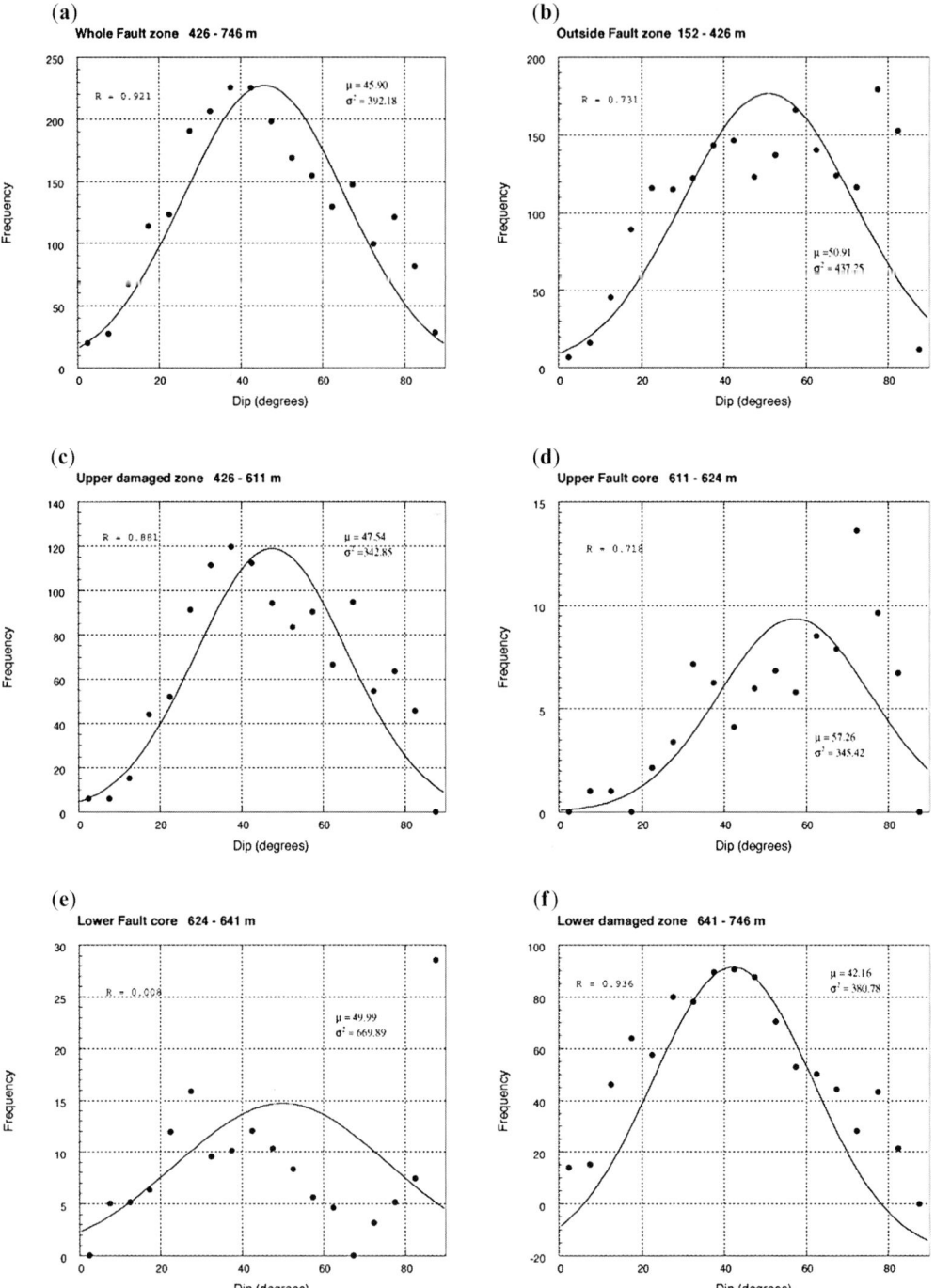

Fig. 9. Comparison between the distribution of the fracture dip (closed circle) and the normal distribution (solid line). The value of *R* in the figure is the correlation coefficient. (**a**) The whole fault zone (426–745 m); (**b**) outside the fault zone (152–426 m); (**c**) the upper damaged zone (426–611 m); (**d**) upper fault core (611–624 m); (**e**) lower fault core (624–641 m); (**f**) lower damaged zone (641–746 m).

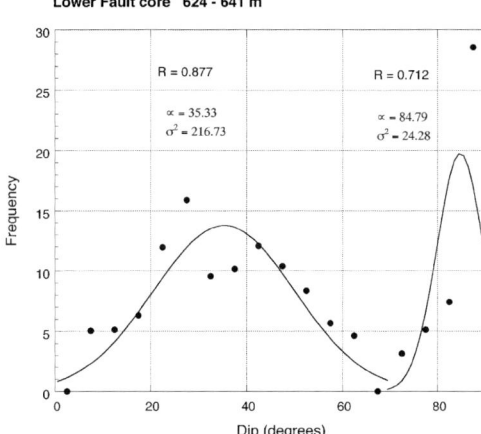

Fig. 10. Comparison between the fracture dip distribution of the lower fault core and a bimodal normal distribution.

Zoback 1992). The clustered distribution means that the high frequency of low spacing values occurs in clusters, and that the low frequency of high spacing values occurs between clusters.

Three theoretical curves are also plotted in Fig. 11. One is the best-fit curve of the exponential distribution, the second is the best-fit curve of exponential function $(f(x) = a \times \exp(-b \times x))$, and the third is the exponential distribution with the λ value determined from the mean fracture frequency per metre. Outside the fault zone, and in the upper and lower damaged zones, the distributions of the fracture spacings have a good fit to a curve of negative exponential distribution, and the correlation coefficient ranges from 0.88 to 0.90. The fracture spacing of the upper and lower fault core does not follow the theoretical curves.

Figure 11 indicates that there is a considerable drop in the frequency for spacing values less than 0.08 m. This drop-off is considered to be largely related to the detection limits of this study, and so the results for spacing values below 0.08 m are discounted. Figure 12 shows three theoretical curves. One is the best-fit curve for the power-law distribution; the second is the best-fit curve of exponential distribution; and the third is the exponential distribution, with the λ value determined from the mean fracture frequency per metre. The negative exponential curve fits the fracture spacing distributions outside the fault zone and in the upper and lower damaged zones better than it fits the power-law curve. The difference in the correlation coefficients between the exponential and power-law

distribution for areas outside the fault zone is larger than that for the damaged zone. This implies that outside the fault zone, fractures are more randomly distributed than within the damaged zone. The small difference in the coefficient values for two of the distributions for the damaged zone indicates that the fracture distribution of the damaged zone has a combination of random and clustered distributions.

Summary and discussion

The Hirabayashi borehole data were defined as being within or outside the fault zone. The fault zone was then classified into four zone from core analysis:

(1) the upper damaged zone (426–611 m);
(2) the upper fault core (611–624 m);
(3) the lower fault core (624–641 m); and
(4) the lower damaged zone (641–746 m).

We analysed fracture frequency, fracture dip and strike, fracture spacing and anisotropy in each zone of the borehole. Table 1 summarizes the results of fracture analysis.

There are large differences in fracture distributions between the zones within the Nojima fault zone and the zones outside the fault zone. The characteristic features of fracture distributions for outside the fault zone (152–426 m) are as follows.

The trend of the strike is N60°W, which is almost perpendicular to the surface trace of the Nojima Fault. A high dip angle (70–80°) of fracture is dominant. From the analysis of seismic anisotropy by the DSI logging data, the fast shear azimuth is localized in the N60°W direction. This seismic anisotropy is thought to be caused by the N60° W aligned fractures with a high dip angle. The seismic anisotropy, determined from natural earthquake observations after the 1995 Kobe earthquake (Mizuno *et al.* 2001) is almost east–west, and this is believed to be parallel to the tectonic stress direction.

The fracture system outside the fault zone is almost consistent with the results of the analysis for shear-wave splitting. The negative exponential distribution provides a good fit to the observed fracture spacing. This indicates that fractures are randomly distributed. This negative exponential distribution of fracture spacing could occur in a homogeneous rock mass (Priest & Hudson 1976).

The features of the whole fault zone (426– 746 m), compared to the fracture distribution outside the fault zone are as follows. The strike of the fractures within the fault zone is rather

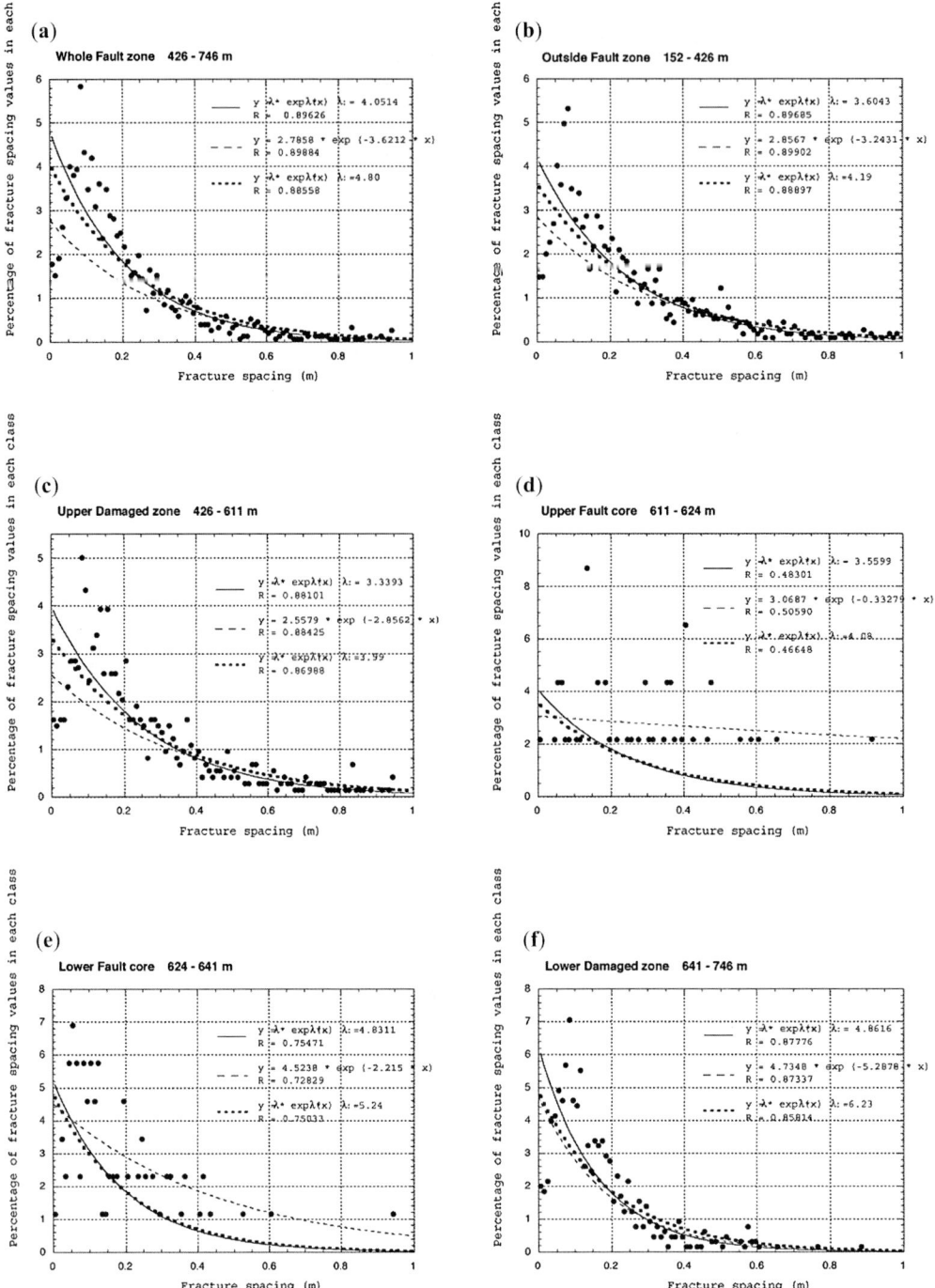

Fig. 11. Distribution of observed fracture spacing (closed circles). The exponential distribution (solid line) and exponential function (dashed lines) fit the data on fracture spacing. The curves of the exponential distribution with the coefficient (λ) value determined from mean fracture frequency per metre are also plotted (bold dotted lines). (**a**) Whole fault zone (426–745 m); (**b**) outside the fault zone (152–426 m); (**c**) upper damaged zone (426–611 m); (**d**) upper fault core (611–624 m); (**e**) lower fault core (624–641 m); (**f**) lower damaged zone (641–746 m).

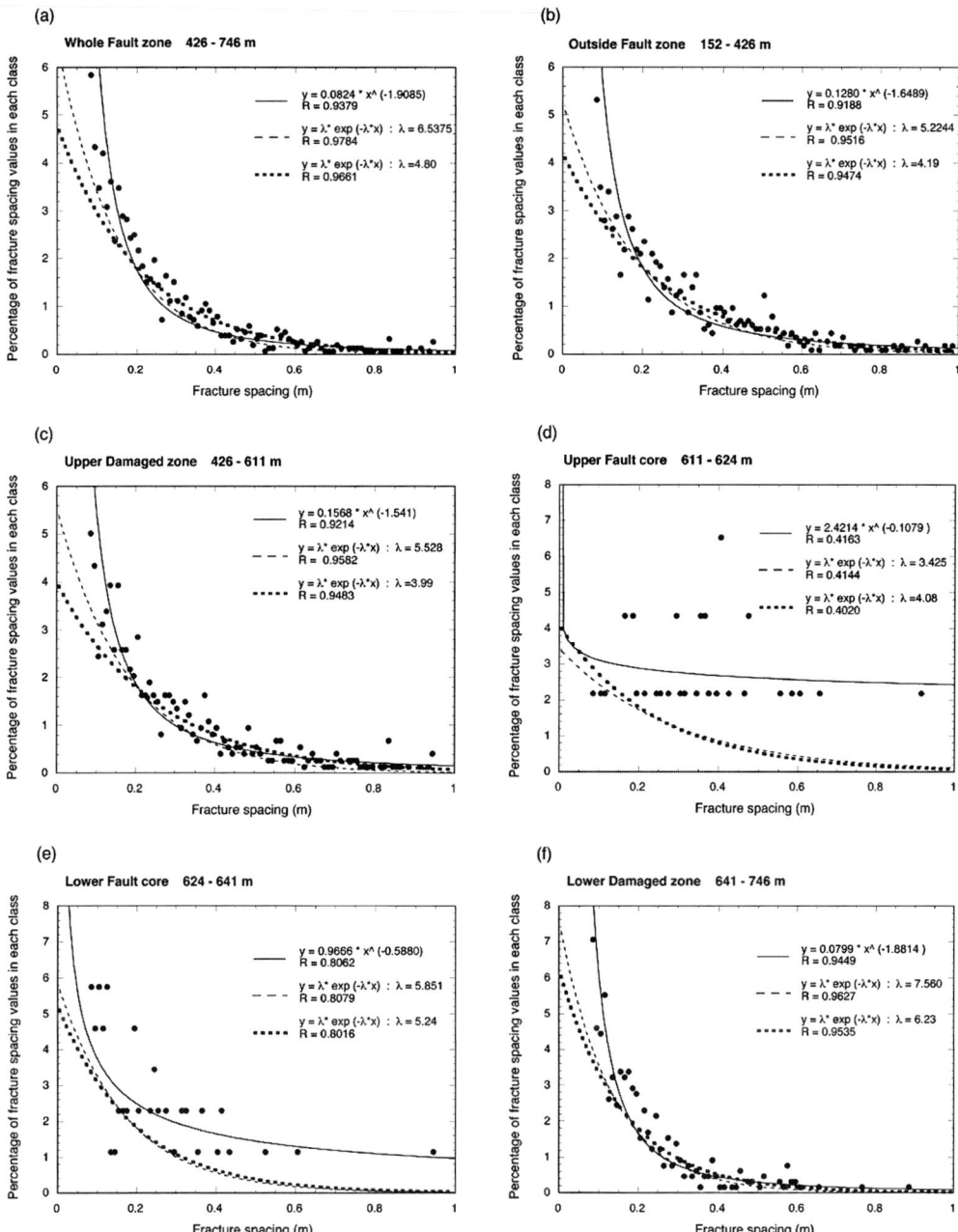

Fig. 12. Distribution of observed fracture spacing after truncating the spacing values lower than 0.08 m (closed circles). The power-law distribution (solid line) and the exponential distribution (dashed line) fit the data on fracture spacing. The curves of the exponential distribution with the coefficient (λ) value determined from mean fracture frequency per metre are also plotted (bold dotted lines). (**a**) Whole fault zone (426–745 m); (**b**) outside the fault zone (152–426 m); (**c**) upper damaged zone (426–611 m); (**d**) upper fault core (611–624 m); (**e**) lower fault core (624–641 m); (**f**) lower damaged zone (641–746 m).

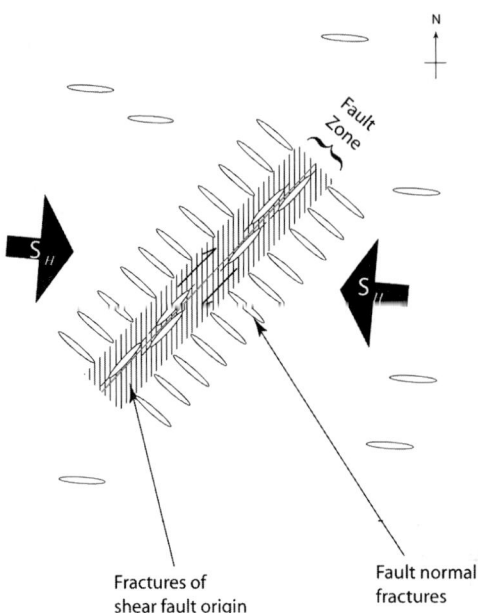

Fig. 13. Fracture model based on the major fracture distribution in the borehole. The fractures close to the fault are fault-normal. In the fault zone, the fault-parallel fault system was observed in a narrow depth interval, otherwise the fracture system in the fault zone is random. The fault-normal fractures indicate the complete reduction in the shear stress of the Nojima Fault.

dispersed, and there is no clear trend. The dominant dip angle is 35–40°, and the fracture dip distribution is described by a normal distribution. The fracture frequency is 4.80 per m, which is about 1.1 times higher than outside the fault zone.

The fault zone is classified into two specific zones. These are the fault core and the damaged zone. The fault core is at the centre of the Nojima Fault zone, and the damaged zone surrounds the fault core. The fracture properties are different between the fault core and damaged zone. Furthermore, the upper and lower parts of the fault core and the damaged zone show several differences. The fractures of both the upper and lower fault core have a trend of N45°E, which is almost parallel to the strike of the Nojima Fault. However, the trend of the fracture dip angle of the lower fault core is not same as that of the upper fault core. The lower fault core is dominated by low-angle fractures (about 30°). The fracture frequency of the lower fault core is about 1.3 times larger than that of the upper fault core. While the seismic anisotropy in the fault core is rather weak, the fast shear azimuths of the lower fault core are localized at N10–40°W. The fractures in the fault core have a N45°E strike trend

Table 1. *Distribution of fractures inside and outside the Nojima Fault zone, within the GSJ borehole*

		Outside fault zone (152–426 m)	Whole fault zone (426–746 m)	Upper damaged zone (426–611 m)	Upper fault core (611–624 m)	Lower fault core (624–641 m)	Lower damaged zone (641–746 m)
Fracture frequency	Total number	1148	1535	739	53	89	654
	Frequency/m	4.19	4.80	3.99	4.08	5.24	6.23
Strike	Dominant trend	N60°W	Dispersed	N45°W	N45°E	N45°E	Dispersed
Dip	Dominant dip angle	75–80°	35–40°	35–40°	70–75°	25–30°, 85–90°	40–45°
	Fit to normal distribution (R: correlational coefficient)	No fit R = 0.731	Good fit R = 0.921	Good fit R = 0.881	No fit R = 0.718	No fit R = 0.008	Good fit R = 0.936
Fracture spacings (fit negative exponential function) (R: correlational coefficient)		Good fit R = 0.896	Good fit R = 0.897	Good fit R = 0.881	No fit R = 0.483	No fit R = 0.754	Good fit R = 0.878
Azimuth of fast S-wave (S-wave anisotropy)		N60°W	Dispersed	Partly N60°W, Partly dispersed	No estimate	Partly N30°W	Partly N70°W

Fractures of shear fault origin

Fault normal fractures

with a low dip angle (25–30°). According to Crampin (1978), a medium with aligned fractures with a low dip angle (less than 60°), has a shear wave anisotropy which is perpendicular to the aligned fractures. Therefore, the shear-wave anisotropy detected by the DSI logging is considered to be caused by the aligned fractures with low dip angles. The trend of the fracture spacing in both the upper and lower fault core is not clear when fitted to the theoretical curves.

The fracture distribution of the damaged zone is different from that of the fault core. The strike of fractures in the upper damaged zone is concentrated at N45°W, which is similar to the trend of the strike outside the fault zone. The dominant fracture-dip angle of both the upper and lower damaged zone is about 40° and the distributions of dip angle fit the normal distribution. These features of fracture-dip angle are quite different from those of the fault core and outside the fault zone. The distributions of fracture spacing show good fit to a negative exponential curve, with the coefficient value determined from the mean fracture frequency per metre. The distributions also fit a power-law distribution with high value of 0.92–0.94 for the correlation coefficient. These features of the fracture spacing in the damaged zones have resulted from the superposition of random and clustered distributions.

There are significant differences in the fracture distribution between the areas within and outside the fault zone. Fractures inside the fault zone have a complicated distribution: the fracture system is almost random, except in a narrow depth interval where the strike is almost parallel to the Nojima Fault.

Tadokoro et al. (1999), Mizuno (2001) and Tadokoro & Ando (2002) studied S-wave splitting from aftershocks near the Nojima Fault, and found that the S-wave anisotropy direction at stations far from the Nojima Fault is parallel to the regional maximum horizontal compressional stress direction (east–west). However, Tadokoro et al. (1999) and Tadokoro & Ando (2002) observed that the S-wave anisotropy direction in the vicinity of the Nojima Fault is parallel to the fault strike (N45–50°E). Tadokoro & Ando (2002) also found a temporal change in S-wave anisotropy. The direction of the S-wave anisotropy rotated from being parallel to the strike of the Nojima Fault to being parallel to the regional tectonic stress direction during the period 33–45 months after the 1995 Kobe earthquake.

Isoyama (2002) analysed the data from a dense seismic array. The array runs across the Nojima Fault surface trace at Hirabayashi. Observations began two months after the 1995 Kobe earthquake, and have continued with essentially the same array. It was found that the S-wave anisotropy depends on the distance from the Nojima Fault. The S-wave direction is east–west far from the Nojima Fault, and is normal to the Nojima Fault when in close proximity to the fault. Also, anisotropy is not clear within the fault zone, because both the S-wave and the trapped wave are observed at almost the same travel time.

Based on the major fracture distribution detected by FMI and comparison with the S-anisotropy from aftershocks, we propose that fractures far from the Nojima Fault (more than 50 m) are aligned east–west, that is, parallel to the tectonic stress direction, whereas the fractures close to the Nojima Fault are in the fault-normal direction. In the fault zone, the fault-parallel fault system was observed in the narrow depth interval, but otherwise the fracture system in the fault zone is random. The major fractures show the fracture system close to the fault, because the distance from the fault core is at most 55 m in the borehole (Fig. 5).

Yamamoto et al. (2002) showed that the friction coefficient is smaller than 0.15, from the stress measurements in several boreholes at the Nojima Fault. Although we do not know the exact mechanism of these fracture systems normal to the fault, our present model suggests that these fracture systems are due to a complete reduction of shear stress. This may be justified from that these are open fracture systems.

Concluding remarks

The major fracture distribution detected by FMI was compared with the S-anisotropy from aftershocks. From the S-wave anisotropy analysis, the fractures far from the Nojima Fault (more than 50 m) are estimated to be aligned in an east–west direction, that is parallel to the tectonic stress direction, whereas the fractures close to the Nojima Fault are in the fault-normal direction. In the fault zone, the fault-parallel fault system was observed in a narrow depth interval; otherwise the fracture system in the fault zone is random. The fault-normal fractures suggest a complete reduction of the shear stress of the Nojima Fault.

Discussions with Y. Kuwahara, T. Ohtani, H. Isoyama and A. Stork were useful.

References

AWATA, Y., MIZUNO, K., SUGIYAMA, Y., IMURA, R., SHIMOKAWA, K., OKUMURA, K., TSUKUDA, E. & KIMURA, K. 1996. Surface fault ruptures on the northwest coast of Awaji Island associated with the Hyogo-ken Nanbu earthquake of 1995, Japan. *Journal of the Seismological Society of Japan*, **49**, 113–124.

BARTON, C. A. & ZOBACK, M. D. 1992. Self-similar distribution and properties of macroscopic fractures at depth in crystalline rock in the Cajon Pass scientific drill hole. *Journal of Geophysical Research*, **97**, 5181–5200.

CRAMPIN, S. 1978. Seismic-wave propagation through a cracked solid: polarization as a possible dilatancy diagnostic. *Geophysical Journal of the Royal Astronomical Society*, **53**, 467–496.

EKSTRUM, M. P., DAHAN, C. A., CHEN, M. Y., LLOYD, P. M. & ROSSI, D. J. 1987. Formation imaging with microelectrical scanning arrays. *The Log Analyst*, **28**, 294–306.

HUDSON, J. A. & PRIEST, S. D. 1979. Discontinuities and rock mass geometry. *International Journal of Rock Mechanics and Mining Sciences and Geomechanics Abstracts*, **16**, 339–362.

HUDSON, J. A. & PRIEST, S. D. 1983. Discontinuity frequency in rock masses. *International Journal of Rock Mechanics and Mining Sciences and Geomechanics Abstracts*, **20**, 73–89.

ISOYAMA, H. 2002. *Spatiotemporal variation of the polarization anisotropy of S waves observed at the Nojima fault.* MSc thesis, Ibaraki University.

ITO, H., KUWAHARA, Y. et al. 1996. *Structure and physical properties of the Nojima Fault, Butsuri–Tansa (Geophysical Exploration)*, **49**, 522–535.

MIZUNO, T., YOMOGIDA, K., ITO, H. & KUWAHARA, Y. 2001. Spatial distribution of shear wave anisotropy in the crust of the southern Hyogo region by borehole observations. *Geophysical Journal International*, **147**, 528–542.

OHTANI, T., FUJIMOTO, K., ITO, H., TANAKA, H., TOMIDA, N. & HIGUCHI, T. 2000. Fault rocks and past to recent fluid characteristics from the borehole survey of the Nojima fault ruptured in the 1995 Kobe earthquake, southwest Japan. *Journal of Geophysical Research*, **105**, 16 161–16 171.

OHTANI, T., TANAKA, H., FUJIMOTO, K., HIGUCHI, T., TOMIDA, N. & ITO, H. 2001. Internal structure of the Nojima fault zone from the Hirabayashi GSJ drill core. *The Island Arc*, **10**, 392–400.

PRIEST, S. D. & HUDSON, J. A. 1976. Discontinuity spacing in rock. *International Journal of Rock Mechanics and Mining Sciences and Geomechanics Abstracts*, **13**, 135–148.

PRIEST, S. D. & HUDSON, J. A. 1981. Estimation of discontinuity spacing and trace length using scanline surveys. *International Journal of Rock Mechanics and Mining Sciences and Geomechanics Abstracts*, **18**, 183–197.

TADOKORO, K. & ANDO, M. 2002. Evidence for rapid fault healing derived from temporal changes in S wave splitting. *Geophysical Research Letters*, **29**, (4), 6-1–6-4.

TADOKORO, K., ANDO, M. & UMEDA, Y. 1999. S wave splitting in the aftershock region of the 1995 Hyogo-ken Nanbu earthquake. *Journal of Geophysical Research*, **104**, 981–991.

TANAKA, H., FUJIMOTO, K., OHTANI, T. & ITO, H. 2001. Structural and chemical characterization of shear zones in the freshly activated Nojima fault, Awaji Island, southwest Japan. *Journal of Geophysical Research*, **106**, 8789–8810.

YAMAMOTO, K., SATO, N. & YABE, Y. 2002. Elastic property of damaged zone inferred from *in-situ* stresses and its role on the shear strength of faults. Earth Planets Space, **54**, 1181–1194

Petrofabric-derived seismic properties of a mylonitic quartz simple shear zone: implications for seismic reflection profiling

G. E. LLOYD & J. M. KENDALL

School of Earth Sciences, The University, Leeds LS2 9JT, UK
(e-mail: g.lloyd@earth.leeds.ac.uk)

Abstract: The link between petrofabric (LPO) and seismic properties of an amphibolite-facies quartzo-feldspathic shear zone is explored using SEM/EBSD. The shear-zone LPO develops by a combination of slip systems and dauphine twinning, with a-maximum parallel to lineation (X) and c-maximum normal to mylonitic foliation (XY). The LPO are used to predict elastic parameters, from which the three-dimensional seismic properties of different shear-zone regions are derived. Results suggest that LPO evolution is reflected in the seismic properties but the precise impact is not simple. In general, the P-wave velocity (V_P) minimum is parallel to the a-axis maximum; the direction of maximum shear-wave splitting (AV_S) is parallel to mylonitic foliation; and the V_P maximum and AV_S minimum are parallel to the c-axis maximum. The seismic anisotropy predicted is significant and increases from shear zone wallrock to mature mylonite. The P-wave anisotropy ranges from 11 to 13%, fast and slow shear waves' anisotropies range from 6 to 15% and the magnitude of shear-wave splitting ranges from 9 to 16%. Nevertheless, such anisotropy requires a considerable thickness of rock with this LPO before it becomes seismically visible (i.e. 100s of m for local earthquakes; 10s of m for controlled source experiments). However, reflections and mode conversions provide much better resolution, particularly across tectonic boundaries. The low symmetry and strong anisotropy due to the LPO suggest that multi-azimuth wide-angle reflection data may be useful in the determination of the deformation characteristics of deep shear zones.

A number of factors may produce elastic anisotropy in rocks, including variations in the spatial distribution of mineral phases, layering, grain-size and shape fabrics, grain boundary properties, and the presence of oriented pores or fractures (e.g. Kern & Wenk 1985; Mainprice *et al.* 2003). In addition, because elastic properties vary with respect to direction in single crystals, the majority of rock-forming minerals are elastically anisotropic (e.g. Christensen 1966). Thus, deformation processes, such as dislocation creep, which produce strong crystal-lattice preferred orientations (LPO) in anisotropic minerals, may also induce bulk elastic anisotropy. It is no surprise therefore that petrophysical properties, such as seismic velocities, are often highly anisotropic in bulk rock aggregates (e.g. Babuška & Cara 1991).

Understanding of the impact of microstructural variables on elastic anisotropy has been obtained typically from experimental measurements of elastic properties of minerals or theoretical models that incorporate microstructural variables (e.g. Burlini & Kern 1994; Wendt *et al.* 2003). Experimental studies have focused primarily on the influence of fractures or LPO on the elastic properties of minerals and rocks. In particular, the seismic properties can be measured via direct laboratory methods at ambient or elevated temperatures and pressures (e.g. Kern 1982), although it is often difficult or impossible to obtain the complete three-dimensional property orientation distribution from a single experiment (however, see Pros *et al.* 2003). Furthermore, experimental investigation of seismic properties usually involves rocks in which all microstructural fabric elements contribute, effectively masking any LPO component.

An alternative approach is to calculate the seismic properties of a rock aggregate from individual crystal orientation measurements, incorporating the single-crystal elastic constants and the LPO for each mineral, weighted according to its modal content (e.g. Mainprice & Humbert 1994). The seismic phase velocities and anisotropies can then be calculated in every direction (e.g. Mainprice & Silver 1993). Recently, the advent of electron backscattered diffraction (EBSD) in the scanning electron microscope (SEM) has made it possible to measure LPO in

From: HARVEY, P. K., BREWER, T. S., PEZARD, P. A. & PETROV, V. A. (eds) 2005. *Petrophysical Properties of Crystalline Rocks.* Geological Society, London, Special Publications, **240**, 75–94.
0305-8719/05/$15.00 © The Geological Society of London 2005.

statistically viable numbers from different minerals within a rock aggregate, whilst maintaining a one-to-one relationship with other microstructural elements (e.g. Mainprice *et al.* 1993; Lloyd 2000; Mauler *et al.* 2000; Bascou *et al.* 2001). In this contribution, the seismic properties of an amphibolite-facies, mylonitic quartz shear zone are predicted by averaging the single-crystal elastic properties according to the SEM/EBSD derived LPO, in order to study the relationship between microstructure, petrofabric and seismic properties. This approach illustrates the usefulness of integrating geological studies of rock microstructures, and particularly petrofabrics, with geophysical studies of seismic properties to obtain a fuller understanding of geodynamic processes.

Specimen details

The sample used in this study was collected from a 30-cm-wide deformed planar quartz vein (Fig. 1a) within Proterozoic gneisses at the head of Upper Loch Torridon, NW Scotland (UK GR NG 840530; Wheeler 1984). It seems to have deformed by bulk simple shear and has been described in detail previously (Law *et al.* 1990; Lloyd *et al.* 1992; Lloyd 2004). The salient aspects of these studies relevant to the present contribution are as follows.

The vein was deformed by crystal–plastic processes to form a dextral shear zone with an intense mylonitic foliation and lineation (Fig. 1a, region MM). The foliation locally displays the orientation variations expected for heterogeneous simple shear (e.g. Ramsay & Graham 1970; Ramsay 1980): X parallel to lineation, XY parallel to foliation and Z normal to foliation. The shear-zone wallrock (Fig. 1a) comprises coarsely crystalline quartz and plagioclase feldspar. Foliation at the shear-zone margins (Fig. 1a) is oriented at $c.45°$ to the vein walls but curves rapidly with a dextral shear sense to become subparallel to the walls. In the XZ section analysed in this contribution, the vein microstructure is that of a classic Type II S-C mylonite (Lister & Snoke 1984) and consists of two planar domains, A (grain size <5 μm) and B (grain size <100 μm), aligned parallel to the macroscopic mylonitic foliation (S_A). The obliquity between grain axes (S_A, S_B) in the two domains is consistent and indicates a dextral shear sense. The microstructures observed are indicative of constant volume, strongly non-coaxial, essentially plane strain deformation that closely approximates to simple shear.

The quartz petrofabric (Figs. 1c–e & 2) is characterized by an $\langle a \rangle$ axis maximum subparallel to X, which occupies a pole position to the corresponding single girdle (c) axis fabric, with a (c) axis maximum oriented subnormal to the XY foliation plane and normal to the average of the quartz basal plane defined by the LPO. These associations suggest a simple relationship between shear-zone geometry, simple shear kinematic framework and orientation of crystal slip systems responsible for shear-zone formation. Most quartz grains are preferentially oriented to exploit slip on systems that utilise $\langle a \rangle$ as the slip direction (i.e. $\{-\pi\}\langle a \rangle$, $\{z\}\langle a \rangle$, $(c)\langle a \rangle$ and $\{m\}\langle a \rangle$). Such observations can be interpreted in terms of a lower resistance to slip on negative forms compared to positive forms, and/or the occurrence of significant dauphiné twinning. The latter is consistent also with the $\{r\}$ point maximum position within the XZ plane at $c.35°$ to the foliation, subparallel to the NW–SE-trending maximum principal stress direction inferred from the simple shear kinematic framework (e.g. Tullis & Tullis 1972). The petrofabric therefore approximates that of a 'dauphiné twinned single crystal'.

SEM studies of the Torridon shear zone have focused attention on to the misorientations between adjacent grains, and hence upon the grain boundary network, as well as the conventional grain microstructure and petrofabric. Misorientations develop by a combination of crystal slip and dauphiné twinning and result in a microstructure dominated by misorientations about the (c) axis. Due to crystal symmetry restrictions, misorientation angles greater than $60°$ about the (c) axis are accommodated apparently by transfer of slip on to systems capable of progressively higher misorientations up to the maximum of $104.5°$ (i.e. successively from $\{m\}\langle a \rangle$, $\{r/z\}\langle a \rangle$, $\{-\pi\}\langle a \rangle$ to (c) $\langle a \rangle$). The resulting grain-boundary network (Fig. 3) consists of tilt boundaries parallel to prism (and hence often YZ tectonic) planes. Although no other tilt boundaries are formed, the individual slip systems can also operate in simple combinations to produce twist boundaries typically parallel to the XY foliation/basal crystal planes and/or XZ tectonic plane.

Methodology

LPO determination

SEM/EBSD (e.g. Venables & Harland 1973; Dingley 1984) provides the precise crystallographic orientation at the point of incidence of the electron beam on the sample surface, with a spatial resolution of $c.1$ μm and angular resolution of $c.1°$ (e.g. Prior *et al.* 1999). A one-to-one correlation is achieved, therefore, between

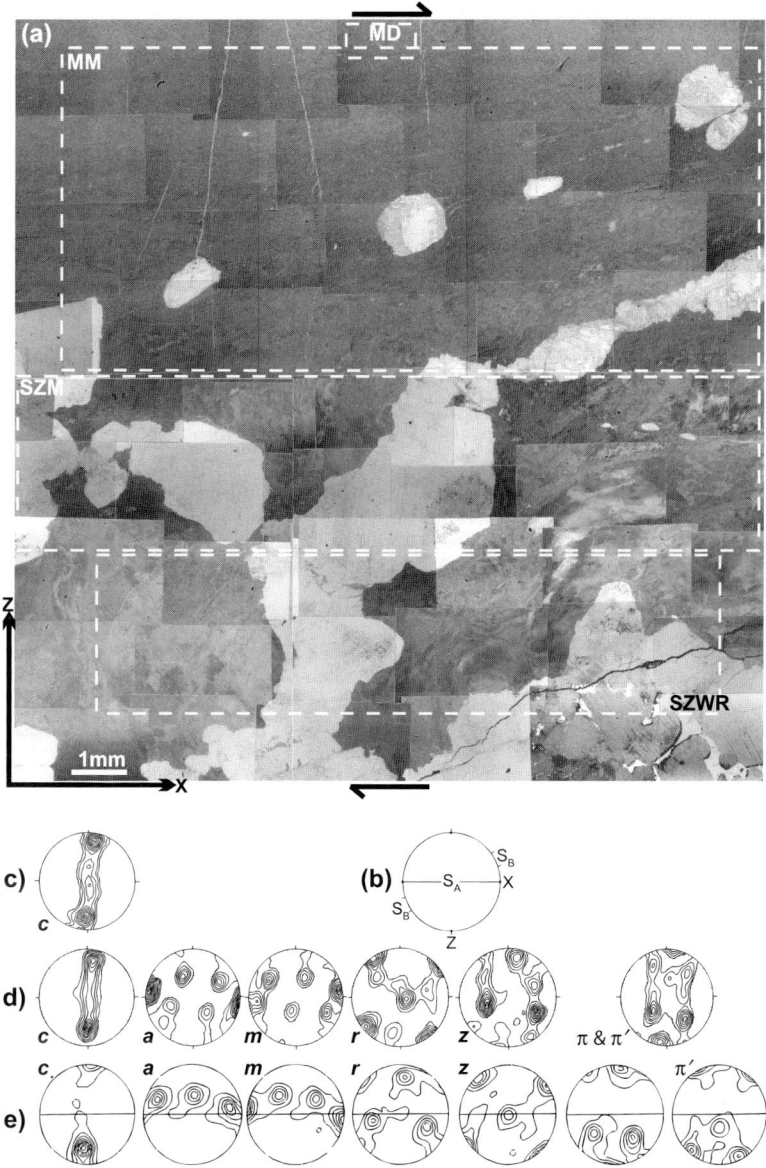

Fig. 1. Summary of Torridon simple shear-zone microstructure and LPO. All pole figures are equal area, upper hemispheres are viewed towards the ENE; contour intervals are multiples of mean uniform distribution. (**a**) SEM electron channelling (see Lloyd 1987 for details) orientation contrast image of microstructure, indicating positions of SEM/EBSD analyses: SZWR, shear-zone wallrock (analysis G030600); SZM, shear-zone margin (G040800); MM, mature mylonite (G270700); and MD, mylonite/shear-zone detail (G050800). Amended from Lloyd (2004). (**b**) Specimen co-ordinates for all LPO diagrams (after Law *et al.* 1990). (**c**) Universal stage optical *c*-axis pole figure; 1815 measurements (after Law *et al.* 1991). (**d**) X-ray pole figures for *c, m, a, r, z* and $\pi + \pi'$ orientations, (after Law *et al.* 1991). (**e**) Manual SEM electron channelling pole figures for *c, m, a, r, z, π* and π' orientations, after Lloyd *et al.* 1992).

crystal orientation and microstructural position. The crystal orientation is defined by three spherical Euler angles (ϕ_1, φ, ϕ_2) with respect to the sample reference frame (Bunge 1982). EBSD analysis can be performed either manually or automatically, but in both cases the diffraction patterns are indexed via computer programs; the system used in this study was the HKL

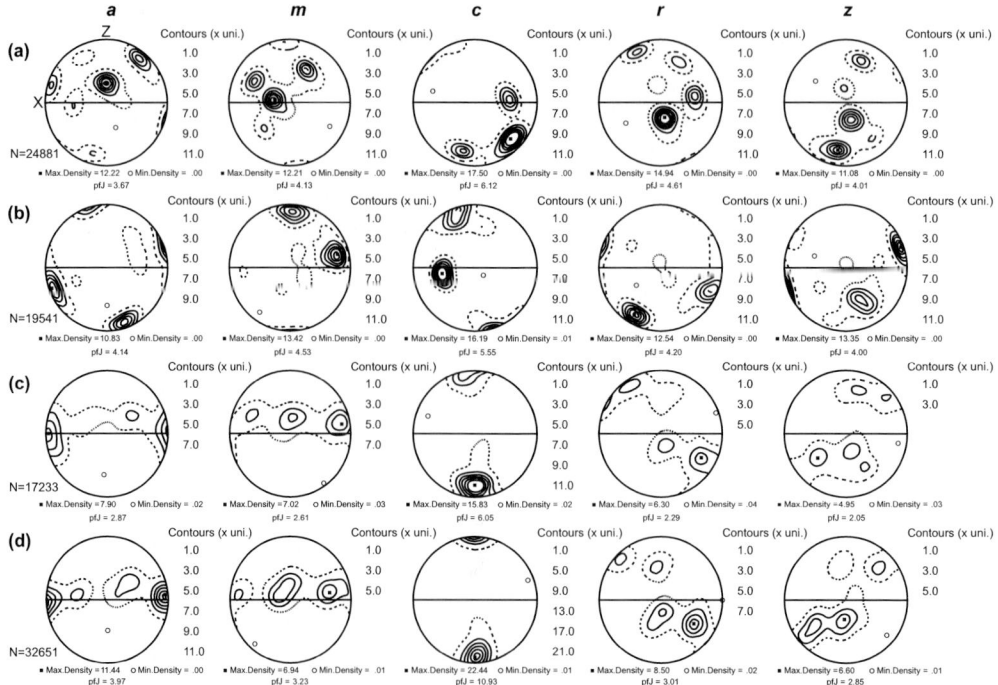

Fig. 2. Summary (after Lloyd, 2004) of SEM/EBSD pole figures for *c, m, a, r* and *z* orientations using the program Pf2k (D. Mainprice). Only wallrock includes quartz and plagioclase, the rest involve quartz alone. All pole figures are equal area, upper hemispheres viewed towards ENE; contours multiples of mean uniform distribution, as indicated (pfJ is an indication of distribution strength, with pfJ = 1 for random; see Bunge 1982). (**a**) Shear-zone wallrock (SZWR). (**b**) Shear-zone margin (SZM). (**c**) Mature mylonite (MM). (**d**) Mylonite details (MD).

Technology 'Channel 5' software (e.g. Schmidt & Olesen 1989). Automated EBSD analysis enables very large (i.e. $>10^6$) orientation data-sets to be acquired from samples $>10 \times 10$ mm area, which can be used to test and/or interpret seismic anisotropy (e.g. Burlini & Kunze 2000; Lloyd 2000; Mauler *et al.* 2000).

Three overlapping large-scale auto-EBSD analyses (G030100, G040800, G270700) were performed, extending from the shear-zone wallrock to the mature mylonite and a detailed auto-EBSD analysis (G050800) of the mature mylonite (Lloyd 2004; see Fig. 1a for locations and Table 1 for experimental details). The LPO derived from the SEM/EBSD experiments (Fig. 2) provide a part of the input data for the seismic properties determinations described below. Important aspects of the LPO are as follows:

1. The shear-zone wallrock (Fig. 2a) and shear-zone margin (Fig. 2b) comprise relatively few, large quartz (and plagioclase in the former) grains, and hence their LPO consist of isolated concentrations associated with individual grain orientations.

2. The mylonite LPO (Fig. 2c & d) approximates that of a 'dauphine twinned single crystal', consistent with results derived from other methods (e.g. Fig. 1c–e), which maintains (strengthens?) the $\langle a \rangle$, (c) and $\{m\}$ positions but interchanges the $\{r\}$ and $\{z\}$ positions.

3. The differences between LPO from the shear-zone margin and mature mylonite indicate a rapid migration of wallrock crystal pole directions towards the mature mylonite LPO, in agreement with micro-structural observations (see Fig. 1a).

4. The shear-zone shear-strain gradient and dynamic recrystallization rates therefore are expected to have been steep and/or rapid.

Seismic properties determination

The propagation of seismic waves generates a short-term deformation in a medium, such that the velocity and polarization direction of the

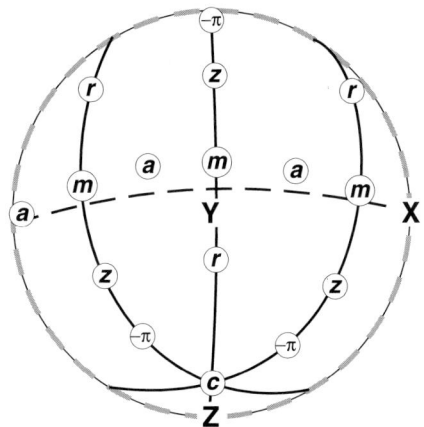

Predicted boundary orientations

—————— dauphine twinning & {m}<*a*> prism-parallel tilt boundaries (TiB), plus higher misoriented prism-parallel TiB due to transfer to (*c*)<*a*> &/or {*r*/*z*}<*a*> &/or {–π}<*a*> slip systems

— — {π'}<*a*> + (*c*)<*a*> twist boundaries (TwB)

— — — {*r*/*z*}<*a*> + {*m*}<*a*> TwB

Fig. 3. Relationship between grain-boundary types and orientations, and shear-zone tectonic (*XYZ*) and crystal ((*a*), (*c*), {*m*}, {*r*} and {*z*}) frameworks (after Lloyd, 2004).

waves depend on the elastic parameters (i.e. the elastic stiffness matrix) of the medium and the nature of the deformation. Thus, knowledge of the elastic parameters can be used to predict seismic velocities and propagation directions. The general relationship between elastic parameters and seismic waves is given by, for example, Nye (1957) and Kendall (2000),

$$(C_{ijkl}X_iX_j - \delta_{kl}\rho V^2)U_\ell = 0 \qquad (1)$$

where C_{ijkl} is the fourth-order proportionality tensor that relates stress to strain, X_i expresses the propagation direction of the wave ($C_{ijkl}X_iX_j$ is the 3×3 Christoffel matrix), δ_{kl} is the Kronecker delta, ρ is the density of the medium, V is the phase velocity of the wave in a given direction and U_ℓ is the displacement.

A non-trivial solution for U_ℓ requires that the determinant of the system in equation (1)

vanishes and results in the Christoffel equation (e.g. Nye 1957),

$$\det\left|C_{ijkl}X_iX_j - \delta_{kl}\rho V^2\right| = 0. \qquad (2)$$

The Christoffel equation has three possible solutions, representing one compressional (*P*) and two shear (S_1, S_2) waves. Due to the symmetry of the elastic tensor, the Christoffel matrix is symmetrical also, which means that the three solutions have mutually perpendicular displacement vectors. In isotropic media, seismic-wave velocities are independent of their propagation direction, and their polarization depends only on the type of wave and the nature of the source. In anisotropic media, seismic velocities depend locally on the propagation direction, and their polarization depends not only on the type of wave, but also on the local symmetry of the elastic properties (i.e. C_{ijkl}).

Although C_{ijkl} has 81 components, these can be reduced to a symmetrical 6×6 matrix, C_{ij}, because of symmetry in the stress and strain matrices (Babuška & Cara 1991). Consideration of the energy function of a strained crystal, which depends only on the strain components, reduces the stiffness matrix to a maximum of 21 *independent* coefficients. Furthermore, elastic parameters of crystalline media depend ultimately on chemical composition and atomic arrangement of the crystal structure, and hence are characteristic for each mineral. Thus, the elastic parameters are closely related to the strength of interatomic bonds in corresponding directions of the crystal structure (e.g. elastic moduli are larger in the direction in which the structure has the strongest bonds). Consequently, because elastic parameters are the same in crystal-symmetry-related directions, the number of independent elastic coefficients for single crystals can be reduced further still, depending on crystal symmetry (see Babuška & Cara 1991, table 2.1).

Single-crystal seismic properties

Understanding the seismic behaviour of the individual constituent crystals is critical to any interpretation of whole-rock seismic behaviour, as exhibited, for example, in mylonitic shear zones. As an example of the impact of crystal structure on seismic properties and the methodology used to predict whole-rock seismic properties, the variations in compressional and shear waves velocities with direction have been calculated for quartz and plagioclase single crystals: the minerals of interest in the Torridon shear zone.

Figure 4a shows common morphological single-crystal forms of quartz and plagioclase,

Table 1. *Summary of Torridon shear-zone auto-EBSD experiments (see Lloyd 2004 for further details)*

Details	Job (format: username–day–month–year)			
	G030600	G040800	G270700	G050800
Specific comments	*Blown filament;	Ended by user break	—	—
Area analysed	Shear-zone wallrock	Shear-zone margin	Mature mylonite	Detail of mature mylonite
Dimensions (mm)	10.0 × 13.0	17.0 × 6.0	13.5 × 7.25	1.5 × 1.0
No. of data points	136 640	120 537	228 825	150 001
Grid step (μm)	50	25	20	1
Time (h:min:s)	11:47:25	29:37:52	41:18:56	22:11:31
Index rate (s/pattern)	1.1	0.88	0.65	0.53
Minerals indexed	Quartz and plagioclase (An$_{16}$)	Quartz	Quartz	Quartz
MAD	<1.5	<1.5	<1.5	<1.5
% Low BC	*46.6	2.2	5.1	1.3
% Low BN	0.1	13.0	0.2	2.1
% Not indexed	27.0	68.5	87.2	74.8
% Indexed	26.19	16.21	7.53	21.77
No. indexed	35 789	19 541	17 233	32 651
No. phases	2	1	1	1
Phase 1	Quartz	Quartz	Quartz	Quartz
Total good points	24 881	19 541	17 233	32 651
Good points %	18.21	16.21	5.53	21.77
Vol. fraction %	69.52	100.00	100.00	100.00
Phase 2	Plagioclase	—	—	—
Total good points	10 908	—	—	—
Good points %	7.98	—	—	—
Vol. fraction	30.48	—	—	—

Key: MAD, mean angular deviation; BC, band contrast; BN, band number; in all but G030100, plagioclase was ignored.

and their relationship to directions of maximum and minimum seismic velocities (*V*) and anisotropy (*A*), based on experimentally determined values summarized by Babuška & Cara (1991). Although the principal crystal directions for quartz (i.e. $a(11\bar{2}0)$, $m(1\bar{1}00)$, $c(0001)$, $r(10\bar{1}1)$, $z(01\bar{1}1)$) and plagioclase (i.e. $a(100)$, $b(010)$, $c(001)$) can be represented on a single stereographic projection, petrofabric analysis typically plots individual 'pole figures' for each form. Appropriate Euler angle triplets (e.g. Casey 1981) for quartz and plagioclase (Table 2, part a) were therefore used to calculate individual pole figure diagrams (Fig. 4b & d), representative of the single-crystal forms via the program 'Pfch5' (Mainprice 2003).

The single-crystal elastic tensor, C_{ijkl}, is generally defined in the crystal reference frame, whereas crystal orientations are generally defined via the Euler angles in the specimen reference frame. The former can be rotated into the latter via (Babuška & Cara 1991),

$$C_{ijkl}(g) = g_{im}g_{jn}g_{ko}g_{lp}C^0_{mnop} \quad (3)$$

where $g_{ij} = g[\phi_1, \varphi, \phi_2]$ is the crystal to sample reference frame rotation matrix, and C^0_{mnop} is

the single-crystal elastic stiffness tensor in the crystallographic reference frame.

The single-crystal orientations, as defined by the Euler angle triplets in equation (3), were combined via (2) with the experimentally determined single-crystal elastic stiffness matrix and density for each mineral (i.e. quartz, McSkimin et al. 1965; plagioclase, Aleksandrov et al. 1974; see Table 2, part b) to calculate the single-crystal seismic-property distributions using the program 'ANISch5' (Mainprice 2003; see also Mainprice 1990; Mainprice & Humbert 1994). Stereographic projections (Fig. 4c & e) of the seismic properties have been plotted in sample coordinates via the program 'VpG' (Mainprice 2003). The specific property distributions calculated are the compressional (V_P) and shear (V_{S1}, V_{S2}) wave phase velocities and the degree of shear-wave splitting for a given direction, represented as either the absolute difference in shear-wave velocities ($dV_S = V_{S1} - V_{S2}$) or the shear-wave anisotropy (AV_S), which is conventionally calculated via (for example, Mainprice & Silver 1993),

$$AV_S\% = 100(V_{S1} - V_{S2})/[(V_{S1} + V_{S2})0.5]. \quad (4)$$

Table 2. *Seismic properties analysis for single-crystal quartz (left) and plagioclase (right)*

Quartz	Plagioclase

(a) Input crystallographic parameters

Symmetry:

Trigonal, space group P3221   Triclinic, space group P-1

Unit-cell dimensions (Å):

$a = 4.9130, b = 4.9130, c = 5.5040$ $a = 8.1553, b = 12.8206, c = 7.1397$

Unit-cell angles (degrees):

$\alpha = 90.0, \beta = 90.0, \gamma = 120.0$ $\alpha = 93.95, \beta = 116.47, \gamma = 89.62$

Euler angle triplet:

$\phi_{1\text{-max}} = 180°, \varphi_{max} = 180°, \phi_{2\text{-max}} = 120°$ $\phi_{1\text{-max}} = 360°, \varphi_{max} = 180°, \phi_{2\text{-max}} = 360°$

(b) Input single-crystal density and elastic stiffness matrix, C_{ij} (Mbar)

Density $= 2.6473$ g cm^{-3} Density $= 2.6100$ g cm^{-3}

McSkimin *et al.* (1965)   Aleksandrov *et al.* (1974)

0.8680	0.0704	0.1191	−0.1804	0.0001	0.0001	0.8200	0.3980	0.4100	0.0001	−0.0840	0.0001
0.0704	0.8680	0.1191	0.1804	0.0001	0.0001	0.3980	1.4500	0.3370	0.0001	−0.0630	0.0001
0.1191	0.1191	1.0575	0.0001	0.0001	0.0001	0.4100	0.3370	1.3280	0.0001	−0.1870	0.0001
−0.1804	0.1804	0.0001	0.5820	0.0001	0.0001	0.0001	0.0001	0.0001	0.1810	0.0001	−0.0100
0.0001	0.0001	0.0001	0.0001	0.5820	−0.1804	−0.0840	−0.0630	−0.1870	0.0001	0.3100	0.0001
0.0001	0.0001	0.0001	0.0001	−0.1804	0.3988	0.0001	0.0001	0.0001	−0.0100	0.0001	0.3350

In addition, the absolute anisotropies of V_P, V_{S1} and V_{S2} have been calculated by substituting their appropriate maximum and minimum values for V_{S1} and V_{S2} respectively in equation (4).

Finally, as shear-wave splitting analysis of real data estimates the degree of splitting and the orientation of the fast shear-wave for a given ray direction (e.g. Kendall 2000), the

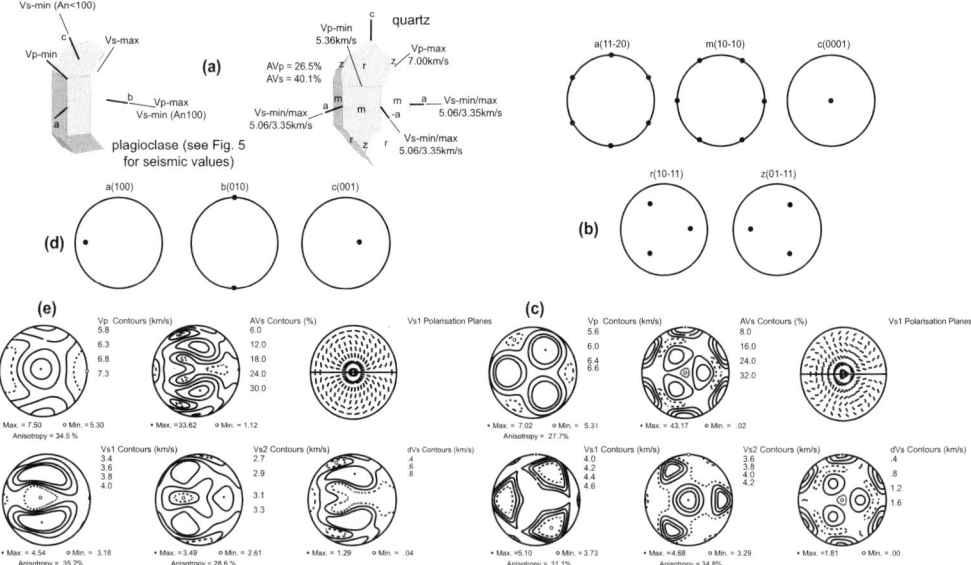

Fig. 4. Seismic property distributions (V_P, V_{S1}, V_{S2}, AV_S, dV_S, V_{S1} polarization planes – see text for details) of quartz and plagioclase single crystals based on LPO-averaging (see Table 2 for input data). All pole figures are equal area, upper hemisphere projections; contours multiples of mean uniform distribution, as indicated. (**a**) Quartz and plagioclase single-crystal forms (after Dana & Ford 1951) and seismic-property variations (after Babuška & Cara 1991). (**b**) Quartz single-crystal LPO generated from an individual Euler angle triplet. (**c**) Variation in quartz single-crystal seismic properties with crystal direction. (**d**) Plagioclase single-crystal LPO generated from an individual Euler angle triplet. (**e**) Variation in plagioclase single-crystal seismic properties with crystal direction.

polarizations of the fast shear waves (V_{S1}) are also shown in stereographic projection.

The trigonal crystal symmetry of quartz is reflected in its seismic property distributions, although AV_S (and dV_S) also exhibit sixfold symmetry close to the basal plane (Fig. 4c). Single-crystal quartz is highly anisotropic in terms of all of its seismic properties. The seismic property distributions for plagioclase are more complex, reflecting the triclinic crystal symmetry, although a seismic symmetry plane normal to $b(010)$ is apparent (Fig. 4e). Single-crystal plagioclase is almost as anisotropic as quartz in terms of all of its seismic properties.

Elastic parameters are known to vary with composition. It is likely therefore that seismic properties for plagioclase vary with anorthite (An) content. The experimentally determined density and elastic stiffness matrix data used to derive Figure 4e were for plagioclase An_{16}, specifically because electron microprobe analysis of the plagioclase in the Torridon shear zone showed a composition range of An_{15-23}. Babuška & Cara (1991, table 3-1) have summarized experimentally measured seismic properties for four other plagioclase compositions (An_9, An_{29}, An_{53} and An_{100}). In an attempt to constrain quantitatively the variations in seismic properties with plagioclase composition, the four sets of experimental values, together with the single-crystal values for An_{16} derived above, have been plotted in Figure 5. The velocities vary linearly across the range of compositions, showing slight to moderate increases with An content. In contrast, both the anisotropy of V_P and shear-wave splitting are almost invariant with An content.

Polycrystal seismic properties

Mineral grains in a rock contribute to the overall seismic properties according to their single-crystal elastic parameters, crystallographic orientation distribution (LPO) and volume fraction. However, although the effect of LPO on seismic-wave velocities has long been recognized (e.g. Kaarsberg 1959; Hess 1964), the significance of LPO on the seismic properties of rocks depends on how the seismic velocities are related to the crystal directions and the origin of the LPO. There must be a significant degree of crystal alignment to produce a seismic anisotropy, whilst randomly oriented crystals generate an isotropic bulk rock. Most LPO-causing mechanisms (e.g. dislocation creep) considered in seismic studies are induced by tectonic stresses in highly deformed rocks (e.g. Babuška & Cara 1991; Blackman et al. 2002).

The methodology used to calculate polycrystal seismic property distributions is the same as that employed in the single-crystal calculations described above. For each mineral grain orientation, measured via SEM/EBSD, the single-crystal parameters (Table 2, part b) need to be rotated into the sample reference frame using (3), such that the elastic parameters of the polycrystal are derived by integration over all

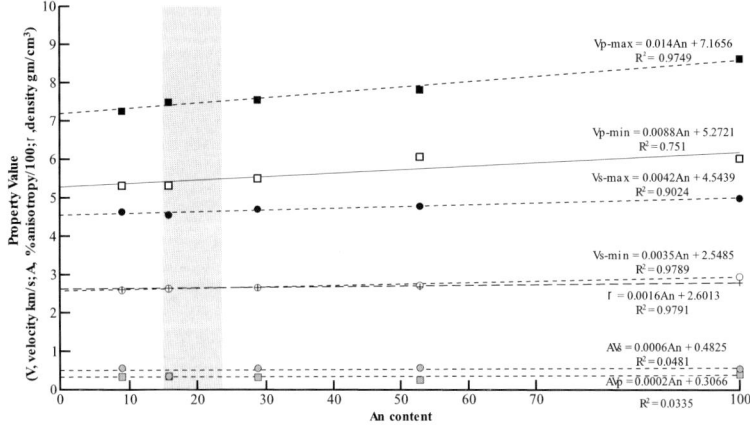

Fig. 5. Variation in seismic properties (V_P, AV_P, V_{S1}, V_{S2}, AV_S – see text for details) and density (ρ) with An content in plagioclase (NB anisotropy represented from 0 to 1 rather than 0–100%). Data for An_9, An_{29}, An_{53}, An_{100} from Babuška & Cara (1991); Data for An_{16} derived from LPO-averaged values using experimentally measured elastic parameters (see Table 2). Broken lines are best-fit trends through data, and R^2 is the square of the correlation coefficient. The shaded box indicates the range of An values measured in the Torridon shear zone.

possible orientations in the three-dimensional orientation distribution function. The three seismic phase velocities (V_P, V_{S1} and V_{S2}), and related anisotropy (AV_P, AV_S), can then be obtained in every direction via solution of equation (2). However, due to stress/strain compatibility assumptions, three different averaging schemes are possible. The *Voigt* (*V*) average (Voigt 1928) assumes a constant strain approximation, whilst the *Reuss* (*R*) average (Reuss 1929) assumes a constant stress approximation in calculating the compliance tensor, the inverse of the stiffness tensor (e.g. Crosson & Lin 1971). The Voigt and Reuss averages represent idealized situations, and provide upper and lower bounds respectively for the real elastic parameters (Bunge *et al.* 2000). Thus, the mean or *Hill* (*H*) average (Hill 1952) of the two values is often taken as the best estimate (VRH) of the average elastic parameters.

Finally, in this section, it must be emphasized that the prediction of seismic properties using the LPO averaging approach requires accurate data on single-crystal elastic parameters of different minerals. Such data are usually determined by measuring the ultrasonic velocities of single crystals in different directions and deriving the elasticity tensor via the Cristoffel equation (e.g. McSkimin *et al.* 1965), although a more recent approach uses Brillioun scattering and is more accurate (e.g. Sinogeikin & Bass 2000). Nevertheless, relevant data for many minerals are still lacking, although the situation is rapidly improving (e.g. Ji *et al.* 2002). Furthermore, single-crystal elastic properties vary with pressure (*P*) and temperature (*T*) but the *P–T*-derivative values are often unavailable. Following the current convention established for mantle minerals (e.g. Mainprice *et al.* 2000), we assume the elastic parameters of quartz and feldspar determined under ambient conditions (McSkimin *et al.* 1965; Aleksandrov *et al.* 1974).

Torridon shear-zone seismic properties

The procedures described in the previous section have been applied to derive the LPO-dependent seismic properties of the Torridon mylonitic simple shear zone. In general, quartz dominates the shear zone region in terms of modal content, although plagioclase is common in the wallrock (see Fig. 1a). Thus, whilst the seismic properties of the mylonitic shear zone are likely to be derived from the quartz LPO alone, those for the wallrock represent a combination of quartz and plagioclase LPO. The relevant LPO have been combined with the single-crystal elastic parameters of quartz and/or feldspar in their appropriate modal proportions, as measured by SEM/EBSD, to determine the seismic property distributions for different regions of the Torridon simple shear zone.

The first step is to derive the elastic parameters (stiffness matrix) for each region from the relevant LPO, using the VRH averaging approach. Results are shown in Table 3. These data are then used to derive the seismic velocities and polarizations in three dimensions for each region by solving equation (2). Results are shown as stereographic projections oriented according to the sample reference frame (Fig. 6). The maximum and minimum phase velocities and anisotropy for each region, and also for single-crystal quartz and feldspar (based on Fig. 4), are summarized in Figure 7, whilst Table 4 summarizes the principal orientation relationships.

Magnitudes

The maximum and minimum velocities in V_P are fairly constant across the shear zone and range from 6.40 to 6.47 km s^{-1} and 5.66 to 5.79 km s^{-1} respectively (Figs 6 & 7a), representing variations of only 1–2%. There is a suggestion of a slight drop in both velocities from the relatively undeformed wallrock and shear-zone margin regions to the mature mylonite. However, the difference between V_P maximum and minimum values (i.e. the compressional-wave anisotropy, AV_P) increases from 11% in the shear-zone wallrock to 12.5% in the mylonite (Figs 6 & 7b), although the maximum value (12.8%) occurs in the shear-zone margin.

Maximum and minimum velocities in V_{S1} range from 4.11 to 4.50 km s^{-1} and 3.83 to 4.02 km s^{-1} respectively, representing variations of 8.6% and 4.7% in each. The difference between V_{S1} maximum and minimum values ranges from 7 to 15%. Maximum and minimum velocities in V_{S2} range from 3.93 to 4.38 km s^{-1} and 3.70 to 3.77 km s^{-1} respectively, representing variations of 10% and 2% in each. The difference between V_{S2} maximum and minimum values ranges from 4 to 15%. Both shear-wave velocities increase from the shear-zone wallrock into the margin and mature mylonite (Figs 6 & 7a). The difference, or anisotropy, between the maximum and minimum values of V_{S1} increases from 7% in the wallrock to 14.7% in the margin, before decreasing to 9–10% in the mylonite (Figs 6 & 7b). In contrast, the anisotropy of V_{S2} increases progressively from 6.0% in the wallrock to 15.0% in the mylonite.

Obviously, the behaviour of shear waves is reflected in the shear-wave splitting, measured as AV_S or dV_S. This is typically large across the

Table 3. *Summary of elastic properties (stiffness matrix) derived via LPO analysis using Voigt–Reuss–Hill (VRH) averaging*

Sample

G030600 – shear-zone wallrock

Agg. density: 2.6359 g cm⁻³

VRH average elastic stiffness for quartz fraction

0.9965	0.0652	0.0964	0.0082	−0.0241	0.0248
0.0652	0.9789	0.0952	−0.0220	0.0002	0.0183
0.0964	0.0952	0.9073	−0.0309	−0.0119	0.0047
0.0082	−0.0220	−0.0309	0.4442	0.0332	−0.0209
−0.0241	0.0002	−0.0119	0.0332	0.4579	−0.0201
0.0248	0.0183	0.0047	−0.0209	−0.0201	0.4534

VRH average elastic stiffness for plagioclase fraction

0.9087	0.4010	0.4426	−0.0315	0.0479	0.0430
0.4010	1.1202	0.4349	−0.0393	0.0000	0.0693
0.4426	0.4349	1.0956	−0.0395	0.0419	0.0637
−0.0315	−0.0393	−0.0395	0.2882	0.0250	−0.0088
0.0479	0.0000	0.0419	0.0250	0.3435	−0.0086
0.0430	0.0693	0.0637	−0.0088	−0.0086	0.2982

VRH average elastic stiffness for each region

0.9453	0.1623	0.1965	−0.0060	0.0005	0.0269
0.1623	1.0062	0.1863	−0.0260	−0.0032	0.0329
0.1965	0.1863	0.9465	−0.0319	0.0032	0.0232
−0.0060	−0.0260	−0.0319	0.3890	0.0304	−0.0159
0.0005	−0.0032	0.0032	0.0304	0.4191	−0.0156
0.0269	0.0329	0.0232	−0.0159	−0.0156	0.3986

G040800 – shear-zone margin

Agg. density: 2.6473 g cm⁻³

VRH average elastic stiffness for quartz fraction

0.9595	0.0787	0.1199	−0.0305	0.0231	0.0185
0.0787	0.9504	0.0875	0.0461	0.0353	−0.0213
0.1199	0.0875	0.9110	0.0332	−0.0421	0.0256
−0.0305	0.0461	0.0332	0.4385	0.0394	0.0404
0.0231	0.0353	−0.0421	0.0394	0.5040	−0.0001
0.0185	−0.0213	0.0256	0.0404	−0.0001	0.4552

VRH average elastic stiffness for plagioclase fraction

Not analysed

VRH average elastic stiffness for each region

0.9595	0.0787	0.1199	−0.0305	0.0231	0.0185
0.0787	0.9504	0.0875	0.0461	0.0353	−0.0213
0.1199	0.0875	0.9110	0.0332	−0.0421	0.0256
−0.0305	0.0461	0.0332	0.4385	0.0394	0.0404
0.0231	0.0353	−0.0421	0.0394	0.5040	−0.0001
0.0185	−0.0213	0.0256	0.0404	−0.0001	0.4552

G270700 – mature mylonite

Agg. density: 2.6473 g cm⁻³

VRH average elastic stiffness for quartz fraction

1.0520	0.1011	0.0805	−0.0106	0.0051	0.0021
0.1011	0.8514	0.1140	0.0082	0.0103	0.0023
0.0805	0.1140	0.9050	0.0008	−0.0517	0.0045
−0.0106	0.0082	0.0008	0.4192	0.0094	−0.0114
0.0051	0.0103	−0.0517	0.0094	0.4919	−0.0117
0.0021	0.0023	0.0045	−0.0114	−0.0117	0.4876

VRH average elastic stiffness for plagioclase fraction

Not analysed

VRH average elastic stiffness for each region

1.0520	0.1011	0.0805	−0.0106	0.0051	0.0021
0.1011	0.8514	0.1140	0.0082	0.0103	0.0023
0.0805	0.1140	0.9050	0.0008	−0.0517	0.0045
−0.0106	0.0082	0.0008	0.4192	0.0094	−0.0114
0.0051	0.0103	−0.0517	0.0094	0.4919	−0.0117
0.0021	0.0023	0.0045	−0.0114	−0.0117	0.4876

G050800 – mylonite detail

Agg. density: 2.6473 g cm⁻³

VRH average elastic stiffness for quartz fraction

1.0605	0.1127	0.0824	0.0077	0.0045	0.0040
0.1127	0.8479	0.1003	−0.0022	0.0101	0.0078
0.0824	0.1003	0.9051	−0.0086	−0.0614	−0.0044
0.0077	−0.0022	−0.0086	0.3942	0.0003	−0.0178
0.0045	0.0101	−0.0614	0.0003	0.4988	0.0060
0.0040	0.0078	−0.0044	−0.0178	0.0060	0.5097

VRH average elastic stiffness for plagioclase fraction

Not analysed

VRH average elastic stiffness for each region

1.0605	0.1127	0.0824	0.0077	0.0045	0.0040
0.1127	0.8479	0.1003	−0.0022	0.0101	0.0078
0.0824	0.1003	0.9051	−0.0086	−0.0614	−0.0044
0.0077	−0.0022	−0.0086	0.3942	0.0003	−0.0178
0.0045	0.0101	−0.0614	0.0003	0.4988	0.0060
0.0040	0.0078	−0.0044	−0.0178	0.0060	0.5097

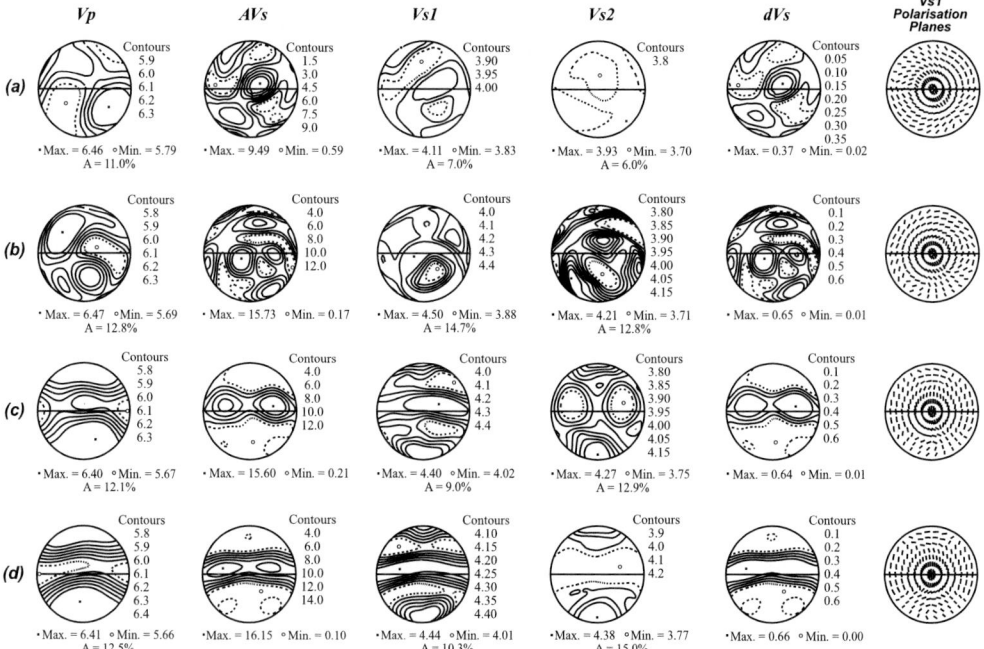

Fig. 6. Seismic-property distributions (V_P, V_{S1}, V_{S2}, AV_S, dV_S, V_{S1} polarization – see text for details) for different parts of the Torridon shear zone, based on LPO-averaging of elastic parameters constructed using Tables 2 and 3 and programs 'Anis2Ck' and 'VpG2k' (Mainprice 2003). Only shear-zone wallrock includes quartz and plagioclase, the rest involve quartz alone. All pole figures equal area, upper hemispheres viewed towards ENE; contours multiples of mean uniform distribution, with tectonic X and Z directions oriented east–west and north–south respectively. See text for discussion. (**a**) Shear-zone wallrock (G030600). (**b**) Shear-zone margin (G040800). (**c**) Mature mylonite (G270700). (**d**) Detail of mature mylonite (G050800).

shear zone, and increases from 9.49% or 0.4 km s^{-1} in the wallrock to 16.15% or 0.7 km s^{-1} in the mylonite (Figs 6 & 7b).

Orientations

The seismic-property orientation distributions, and, in particular, the orientations of their maximum and minimum values, relative to the LPO and tectonic reference frame are crucial in terms of explaining and using seismic properties in geodynamic interpretations (see Figs 6 & 8, and Table 4).

The seismic property orientation distributions for the shear-zone wallrock and margins regions (Fig. 6a & b) result from relatively weak LPO development (Fig. 2a & b) and, in the case of the wallrock, the contrasting impact of quartz and plagioclase LPO. Nevertheless, the seismic properties do show some consistent orientation patterns, particularly relative to the quartz LPO and tectonic reference frame. The maximum in V_P is aligned subparallel to the quartz c-axis

maximum and inferred position of σ_1, whilst the minimum in V_P is subparallel to the tectonic Y direction (and perhaps also the inferred position of σ_3 in the wallrock). The maximum in V_{S1} has a broad distribution that encompasses the quartz c and r maximum directions, the tectonic Y direction and the inferred position of σ_1, but the orientation of the minimum in V_{S1} is poorly defined relative to crystal or tectonic directions. The maximum in V_{S2} also aligns subparallel to the quartz c-axis maximum (and perhaps the inferred position of σ_1), whilst the minimum in V_{S2} appears to align parallel to the quartz a-axis maximum and subparallel to the tectonic Y direction. The orientation distributions of AV_S (and dV_S) have maxima oriented subparallel to quartz a-axis concentrations and the tectonic Y direction, and minima oriented subparallel to the tectonic X direction and perhaps the quartz m direction concentrations.

In contrast to the shear-zone wallrock and margin, seismic-property orientation distributions in the mature mylonite are exceptionally well

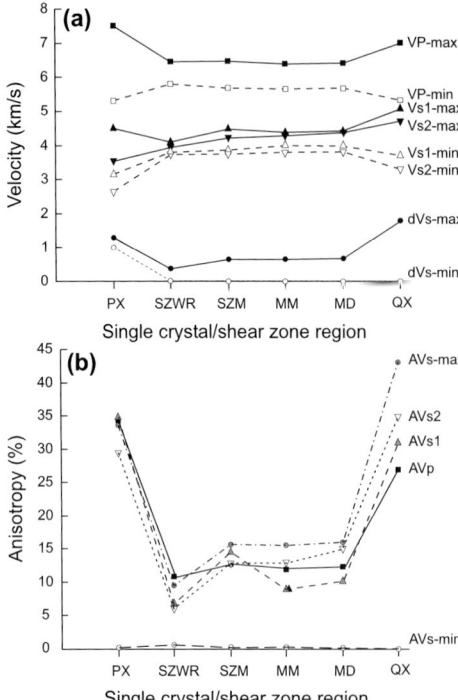

Fig. 7. Comparison of calculated variations in seismic properties (V_P, AV_P, V_{S_1}, V_{S_2}, AV_S, dV_S – see text for details) for single-crystal quartz (QX) and plagioclase (PX) and shear-zone wallrock (SZWR), margin (SZM), mature mylonite (MM) and mylonite detail (MD) regions. See text for discussion. (a) Velocity values. (b) Anisotropy values.

ordered (Fig. 6c & d), and clearly reflect the quartz LPO (Fig. 2c & d). Several of the relationships recognized in the shear-zone wallrock and margin regions persist into the mylonite. A broad maximum in V_P encompasses the quartz c-axis maximum and r/z direction small circle distributions, and is subparallel to the tectonic Z direction. V_P has an absolute minimum parallel to the a-axis maximum and tectonic X direction, and a great-circle distribution of low values parallel to the quartz basal plane and subparallel to the XY tectonic plane. Although the absolute maximum in V_{S_1} is subparallel to the tectonic Y direction, similar values occur parallel to both the quartz basal plane and the c-axis maximum, and hence encompass all three tectonic directions. The minima in V_{S_1} form small-circle distributions that appear to reflect the r/z LPO (the absolute minimum in V_{S_1} may indicate the inferred directions of both σ_1 and σ_3). The maximum in V_{S_2} is parallel to the quartz c-axis

maximum and subparallel to the tectonic Z direction, whilst minimum values in V_{S_2} occupy a great-circle distribution parallel to the quartz basal plane and subparallel to the XY tectonic plane. Maximum values in AV_S and dV_S occupy great-circle distributions parallel to the quartz basal plane and subparallel to the XY tectonic plane, whilst minimum values occur over broad areas that encompass the quartz c-axis maximum and r/z small-circle distributions.

The polarization behaviours of the fast shear waves (V_{S_1}) for the three shear-zone regions are somewhat similar, particularly for vertical wave propagation. More specifically, those for the shear-zone margin rock (Fig. 6b) and the mylonite (Fig. 6c–d) are very similar.

Discussion

The results of the LPO-averaged seismic-property determinations for different regions in the Torridon shear zone have significant implications for the interpretation and use of seismic velocities and anisotropy in geodynamic analysis. Firstly, it is worth comparing the single-crystal seismic properties of quartz and feldspar with the LPO-averaged results obtained from the shear-zone rocks to see how the former become modified in the latter. Secondly, the relationships between the LPO-averaged seismic properties and shear-zone structure and tectonics are briefly considered, including the potential for using LPO-averaged seismic property results in seismic waveform modelling. Finally, the potential impact of the grain-boundary microstructure elements on shear-zone seismic properties is considered briefly.

Comparison between single-crystal and polycrystal seismic properties

There are significant differences between the seismic properties of quartz and plagioclase single crystals and those of the three shear-zone regions. The trigonal symmetry anisotropy parallel to the c-axis, clearly seen in single-crystal quartz seismic properties (see Fig. 4a & c) is dispersed by the LPO-averaging effect in polycrystal aggregates (Fig. 6). The compressional and shear-wave velocities show wider variations (higher anisotropy) for the single crystals compared to the shear-zone regions (Fig. 7). This is the most important difference between single-crystal and polycrystal behaviours.

In terms of orientation, the overall effect of shear-zone LPO (Fig. 2) is to either displace and/or disperse (see Fig. 6 & Table 4) the

Table 4. *Summary of single-crystal and Torridon shear-zone seismic properties (see Figs 6 & 8)*

Property		Single crystal	SZ wallrock	SZ margin	Mature mylonite	Mylonitic detail
V_P	Max.	Q:||z P: sub-||(101)	Q: broad||c P: broad||(001)	Q: broad sub-||c	Q: sub-||c, r, z	Q: sub-||c, r, z
	Min.	Q: sub-||r P:||(001)	T: broad sub-||σ_1 Q: sub-||a, z? P: sub-||(100)	T: broad sub-||σ_1 Q:?	T: sub-||Z Q:||a and basal plane	T: sub-||Z Q:||a and basal plane
V_{S1}	Max.	Q:||a P: sub-||(110)	T: sub-||Y, σ_3 Q: broad sub-||c, r P: sub-||(010)	T: sub-||Y Q: broad sub-||c, r	T:||X, sub-||XY Q:||basal plane, sub-||c	T:||X, sub-||XY Q:||basal plane, sub-||c
	Min.	Q:||z P: sub-||(101)	T:||σ_1, sub-||Y Q: broad sub-||a, m? P: broad sub-||(010)	T: sub-||σ_1, Y Q:?	T:||X, Y, Z Q:||z small circle	T:||X, Y, Z Q:||r, z small circle
V_{S2}	Max.	Q:||c P:||(100)	Q:||c P: sub-||(001)	T:? Q: sub-||c?	T:||σ_3? Q:||c	T:||σ_1 and σ_3? Q:||c
	Min.	Q:||a P:||(101)	T:||σ_1? Q:||a P:||(010)	T:? Q:||a?	T: sub-||Z Q:||basal plane	T: sub-||Z Q:||basal plane
AV_S and dV_S	Max.	Q:||a P: sub-||(110)	T: sub-||Y Q: sub-||a P: sub-||(010)	T: sub-||Y? Q: sub-||a, c?	T: between X and Y Q:||basal plane	T: sub-||XY Q:||basal plane
	Min.	Q:||c P:||(100)	T: sub-||Y Q: sub-||m? P: sub-||(100), (001)? T: sub-||X?	T: sub-||Y? Q: sub-||m? T: sub-||X?	T: sub-||XY Q: sub-||c and r/z Small circle T: sub-||Z?	T: sub-||XY Q: sub-||c and r/z Small circle T: sub-||Z?

Q, quartz crystal directions (c, m, a, r, z); P, plagioclase crystal directions ((100), (010), (001)); T, tectonic axes (X ≥ Y ≥ Z); σ_1 and σ_3, inferred maximum and minimum principal stress directions, parallel to Qr (NW—SE) and Qz (NE—SW) respectively; \|\|, parallel to.
Note, only shear-zone wallrock contains significant plagioclase.

unique maximum and minimum seismic orientations recognized for the single crystals (Fig. 4). For example, the sharp orientation maximum in V_P parallel to the quartz single-crystal $\{z\}$ pole becomes broadened and parallel to the c-axis and/or $\{z\}$. Similarly, the quartz single-crystal maxima in V_{S^1}, AV_S and dV_S parallel to the a-axes are broadened into a great-circle distribution parallel to the quartz basal plane.

The polarizations of the fast shear waves (V_{S^1}) for all three shear-zone regions (Fig. 6) are very different to those of the single crystals and particularly quartz (Fig. 4c & e). This is due to the fact that although the c-axes of quartz crystals are highly aligned (Fig. 2c–d), the other crystallographic directions are more dispersed in orientation. The polycrystal elasticity therefore is not trigonal in symmetry.

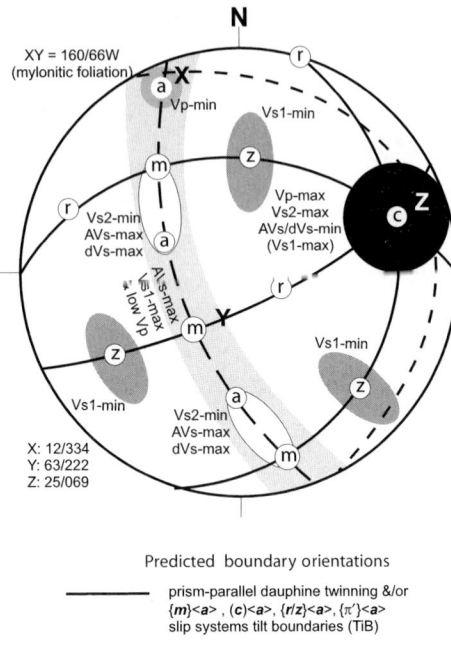

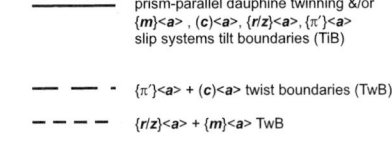

Predicted boundary orientations

———— prism-parallel dauphine twinning &/or $\{m\}<a>$, $(c)<a>$, $\{r/z\}<a>$, $\{\pi'\}<a>$ slip systems tilt boundaries (TiB)

— — — · $\{\pi'\}<a> + (c)<a>$ twist boundaries (TwB)

– – – – $\{r/z\}<a> + \{m\}<a>$ TwB

Fig. 8. Summary of orientations of seismic properties (V_P, V_{S^1}, V_{S^2}, AV_P, AV_S, dV_S – see text for details) with respect to Torridon shear-zone tectonic (X, Y, Z) and crystallographic (quartz $c-m-a-r-z$ LPO) frameworks in geographical (N) co-ordinates. Also shown are predicted orientations of the principal quartz grain boundaries (Fig. 3). See text for discussion.

Relationship between shear-zone structure, tectonics and seismic properties

In general, the results of the LPO-averaging of shear-zone seismic properties suggest that deformation (increasing strain) during shear-zone formation and mylonitization modifies the absolute seismic-velocity values, and increases the seismic anisotropy of the original wallrock (Fig. 7). To consider further the relationship between shear-zone structure, tectonics and seismic properties, a summary stereographic projection of the orientations (see Table 4) of structural (i.e. LPO), tectonic (i.e. $X \geq Y \geq Z$) and seismic (i.e. V_P, V_{S^1}, V_{S^2}, AV_S, dV_S) properties in the field (i.e. geographical) reference frame is shown in Figure 8.

Recall that the mylonitic shear zone has a strong LPO (Fig. 2c & d) that approximates a dauphine twinned single-crystal configuration, with c-axis maximum subparallel to Z and subnormal to the foliation (XY) plane, and the a-axis maximum parallel to X. The results of the LPO-averaging of the seismic properties (Figs 6 & 7) suggest that within the mylonitic shear zone the maximum in V_P and the minima in V_{S^2}, AV_S and dV_S tend to align parallel to the c-axis maximum (i.e. normal to the mylonitic foliation), whilst the minimum in V_P aligns parallel to the a-axis maximum and hence defines the extension (X) direction. High values of V_{S^1}, AV_S and dV_S and low values of V_P define the foliation (XY) plane, whilst the minimum in V_{S^1} either aligns subparallel to $\{z\}$ or defines a weak small-circle distribution about the c-axis maximum that includes $\{z\}$.

The relationships between structure, tectonics and seismic properties described above may

prove useful in the tectonic interpretation of seismic measurements. However, it is clear from the geographical representation of the various parameters in Figure 8 that any analysis of seismic waves (e.g. see below) must be interpreted in terms of the appropriate structural and tectonic orientations in the field.

Seismic waveform modelling

Wave propagation is considerably more complicated in anisotropic media than it is in isotropic media. For example, wavefronts in homogeneous anisotropic media are no longer spherical, and the direction of particle motion, ray direction and wavefront normal are in general not aligned. Furthermore, two shear waves propagate in anisotropic media, and they may exhibit folding, which leads to travel-time triplications (e.g. the fast shear wave arrives at three different times at a receiver oriented parallel to the a-axis

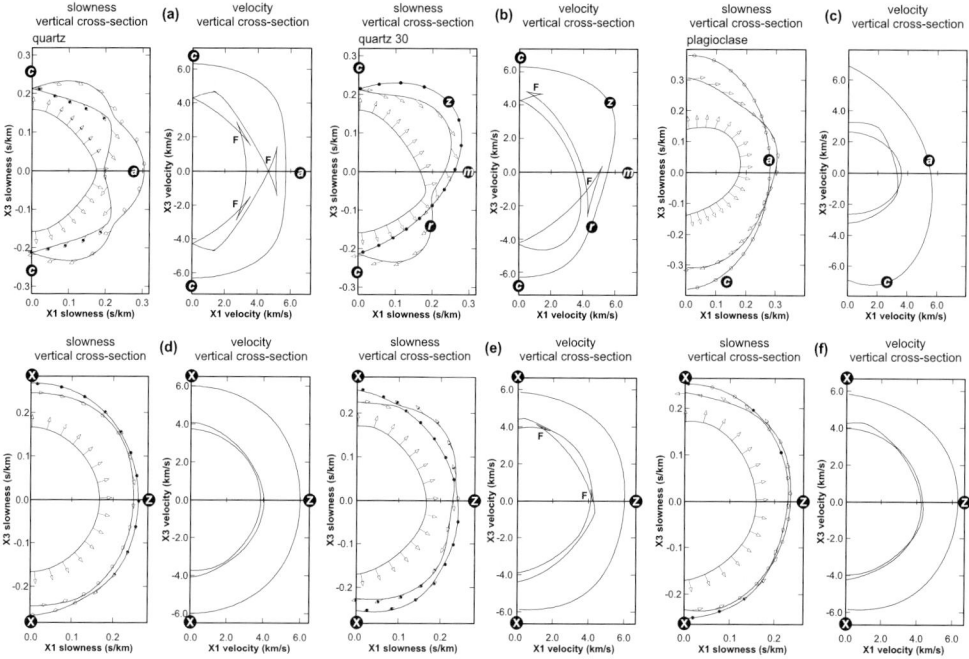

Fig. 9. Cross-sections through slowness and wave surfaces for single crystals and shear-zone regions. In each, the left-hand side shows slowness surfaces for each wave, including directions of particle motion (arrows); note, the P-wave is the innermost surface. The right-hand side shows wave surfaces, and can be viewed as a snapshot of the wavefront after one second; note 'folding' (F) on some S-wavesheets, which is *never* seen for P-waves. (**a**) Quartz single crystal parallel to plane normal to $m\{10\bar{1}0\}$; note symmetry about quartz basal plane parallel to $\langle a \rangle$ direction. (**b**) Quartz single crystal parallel to plane normal to $a\{11\bar{2}0\}$; note symmetry about quartz basal plane that reflects maximum in V_P parallel to $\{z\}$ and minimum in V_P between $\{r\}$ and $\{m\}$. (**c**) Plagioclase single crystal parallel to plane normal to $b\{010\}$; note overall (triclinic) asymmetry. (**d**) Shear-zone wall rock, XZ section. (**e**) Shear-zone margin, XZ section (note folding). (**f**) Mature mylonite, XZ section.

of quartz, see Fig. 9a). Figure 9 illustrates these effects, showing cross-sections of the slowness and wave surfaces for the minerals and rocks of the Torridon shear zone parallel to specific crystal (i.e. $\langle 1\bar{1}00 \rangle$ and $\langle 11\bar{2}0 \rangle$) and tectonic (i.e. XZ) planes.

It is interesting to consider how seismic waves propagate through the shear zone and how they are affected by the enhanced deformation and hence seismic anisotropy. Teleseismic and regional earthquakes provide a passive means of imaging the deep crust with relatively low frequencies and hence low resolution. In contrast, wide-angle controlled-source surveys give better resolution. Shear-wave splitting is the most unambiguous indicator of anisotropy. However, whether or not this is observable depends on the magnitude of the anisotropy and its spatial extent. The rocks of the Torridon shear zone show varying degrees of splitting (e.g. see Figs 6 & 7), but in general the anisotropy must be uniform and persist over a large region with

respect to seismic wavelength, in order to accrue a measurable level of splitting. In earthquake studies such a region must be kilometres in extent, and therefore is only likely to be observed for regional-scale shear zones or rock masses with constant petrofabric characteristics (e.g. gneisses). Fortunately, reflections and mode conversions at interfaces provide better vertical resolution (perhaps tens of metres for controlled-source experiments) and can be very sensitive to changes in anisotropy. High degrees of anisotropy can enhance the seismic contrast across a boundary, both in terms of absolute values of anisotropy and also if the directional characteristics of the anisotropy change across the boundary.

If the crystals in the shear-zone rocks are randomly oriented, their elastic parameters are isotropic. Figure 10a & b shows the P-wave reflections and P-to-S-wave conversions at boundaries between such isotropic assemblages. These averages are based on the LPO-averaged elastic parameters for the Torridon rocks (see Tables 2,

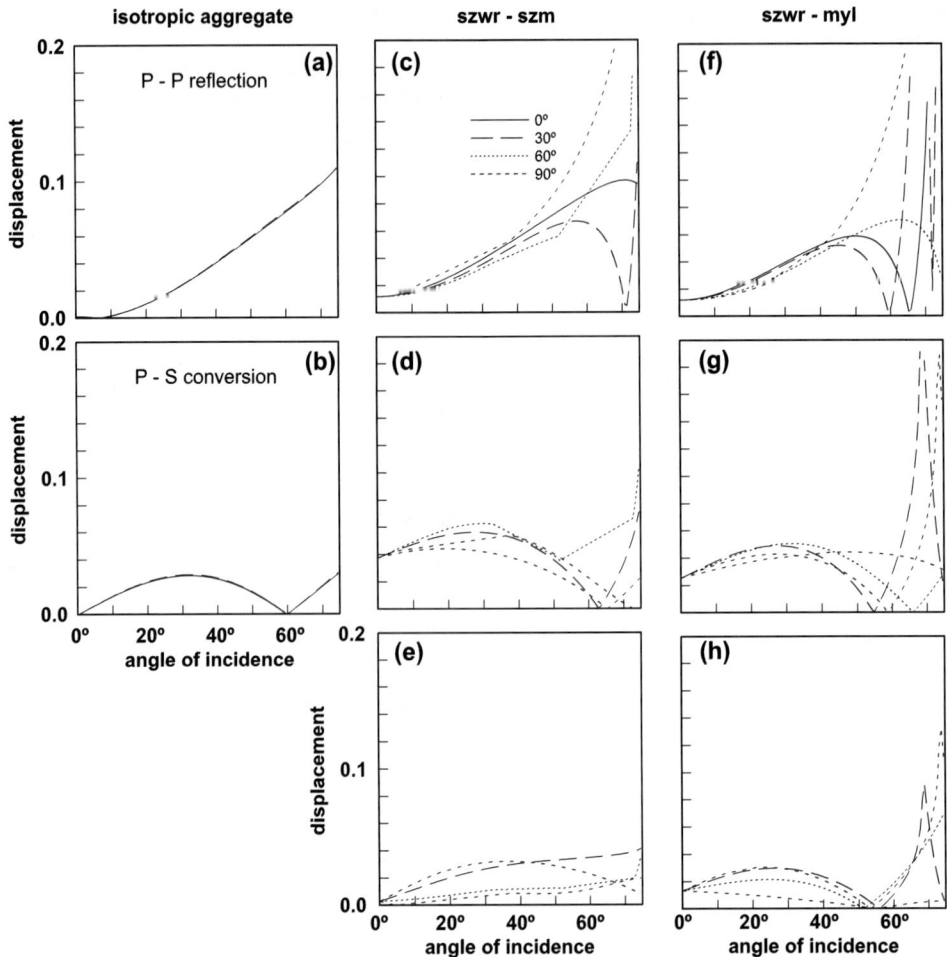

Fig. 10. Seismic-reflection coefficients. (**a**) and (**b**) Consider interfaces between isotropic aggregates (see Table 5) for P-wave reflections and P–S mode-converted reflections respectively. Solid line shows coefficients for isotropic shear-zone wallrock and isotropic shear-zone margin interface, while the (barely visible) dashed line shows coefficients for interface between the shear-zone wallrock and the isotropic mylonite. (**c**)–(**e**) Consider an interface between anisotropic shear-zone wallrock and shear-zone margin (NB no azimuthal variation in reflection coefficient in isotropic cases; see (**c**) for key to azimuthal directions). (**c**) P-wave reflections as a function of the incidence angle and azimuthal angle. (**d**) Mode-converted reflections for P-S1-waves. (**e**) Mode-converted reflections for P-S2-waves. (**f**)–(**h**) Consider an interface between anisotropic shear-zone wallrock and mature mylonite (NB no azimuthal variation in reflection coefficient in isotropic cases; see (**c**) for key to azimuthal directions). (**f**) P-wave reflections as a function of incidence angle and azimuthal angle. (**g**) Mode-converted reflections for P-S1-waves. (**h**) Mode-converted reflections for P-S2-waves.

3 & 5). The reflections are very weak, especially at near offsets (small angles of incidence).

Crystal alignment and hence anisotropy affects the reflection properties in a number of interesting ways. Figure 10c–h shows the reflection coefficients for anisotropic interfaces as predicted from the petrophysical analysis described above. In general, more energy is reflected in the anisotropic case. This is due to higher vertical

velocities in the underlying more-deformed layers. There is a significant level of mode-converted reflections at these interfaces. Both the fast and slow shear waves are generated in the conversion, due to the complex anisotropic symmetries. There is a surprising amount of reflected converted-wave energy at normal incidence, which never exists in the isotropic case. This is due to the strong difference in the

Table 5. *Isotropic aggregate properties of Torridon shear-zone rocks calculated from the measured elastic properties in Tables 2 and 3*

Property	SZWR	SZM	MM
P-wave velocity (km s^{-1})	4.298	4.291	4.290
S-wave velocity (km s^{-1})	2.748	2.911	2.906
Density (g cm^{-3})	2.636	2.647	2.647

SZWR, shear-zone wallrock; SZM, shear-zone margin; MM, mature mylonite.

directions of group velocity (ray direction) and phase velocity (normal to the wavefront), again due to the low order of symmetry and strong anisotropy. There are also clear azimuthal variations in the reflections. This suggests that multi-azimuth wide-angle reflection data potentially can be used to study the sense of deformation in deep-rooted shear zones (see also Burlini *et al.* 1998; Khazanehdari *et al.* 1998). Such reflection effects have been used also to map fracture patterns in oil reservoirs (Hall & Kendall 2003).

Grain-boundary microstructure

In addition to LPO, a number of other microstructural elements are known to impact on elastic anisotropy, such as cracks, fractures, pores, grain morphology and shape-preferred orientation, and layering. The impact of many of these elements on elastic anisotropy has recently been considered by Wendt *et al.* (2003) and further discussion here is largely inappropriate. However, it is worth considering one particular microstructural element, the role of grain boundaries, that impacts directly on the present study.

Grain boundaries are important microstructural elements in rocks, and may contribute a significant volume fraction, particularly in fine-grained rocks such as mylonites. They represent regions of relatively high lattice distortion and defect concentration that form extended three-dimensional networks, and are likely to have different elastic parameters from those of the grains. In addition, as rock elasticity is also dependent on cohesion forces between grains (Kaarsberg 1959), the contact area between grains should be considered. Furthermore, grain boundaries may form crack-like discontinuities in some rocks, and consequently influence seismic anisotropy if the boundaries are preferentially aligned (e.g. Babuška and Cara 1991). Thus, many rock properties, including the elastic parameters, may be influenced by the grain-boundary network.

Unfortunately, there is little known about the exact impact of grain boundaries on seismic anisotropy (e.g. Wendt *et al.* 2003). However, a recent interpretation of the microstructural and petrofabric evolution of the Torridon shear zone (Lloyd 2004) considered the formation and orientation of quartz grain-boundaries. It was suggested that the grain boundary network consists of tilt boundaries parallel to quartz prism (and hence often *YZ* tectonic) planes and twist boundaries parallel to the *XY* foliation/basal-crystal planes and *XZ* tectonic plane (see Fig. 3). The *physical* presence of grain boundaries was not considered in the LPO-averaging analysis described here, and hence any *direct* impact due to grain-boundary configuration on seismic properties cannot be assessed. However, comparison between grain-boundary and seismic property orientations (Fig. 8; see also Fig. 6) may provide significant indications, as follows.

Twist boundaries that developed parallel to the *XY* foliation/basal crystal planes due to a combination of $\{\pi'\}\langle a \rangle + (c)\langle a \rangle$ slip could contribute to the great circle distribution of high AV_S, dV_S and V_{S1} and low V_P values. In addition, the broad concentration of high/maximum V_P, V_{S2} and V_{S1} and low/minimum AV_S and dV_S values parallel to the quartz *c*-axis maximum and subparallel to the tectonic *Z* direction may be influenced by two characteristics of the grain boundary network:

(1) the intersection of tilt boundaries that develops due to a combination of dauphine twinning and/or slip on systems that exploit the quartz $\langle a \rangle$ axis as the slip direction; and

(2) twist boundaries that develop parallel to the *XZ* tectonic plane due to combined slip on $\{r/z\}\langle a \rangle + \{m\}\langle a \rangle$ systems.

Similarly, the tendency for the minimum in V_P to occur parallel to the quartz *a*-axis maximum and subparallel to the tectonic *X* direction may be augmented by the intersection of twist boundaries that develop parallel to the *XY* foliation/basal crystal planes and the *XZ* tectonic plane, due to combined slip on $\{\pi'\}\langle a \rangle + (c)\langle a \rangle$ and $\{r/z\}\langle a \rangle + \{m\}\langle a \rangle$ systems respectively. Thus, it appears that grain-boundary configuration has the *potential* to impact on seismic property characteristics, but much further work (i.e. direct measurements) is needed.

Conclusions

This contribution has explored the link between petrofabric (LPO) and petrophysical (seismic) properties of a quartzo-feldspathic shear zone developed under conditions typical of the mid–lower crust. Although it is clear that shear-zone formation and petrofabric evolution from an initially quartzo-feldspathic gneissic wallrock to a mature quartz mylonite is reflected in the seismic properties, the precise impact of LPO on seismic properties is not simple. Several conclusions can be drawn, based on these observations:

(1) The most obvious control of LPO on shear zone seismic properties results in the following relationships: $V_{P\text{-min}}$ parallel to the a-maximum (i.e. tectonic X); $AV_{S\text{-max}}$ parallel to mylonitic foliation (i.e. XY mylonitic foliation plane); $V_{P\text{-max}}$ and $AV_{S\text{-min}}$ parallel the c-maximum (i.e. foliation normal, Z); and $V_{S1\text{-min}}$ parallel to the z-maxima.

(2) The development of a strong LPO during shear-zone evolution, and particularly during mylonitization, is responsible for considerable seismic anisotropy. P-wave anisotropy varies between 11–13%, the fast and slow shear-waves show 6–15% anisotropy, and the magnitude of shear-wave splitting is 9.5–16%.

(3) Although the degree of shear-wave splitting exhibited by the shear zone due to LPO requires considerable thicknesses of rock (depending on the dominant wavelength of the source) with constant LPO characteristics before it becomes seismically visible, reflections and mode conversions provide much better resolution, particularly across boundaries (e.g. between the shear-zone margin and the mature mylonite).

(4) The low symmetry and strong anisotropy due to the petrofabric suggest that multi-azimuth wide-angle reflection data may be useful in the determination of the deformation characteristics of deep shear zones.

(5) Ultimately, how well seismic data can be used to study anisotropy, and hence lower-crustal deformation, remains to be seen. Data quality and spatial coverage will be crucial.

We thank D. Mainprice for use of his LPO and seismic property calculation and plotting programs. The original manuscript was considerably improved by the comments of two anonymous referees. Part of the automated EBSD system used was funded by UK NERC Grant GR9/3223 (GEL).

References

ALEKSANDROV, K. S., ALCHIKOV, U. V., BELIKOV, B. P., ZASLAVSKI, B. I. & KRUPNYI, A. I. 1974. Velocities of elastic waves in minerals at atmospheric pressure and increasing precision of elastic constants by means of EVM. *Izvestiya Acad. Science USSR Geol Ser.* **10**, 15–24.

BABUŠKA, V. & CARA, M. 1991. *Seismic Anisotropy in the Earth*. Kluwer Academic, Dordrecht.

BASCOU, J., BARRUOL, G., VAUCHEZ, A., MAINPRICE, D. & EGYDIO-SILVA, M. 2001. EBSD-measured lattice preferred orientations and seismic properties of eclogites. *Tectonophysics*, **342**, 61–80.

BLACKMAN, D. K., WENK, H.-R. & KENDALL, J.-M. 2002. Seismic anisotropy of the upper mantle: 1, factors that affect mineral texture and effective elastic properties. *Geochemical and Geophysical Geosystems*, **3**, 8601, doi:10.1029/2001GC000248, 2002.

BUNGE, H. J. 1982. *Texture Analysis in Materials Science*. Butterworths, London, 599 pp.

BUNGE, H. J., KIEWEL, R., REINERT, TH. & FRITSCHE, L. 2000. Elastic properties of polycrystals – influence of texture and stereology. *Journal of the Mechanics and Physics of Solids*, **48**, 29–66.

BURLINI, L. & KERN, H. 1994. Special Issue: Seismic properties of crustal and mantle rocks – Laboratory measurements and theoretical calculations. *Surveys in Geophysics*, **15**, 439–672.

BURLINI, L. & KUNZE, K. 2000. Fabric and seismic properties of Carrara marble mylonite. *Physics and Chemistry of the Earth*, **25**, 133–139.

BURLINI, L., MARQUER, D., CHALLANDES, N., MAZZOLA, S. & ZANGARINI, N. 1998. Seismic properties of highly strained marbles from the Splugenpass, central Alps. *Journal of Structural Geology*, **20**, 277–292.

CASEY, M. 1981. Numerical analysis of X-ray textural data – an implementation in FORTRAN allowing triclinic or axial specimen symmetry and most crystal symmetries. *Tectonophysics*, **78**, 51–64.

CHRISTENSEN, N. I. 1966. Elasticity of ultrabasic rocks. *Journal of Geophysical Research*, **71**, 5921–5931.

CROSSON, R. S. & LIN, J. W. 1971. Voigt and Reuss prediction of anisotropic elasticity of dunite. *Journal of Geophysical Research*, **76**, 570–578.

DANA, E. S. & FORD, W. E. 1951. *A Textbook of Mineralogy*. John Wiley & Sons, Chichester.

DINGLEY, D. J. 1984. Diffraction from sub-micron areas using electron backscattered diffraction in a scanning electron microscope. *Scanning Electron Microscopy*, **2**, 569–575.

HALL, S. & KENDALL, J.-M. 2003. Fracture characterisation at Valhall: application of P-wave AVOA analysis to a 3D ocean-bottom data set. *Geophysics*, **68**, 1150–1160.

HESS, H. H. 1964. Seismic anisotropy of the uppermost mantle under oceans. *Nature*, **203**, 629–631.

HILL, R. 1952. The elastic behaviour of a crystalline aggregate. *Proceedings of the Physical Society, London A*, **65**, 351–354.

JI, S., WANG, Q. & XIA, B. 2002. *Handbook of Seismic Properties of Minerals, Rocks and Ores*. Polytechnic International Press, Canada.

KAARSBERG, E. A. 1959. Introductory studies of natural and artificial argillaceous aggregates by sound-propagation and X-ray diffraction methods. *Journal of Geology*, **67**, 447–472.

KENDALL, J.-M. 2000. Seismic anisotropy in the boundary layers of the Earth's mantle. *In*: KARATO, S.-I., FORTE, A. M., LIEBERMAN, R. C., MASTERS, G. & STIXRUDE, L. (eds) Earth's deep interior; mineral physics and tomography from the atomic to the global scale. *Geophysical Monograph*, **117**, 149–175.

KERN, H. 1982. P- and S-wave velocities in crustal and mantle rocks under the simultaneous action of high confining pressure and high temperature and the effect of microstructure. *In*: SCHREYER, W. (ed.), *High Pressure Researches in Geoscience*. H. Schweizerhart'sche Verlagsbuchhandlung, Stuttgart, 15–45.

KERN, H. & WENK, H.-R. 1985. Anisotropy in rocks and the geological significance. *In*: WENK, H.-R. (ed.) Preferred orientation in deformed metals and rocks: an introduction to modern texture analysis, Academic Press, Orlando, 537–555.

KHAZANEHDARI, J., RUTTER, E. H., CASEY, M. & BURLINI, L. 1998. The role of crystallographic fabric in the generation of seismic anisotropy and reflectivity of high strain zones in calcite rocks. *Journal of Structural Geology*, **20**, 293–300.

LAW, R. D., SCHMID, S. M. & WHEELER, J. 1990. Simple shear deformation and quartz crystallographic fabrics – a possible natural example from the Torridon area of NW Scotland. *Journal of Structural Geology*, **12**, 29–45.

LISTER, G. S. & SNOKE, A. 1984. S-C mylonites. *Journal of Structural Geology*, **6**, 617–638.

LLOYD, G. E. 1987. Atomic number and crystallographic contrast images with the SEM: a review of backscattered electron techniques. *Mineralogical Magazine*, **51**, 3–19.

LLOYD, G. E. 2000. Grain boundary contact effects during faulting of quartzite: an SEM/EBSD analysis. *Journal of Structural Geology*, **22**, 1675–1693.

LLOYD, G. E. 2004. Microstructural evolution in a mylonitic quartz simple shear zone: the significant roles of dauphine twinning and misorientation. *In*: ALSOP, G. I., HOLDSWORTH, R. E., MCCAFFREY, K. & HAND, M. (eds) Flow Processes in Faults and in Shear Zones, Geological Society, London, Special Publications **224**, 39–61.

LLOYD, G. E., LAW, R. D., MAINPRICE, D. & WHEELER, J. 1992. Microstructural and crystal fabric evolution during shear zone formation. *Journal of Structural Geology*, **14**, 1079–1100.

MCSKIMIN, H. J., ANDREATCH, JR, P. & THURSTON, R. N. 1965. Elastic moduli of quartz versus hydrostatic pressure at 25°C and −195.8°C. *Journal of Applied Physics*, **36**, 1624–1632.

MAINPRICE D. 1990. An efficient Fortran program to calculate seismic anisotropy from the lattice preferred orientation of minerals. *Computers and Geosciences*, **16**, 385–393.

MAINPRICE, D. 2003. World Wide Web Address: http://www.isteem.univ-montp2.fr/TECTONOPHY/pet rophysics/software/petrophysics_software.html.

MAINPRICE, D. & HUMBERT, M. 1994. Methods of calculating petrophysical properties from lattice preferred orientation data. *Survey Geophysics*, **15**, 575–592.

MAINPRICE, D. & SILVER, P. G. 1993. Interpretation of SKS waves using samples from the subcontinental lithosphere. *Physics of Earth and Planetary Interiors*, **78**, 257–280.

MAINPRICE, D., LLOYD, G. E. & CASEY, M. 1993. Individual orientation measurements in quartz polycrystals – advantages and limitations for texture and petrophysical property determinations. *Journal of Structural Geology*, **15**, 1169–1187.

MAINPRICE, D., BARRUOL, G. & BEN ISMAIL, W. 2000. The seismic anisotropy of the Earth's mantle; from single crystal to polycrystal. *In*: KARATO, S.-I., FORTE, A. M., LIEBERMAN, R. C., MASTERS, G. & STIXRUDE, L. (eds) *Earth's Deep Interior; Mineral Physics and Tomography from the Atomic to the Global Scale*. Geophysical Monographs, **117**, 237–264.

MAINPRICE, D., POPP, T., GUEGUEN, Y., HUENGES, E., RUTTER, E. H., WENK, H.-R. & BURLINI, L. 2003. Physical properties of rocks and other geomaterials, a Special Volume to honour Professor H. Kern. *Tectonophysics*, **370**, 1–311.

MAULER, A., BURLINI, L., KUNZE, K., BURG, J. P. & PHILIPPOT, P. 2000. P-wave anisotropy in eclogites and relationship to omphacite crystallographic fabric. *Physics and Chemistry of the Earth*, **25**, 119–126.

NYE, J. F. 1957. *Physical Properties of Crystals*. Clarendon Press, Oxford, 322 pp.

PESELNICK, L. A., NICOLAS, A. & STEVENSON, P. R. 1974. Velocity anisotropy in a mantle peridotite from the Ivrea Zone: aplication to upper mantle anisotropy. *Journal of Geophysical Research*, **79**, 1175–1182.

PRIOR, D. J., BOYLE, A. P. *et al.* 1999. The application of electron backscatter diffraction and orientation contrast imaging in the SEM to textural problems in rocks. *American Mineralogist*, **84**, 1741–1759.

PROS, Z., LOKAJICEK, T., PŘIKRYL, R. & KLIMA, K. 2003. Direct measurement of 3D elastic anisotropy on rocks from the Ivrea zone (Southern Alps, NW Italy). *Tectonophysics*, **370**, 31–47.

RAMSAY, J. G. 1980. Shear zone geometry: a review. *Journal of Structural Geology*, **2**, 83–100.

RAMSAY, J. G. & GRAHAM, R. H. 1970. Strain variation in shear belts. *Canadian Journal of Earth Sciences*, **7**, 786–813.

REUSS, A. 1929. Berechnung der Fließgrenze von Mischkristallen auf Grund der Plastizitätsbedingungen für Einkristalle. *Zeitschrift für Angewandte Mathematik und Mechanik*, **9**, 49–58.

SCHMIDT, N. H. & OLESEN, N. Ø. 1989. Computer-aided determination of crystal-lattice orientation from electron channelling patterns in the SEM. *Canadian Mineralogist*, **27**, 15–22.

SERONT, B., MAINPRICE, D. & CHRISTENSEN, N. I. 1993. A determination of the three dimensional seismic properties of anorthosite: comparison between values calculated from the petrofabric and direct laboratory measurements. *Journal of Geophysical Research*, **98**, 2209–2221.

SINOGEIKIN, S. V. & BASS, J. D. 2000. Single-crystal elasticity of pyrope and MgO to 20 GPa by Brillouin scattering in the diamond cell. *Physics of the Earth and Planetary Interiors*, **120**, 43–62.

VENABLES, J. A. & HARLAND, C. J. 1973. Electron backscattering patterns – a new technique for obtaining crystallographic information in the scanning electron microscope. *Philosophical Magazine*, **27**, 1193–1200.

VOIGT, W. 1928. *Lehrbuch der Kristallphysik: mit Ausschluss der Kristalloptik*. B. G. TEUBNER, Leipzig, 978 pp.

WENDT, A. S., WIRTH, R., BAYUK, I. O., COVEY-CRUMP, S. J. & LLOYD, G. E. 2003. An experimental and numerical study of the microstructural parameters contributing to the seismic anisotropy of rocks. *Journal of Geophysical Research*, **108** (B8), 2365, doi:10.1029/2002JB001915.

WHEELER, J. 1984. A new plot to display the strain of elliptical markers. *Journal of Structural Geology*, **6**, 417–423.

Fractured reservoir analysis using modern geophysical well techniques: application to basement reservoirs in Vietnam

S. M. LUTHI

Department of Geotechnology, Mijnbouwstraat 120,
Delft University of Technology, 2628RX Delft, The Netherlands
(e-mail: s.m.Luthi@ta.tudelft.nl)

Abstract: Recent geophysical well techniques have significantly improved the analysis of fractured reservoirs. These methods include electrical and ultrasonic scans and, in some cases, optical video images, that provide azimuthal high-resolution images of the borehole wall on which fractures are prominently visible. Additionally, fractures produce reflections and attenuations of the Stoneley wave, a borehole mode recorded by the array sonic wireline tool. A fracture identified with these methods can be individually probed with a new wireline formation tester featuring a dual-packer module that hydraulically isolates it from the surrounding formation. The combination of these techniques can provide information on fracture locations, dip, azimuth, aperture, permeability and fluid content. Seismic data can be used to extrapolate this information away from the wells. A case study on basement reservoirs from offshore Vietnam exhibits foliations, borehole breakouts, hydraulic and tectonic fracturing. Oil production comes from a small number of point entries that correspond to fractures, most of which produce more than 1000 barrels of oil per day. Two intersecting fracture sets were found, which may explain the high sustained production. Properly planned horizontal wells may increase production and decrease the chance of water breakthrough.

Oil and gas accumulations in basement rocks are generally ranked among the more unusual reservoirs. The fact that a crystalline or metamorphic rock contains hydrocarbons, which are generally considered to be sourced from organic-rich sedimentary rock that is younger than the reservoir rock, seems still somewhat unusual to many geoscientists. Nevertheless, hundreds of such reservoirs have been found in many parts of the world, and many of them are prolific producers (GeoScience Ltd, 2000).

We consider here as basement rocks only crystalline and metamorphic rocks, a definition that is in line with Landes *et al.* (1960) but at variance with P'an (1982), who included tight Lower Palaeozoic sedimentary rocks as well (such as the Hassi Messaoud field in Algeria). Oil accumulations in basement rocks have been discovered since the early decades of the 20th century, but, according to most accounts, this has happened largely by accident. Among the best-known examples are the La Paz field in Lake Maracaibo, Venezuela (Smith 1956), the Hurghada and Zeit Bay fields in the Gulf of Suez, Egypt (P'an 1982; Zahran & Askary 1988), the Nafoora–Augila field in Libya (Belgasem *et al.* 1990), and several fields in the central Kansas uplift (Hubbert & Willis 1955), as well as in the former Soviet Union and China (P'an 1982). More recently, the Clair field west of Shetland was found to contain significant quantities of oil in the Permian red beds and the underlying granite basement (Coney *et al.* 1993); however the field is still not under production. In the 1980s, Russian explorers found significant quantities of oil in basement fault blocks offshore of Vietnam, including the White Tiger field that contributes the bulk of the country's present-day oil production (Areshev *et al.* 1992). This find triggered significant drilling activity offshore of South Vietnam (Areshev *et al.* 1992; Tandom *et al.* 1997, 1999) that led to several additional but smaller discoveries. Since evaluating the reserves and production capacities of these non-conventional reservoirs is difficult, new technologies were evaluated, and some were found to be exceptionally useful. Among these are new logging and borehole imaging techniques that are specifically suited for the analysis

From: HARVEY, P. K., BREWER, T. S., PEZARD, P. A. & PETROV, V. A. (eds) 2005. *Petrophysical Properties of Crystalline Rocks.* Geological Society, London, Special Publications, **240**, 95–106.
0305-8719/05/$15.00   © The Geological Society of London 2005.

of fractured systems. These techniques will be
briefly described below.

Hydrocarbon accumulation in
basement rocks

The accumulation of hydrocarbons in basement
rocks happens under specific circumstances,
which have been summarized by P'an
(1982). These include the following common
characteristics:

(1) Basement reservoirs always occur on highs
     or uplifts within a basin that were subject to
     long periods of weathering or erosion prior
     to being submerged and covered with
     marine sediments that formed the seal and
     the source rock for the reservoir.
(2) Basement reservoirs always occur below a
     regional unconformity that is considered
     important in the migration of oil towards
     the reservoir.
(3) All source beds of basement reservoir lie
     above the reservoir rock, implying that
     migration took place laterally upwards or
     vertically downwards. The latter can be
     achieved by dilatancy in the basement
     rocks due to shearing that reduces the
     hydrostatic pressure in the fractures and
     thus creates a pressure gradient favourable
     for downward migration (McNaughton
     1953).
(4) Differential compaction of argillaceous
     sediment over the buried high has created
     a draping effect that is considered essential
     in the trapping of hydrocarbons.
(5) Gas is rarely found in basement reservoirs.
(6) In general, the pore space in basement
     reservoirs consists of tectonic fissures and
     faults, dissolved interstices, and caverns.
     Areshev et al. (1992) add shrinkage vugs
     as potentially important elements in the
     Vietnam reservoirs. These pore types are
     distributed very irregularly, and hence the
     porosity and permeability distribution is
     highly heterogeneous.
(7) Towards the top of the basement reservoirs,
     an increasing degree of weathering is
     found, that can lead, in some cases, to a
     gradual transition to the overlying sedimen-
     tary rocks. Intergranular porosity similar to
     that in sandstones can be found here.
(8) Basement reservoirs are characterized by
     thick reservoir rocks and single well
     production that may range from nil to
     very high, depending on the pore type and
     distribution. Water coning and fingering
     is common, but can be overcome by

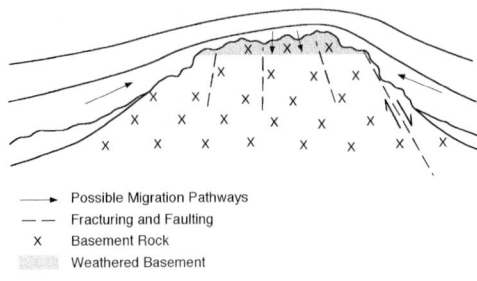

——▶ Possible Migration Pathways
— — Fracturing and Faulting
 x   Basement Rock
▨▨▨ Weathered Basement

Fig. 1. A schematic sketch of a basement rock
reservoir and its accumulation conditions, compiled
from various sources.

     successive completions and production
     from the bottom upwards.
(9) Reserves in basement reservoirs can be very
     large, but they carry a high uncertainty.

Figure 1 shows a schematic sketch of the
migration and accumulation conditions typically
found in basement reservoirs. The occurrence of
oil in basement rocks has been used by some
geologists as an argument to support the hypothesis of an inorganic origin of petroleum (e.g.
Porfirev 1974; Gold & Soter 1980). At present,
however, there is sufficient evidence to think
that oil accumulation conditions in basement
rock reservoirs do not differ from those in more
conventional reservoirs. The potential of basement rock reservoirs has been recognized early
(Eggleston 1948), and there were always propo-
nents of increased efforts to explore for them in
a more targeted manner. Thus, Landes et al.
(1960) state that: 'In suitable areas exploratory
wells should be located deliberately in order to
probe upward bulges of the basement rock as
well as the sedimentary veneer.'

Characterization of basement reservoirs

The most difficult problem after finding
basement rock reservoirs is to evaluate them,
particularly their production capacity and their
reserves. Nelson (1985), North (1990) and
Aguilera (1995) describe relatively traditional
methods, that include core analysis, well tests
and log evaluation. It is generally agreed that
an accurate description of the fracture system is
of paramount importance. This includes the
shape, orientation, spacing and apertures of
the fractures, from which their volume and interconnectedness can be obtained and, therefore,
the fracture porosity and the permeability of the
system. However, such complete information is
never available, and simplifying assumptions

have to be made. Thus, for example, fractures are often modelled as regularly spaced and arranged in an orthogonal pattern. Such simple fracture systems are also easier to use in fluid flow simulators that forecast the production behaviour with time. In some cases, nearby outcrops allow inspection of the fracture system and, by analogy, application to the subsurface. As an example, the area around Gebel Zeit in Egypt, an outcrop of tilted basement and associated sedimentary cover (Fig. 2), has been used as an analogue for the Zeit Bay basement rock reservoir in the Gulf of Suez (Luthi 1983; Younes et al. 1998).

At a meeting on fractured reservoirs, held in 2000 by the Society of Petroleum Engineers in Sainte Maxime (France), recent progress in the field was reviewed. An important conclusion at this meeting was that new technologies in the fields of seismic surveying, borehole logging and fluid-flow simulation now provide geoscientists with powerful tools to characterize fracture systems better than ever before. In the field of seismic surveying, this happens through

Fig. 2. Outcrop photograph of fractured rhyolite in the northern part of Gebel Zeit, Gulf of Suez, Egypt. This fracture system can be used as an analogue to that of the nearby Zeit Bay oilfield in basement rock.

improved data acquisition, particularly with multi-component geophones in 3D, and through advanced processing such as seismic attribute analysis and variance cubes that provide information on the degree and directionality of fracturing. In the field of fluid-flow simulation, discrete fracture network simulators can now model realistic fractures, rather than their equivalent mathematical representations (proxies) used in conventional simulators.

New geophysical borehole measurements

In the field of borehole logging, the analysis of fractured reservoirs has been based on circumstantial evidence, at best, for a long time. A common method consisted of evaluating the fracture probability by looking at all indicators on the various logs. As shown in Figure 3, each log can be affected to some degree by the presence of a fracture. Thus, the caliper might show a slight enlargement in front of a fracture. The deep and shallow resistivity logs might show a separation larger than usual, because of their different current geometry in the presence of a fracture (Sibbit 1995). Barium in the drilling mud might invade a fracture and result in a higher than usual photo-electric absorption or density reading. The dip-meter tool rotation can slow down because a pad follows a vertical fracture trace, or the dip-meter curves show relative differences because one pad is in front of a fracture while the others are not. And finally, the sonic velocities may decrease in front of a fracture, an effect seen particularly well on the shear-wave velocity log. If such evidence occurs on multiple logs at a given depth, the chance increases that a fracture exists at that depth. However, no information is gleaned on the orientation or the width of the fracture, and thus the fracture characterization is essentially reduced to a probability estimation of the fracture density.

Borehole imaging

Stearns & Friedman (1972) found that borehole imaging – at that time with the acoustic borehole televiewer (Zemanek et al. 1969) – has the potential of becoming an important tool in fractured reservoir characterization. This ultrasonic borehole scannning measurement took a considerable time to reach commercial maturity (Faraguna et al. 1989; Hayman et al. 1998), but today all major oil service companies provide this service in a variety of implementations. Their principle consists of a rotating ultrasonic transducer, mounted on a tool that is pulled up

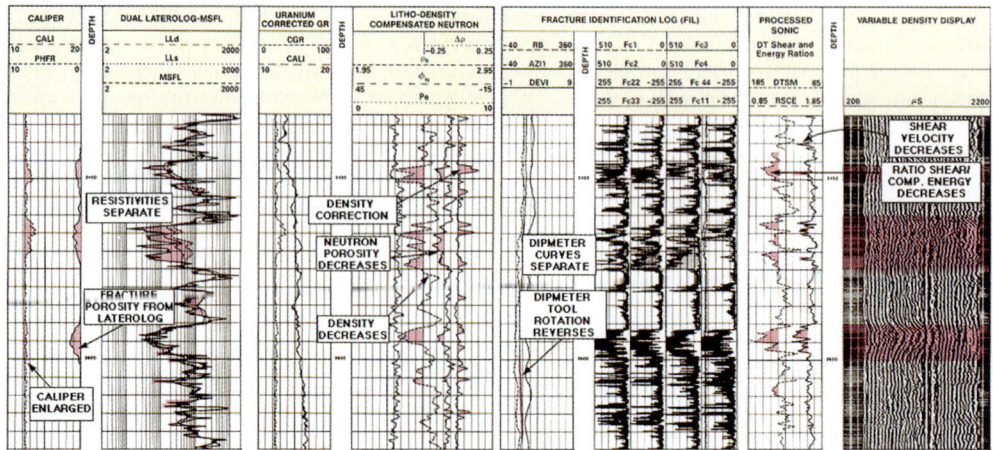

Fig. 3. A suite of conventional well logs through a basement reservoir in Egypt. None of the logs is a direct indicator of fractures, but each is, to some degree, affected by the presence of fractures (shaded areas). If at a given depth several such indicators are present, the probability of a fracture to occur is considered high. From left to right the tracks show the following logs: caliper; resistivity logs; natural gamma ray; density/neutron logs; dipmeter curves including auxiliary measurements on left; sonic velocities; sonic waveforms (from Schlumberger 1984).

the borehole, emitting short acoustic pulses in the direction of the borehole wall. The same device, which acts as a transmitter/receiver, measures their reflections, and the transit time as well as the amplitude of the reflected signal is recorded versus depth and azimuth. A map of the amplitude is then made on a co-ordinate system of depth versus azimuth. These planar projections of borehole wall scans, or borehole images, are nowadays widely used by geologists, because they show bedding and structural features – including fractures – in great detail. The dip and azimuth of these features can be measured, providing valuable information for building a reservoir model (Luthi 2001).

Ultrasonic imaging was later confronted with heavy competition by a group of electrical borehole imaging devices. Among these, the Formation MicroImager (FMI, Safiniya *et al.* 1991), has the capability of identifying fractures as thin as a few micrometres (10^{-3} millimetres) with a resolution of one-half centimetre. Fractures are generally filled with drilling fluid in the vicinity of the borehole and, if the mud is water based, exhibit much lower resistivities than the usually tight rock matrix surrounding them. The excess current injected into the fracture is a function of the fracture width, for which an inversion scheme has been developed (Luthi & Souhaité 1990). The method is also sensitive to differences in layer resistivities, and usually shows the bedding in more detail than the ultrasonic borehole images. This is because of the diffusive, penetrating nature of the

technique, in contrast to the reflection method that depends on the geometry of the borehole surface. The tool contains 192 small electrodes mounted on eight pads that are pressed against the borehole wall while the tool is pulled up-hole. Electrodes record variations in the local resistivities of the formation immediately in front of them. Other electrical imaging devices were developed to obtain lower-resolution images further away from the borehole, and in logging-while-drilling mode (Davis *et al.* 1992; Bonner *et al.* 1994; Rohler *et al.* 2001).

In wells with clear drilling fluids, such as research and some mining wells, optical borehole imaging can be made with video cameras adapted for down-hole conditions. Video systems are available using a fibre-optic cable that can transmit live videos as well as still pictures to the surface at a very high rate. The resolution of these images can be much higher than the two methods mentioned above, but application is limited in the oil industry because the drilling muds are generally opaque (Whittaker & Linville 1996).

Other borehole measurements

An important contribution to fracture analysis comes from the Stoneley wave measurement of array sonic logging tools. This acoustic borehole mode travels at comparatively low speeds, but with high amplitudes, and is affected by changes in the borehole volume. Thus, in the presence of a fracture above the tool, the upgoing Stoneley wave partially enters the fracture and is reflected back into the borehole, where one part travels downwards to the array of sonic receivers. This characteristic signal – which resembles reflections in vertical seismic profiling – can be extracted, and its amplitude has been shown to be a function of the fracture width (Hornby *et al.* 1989). Borehole washouts and other surface irregularities may also cause Stoneley wave reflections, and the technique is therefore best used in combination with borehole imaging tools (Hornby *et al.* 1992).

The combination of these tools can provide the accurate locations of fractures, their dips and azimuths, as well as their local apertures. In practice, it has been found that numerous types of fractures and fracture-related features occur in boreholes. These include drilling-induced features such as breakouts caused by shear failure in places of maximum stress concentration, and tensional (hydraulic) fracturing at the borehole wall. Additionally, in most fractured reservoirs, a variety of different fractures from separate tectonic phases may occur. These can often be distinguished based on their geometry and morphology on borehole images.

Once a fracture of interest is identified, a wireline formation tester with a dual packer module can be placed in front of a particular fracture, and its permeability can be examined with pressure draw-down of build-up tests (Fig. 4). Additionally, fluid samples can be taken to investigate whether the fracture system is oil bearing, a notoriously difficult task. The method can be extended to drill-stem tests, whereby an entire interval with fractures of interest can be probed in a similar manner. This approach of directly identifying fractures and testing them individually or in groups is a considerable improvement over earlier methods, and it leads to a significantly better understanding of the fracture system and its fluid content.

Application to basement reservoirs in Vietnam

These Vietnam basement reservoirs are located on the continental shelf of Southern Vietnam, in block-faulted basement highs of the Mekong and the Nam Con Son basins. The water depth reaches 120 metres, and the basement structures are generally encountered at depths of 2500 to 4000 metres. The basement rocks are predominantly granodiorites to granites and diorites, and they range in age from Late Jurassic to Early Cretaceous. According to Areshev *et al.* (1992) and Tandom *et al.* (1997, 1999) the geological history can be summarized as follows:

(1) End Jurassic–Early Cretaceous: acidic magmas intruded as granitoid batholiths into Jurassic and older sedimentary rocks. Subsequent uplifting and cooling led to the formation of contraction voids.

(2) Middle Cretaceous: intensive tectonization led to batholith fracturing and formation of cataclasites and mylonites, through which hydrothermal water circulated.

(3) End Cretaceous–Early Palaeogene: extensional faulting in a WNW–ESE direction led to the formation of lithospheric blocks with a mountainous palaeo-relief. Erosion and weathering took place on the highs of the granitoid bodies.

(4) Middle–End Palaeogene: regional subsidence occurred, and basic magmas were emplaced at the surface. Weathering continued on topographic highs, while lows were filled with terrigenous sediments that include lacustrine shales.

(5) Neogene–Pleistocene: after a short regional uplift, marine transgression set in, with the deposition of a thick wedge of sediments. In the Middle Miocene a regional NNE–SSW compression caused wrench-related fracturing in the basement. In the Late Miocene and Early Palaeocene, oil maturation of Oligocene organic shales and subsequent migration filled the basement highs. The present-day maximum horizontal stress is in the NW–SE direction in an essentially extensional regime.

The main porosity types in the White Tiger field (Mekong Basin) and the Big Bear field (Nam Con Son Basin) are fractures, caverns, leaching pores and possibly contraction voids. In the White Tiger field, the oil column exceeds one kilometre, and production is in excess of 10 000 barrels per day.

Tandom *et al.* (1997, 1999) confirm that the weathering crust is important, as most of the production in the Ruby field comes from the uppermost part of the basement where vugs and solution-enhanced fractures are most common. Additional production comes from Oligocene and Miocene arkosic sandstone overlying the basement. Interestingly, they note that when the matrix porosity in the basement rock exceeds 4%, the production seems to deteriorate. They believe that increased weathering led to an increase in clay and zeolite formation then filled the fractures and thus reduced the reservoir quality. Development wells are slanted and drilled in a direction orthogonal to the maximum horizontal stress, which is SE–NW. This strategy seems to work, because most of

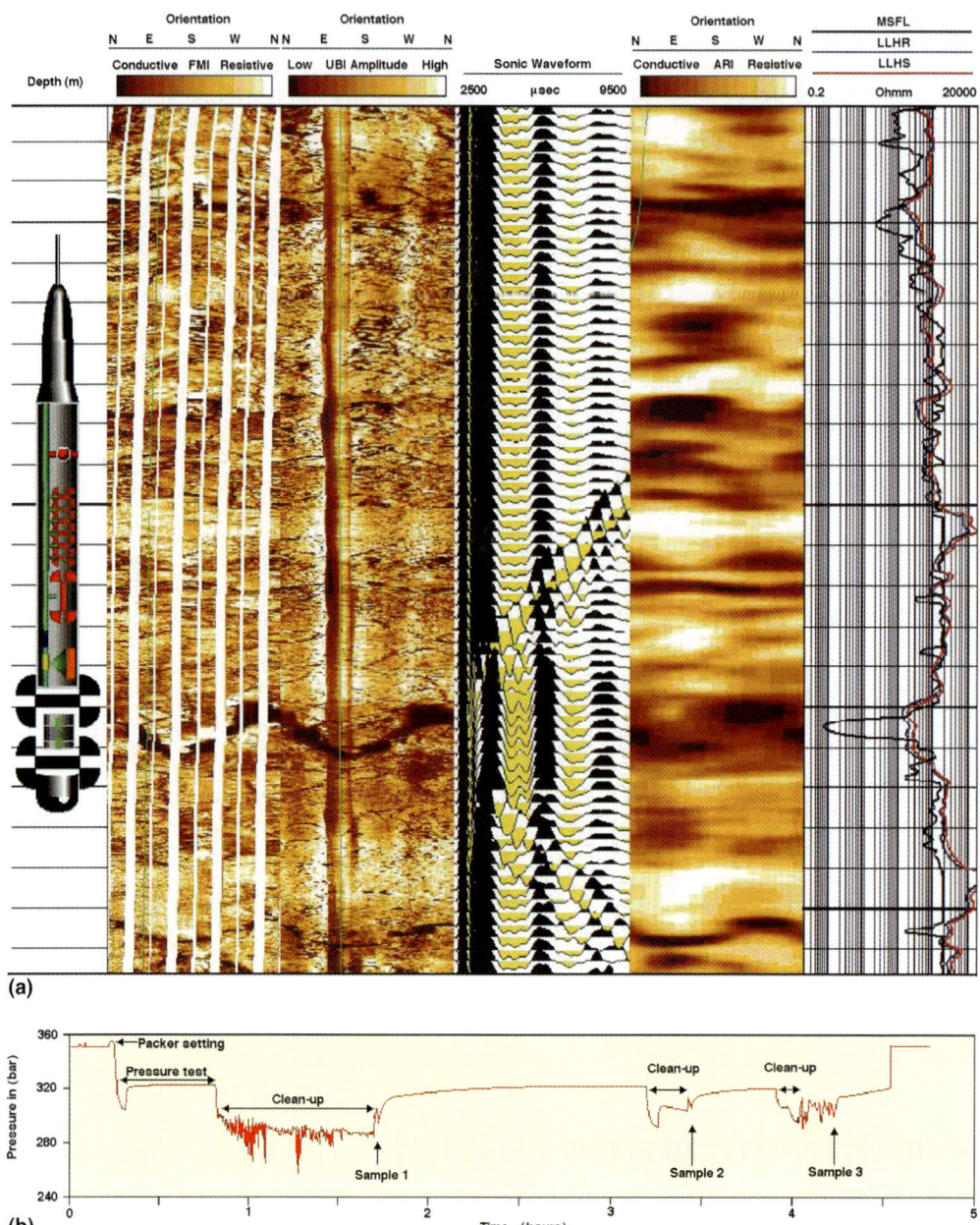

Fig. 4. (a) A new suite of borehole measurements specifically designed for fracture characterization. From left to right the tracks show the following measurements: FMI electrical borehole images; UBI ultrasonic borehole images; sonic waveforms showing the Stoneley wave direct arrivals and reflections; ARI deep-resistivity borehole images; dual laterolog and microspherically focused log. The tool shown on the left is a modular dynamic formation tester with two packers that can isolate a fracture and perform a series of pressure tests on it as well as sample the formation fluid in it. (b) Example of a series of operations performed by the modular dynamic formation tester to determine the formation fluid pressure and the permeability of the fracture, and to take fluid samples. The clean-ups serve to remove fluids in the invaded zone. The largest sample is taken at the end, and has a volume of about 4 litres.

the wells are reported to produce between 1500 and 3000 barrels (2.8 to 5.5 litres per second) of oil per day in the initial stages.

Tandom *et al.* (1999) also analysed FMI images and Stoneley wave reflection data, and found a variety of different fracture types. They concluded that productive intervals are characterized by a strong Stoneley response, with fracture apertures in excess of one millimetre, by vuggy fractures and visible vugginess on the FMI images, and by matrix porosities between 2% and 4%. The term 'vuggy fractures' is used for fractures that have been affected by leaching, which happened most likely during the weathering phase. These authors also related the well data to directional patterns on the seismic incoherence data, and from this interpreted the likely regional distribution of the fractures.

Figure 5 shows data from another basement well in the same basin, in a setting that is in most aspects similar to the one described above (Luthi 2001). The data include Stoneley waveforms, electrical borehole images (FMI), the ultrasonic borehole images (UBI), and the deep resistivity borehole images (ARI). The example discussed here is from a granitic interval in the water zone of this well – which is vertical – and it serves to illustrate the features typically observed:

- Hydraulically induced fracturing of the borehole wall (A), caused by excessive mud weight. Tensional failure of the rock occurred in the direction of maximum horizontal stress (i.e. perpendicular to the

minimum horizontal stress). This direction is seen to be SE–NW, and is thus in line with the regionally observed maximum stress direction mentioned above. One notices that these induced fractures are generally vertical, but in places seem to deviate in short wisps. This happens where foliations or microfractures weaken the rock in a preferred direction, which the induced fractures locally follow.

- Breakouts of the borehole wall (B). These local enlargements are caused by shear failure in directions where the tangential stress on the borehole wall exceeds the shear strength of the rock (Zoback *et al.* 1985; Bell 1990). Such local stress concentrations are caused by a large deviatoric stress (the difference between the maximum and minimum horizontal stress) and occur at azimuths corresponding to the minimum stress direction, which in this case is in the NE–SW direction and at 90° to the maximum stress direction obtained above. The interval where breakouts are observed has probably concentrated the regional stresses more than the under- and overlying sections. Notice that the ultrasonic image shows the breakouts better than the electrical image, because, as a reflection method, it is more sensitive to the surface geometry of the borehole.

- Foliations with delaminations (C). Such foliations are common in most basement rocks, and are usually caused by local concentrations of aligned mica or other platy minerals. These not only cause a conductivity

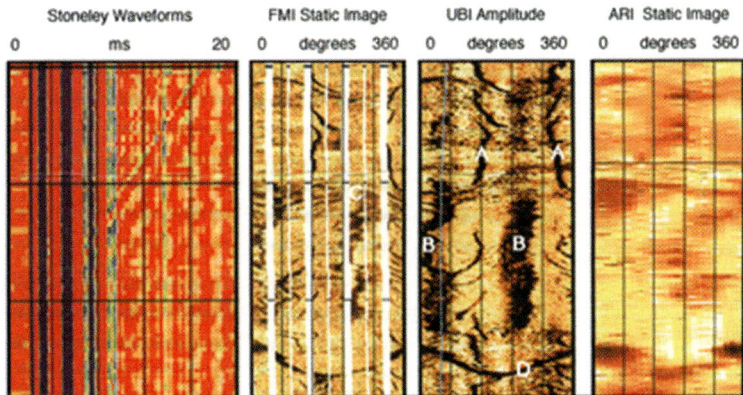

Fig. 5. Stoneley waveforms, FMI, UBI and ARI borehole images over an interval covering 15 metres from a well into granodioritic basement rock, Vietnam. Prominent features include induced fracturing (**A**), borehole breakouts, (**B**), foliations with delaminations (**C**), and tectonic fractures (**D**). Induced fracturing is generally tensional failure caused by high mud-weight, while breakouts are caused by shear failure in compressionally stressed zones. Both indicate a maximum horizontal stress in the NW–SE direction.

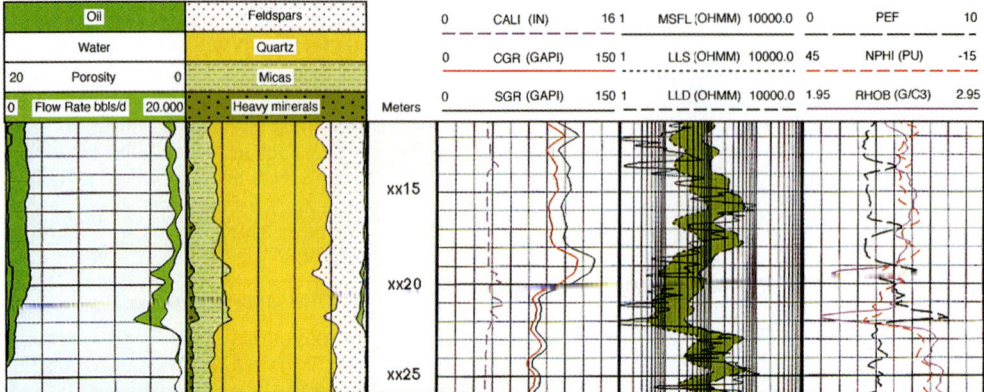

Fig. 6. A complete logging suite over the lowermost oil-producing interval of a basement well in Vietnam. From left to right, the tracks include the following measurements: flow rates from production log and porosity as well as oil saturation; mineralogical analysis based on open-hole logs; caliper and gamma-ray log; shallow and deep resistivity logs; density, neutron porosity, and photo-electric absorption factor; sonic logs; Stoneley wave reflection coefficient and energy loss; Stoneley waveforms; FMI borehole images; UBI borehole images; and ARI borehole images. Prolific production comes from several fractures seen on the borehole images, most prominently on the UBI.

contrast measurable by the imaging tools, but they also constitute points of weakness that induced fractures sometimes follow (see A).

• Natural fracturing (D) caused by tectonic forces is usually characterized by a planar signature (and thus a sinusoid on these planar projections of the borehole) that cross-cuts the entire well-bore. Both the FMI as well as the UBI image show this fracture clearly, and the deep-resistivity image of the ARI illustrates that this fracture extends much farther away from the wellbore than the induced fractures and the breakouts, both of which are not seen by the ARI. A weak Stoneley wave reflection from this feature can be seen, suggesting that it is an open fracture.

From this analysis it appears that the well was drilled with a mud that was sufficiently heavy to cause tensional failure of the borehole wall at certain depth intervals. The deviatoric stress is high and caused shear failure on the borehole wall at other intervals that probably differ in lithology and rock strength from those where hydraulic fracturing occurred. Knowing the overburden stress, the mud weight and the rock strengths, the downhole horizontal stresses can be evaluated with an adequate geomechanical model and suitable failure criteria. The rock shows natural tectonic fractures that seem to be open, and whose dip, azimuth and aperture can be determined.

For the second interval discussed here (Fig. 6) we show the complete suite of logs, including all borehole images. The logs have been processed to obtain the porosities and fluid saturations, shown on the leftmost column together with open-hole flow rates measured with the spinner of the production logging tool PLT run during a production test in barefoot completion. This interval covers about 15 metres and is located at the first oil entry into the well. The flow rates increase, over an interval of about 5 metres, from zero to over 2000 barrels per day. The average porosity is about 2%, with a peak of slightly above 5%. Inspection of the borehole images on the right-hand three tracks suggests that the flow comes from a series of perhaps five parallel fractures. These are particularly well seen on the UBI images, but the FMI images could be processed to enhance the fractures. They dip towards NNW at about 70°, and show strong Stoneley wave reflections. The caliper log shows some borehole enlargements, and the density and photoelectric factor logs show strong excursions caused by the baryte-doped mud invading the fractures. The response of the other open-hole logs is much subtler, and it would be difficult to gain further insights from them regarding the fracturing system. The mineralogical composition, shown in the second column from the left, was calculated from these logs using a mineral model that includes heavy minerals, mica, quartz and feldspars. At the depth of the fluid entry, an increase in the amount of micas (or perhaps clays, which are

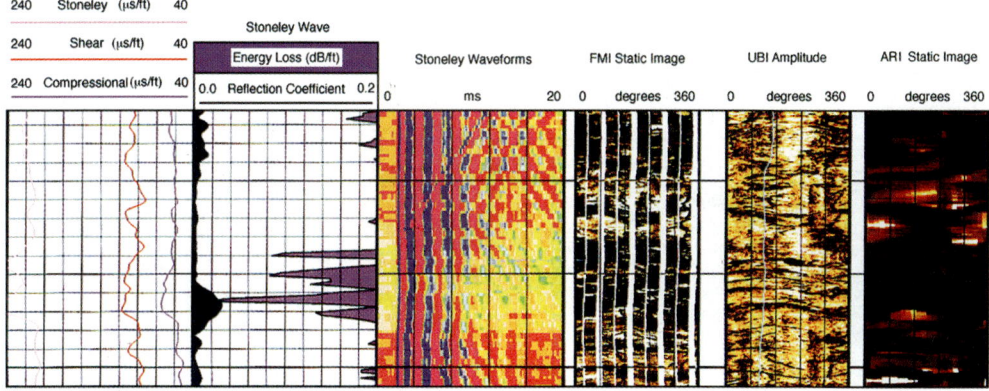

Fig. 6. (cont.)

indistinguishable from each other with this suite of logs) is seen, which may indicate some amount of weathering due to fluid circulation in the fracture system at an early stage.

Figure 7 shows an interval higher up in the same well, where the cumulative flow rate is already substantially higher. Only a subset of the data is shown, including production flow rates, the Stoneley wave data, and the borehole images. The flow profile shows a very localized increase by about 3000 barrels per day, which comes ostensibly from a fracture (highlighted by an arrow on the UBI images) that dips at 70° towards the NNW. It may be connected with a similar fracture deeper down in the well through one or more fractures that dip in the opposite (SSE) direction. The fracture shows considerable damage on the borehole wall, as seen by the dark patches on FMI and UBI images, an effect perhaps caused by the intersection of the two fracture systems. The strong

response of the Stoneley wave is at least partially attributed to this, and a fracture width can therefore not be computed. This fracture may indeed be a shear zone with a total width in the order of millimetres or even centimetres – considerably larger than the range for which an aperture from the FMI can be computed. It is an example of a point entry for oil production, but it is difficult to assess why only one of the fractures is producing.

The last interval discussed here is shown in Figure 8. It is in the uppermost part of the well, but still below the weathered basement. The same reduced number of measurements as in the previous example is shown. In the middle of the interval, the flow rate is seen to increase by almost 2000 barrels per day, and this corresponds to a single fracture (highlighted by an arrow on the UBI image) that dips steeply towards the east. A pronounced peak of the Stoneley wave reflection in the uppermost part

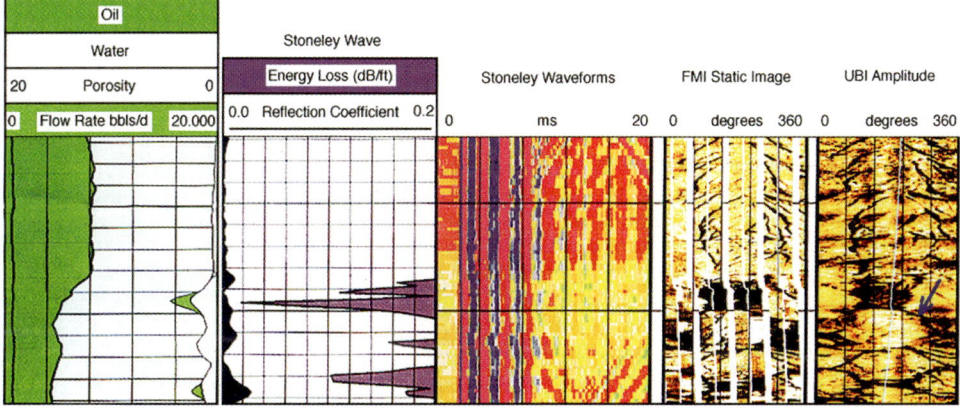

Fig. 7. Another productive interval, shown with only a subset of all measurements. Prolific production is seen to come from a single fracture shown by the arrow on the UBI borehole images.

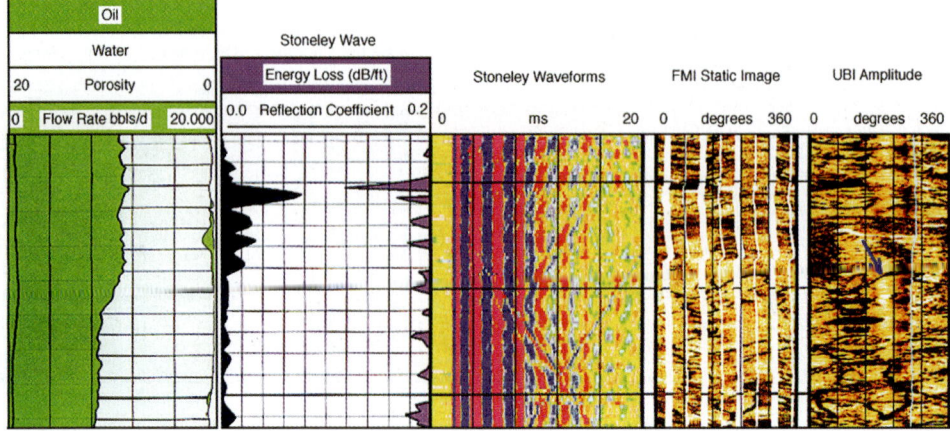

Fig. 8. The most productive interval in this case study well. Production is seen to come from a single fracture (arrow). The strong Stoneley wave reflection higher up in the well is attributed to a borehole washout, not a fracture.

of the interval corresponds to a local borehole enlargement, as evidenced by the caliper (not shown). This seems to be associated with another fracture that is similar in geometry to the producing one, but that is tight, as the flow rate does not increase at that depth. The productive fracture is again a single-entry point, with the major difference to the previous example being the dip direction of the fracture.

Fracture geometry and production

This well produced a total of about 12 000 barrels of oil per day during the time of the well test, with no water at all. This production comes from the topmost 100 metres of the reservoir, and, after our analysis, can be traced back to a few point entries (from the production log) that correspond to fractures (from the borehole images). In order to build a reservoir model that can serve as a basis for a field development plan, the relationship of production versus fracture geometry needs to be analysed. The flow rates for each producing fracture are plotted versus fracture dip and azimuth in Figure 9. Although there is considerable scatter and relatively few data points, two clusters of productive fractures seem to be discernible, with average strikes of ENE–WSW, and SSE–NNW respectively. It is not possible to relate them directly and unambiguously to one of the tectonic phases mentioned earlier, and it is quite probable that they have been reactivated, perhaps even several times. One possible interpretation is that they are conjugate shear fractures formed during the Mid-

Miocene NNE–SSW compression, perhaps superposed on to pre-existing fractures from the

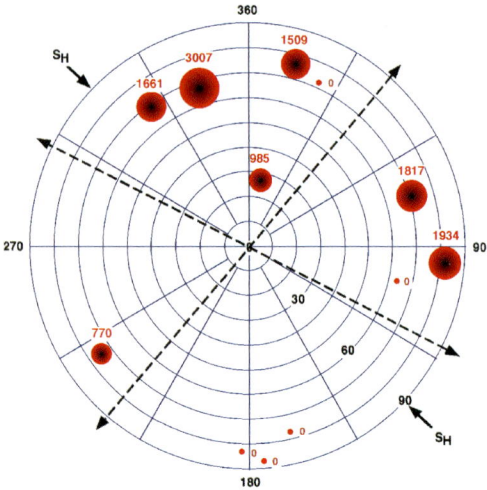

Fig. 9. A radial plot of the dip and azimuth of prominent fractures in the case study well of a basement reservoir in Vietnam. Production (in barrels per day) is indicated by the numbers as well as the size of circle area, which is proportional to production. Production rates have been obtained by averaging five different production log runs. Despite a considerable scatter, two fracture sets seem to be present, one dipping towards NNW, the other towards the east. This may cause a large-scale interconnectedness of the fractures, explaining their prolific production. Dashed lines indicated two possible drilling directions for a slanted well, which could increase production and decrease the chance for early water breakthrough. Maximum present horizontal stress direction is indicated by S_H.

pre-Miocene extensional phase. Leaching may have led to irregular fracture surfaces that prevented them from completely closing at later stages. The dips of these fractures typically lie between 60 and 80°. These relatively steep dips explain why relatively few fractures were encountered in the well, which was drilled vertically and thus had a smaller probability to penetrate steeply dipping fractures than a horizontal well would have had. The variety in fracture geometry implies that the likely explanation for the prolific production of this well is the interwoven network that the fractures form, providing the principal production conduit but also the reservoir. A rough assessment of the reservoir volume contained in the fracture system of this fault block can be based on an average porosity of 3%, a block size of 1.5 × 3.0 km and an oil column of 1 km. The pore volume then amounts to almost 900 million barrels. If most of the pore space is interconnected and water saturations are negligible, the recoverable reserves from such a block can exceed half a billion barrels and sustained high flows can be achieved, as is observed in some of the basement reservoir fields mentioned earlier. Talus sediments on the fringes of the basement blocks or clastics overlying them might contain additional reserves. However, water breakthrough is a problem in fractured reservoirs, and water entries are likely to be as localized as oil entries are. The capability of selective water shut-off is therefore important, as is fluid flow monitoring to determine the type of fluid and the exact entry depths. Deviated or horizontal wells offer longer producing intervals and lower draw-down pressures, which reduces the chance of water breakthrough. In order to be efficient, these wells have to be drilled in a direction that maximizes the amount of fractures intersected, which, for the case of a single fracture system is orthogonal to it (Tandom *et al.* 1997, 1999). In our case there are two main fracture sets, and an interesting option might be to drill a slanted or horizontal well in the directions exactly halfway between the fracture strikes, thereby penetrating both fracture sets at equal angles. These two well directions are indicated in Figure 9. Analysis of data collected in such wells can then provide additional refinement of the structural–geological model and further improvement in well design.

Conclusions

Modern down-hole geophysical techniques that include borehole imaging, Stoneley wave reflections, production data, and 3D seismic surveys can considerably improve the characterization of fractured basement reservoirs. These techniques far surpass the conventional analysis of standard logs. A case study from a highly productive well in Vietnam shows that the oil flows into the wellbore mainly through point entries, which upon inspection of the borehole images are identified as single or multiple fractures. They vary in dip and azimuth, but can be grouped into two sets that can be related tentatively to a compressional tectonic event. Induced hydraulic fracturing of the well-bore is also observed, but does not contribute to production. Since the two sets interconnect, the fracture system is likely to be an interconnected network, which may explain the high sustained flow rates. Deviated drilling can be used to enhance well productivity by intersecting an optimum number of fractures using a well trajectory based on this geological model. This will also reduce the draw-down pressure and therefore the chance of early water breakthrough, a typical problem encountered in fractured reservoirs. Fluid flow monitoring and the ability for selective shut-off of parts of the well are additional methods to guarantee efficient production.

The help of Bill Borland, Tom McDonald, Brian Hornby and Katsuei Saito is greatly appreciated. FMI, UBI, ARI and PLT are trademarks of Schlumberger.

References

AGUILERA, R. 1995. *Naturally Fractured Reservoirs.* Pennwell, Tulsa Oklahoma.

ARESHEV, E. G., DONG, L. E., SAN, N. T. & SHNIP, O. A. 1992. Reservoirs in fractured basement on the continental shelf of Vietnam. *Journal of Petroleum Geology*, **15** (4), 451–464.

BELGASEM, B. A., BARLAI, Z. & REZ, F. 1990. Interpretation of well logs in the oil-bearing reservoir, Nafoora–Augila Field, Libya. *Society of Professional Well Log Analysts, Proceedings of 13th European Formation Evaluation Symposium*, paper OO.

BELL, J. S. 1990. Investigating stress regimes in sedimentary basins using information from oil industry wireline logs and drilling records. *In*: HURST, A., LOVELL, M. A. & MORTON, A. C. (eds) *Geological Application of Wireline Logs.* Geological Society, London, Special Publications, **48**, 305–325.

BONNER, S. & BAGERSH, A., *et al.* 1994. A new generation of electrode resistivity measurements for formation evaluation while drilling. *Society of Professional Well Log Analysts, Proceedings of 35th Annual Symposium*, paper OO.

CONEY, D., FYEE, T. B., REATIL, P. & SMITH, P. J. 1993. Clair appraisal: the benefits of a cooperative approach. *Proceedings, 4th Conference of Petroleum Geology of Northwest Europe*, 1409–1420.

DAVIES, D. H., & FAIVRE, O. 1992. Azimuthal resistivity imaging: a new generation laterolog. *Society of Petroleum Engineers, 67th Annual Conference*, paper **24676**.

EGGLESTON, W. S. 1948. Summary of oil production from fractured rock reservoirs in California. *American Association of Petroleum Geologists, Bulletin*, **32/7**, 1352–1355.

FARAGUNA, J. K., CHACE, D. M. & SCHMIDT, M. G. 1989. An improved borehole televiewer system – image acquisition, analysis and integration. *Society of Professional Well Log Analysts, Proceedings of 30th Annual Symposium*, paper UU.

GEOSCIENCE LTD 2000. Hydrocarbon production from fractured basement formations. Worldwide Web Address: *http://www.geoscience.co.uk*.

GOLD, T. & SOTER, S. 1980. The deep-earth gas hypothesis. *Scientific American*, **242/6**, 130–137.

HAYMAN, A. J., PARENT, P., CHEUNG, P. S. & VERGES, P. 1998. Improved borehole imaging by ultrasonics. *Society of Petroleum Engineers, Production & Facilities, February 1998*, 5–13.

HORNBY, B. E., JOHNSON, D. L., WINKLER, K. W. & PLUMB, R. A. 1989. Fracture evaluation using reflected Stoneley-wave arrivals. *Geophysics*, **54**, 1274–1288.

HORNBY, B. E., LUTHI, S. M. & PLUMB, R. A. 1992. Comparison of fracture apertures computed from electrical borehole scans and reflected Stoneley waves: an integrated interpretation. *The Log Analyst*, **33** (1), 50–66.

HUBBERT, M. K. & WILLIS, D. G. 1955. Important fractured reservoirs in the United States. *Proceedings, 4th World Petroleum Congress, Rome, section I/A-1*, 57–81.

LANDES, K. K., AMORUSO, J. J., CHARLESWORTH, L. J., HEANY, F. & LESPERANCE, J. 1960. Resources in basement rock. *American Association of Petroleum Geologists, Bulletin*, **44** (10), 1682–1691.

LUTHI, S. M. 1983. Shear zones indicated by a dipmeter survey in the Gulf of Suez, Egypt. *International Basement Tectonics Association, Proceedings of 5th International Conference on Basement Tectonics*, 167–173.

LUTHI, S. M. 2001. *Geological Well Logs – Their Use in Reservoir Modeling*. Springer-Verlag.

LUTHI, S. M. & Souhaité, P. 1990. Fracture apertures from electrical borehole scans. *Geophysics*, **55**, 821–833.

MCNAUGHTON, D. A. 1953. Dilatancy in migration and accumulation of oil in metamorphic rock. *American Association of Petroleum Geologists, Bulletin*, **37** (2), 217–231.

NELSON, R. A. 1985. *Geologic Analysis of Naturally Fractured Reservoirs*. Gulf Publishing Co.

NORTH, F. K. 1990. *Petroleum Geology*. Unwin Hyman (2nd edition), Boston, MA.

P'AN, C. H. 1982. Petroleum in basement rock. *American Association of Petroleum Geologists, Bulletin*, **66/10**, 1597–1643.

PORFIREV, V. B. 1974. Inorganic origin of petroleum. *American Association of Petroleum Geologists, Bulletin*, **58** (1), 3–33.

ROHLER, H., BORNEMANN, T., DARQUIN, A. & RASMUS, J. 2001. The Use of real-time and time-lapse logging-while-drilling images for geosteering and formation evaluation in the Breitbrunn Field, Bavaria, Germany. *Society of Petroleum Engineers, 76th Annual Conference*, paper **71331**.

SAFINIYA, K. A., LE LAN, P., VILLEGAS, M. & CHEUNG, P. S. 1991. Improved formation imaging with extended microelectrical arrays. *Society of Petroleum Engineers, 66th Annual Conference*, paper **30550**.

SCHLUMBERGER, 1984. *Egypt Well Evaluation Conference*. Schlumberger Middle East S. A., 270p.

SIBBIT, A. M. 1995. Quantifying porosity and estimating permeability from well logs in fractured basement reservoirs. *Society of Petroleum Engineers, Conference 1995*, Ho Chi Minh City, Vietnam, paper **30157**.

SMITH, J. E. 1956. Basement reservoir of La Paz-Mara Fields, Western Venezuela. *American Association of Petroleum Geologists, Bulletin*, **40** (2), 380–385.

STEARNS, D. W. & FRIEDMAN, M. 1972. Reservoirs in fractured rock. *American Association of Petroleum Geologists, Memoir*, **16**, 82–100.

TANDOM, P. M., NHUAN, T. X., TIJA, H. D. & SPAGNUOLO, S. A. 1997. Fractured basement reservoir characterization, offshore Vietnam. *Society of Professional Well Log Analysts, Asian Conference in Japan*, paper XXX.

TANDOM, P. M., NGOC, N. H., TIJA, H. D. & LLOYD, P. M. 1999. Identifying and evaluating producing horizons in fractured basement. *Society of Petroleum Engineers, Conference on Asia Pacific Improved Oil Recovery*, paper **57324**.

WHITTAKER, J. L. & LINVILLE, G. D. 1996. Well preparation – essential to successful video logging. *Society of Petroleum Engineers, Transactions of 66th Western Region Meeting*, paper **35680**, 297–308.

YOUNES, A. I., ENELGER, T. & BOSWORTH, W. 1998. Fracture distribution in faulted basement blocs: Gulf of Suez, Egypt. *In*: COWARD, M. P., DALBATAN, T. S. & JOHNSON, H. (eds) *Structural Geology in Reservoir Characterization*. Geological Society, London, Special Publications, **127**, 167–190.

ZAHRAN, I. & ASKARY, S. 1988. Basement reservoir in Zeit Bay oilfield, Gulf of Suez. *American Association of Petroleum Geologists, Bulletin*, **72** (2), 261–280.

ZEMANEK, J., CALDWELL, R. I., GLENN, E. E., HOLCOMB, S. V., NORTON, I. J. & STRAUS A. J. D. 1969. The borehole televiewer – a new logging concept for fracture location and other types of borehole inspection. *Journal of Petroleum Technology*, **21**, 762–774.

ZOBACK, M. D., MOSS, D., MARTIN, L. & ANDERSON, R. N. 1985. Wellbore breakouts and *in-situ* stress. *Journal of Geophysical Research*, **90**, 5523–5530.

Fracture mapping with electrical core images

M. LOVELL[1], P. JACKSON[2], R. FLINT[3] & P. K. HARVEY[1]

[1]*Department of Geology, University of Leicester, Leicester, LE1 7RH, UK*
(e-mail: mike.lovell@leicester.ac.uk)
[2]*British Geological Survey, Keyworth, Nottingham, NG12 5GG, UK*
[3]*Department of Aeronautical and Automotive Engineering,*
Loughborough University, LE11 3TU, UK

Abstract: Naturally fractured reservoirs often contain a range of different fracture types and networks; fractures that are relatively permeable and relatively impermeable, unconnected and connected to the part of the fracture network that carries fluid flow, and naturally occurring or drilling induced. Consequently, in terms of their fluid connectivity, fractures may be open or closed, while individual fractures may be isolated or well connected.

We have adapted our approach to imaging sedimentary fabric in the laboratory, where we related electrical core images to properties such as porosity, permeability, grain size and cementation, to enable electrical imaging of fractures in core. Our approach uses similar principles to those employed in down-hole electrical imaging. The results demonstrate an ability to image conductive fractures in fully saturated low-porosity water-bearing core: these fractures being electrically connected from the flat measurement surface through to the outer surface of the core.

Published results for numerical modelling of down-hole electrical imaging tools show the electrical response is related to fracture depth and fracture aperture. Our experimental results on fractured core in the laboratory support these numerical observations, increased current flowing into the fracture as the aperture increases. The finite size of the electrode, however, means that this technique cannot distinguish between a single fracture and smaller groups of fractures intersecting the electrode.

Down-hole electrical images are used to assess the nature, variability and distribution of subsurface formations. The images can be used to describe and delineate sedimentary features; to provide input into depositional models; to describe and quantify fracture occurrence and orientation; to identify and distinguish breakouts and hydraulic (induced) fractures; and to provide indications of local variability and heterogeneity in terms of porosity and permeability variations (with calibration to core). In parallel to these down-hole images, detailed laboratory studies have provided improved understanding of the variability of petrophysical properties at the core and pore scale. These variations are often related to the fine-scale sedimentary (depositional and/or diagenetic) and structural (gross geometry and stress related) fabric, although stratigraphic descriptions based on visual (optical) observations do not always agree with petrophysical parameters. The use of electrical resistivity measurements to generate down-hole images from which geological characteristics can be interpreted is now well established. This technique is based on the underlying principle that electrical flow in rocks takes place predominantly through electrolytic conduction in the pore fluid, with the individual matrix components acting as relative insulators. Thus low resistivity (high conductivity) is often seen to correspond to increased porosity (Archie 1942). One exception to this can be the presence of clays when an excess conductivity is observed above that contributed solely by the fluid-filled porosity of the rock. This excess clay-conductivity reduces the resistivity further. The effect occurs in shaly sands (Worthington 1985) and declines as the salinity (and hence conductivity) of the pore fluid increases and the fluid becomes the dominant conductor.

The pore fluid can be contained within the primary (matrix) pore space, secondary (diagenetic) pore space, or within the confines of fractures. Regardless of location, however, the effect on the resistivity response remains consistent, with increasing porosity reducing the electrical resistivity of the rock for water-saturated formations. Archie (1942) described a trend between porosity

From: HARVEY, P. K., BREWER, T. S., PEZARD, P. A. & PETROV, V. A. (eds) 2005. *Petrophysical Properties of Crystalline Rocks*. Geological Society, London, Special Publications, **240**, 107–115.
0305-8719/05/$15.00 © The Geological Society of London 2005.

and electrical resistivity for sedimentary rocks with an exponent '*m*' defined by the slope of the relationship on a log–log scale; this exponent is often described as a cementation factor, but actually relates to the shape of the pore space. Recent numerical modelling (Jackson *et al.* 2002) of different pore shapes supports Archie's observations of the variability of '*m*' with changing pore shape. In fractured rocks, however, the resistivity contrast between the fluid in an open fracture and the surrounding rock may be several orders of magnitude, and consequently open fractures are amongst the most prominent features seen on electrical images (Luthi & Souhaité 1990).

The down-hole electrical images are generated from simultaneous, multiple closely spaced micro-resistivity logs (Lloyd *et al.* 1986; Ekstrom *et al.* 1987; Boyeldieu & Jeffreys 1988). Figure 1 shows a typical, generic tool with one shaded button identified. The current from each button is passively focused into the formation through the use of an isolated surrounding conducting plate that is held at the same potential, and returns to a second electrode in the remote upper part of the tool. These data are presented as visual images that are optimized to respond to variations in the electrical conductivity of the rock, on the millimetre scale. These electrical images, while not equivalent to optical images, provide the geologist with an opportunity to view the subsurface formations in their *in situ* state (e.g. Bourke *et al.* 1989; Luthi 1990; Luthi & Souhaité 1990; Bourke 1992; Lovell *et al.* 1998; Prensky 1999; Cheung 1999;

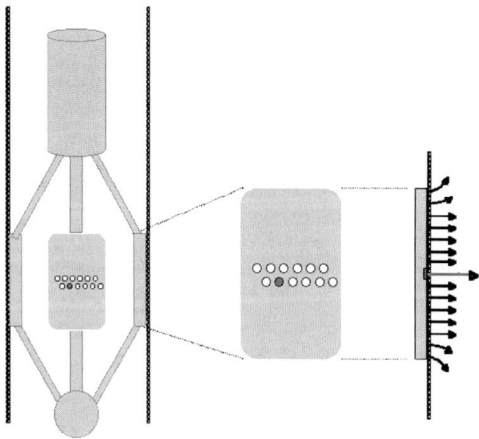

Fig. 1. Principles of a generic down-hole electrical imaging tool, showing passive focusing of electrical current at the pad into the formation.

Lofts & Bourke 1999). The problems of attempting to compare the quantitative values of conductance–depth traces from adjacent electrodes of the down-hole imaging devices are discussed in detail by Lovell *et al.* (1998). They concluded that, while the current flowing from each button-electrode can be used to produce images depicting geological structure, they do not necessarily represent quantitative estimations of formation resistivity.

Electrical core imaging

Electrical down-hole measurements are designed to produce borehole images that respond to geological characteristics such as rock structure and texture; quantitative comparison of these down-hole measurement traces must be done with caution (Jackson *et al.* 1998). In contrast, the development of electrical resistivity imaging for the quantitative characterization of cores using four-electrode contact measurement technology is well documented (e.g. Jackson *et al.* 1990, 1991, 1992; Lovell & Jackson 1991; Harvey *et al.* 1994; Lovell *et al.* 1994, 1997). The technique aims to produce quantitative resistivity data on core at a similar vertical and horizontal resolution to that of a down-hole imaging device. Thus, the heterogeneity within the rock, which can influence the electrical resistivity measurement at any particular scale, should affect both core and down-hole measurements in a similar manner. The core images may be used to quantitatively constrain the millimetre-scale structural changes seen on down-hole images and provide a means of defining the variability of the electrical resistivity structure of a formation. Furthermore, these core images should provide for the possible interpretation of down-hole images in terms of fine-scale petrophysical variation.

Electrical resistivity mapping of sedimentological and structural elements of core in the laboratory enables detailed study of the fine-scale variability of formations. This can include spatial heterogeneity or dynamic fluid-flow anisotropy of high-porosity sediments or low-porosity fractured reservoirs.

Laboratory measurements of electrical resistivity require relatively simple measurements and for investigations on rocks a four-electrode method is preferred and this approach is deemed inherently more accurate than two-electrode measurements (Rust 1952; Brace *et al.* 1965; Jackson *et al.* 1990). We have adapted the successful down-hole approach to electrical imaging, to provide similar core images on low-porosity fractured rock through laboratory

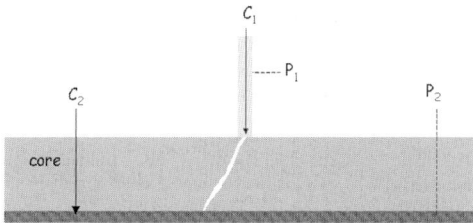

Fig. 2. Laboratory measurement technique simulating a down-hole electrical imaging tool measurement with a four-electrode array. Current is passed from electrode C_1 on the core to electrode C_2 connected to the metal split-core liner, and the potential difference is measured between electrodes P_1 and P_2.

measurements of electrical resistivity on core. This required a novel approach combining the advantages of four-electrode measurements

with the principles of the down-hole tool, with the core sample being resaturated under vacuum, and contained in a metallic split-core liner containing conductive formation fluid (Fig. 2). The technique uses precise control of the current flow, while voltage measurements are not degenerated by surface impedances and electrochemical effects. Using this approach it is possible to generate an electrical resistivity image of a core, as shown in Figure 3. In this image, the bright feature represents the electrical response to a conductive fluid-filled fracture in which the current flows from the electrode into the rock. The image covers only the central part of the sample surface. The core is a low-porosity crystalline marble, and contains several minor fractures visible on the photo, which run obliquely from the main fracture; these are less clearly visible on the electrical image, suggesting that they are partially closed,

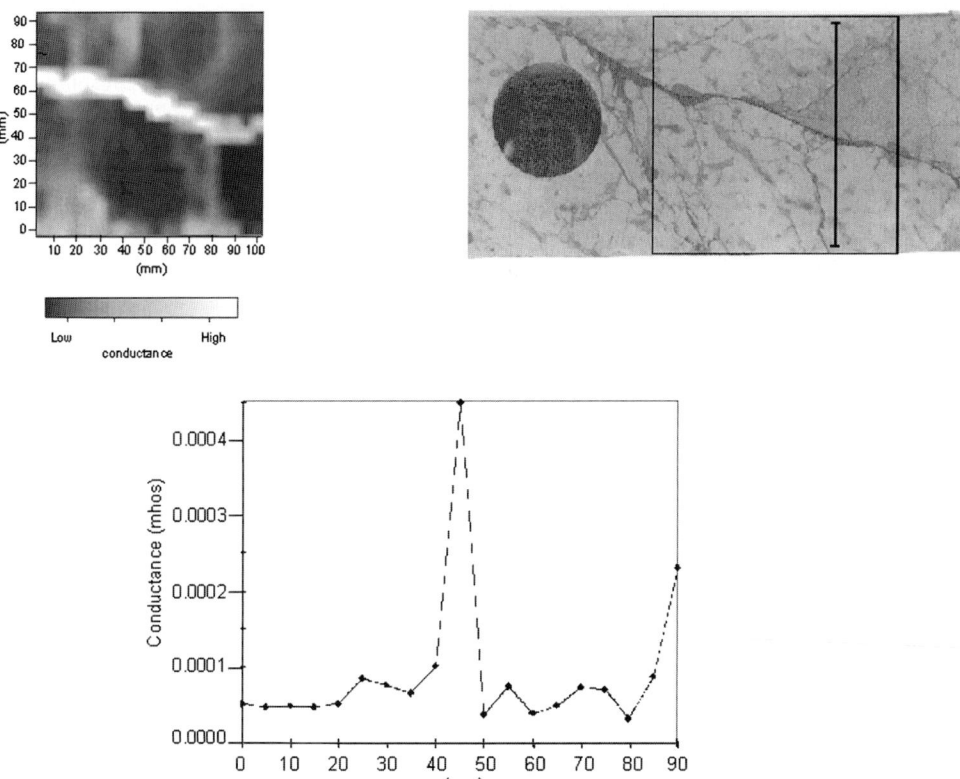

Fig. 3. Electrical resistivity core image and core photograph showing a dominant single fracture in a low-porosity marble; note that the shading used in these laboratory images is the inverse of that used down-hole, thus the conducting features occur as bright/light/white shades, while resistive features are assigned the dull/dark/grey shades. The location of the electrical resistivity profile across the single fracture is shown in the core photograph. Units of conductance are mhos.

or have a smaller aperture with depth profile, and, consequently, current does not flow preferentially through them. Data from the same image are plotted as a conductance profile across the fracture from left to right (Fig. 3). The main fracture is clearly identified as a conductive anomaly, demonstrating the ability of the core imaging technique to image conductive fractures in a low-porosity rock matrix.

Fracture characterization

Trice (1999) defines a naturally fractured reservoir as a large volume of rock and a network of fractures. Within the fracture system there will be relatively permeable and relatively impermeable fractures. There will also be fractures unconnected and connected to the part of the fracture population that carries fluid flow.

Laboratory characterization of fractures is conventionally done using visual mapping of fractures and fracture sets. These observations can be done manually using overlays, or through the use of high-resolution digital imaging. This approach provides detailed fine-scale mapping of the fractures present in the core, plus separation of fractures into different sets using specific criteria such as hand-specimen-scale description of the likely fracture infill and mineralogy. Such an approach is feasible for larger fractures of the order of several millimetres width, but is increasingly difficult for smaller fractures of less than a millimetre. Furthermore, visual description of fractures in core rarely is able to demonstrate the continuity of fractures through the core, or to assess the connectivity with other fracture (sets).

Borehole-wall electrical images can be used to identify, map and characterize fractures downhole where a contrast exists between the electrical resistivity of the fracture and the host rock. In low-porosity formations drilled with a conducting mud, low-resistivity (high-conductivity) fractures tend to be filled with conductive fluids or clay minerals, whereas high-resistivity (low-conductivity) fractures tend to be tightly cemented. The role of clays as a secondary conductance mechanism within the rock affects the analysis of shaly sands, and is described in the literature (e.g. Worthington 1985). The effect dominates in the presence of low-salinity pore fluids, and under these circumstances a clay-filled fracture would provide a positive conductance anomaly similar in appearance to an open fluid-filled fracture on an electrical resistivity image.

The appearance of a fracture on a down-hole electrical image relies on the contrast in resistivity between the fracture and the host rock; thus the background or matrix porosity of the host formation is critical. The vertical extent of the intersection of the fracture with the borehole is a function of the borehole diameter and the angle of dip of the fracture. *In situ* planar fractures will intersect the borehole as a sinusoidal feature, with the amplitude of the feature proportional to the dip of the fracture relative to the borehole inclination; in deviated wells the borehole inclination must be noted if fracture orientations are to be related to regional stress fields. A simple model of a planar dipping feature (Fig. 4) intersecting the borehole is thus often considered. While this may adequately model larger-scale fractures, it often fails to match frequently observed fine-scale fracture patterns observed in both outcrop and core. These detailed fracture patterns can be grouped on the basis of orientation, although they often exhibit more than one preferential orientation direction. Where electrical resistivity contrasts exist, these different orientations can be identified on down-hole images. The size of the resistivity feature on the down-hole electrical image cannot be directly related to the dimensions of the fracture, because the vertical resolution is controlled by the physical size of the button. A range of factors other than the fracture shape

Fig. 4. Simple fracture aperture models for horizontal and dipping fractures, shown in schematic core sample.

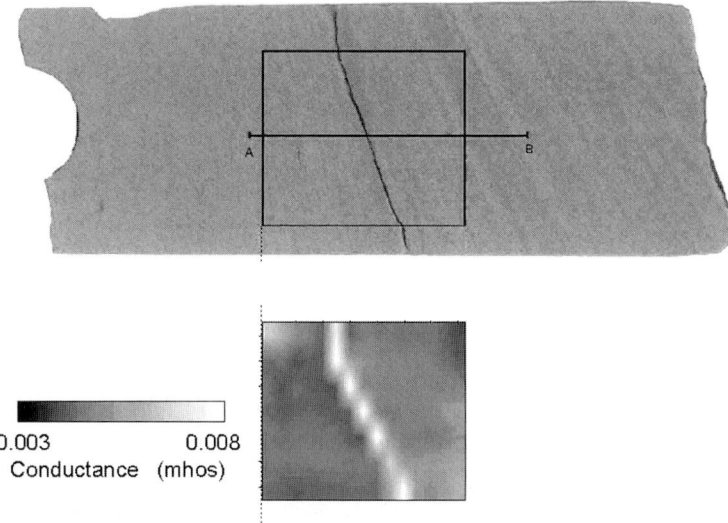

0.003   0.008
Conductance (mhos)

Fig. 5. Electrical image and core photograph for sandstone core with fracture at grain contacts.

and size controls the electrical resistivity response. The simple model of a planar dipping fracture (Fig. 4) can be used to calculate the fracture dimensions, particularly the aperture or width (Luthi & Souhaité 1990). This usually requires a range of assumptions to be made, including the planarity of the fracture across the borehole; the parallel-sided nature of the fracture for its infinite extent away from the borehole; as well as the resistivity of the host formation and the fracture material. Some of these assumptions may be valid, since it is the properties of the fracture closest to the borehole that primarily influence the measurement response. Typical practice is to estimate the resistivity of the formation by calibrating the down-hole image response curves to a resistivity log at a scale best approaching the image log.

In the laboratory, we examined the resistivity response to a single fracture, using a core sample (Fig. 5) in which the fracture has completely broken the core into two separate pieces, and these two pieces of core, which break along grain boundaries, have been reassembled. Thus the fracture is known to be both open and fluid filled. This feature of the core was used to demonstrate the variation in electrical response to increasing fracture aperture as the two pieces are moved physically apart. Figure 6 shows results drawn from a series of measurements on the core as the fracture aperture is increased. The changing shape of the excess conductivity peak (reduced resistivity) across the open, fluid-filled fracture supports the concept that

fracture aperture estimates from laboratory measurements should be feasible. The spatial distribution of the measurement data shown in Figure 3, and the physical separation of the core pieces defining the fracture are, however, insufficiently constrained in this experiment to provide fracture aperture estimates; such work requires exact positioning of both the electrodes and the core pieces, and is the subject of further work in progress.

The experimental results presented in Figures 5 & 6 relate to a fracture along grain boundaries in a core. This fracture enabled the investigation of changing aperture on the electrical response, and is instructive in supporting the concept of

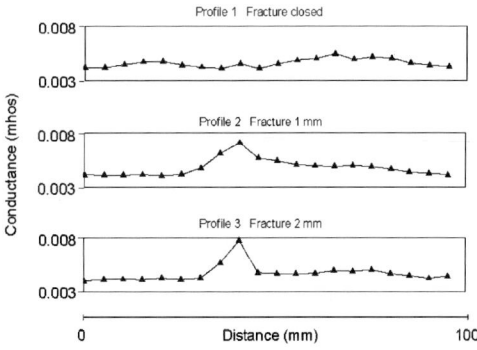

Fig. 6. Electrical resistivity profiles showing variation in response to changing fracture aperture for sample in Figure 5.

fracture aperture evaluation from electrical resistivity measurements. *In situ* fractures and sets of fractures are not necessarily as simple. Fractures often follow different orientations and intersect at varying angles related to the stress history of the formation. Fractures may be sealed or open, connected or closed. While large fractures may offer major fluid flow paths it is often the minor fracture sets that provide for both storage and drainage of fluids. Thus the investigation of fine-scale fractures is important to the understanding of fluid dynamics.

Figure 7 shows a photo and laboratory electrical image of a carbonate core with intersecting fine-scale fractures. In contrast to identifying an individual fracture, this image shows a fracture map with clear variability between conductive and resistive fractures. Comparison of the optical fractures with those exhibiting low resistivity (high conductivity) quickly identifies only a subset of fractures open to electrical current flow. Furthermore, those that are electrically conducting are not necessarily consistent with those that are optically dominant. Similarly, Figure 7 highlights the difficulty of identifying

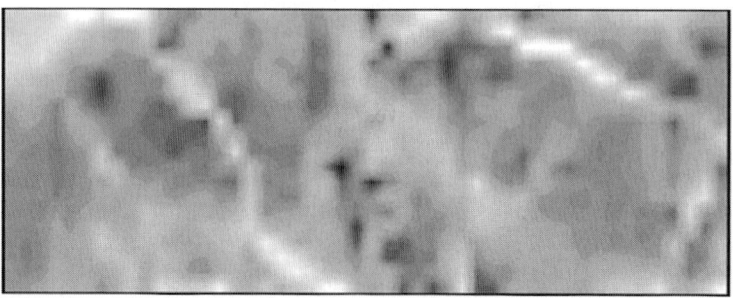

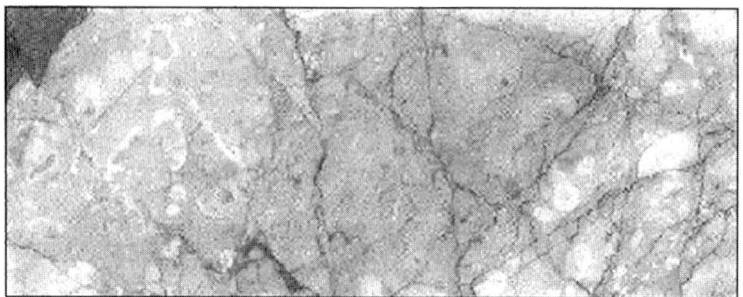

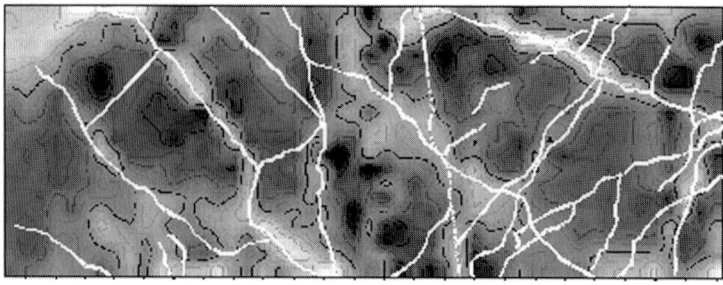

Fig. 7. Core photo (centre) and electrical map (top) of fracture network in low-porosity marble showing both open and closed fractures. Fractures are highlighted in electrical image (bottom) emphasizing fracture network. Vertical (across core) scale of images is 10 cm.

electrically conductive fractures (and hence in this case probably open fractures) from optical observations.

Discussion

In this paper we have demonstrated the ability to obtain electrical resistivity images of fine-scale fractures in the laboratory. This approach simulates a down-hole generic electrical imaging tool, but uses the four-electrode approach to provide quantitative resistivity data. It also provides for the further study of fracture aperture in a well-constrained environment that can be readily related to down-hole images. At the outset, we discussed briefly the general premise that electrical resistivity is related to the porosity of the rock; thus, an open fluid-filled fracture appears as a conductive feature. We also mentioned, however, that the presence of clays (in a fracture) could also lead to a conductive feature. Thus, given a piece of core containing fractures, which are conductive features (in the laboratory or down-hole), how can we be certain that the fractures are fluid pathways rather than simply clay filled? It is possible to draw up various conditions or rules based on the magnitude of the anomaly, the background porosity, the conductivity of the fluid, and the nature of the clays – but these would at best only constrain the problem. The alternative is to flow tracer fluids through the fractures while monitoring the electrical resistivity of the core. Changing the salinity of the fluid would lead to a change in the resistivity of any open fluid-filled fracture connected to the tracer pathway. This has been demonstrated for fluid flow through the matrix porosity of an unconsolidated sandstone and an aeolian sandstone (e.g. Jackson et al. 1990; Lovell et al. 1994). Applying the same principle to fractured core to separate fluid pathways from clay-filled fractures is feasible.

These results on core simulate the down-hole measurement described earlier, and are supported in general by numerical modelling. Luthi & Souhaité (1990) modelled a generic down-hole electrical imaging tool (Fig. 8) and demonstrated a clear relationship between fracture width and additional current flowing in the fracture (for a known resistivity contrast between the fluid and the formation matrix). Furthermore, they demonstrated that the influence of fracture dip was insignificant in the range 0 to 40 degrees dip. Williams (1996) also modelled a generic down-hole electrical imaging device. The results (Figs 9 & 10) show how increasing fracture depth away from the borehole causes

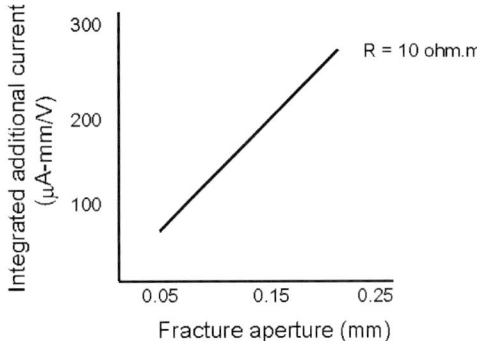

Fig. 8. Results (after Luthi & Souhaité 1990), showing a clear dependence of the integrated additional current on the fracture aperture. *R* is the background formation resistivity.

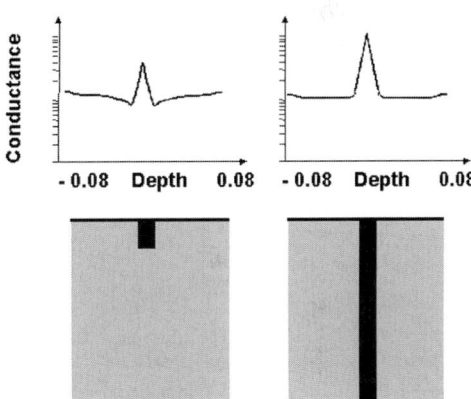

Fig. 9. Numerical modelling of electrical response (for a generic electrical borehole imaging tool) to different depths of fracture into the formation, away from the borehole (after Williams et al. 1997).

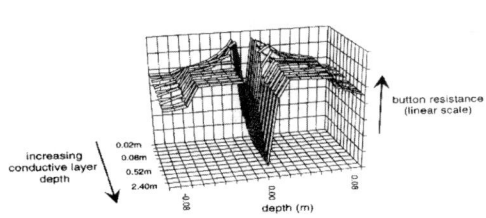

Fig. 10. Numerical modelling of down-hole electrical response for a generic electrical borehole imaging tool to different depths of fracture into the formation (away from the borehole) (after Williams et al. 1997).

the shape of the individual button trace on the imaging tool pad to change. In particular, the shoulders adjacent to the fracture become more pronounced as the fracture depth is limited close to the borehole wall (Williams 1996; Williams *et al.* 1995, 1997). Note that the falling off of the response for all traces at some distance away from the fracture relates to the boundary of the model and not the fracture parameters.

There has also been reference to the determination of fracture aperture from electrical measurements, as well as to connectivity of fractures. The technique described here is applicable to both, and theoretically enables the estimation of aperture for a single fracture and the mapping of connected conductive features through the core in three-dimensions. Furthermore, however, the technique is also applicable to the monitoring of the compressibility of fractures during stress experiments. Combined with down-hole images, this could lead to improved understanding of the fracture–fluid-flow characterization of reservoirs and their behaviour during reservoir production.

Using solely the static electrical images it is impossible to determine whether the fractures are open or closed to fluid flow. Our work with sediments, however, has demonstrated that passing a tracer fluid of different resistivity to the saturating fluid can identify fluid pathways through the core. Applying this technique to fractured core will enable the investigation of open and closed fractures and the effects of stress on fracture compressibility.

Conclusions

(1) Individual fractures and fracture networks that are open and electrically connected can be delineated on core.
(2) Electrical measurements on core can produce detailed images of fractured formations.
(3) The electrical images do not always correlate with optically observed obvious fractures; i.e. they can provide additional information.
(4) The electrical response varies with changing fracture morphology (e.g. aperture).
(5) The electrical images on core presented here concur with published results.

Electrical Core Imaging is the result of research over several years, funded at various stages by the Natural Environment Research Council, the Department of Trade and Industry, Enterprise Oil plc, LASMO plc, Mobil North Sea Ltd, and Shell UK through collaboration between the University of Leicester and the British Geological Survey. The fracture-imaging aspect of the work was particularly supported by LASMO Italia SUD S.p.A. This paper is published with the permission of the Director of the British Geological Survey.

References

ARCHIE, G. E. 1942, The electrical resistivity log as an aid in determining some reservoir characteristics. *Transactions of the AIMME*, **146**, 54–62.

BOURKE, L. T. 1992. Sedimentological borehole image analysis in clastic rocks: a systematic approach to interpretation. *In*: Hurst, A., Griffiths, C. M. & Worthington, P. F. (eds) *Geological Applications of Wireline Logs II*, Geological Society. London, Special Publications, **65**, 31–42.

BOURKE, L. T. & DELFINER, P. *et al.* 1989. Using formation MicroScanner images. *The Technical Review*, **37** (1), January, 16–40.

BOYELDIEU, C. & JEFFREYS, P. 1988. Formation MicroScanner: new developments. *Society of Professional Well Log Analysts, 11th European Formation Evaluation Symposium*. Oslo, Norway, Paper WW.

BRACE, W. F., ORANGE, A. S. & MADDEN, T. R. 1965. The effect of pressure on the electrical resistivity of water saturated crystalline rocks. *Journal of Geophysical Research*, **70**, 5669–5678.

CHEUNG, P. 1999. Microresistivity and ultrasonic imagers: tool operations and processing principles with reference to commonly encountered image artifacts. *In*: LOVELL, M. A., WILLIAMSON, G. & HARVEY, P. K. (eds) *Borehole Images: Case Histories and Applications*. Geological Society, London, Special Publications, **159**, 45–57.

EKSTROM, M. P., DAHAN, C., CHEN, M.-Y., LLOYD, P. & ROSSI, D. J. 1987. Formation imaging with microelectrical scanning arrays. *The Log Analyst*, **28**, 294–306.

HARVEY, P. K. JACKSON, P. D., LOVELL, M. A. *et al.* 1994. Structural implications from fluid flow and electrical resistivity images in aeolian sandstones. *SPWLA 35th Annual Logging Symposium*, Tulsa, OK, Paper LL, 1–7.

JACKSON, P. D. & ODP Leg 133 Shipboard Party 1991. Electrical resistivity core scanning: a new aid to the evaluation of fine scale sedimentary structure in sedimentary cores. *Scientific Drilling*, **2**, 41–54.

JACKSON, P. D., BRIGGS, K. B., FLINT, R. C., HOLYER, R. J. & SANDIDGE, J. C. 2002. Two- and three-dimensional heterogeneity in carbonate sediments using resistivity imaging. *Marine Geology*, **182**, 55–76.

JACKSON, P. D., HARVEY, P. K., LOVELL, M. A. & WILLIAMS, C. G. (1998) Measurement scale and formation heterogeneity: effects on the integration of resistivity data. *In*: *Core-Log Integration*, HARVEY, P. K. & LOVELL, M. A. (eds) Geological Society. London, Special Publications, **136**, 261–272.

JACKSON, P. D. & LOVELL, M. A. *et al.* 1992. Electrical resistivity core imaging: theoretical and practical experiments as an aid to reservoir characterisation. *SPWLA 35th Annual Logging Symposium, Tulsa, OK.* Paper VV, 1–13.

JACKSON, P. D., LOVELL, M. A., PITCHER, C. A., GREEN, C. A., EVANS, C. J., FLINT, R. C. & FORSTER, A. 1990. Electrical resistivity imaging of core samples. *Transactions of the European Core Analysis Symposium (EUROCAS I),* London, May, 365–378.

LLOYD, P. M., DAHAN, C. & HUTIN, R. 1986. Formation imaging with electrical scanning arrays. A new generation of stratigraphic high resolution dipmeter tool. *Transactions of the Society of Professional Well Log Analysts, 10th European Formation Evaluation Symposium.* Paper L.

LOFTS, J. C. & BOURKE, L. T., 1999. The recognition of artefacts from acoustic and resistivity borehole imaging devices. *In*: LOVELL, M. A., WILLIAMSON, G. & HARVEY, P. K. (eds) *Borehole Images: case histories and applications.* Geological Society, London, Special Publications, **159**, 59–76.

LOVELL, M. A. & JACKSON, P. D. 1991, Electrical flow in rocks: the application of high resolution electrical resistivity core measurements, in *32nd Annual Logging Symposium Transactions: Society of Professional Well Log Analysts, Midland, Texas,* Paper WW.

LOVELL, M. A., HARVEY, P. K., JACKSON, P. D., BALL, J. K., ASHU, A. P., FLINT, R. C. & GUNN, D. 1994. Electrical resistivity core imaging: towards a 3-dimensional solution. *SPWLA 35th Annual Logging Symposium, Tulsa, OK.* Paper JJ, 1–6.

LOVELL, M. A., HARVEY, P. K., BREWER, T. S., WILLIAMS, C., JACKSON, P. D. & WILLIAMSON, G. 1998. Application of FMS images in the Ocean Drilling Program: an overview. *In*: CRAMP, A., MACLEOD, C. J., LEE, S. V. & JONES, E. J. W. (eds) *Geological evolution of ocean basins: results from the Ocean Drilling Program.* Geological Society, London, Special Publications, **131**, 287–303.

LOVELL, M. A., HARVEY, P. K., WILLIAMS, C. G., JACKSON, P. D., FLINT, R. C. & GUNN, D. A. 1997. Electrical resistivity core imaging: a petrophysical link to borehole images. *The Log Analyst*, **38** (6), 45–53.

LUTHI, S. M. 1990. Sedimentary structures of clastic rocks identified from electrical borehole images. *In*: HURST, A., LOVELL, M. A. & MORTON, A. C. (eds) *Geological Applications of Wireline Logs.* Geological Society, Special Publications, **48**, 3–10.

LUTHI, S. M., & SOUHAITÉ, P. 1990. Fracture apertures from electrical borehole scans. *Geophysics*, **55** (7), 821–833.

PRENSKY, S. 1999. Advances in borehole imaging technology and applications. In: LOVELL, M. A., WILLIAMSON, G. & HARVEY, P. K. (eds) *Borehole Images: Case Histories and Applications.* Geological Society, London, Special Publications, **159**, 1–43.

RUST, C. F. 1952. Electrical resistivity measurements on reservoir rock samples by the two-electrode and four-electrode methods. *Transactions of the American Institution of Mining and Metallurgical Engineers*, **195**, 217–224.

TRICE, R. 1999. Application of borehole image logs in constructing 3D static models of productive fracture networks in the Apulian Platform, Southern Apennines. *In*: LOVELL, M. A., WILLIAMSON, G. & HARVEY, P. K. (eds) *Borehole Imaging: Applications and Case Histories.* Geological Society, London, Special Publications, **159**, 155–176.

WILLIAMS, C. G. 1996. Assessment of electrical resistivity properties through development of three-dimensional numerical models. PhD thesis, Leicester University.

WILLIAMS, C. G., JACKSON, P. D., LOVELL, M. A., HARVEY, P. K. & REECE, G. 1995. Numerical simulation of electrical imaging tools.*4th International Conference of the Brazilian Geophysical Society – 1st Latin American Geophysical Conference, Rio de Janeiro, Volume II*, 744–746.

WILLIAMS, C. G., JACKSON, P. D., LOVELL, M.A. & HARVEY, P. K. 1997. Assessment of and interpretation of electrical borehole images using numerical simulations. *The Log Analyst*, **38** (6), 34–44.

WORTHINGTON, P. F. 1985. The evolution of shaly sand concepts in reservoir evaluation. *The Log Analyst (Jan–Feb)*, **26** (1), 23–40.

Shear-wave anisotropy from dipole shear logs in oceanic crustal environments

G. J. ITURRINO,[1] D. GOLDBERG[1], H. GLASSMAN[2], D. PATTERSON[2],
Y.-F. SUN[1], G. GUERIN[1] & S. HAGGAS[3]

[1]*Lamont–Doherty Earth Observatory, Borehole Research Group, Route 9W,
Palisades, NY 10964, USA (e-mail: Iturrino@ldeo.columbia.edu)*
[2]*Baker-Hughes, 2001 Rankin Road, Houston, TX 77073-5114, USA*
[3]*Leicester University, Department of Geology, University Road,
Leicester LE1 7RH, UK*

Abstract: The deployment of a down-hole dipole shear sonic tool in Hole 395A and Hole 735B marked the first two opportunities to measure high-resolution shear-wave velocity and V_S anisotropy profiles in oceanic crustal rocks. In Hole 395A near the Kane Fracture Zone, dipole sonic logs were recorded from 100–600 mbsf, and allow azimuthal anisotropy to be determined as a function of depth in the crust. The magnitude of V_S anisotropy varies with depth, from less than 3.2% in low-porosity flows at the bottom of the hole, to approximately 15.5% in highly fractured pillow basalts and breccias. The orientation of the fast V_S direction also varies over depth, with a mean value between 75°N and 80°E, and aligns with the strike of steeply dipping structures observed by down-hole electrical and acoustic images. This fast V_S angle orientation is locally oblique to the plate-spreading direction and to the Mid-Atlantic Ridge axis. In Hole 735B, drilled near the Atlantis Fracture Zone, dipole sonic logs from 23 to 596 mbsf indicate that V_S anisotropy varies with depth, with averages of 5.3% in the foliated and deformed gabbros recovered at the bottom of the hole; 4.5% in undeformed olivine and oxide-rich gabbros around 300 mbsf; and 6.8% in highly deformed mylonitic zones at shallow depths. The fast V_S angle also varies with depth, giving a mean orientation of approximately S45°E for well-resolved estimates in the upper interval of the hole. This direction aligns with the strike of steeply dipping fractures observed by down-hole imaging, and is locally oblique to the Southwest Indian ridge axis. Although the effects of regional stresses and local deformation of these holes may introduce anisotropy in the dipole sonic data, we conclude that crustal morphology in the vicinity of the holes contributes significantly to the magnitude and orientation of V_S anisotropy.

Over the last three decades, one of the main scientific goals of both the Deep Sea Drilling Project (DSDP) and the more recent Ocean Drilling Program (ODP) has been to drill deep into the Earth to study the composition and structure of the oceanic crust and upper mantle. Scientific expeditions to the Mid-Atlantic Ridge, the East Pacific Rise, and the Southwest Indian Ridge have increased our understanding of the variations in geophysical properties of the crust as a function of depth, age and spreading rate. The anisotropy of seismic velocity, in particular, has been reported to exist in the crust near these locations (Stephen 1981; Ando *et al.* 1983; White & Whitmarsh 1984; Shearer & Orcutt 1985, 1986). These authors, among others, have typically attributed the observed seismic velocity anisotropy to preferred crack or mineral orientation in crustal and upper mantle rocks. In this paper, we present velocity anisotropy results determined from down-hole logging data at two sites in slow–intermediate crustal spreading rate environments. For both cases, one near the Mid-Atlantic Ridge and one near the Southwest Indian Ridge, shear-wave splitting is measured in a vertical hole using a dipole shear-logging tool.

Hole 395A was drilled at 22°45.352′N, 46°4.861′W in a sediment pond along the western flank of the Mid-Atlantic Ridge and south of the Kane Fracture Zone (Fig. 1a). The hole was originally drilled to a depth of 664 metres below the sea-floor (mbsf), or 571 m into basement (Melson *et al.* 1978). Drilling results show that below 11.5 m of Pleistocene–Pliocene sediments, Hole 395A recovered cores

From: HARVEY, P. K., BREWER, T. S., PEZARD, P. A. & PETROV, V. A. (eds) 2005. *Petrophysical Properties of Crystalline Rocks.* Geological Society, London, Special Publications, **240**, 117–131.
0305-8719/05/$15.00   © The Geological Society of London 2005.

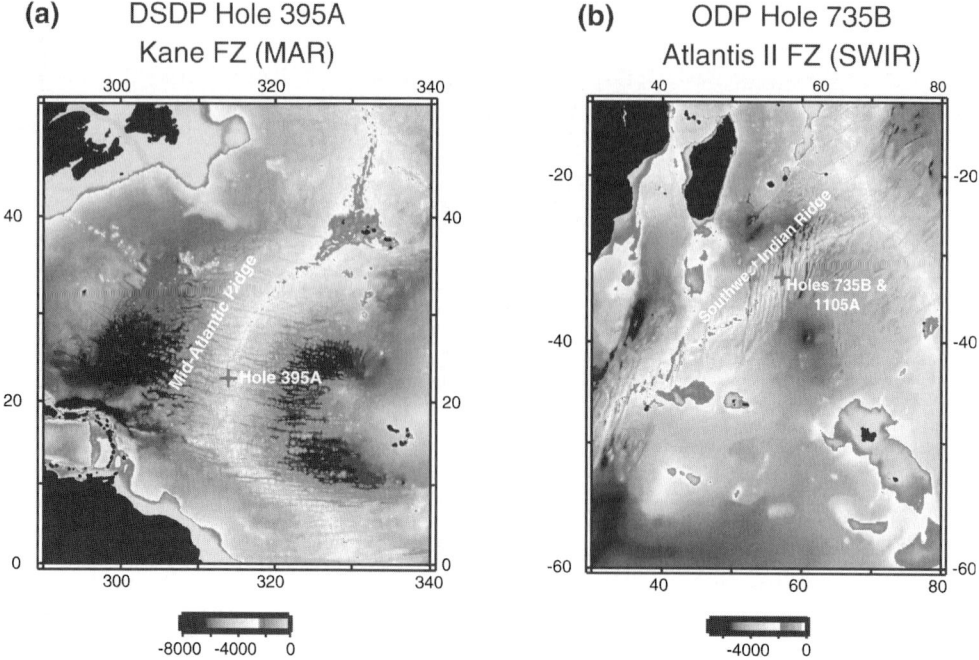

(a) DSDP Hole 395A
Kane FZ (MAR)

(b) ODP Hole 735B
Atlantis II FZ (SWIR)

Fig. 1. (a) The location of Hole 395A near the Mid-Atlantic Ridge is marked with a cross. (b) The location of ODP Hole 735B in the Atlantis II Fracture Zone of the Southwest Indian Ridge.

of alternating layers of aphyric basalts, phyric basalts, breccias (mainly basaltic), and intrusive dolerites (Fig. 2 & Table 1). Core recovery in these basalts averaged 17.6%, and the crustal age is approximately 7 Ma (Melson *et al.* 1978). Numerous return visits to Hole 395A conducted logging and borehole investigations to augment our knowledge of the composition and physical processes occurring in shallow crustal spreading at intermediate rates (Hyndman & Salisbury 1984; Bryan *et al.* 1988; DIANUT 1992; Becker *et al.* 1998). These subsequent expeditions encountered poor hole conditions in the deepest 50 m, and the current depth of Hole 395A penetrates approximately 513 m into basaltic rocks, or 603 mbsf. Down-hole logs were recorded over this entire basement interval.

Hole 735B achieved the deepest penetration of the lower oceanic crust, to a depth of 1508 mbsf during two ODP cruises to this site. The high recovery of gabbroic core (86.3%) at this site provides an excellent data-set to study lower-crustal rocks that are analogous to ophiolitic sequences. They have been used to describe the physical properties and the magmatic, structural, and metamorphic history of a block of the lower ocean crust that formed at a slow-spreading ridge

(Robinson *et al.* 1989; Dick *et al.* 1999). The recovered core from Hole 735B also allows the laboratory assessment of the seismic anisotropy of Layer 3 rocks (Iturrino *et al.* 1991, 2002).

Hole 735B is located at 32°43.40'S, 57°16.00'E (Fig. 1b) on the rift mountains along the eastern transverse ridge of the Atlantis II Fracture Zone, which is a 210-km-long left-lateral offset of the Southwest Indian Ridge (SWIR). The hole was drilled on a shoal platform, known as the Atlantis Bank, in approximately 700 m of water. The platform is believed to be part of a series of uplifted horst blocks forming a ridge parallel to the transform valley. Based on magnetic anomalies and zircon U–Pb isotopic dating, the age of the crust at this site is approximately 11.5 Ma (Dick *et al.* 1991; Stakes *et al.* 1991). The lithostratigraphic units recovered from Hole 735B (Fig. 3 & Table 2) are mainly metagabbros, olivine gabbros, gabbros, oxide-rich gabbros, gabbronorites, and troctolitic gabbros (Robinson *et al.* 1989; Dick *et al.* 1999). Several prominent mylonitic shear zones are present within the uppermost 40 m of the hole (Unit I) and show the most intense deformation, whereas the bottom-most 900 m interval of the hole has

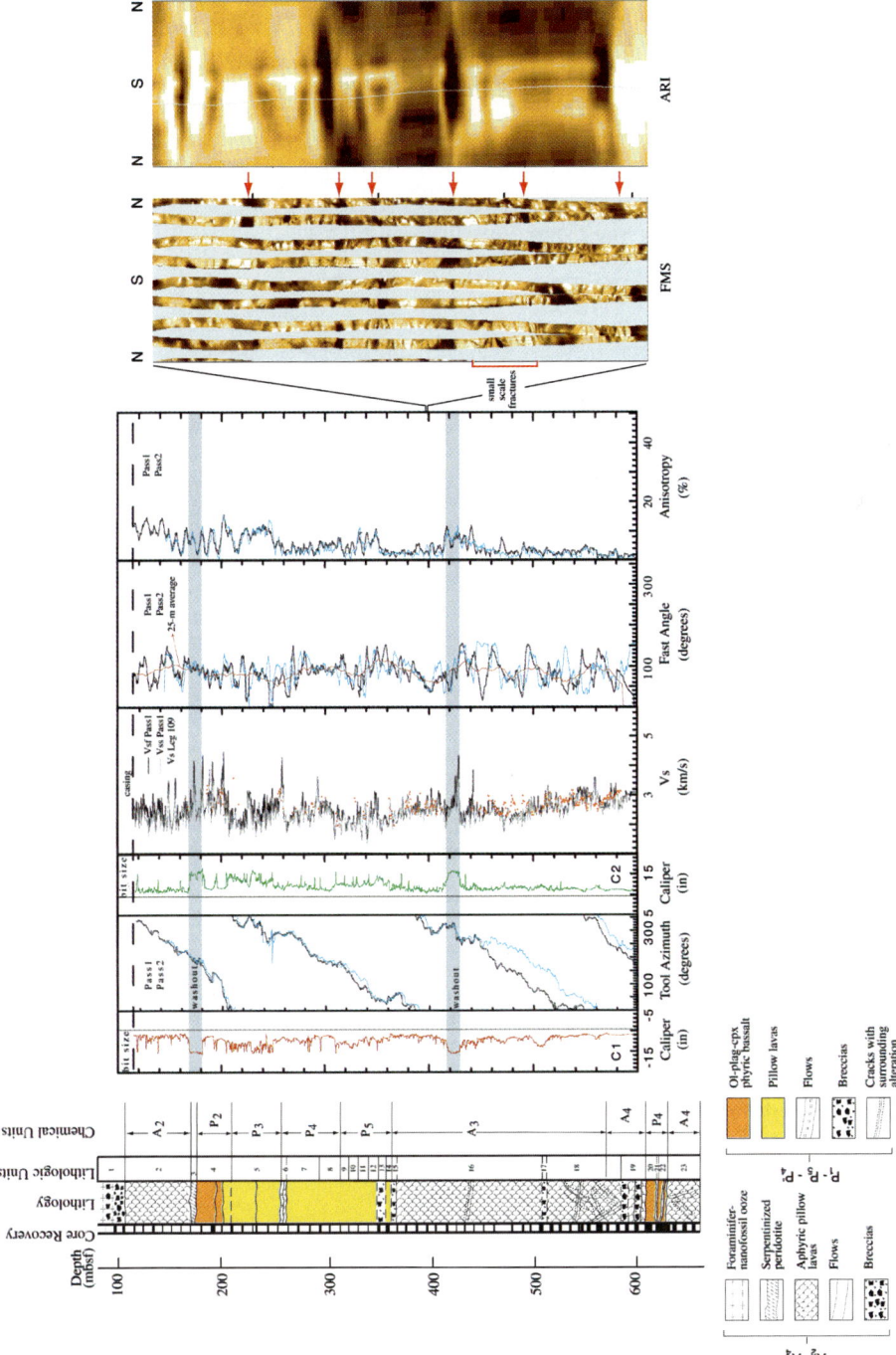

Fig. 2. Lithostratigraphy, core recovery, and down-hole logs from Hole 395A. The V_{s} (V_{sf} = fast shear-wave velocity and V_{ss} = slow shear-wave velocity), fast angle, and anisotropy logs were obtained from processing of the dipole sonic recorded waveforms. The V_{S} data (dots) from Moos (1990) is shown for comparison. FMS and ARI images show predominantly subhorizontal fractures and pillow lavas, with arrows indicating the locations of large, subhorizontal fractures and a bracket showing small-scale fractures.

Table 1. *Lithological unit descriptions of DSDP Hole 395A*

Unit	Lithological description
Unit 1	Sedimentary breccia with subrounded to angular fragments of aphyric basalt, and ultramafic to mafic plutonic rocks in foraminifer–nannofossil ooze
Unit 2	Aphyric basalt
Unit 3	Sedimentary breccia, subangular fragments of serpentine and aphyric basalt
Unit 4	Plagioclase–olivine-phyric fine-grained basalt
Unit 5	Plagioclase–olivine–clinopyroxene-phyric basalt
Unit 6	Plagioclase–olivine–clinopyroxene-phyric basalt with fewer plagioclase phenocrysts than Unit 5
Unit 7	Plagioclase–olivine-phyric basalt with rare clinopyroxene phenocrysts
Unit 8	Plagioclase–olivine–clinopyroxene-phyric basalt
Unit 9	Plagioclase–olivine-phyric basalt
Unit 10	Plagioclase–olivine–clinopyroxene and plagioclase–olivine-phyric basalt
Unit 11	Olivine–plagioclase-phyric basalt
Unit 12	Mixed fragments of plagioclase–olivine–clinopyroxene and plagioclase–olivine-phyric basalt
Unit 13	Basaltic breccia with angular clasts of fine- to medium-grained plagioclase–olivine-phyric basalt in carbonate clay-rich matrix
Unit 14	Mixed fragments of plagioclase–olivine–clinopyroxene and plagioclase–olivine-phyric basalt
Unit 15	Hyaloclastite, fine-grained basalt and basaltic glass in recrystallized carbonate ooze
Unit 16	Aphyric basalt with rare olivine and plagioclase xenocrysts, highly fractured, abundant veins
Unit 17	Basaltic breccia with angular and brecciated fine- to medium-grained basalt clasts
Unit 18	Aphyric basalt with rare olivine and plagioclase
Unit 19	Glass-rich basaltic breccia and aphyric basalt; abundant basaltic glass altered to numerous secondary minerals
Unit 20	Dolerite, plagioclase–olivine–clinopyroxene medium-grained basalt
Unit 21	Aphyric basalt, thin zone, surfaces sheared and coated with alteration minerals
Unit 22	Dolerite, plagioclase–olivine–clinopyroxene medium-grained basalt with quenched contact at the base
Unit 23	Aphyric basalt with some glassy breccia zones, highly fractured with abundant soft clays in fractures

significantly less deformation and more uniform composition. Down-hole logs were only recorded in the upper 600 m of Hole 735B because the lower 900 m interval remains obstructed by a severed drill pipe.

Anisotropy

In its simplest form, there are two styles of alignment in Earth materials that cause velocity anisotropy, those having principal planes with horizontal and those with vertical orientations.

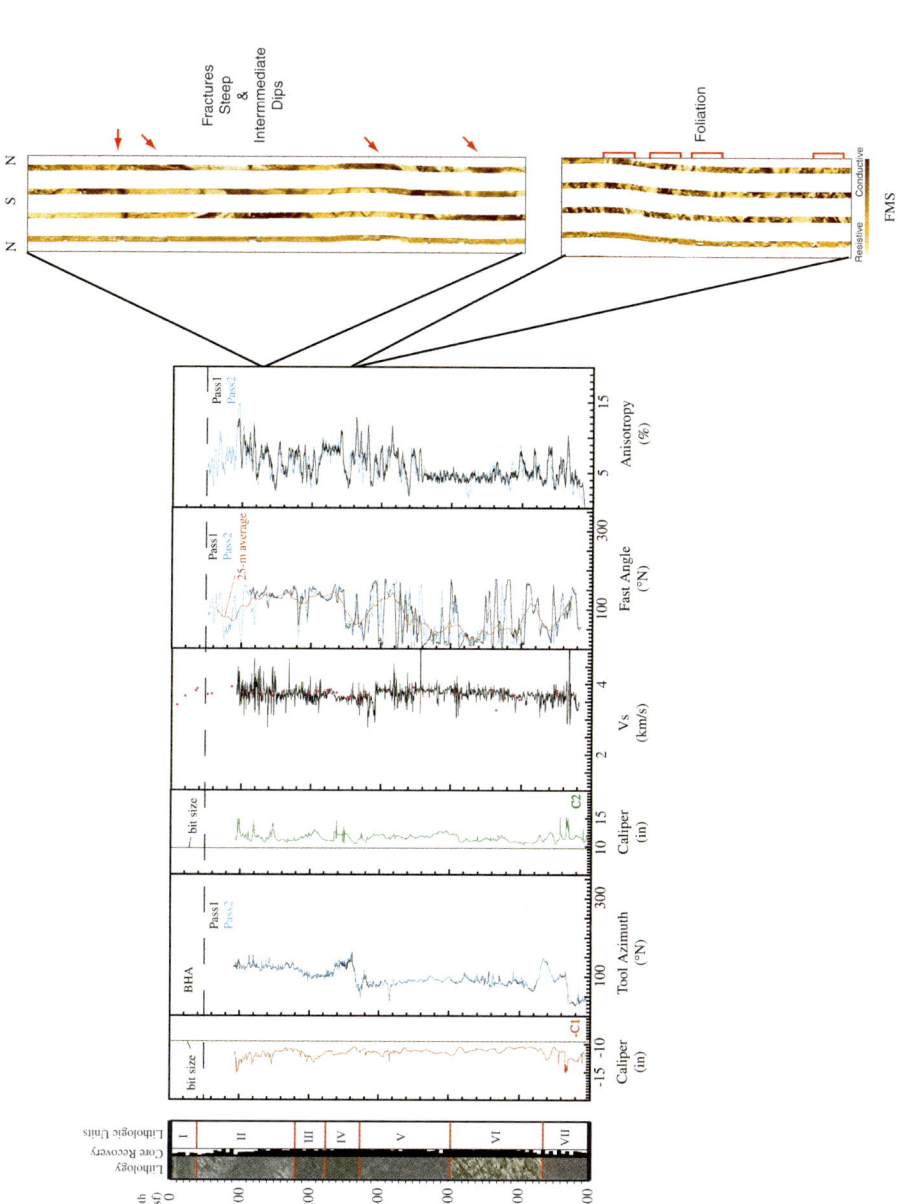

Fig. 3. Lithostratigraphy and down-hole logs from the upper 600 m of Hole 735B. The V_S, fast angle, and anisotropy logs were obtained from processing of the dipole sonic recorded waveforms. Laboratory average shear-wave velocity data (dots) are shown, corrected for *in situ* pressure effects (Iturrino *et al.* 1991, 2001). FMS images show steep and intermediate dipping fractures, as well as foliation within the oxide-rich gabbros.

Table 2. *Lithological unit descriptions of the upper 600 m of ODP Hole 735B*

Unit	Lithological description
Unit I (0–39.5 mbsf)	Consists of porphyroclastic to mylonitic metagabbro. Well-developed foliations with dips ranging from 25 to 35 degrees are observed in the majority of the metagabbros. Porphyroclastic textures are quite abundant. Relict igneous minerals such as plagioclase, cpx, opx, and Fe–Ti oxides occur as porphyroclasts in a mainly recrystallized matrix controlled by slightly oriented plagioclase.
Unit II (39.5–180 mbsf)	Defined by abrupt decrease in the degree of plastic deformation and appearance of distinctive igneous textures. Composed of olivine-bearing (<5% olivine) to olivine gabbros (>5% olivine). Primary mineralogy: olivine, plagioclase, cpx, opx (trace amounts), small amounts of opaques, brown amphiboles and sulphides. Most of the rocks are mesocumulates. Small basaltic dykes with chilled margins intrude the olivine gabbros.
Unit III (180–224 mbsf)	Composed of olivine gabbros and intervals of Fe–Ti oxide gabbros. Steeply dipping and well-defined laminations present. Contact between Unit II and Unit III is defined by a change in chemical composition of the olivine gabbros from 12–13% MgO to 8–9% MgO (Robinson *et al.* 1989). Boundary with Unit IV is marked by a 60-cm-thick mylonitic zone and flattening of laminations.
Unit IV (224–272 mbsf)	Iron–titanium oxide gabbros are the predominant rock type. Gabbro characterized by an abundance of opaques (usually >10%), deficiency in olivine and presence of a more sodic plagioclase (andesine). Primary layering expressed as strong lamination of clinopyroxenes and plagioclase, with most of the igneous laminations subhorizontal to horizontal. Olivine gabbros occur in small and sparse intervals. Brecciated zones filled with felsic intrusions and trondhjemite veins are found in small intervals. A 3-m-thick layer of mylonitic oxide-rich gabbros defines the lower boundary of this unit.
Unit V (272–403.5 mbsf)	This is the most massive and homogeneous section. Composed of massive olivine gabbros. Brecciated zone at the top of the unit. Samples are mesocumulates characterized by abundance of olivine (>5%) and the scarcity of oxides and low-calcium pyroxenes. Primary layering defined by changes in grain size instead of laminations of the plagioclase. Plagioclase is more calcic (labradorite) and the MgO content averages 11%. Brecciated zones associated with felsic veins and growths of epidote and albite are abundant. Extensive amphibolitization of clinopyroxenes and small troctolite intervals are also present.
Unit VI (403.5–536 mbsf)	Olivine-rich interval (>10%) with frequent layers of troctolite and lenses of microgabbros with equigranular textures. Olivine-rich gabbros similar to gabbros of Unit V, except that higher modal proportions are present. This unit is characterized by intervals of plastic deformation in which mylonitic and porphyroclastic textures are abundant. Iron-Ti oxide-rich zones are found within small shear and amphibolitized zones.
Unit VII (536–599 mbsf)	The transition to lithological Unit VII is marked by the occurrence of orthopyroxene-bearing gabbro and gabbronorite, with and without oxides, as well as the disappearance of troctolitic gabbro. Intervals of oxide-bearing rocks are more abundant, and the average grain size decreases relative to the Unit VI rocks. This unit is composed of *c.*70% orthopyroxene-bearing rocks, which are relatively rare in the units above. The oxide-rich rocks clearly cross-cut the gabbros, and some have sheared contacts. Unit VII is divided into two subunits, VIIA and VIIB, at 560 mbsf. The bottom of Subunit VIIA at 560 mbsf coincides with a major fault, the disappearance of olivine gabbro, and the occurrence in Subunit VIIB of intervals with subophitic texture.

These two styles combine in complex arrangements to form the three-dimensional anisotropy observed in many realistic Earth materials. In the simple horizontal layered case, elastic properties vary vertically, for example from layer to layer, but not horizontally (Fig. 4a). Such a material is called transversely isotropic with a vertical axis of symmetry (TI-V). Seismic waves generally travel faster horizontally in these environments than vertically. A material with aligned vertical weaknesses such as cracks or fractures, or with unequal horizontal stresses, is called transversely isotropic with a horizontal axis of symmetry (TI-H). This forms azimuthal anisotropy in the formation, and seismic waves travelling along the fracture direction generally travel faster than waves crossing the plane of the fracture (Thomsen 1977; Schoenberg & Sayers 1995). Laboratory velocity measurements can assess the effects of both TI-H and TI-V symmetry on anisotropy by careful sampling along the orientations of planes of weakness, such as cracks, and, in particular, when shear waves are measured (Crampin 1981; Kern *et al.* 1997). Laboratory velocity studies on oceanic, mafic and ultramafic rocks have indicated a decrease in shear-wave splitting and elastic wave anisotropy with increasing confining pressure, due to progressive crack closure (Iturrino *et al.* 1991, 1995; Kern *et al.* 1997). As a consequence of

this behaviour, anisotropy is often important in the interpretation of oceanic seismic data, especially at the relatively shallow depths investigated by seismic surveys and drilling studies, where fracturing and deformation is prevalent (Stephen 1981; White & Whitmarsh 1984; Shearer & Orcutt 1985, 1986).

Methods

Using logging tools, formation anisotropy can only be measured with respect to the axis of the borehole. In a vertical borehole, shear-wave splitting is largely controlled by TI-H anisotropy. A maximum value of anisotropy is produced at the preferred crack or mineral orientation with respect to the borehole axis (Fig. 4b). In this study, the deployment of the Schlumberger[TM] Dipole-Shear Sonic Imaging (DSI) tool in Holes 395A and 735B marked the first two opportunities to measure high-resolution V_S anisotropy profiles using logs in the ocean-crustal environment. The DSI utilizes directional sources and receivers which behave much like a piston, creating a pressure increase on one side of the hole and a decrease on the other (Sinha *et al.* 1994). This generates flexural waves travelling along the borehole wall that directly excite compressional and shear waves in the formation (Fig. 4b). The source operates

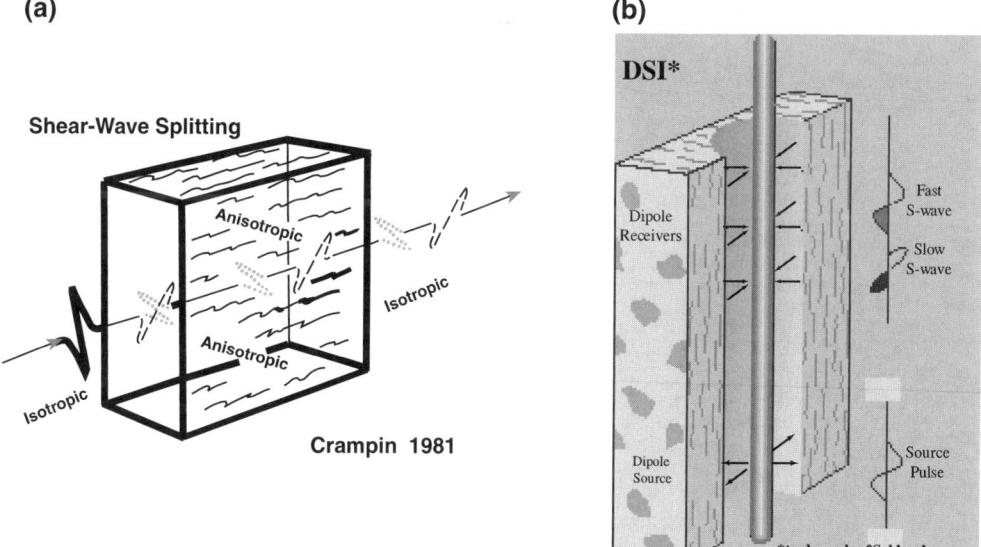

Fig. 4. Examples of shear-wave velocity anisotropy. (**a**) In a layered case, elastic properties may vary vertically, such as from layer to layer, but not horizontally. (**b**) Maximum shear-wave splitting in a borehole is controlled by transverse anisotropy with a horizontal axis of symmetry (azimuthal anisotropy). Dipping structures also may introduce azimuthal anisotropy.

at low frequencies, usually below 4 kHz, where excitation of these flexural waves is optimal (Kimball 1998). Two receiving transducers orthogonal to each other are used to record these waves for every source firing. When no TI-H anisotropy is present, propagation of the flexural wave along the wall surface is coaxial with the borehole and in line with the transducers. The presence of TI-H anisotropy introduces partitioning of the flexural wave energy between the two orthogonal receivers. This shear-wave splitting may be used to determine the magnitude and orientation of the anisotropy.

Shear-wave anisotropy measurements from dipole logs have previously been analysed to determine fracture intensity and orientation (Joyce *et al.* 1998; Wade *et al.* 1998), infer bedding-induced anisotropy (Müller *et al.* 1994), as well as the orientation of regional stresses (Sinha *et al.* 1994). These studies use an inversion technique, common in vertical seismic profiling and shear-wave seismic exploration, that is based on the numerical rotation of fast and slow shear waveforms in the principal planes using data from orthogonal receivers (Alford 1986). Using a software package developed by Baker Atlas, we analyse the dipole waveforms from Holes 395A and 735B for V_S anisotropy as indicated by shear-wave splitting between orthogonal receivers on the DSI tool (Tang & Chunduru 1999). Four waveforms are analysed simultaneously at each depth from two source–receiver pairs to obtain a slow and fast shear-wave velocity (Patterson & Shell 1997). When a source and a receiver align with one of the principal axes, the transmitted shear-wave energy does not split and the maximum fast and slow waveforms are recorded directly. When the source–receiver pair does not align with the principal axes, the energy splits and numerical rotation into the principal planes is required. This may be a difficult process if all received waveforms do not have a high signal-to-noise ratio, or if the depth ambiguity of the tool is large. Fortunately, the technique used is based on waveform matching between receivers, and data stacking to suppress noise effects. Thus, reliable anisotropy estimates and the associated fast shear azimuth can be determined in one numerical process and with greater accuracy (Tang & Chunduru 1999). Three parameters are simultaneously derived: anisotropy, fast shear azimuth, and shear-wave slowness. However, a 90° ambiguity in determining orientations may occur if the difference between the fast and slow shear-wave velocities is small.

By numerically identifying shear slowness, fast azimuth, and anisotropy, quantitative error estimates are also produced. These measure the relative differences between the fast and slow shear-wave angles with respect to an isotropic model, and their ratio, which represents a data-fitting error for the inversion. Larger ratio values indicate higher confidence in the estimated magnitude of the anisotropy. If the ratio is low, however, it is difficult to distinguish the fast from the slow shear-wave velocity. Another indicator (S1S2) represents the relative difference between the fast and slow velocity residual errors. Larger values indicate more reliable fast shear orientations. Figure 5 illustrates the ratio and the residual error curves for the V_S anisotropy computation in Holes 395A and 735B. Hole 395A shows relatively high ratios and S1S2 values throughout most of the interval, indicating reliable anisotropy and fast angle estimates. Hole 735B has large variations over the transition between the oxide-rich and olivine gabbros (240–300 mbsf) and in the interval below 380 mbsf. The confidence in fast shear orientations in Hole 735B therefore has 90° ambiguity over the lower interval.

Results

DSDP Hole 395A – Kane Fracture Zone, Mid-Atlantic Ridge

Down-hole logs obtained from Hole 395A are shown in Figure 2. Orthogonal caliper logs of the borehole diameter and tool rotation profiles are useful for assessing log data quality. Borehole diameter is consistently between 25 and 40 cm (10–16 inches) with the exception of two severely enlarged intervals at 163–176 and 418–430 mbsf that have severely deteriorate the waveform data quality. The similarity of the two caliper curves and the relatively constant rate of rotation of the logging tool as it was pulled upwards indicates that Hole 395A is relatively circular in cross-section. High-quality images of the borehole wall from the formation microscanner (FMS) and azimuthal resistivity imaging (ARI) tools show a thin intercalation of pillow basalt and massive flows and cm-scale fracturing within them (Fig. 2). Porosity is often the most important determining factor for electrical conduction in these rocks. High-conductivity voids and fractures correspond to more porous (dark) features in Figure 2. This interval is typical of the pillow basalts encountered in the hole. Not all these porosity changes have a systematic orientation with respect to the borehole axis, however. Fractures and voids may introduce heterogeneity at oblique angles to the borehole. Structural analyses based on

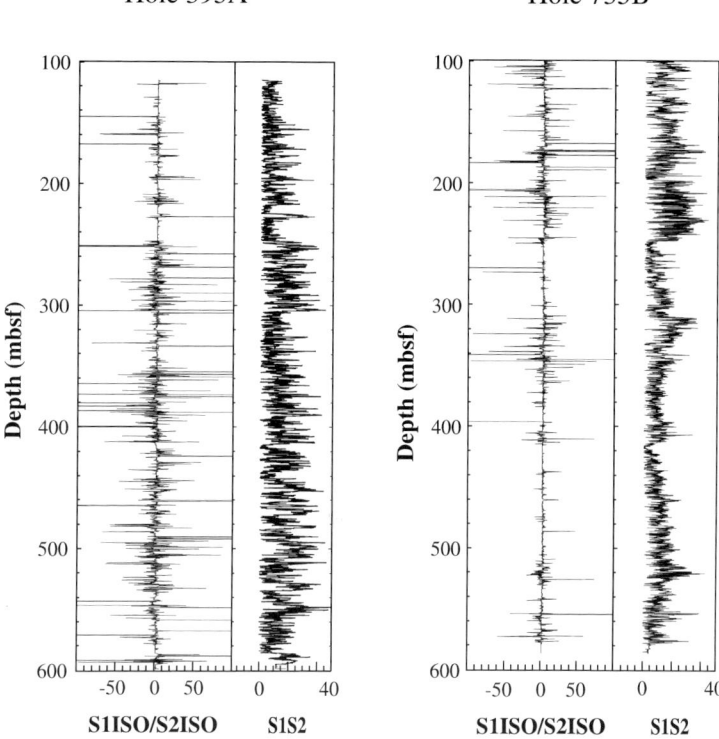

Hole 395A   Hole 735B

Fig. 5. Quality curves derived from dipole sonic waveform analyses from Holes 395A and 735B. The relative differences between the fast shear-wave and slow shear-wave polarization angles and their ratio represents the data-fitting residual error of an equivalent isotropic formation (S1ISO/S2ISO). If the S1ISO and S2ISO values are comparable it is difficult to distinguish the fast from the slow shear waves. Low values of S1S2 indicate uncertainty in the determination of the fast angle of V_S anisotropy.

the FMS images show that most of the intermediate–steeply dipping features (30–60° and 60–90°) occur below 483 mbsf and strike approximately N75°E and dip north to NW (Fig. 6).

Figure 2 shows V_S and anisotropy logs computed using the method described above, from two passes of the DSI tool. The V_S logs correlate well with V_S estimates derived from a monopole sonic tool deployed during an earlier visit to this site (Moos 1990). The offset between the dipole and monopole V_S values increases slightly with depth and may, in part, be due to greater dispersion of the dipole flexural wave with increasing velocity (Lang *et al.* 1987; Sinha *et al.* 2000). The V_S anisotropy log has a mean value of 5.3% ± 3.2%, and tends to decrease with depth from a peak value of 15.5% at the top of the basement to 3.2% below 260 mbsf. Increases in anisotropy are observed in zones where more fracturing is present. The quality of these anisotropy estimates, as computed by the inversion, yields an average error of 3.7% in the hole. The

fast V_S angle was determined from both passes of the tool. These were averaged using 20-point and 25-m sliding windows, and agree over the entire logged interval. These orientations are corrected for the declination (latitude) of the site and for rotation of the tool during data acquisition. Local variations in fast angle are pervasive with a mean orientation of N83°E over the entire hole, and N75°E over the lowermost 120 m where steeply dipping features are more prevalent. This geographical azimuth is approximately parallel to the mean orientation of these features observed in the FMS images.

ODP Hole 735B – Atlantis II Fracture Zone, Southwest Indian Ridge

The down-hole logs recorded in Hole 735B are shown in Figure 3. The V_S log varies with depth and lithological composition over the logged interval, and the caliper shows a regular

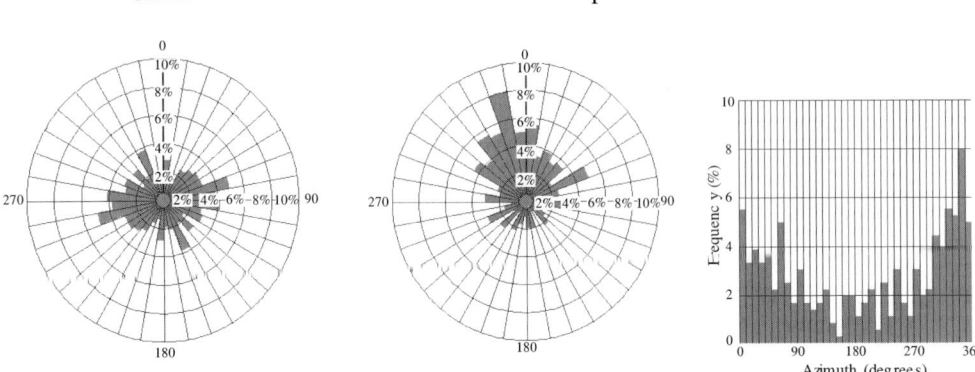

362 FMS dips (30- 90°) from 483- 603 mbsf

Fig. 6. Analyses of FMS images from the bottom 119 m of Hole 395A showing predominantly north- to NW-dipping structures.

borehole size with significant enlargements only near 100 mbsf and below 560 mbsf. FMS images in this area confirm that fractures are present and have dips ranging from 20 to 70° north–south and strikes generally in the WNW–ESE direction (Fig. 7). The vertical and steeply dipping fractures contribute to TI-H anisotropy in Hole 735B. Average V_S anisotropy estimated from the dipole data is $5.8 \pm 4.8\%$ over the upper 600 m of the hole. The V_S anisotropy tends to decrease with depth where the overburden pressure appears to close cracks and alteration minerals infill fractures. Based on the error computations in the inversion, V_S anisotropy estimates above 340 mbsf are believed to be the most reliable (see Fig. 5).

In Figure 3, the computed fast V_S angle is averaged using 20-point and 25-m sliding windows. The mean orientation of the fast angle is 89°; however, the distribution is strongly bimodal. From the sea-floor to 245 mbsf, the orientation is S45°E. Below this depth, it rotates c.90° to N 40°E. Large variations in fast angle are also observed from 224 to 272 mbsf, where the magnetization of oxide-rich gabbros may affect the measurements. The bimodal distribution may reflect the difficulty in distinguishing the fast and slow V_S angles where the inversion error ratio is low towards the bottom of the hole (Fig. 5). We believe that the fast V_S angle of S45°E is accurate above 340 mbsf in Hole 735B.

Discussion

Several factors contribute to the azimuthal V_S anisotropy observations in these holes. Intrinsic anisotropy may arise from the alignment of fractures, foliation and deformation features, as well as the morphology of crustal structures. In addition, under horizontal stresses that exceed the strength of the rock, brittle deformation of the formation and the borehole may preferentially occur at orientations associated with the principal stress directions (Zoback *et al.* 1985). At low frequencies, dipole waves penetrate relatively deeply into the formation and are not significantly affected by fracturing near the borehole, but at higher frequencies, the velocity of these waves changes considerably around the azimuth of the damaged borehole (Winkler *et al.* 1998). Thus, local and regional stresses may generate an induced anisotropy in addition to the intrinsic or structural anisotropy present in the formation.

Seismic anisotropy in Layer 2 of the crust has been studied previously using vertical seismic profiles. Shear-wave splitting has been observed in these data, and estimates of up to 30% V_S anisotropy in the upper 1500 m have been measured (e.g. Stephen 1981). Although somewhat poorly constrained in azimuth, these authors suggest that the anisotropy indicates ridge-parallel, laterally variable fracturing in the upper oceanic crust normal to the plate motion vector. Velocity profiles from ocean-bottom seismic data near Hole 395A also suggest that anisotropy exists due to the preferential closure by compression and chemical filling of ridge-parallel fissures in the uppermost few hundred metres of the crust (Purdy 1987). The V_S anisotropy probably aligns with those fractures and voids which preferentially remain open (Fryer & Wilkens 1990;

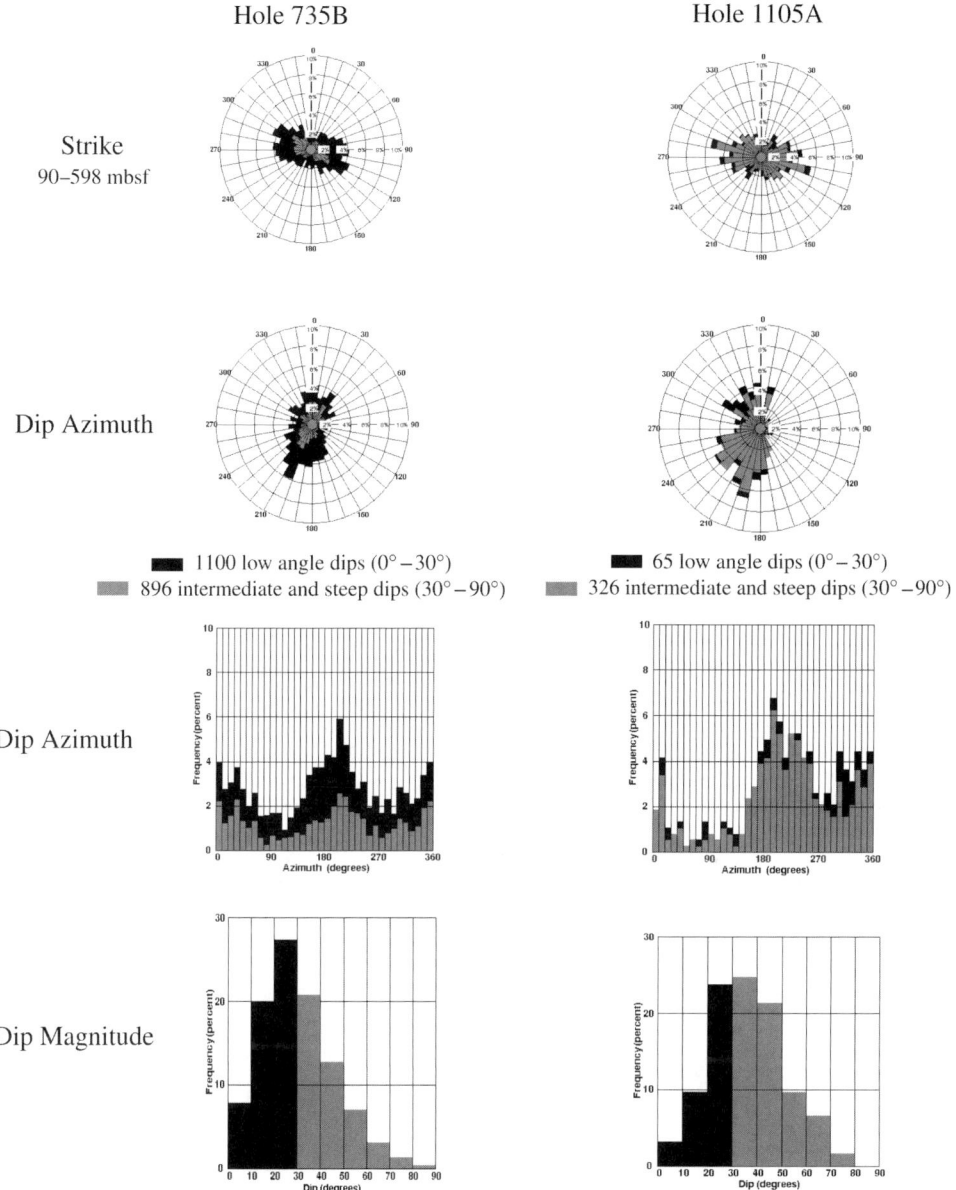

Fig. 7. Analyses of FMS images from Holes 735B and 1105A showing a bimodal distribution of fracture dip in north–south directions with most of the features dipping to the south.

Moos & Marion 1994). The correlation of the fast V_S angle from the dipole log results in Hole 395A with the orientation of fractures, pillow contacts and lava flows, suggests that crustal morphology on the scale of centimetres to metres similarly affects the alignment of the observed anisotropy, at least in the vicinity of the borehole.

Figure 8 shows the bathymetry in the region near Hole 395A and earthquake focal mechanism solutions based on the Harvard Centroid Moment Tensor (CMT) solutions along the Kane Fracture Zone and the Mid-Atlantic Ridge (Dziewonski & Woodhouse 1983; Cornell University GIS Group 1998). Studies of global earthquake focal mechanism solutions suggest that spreading-parallel

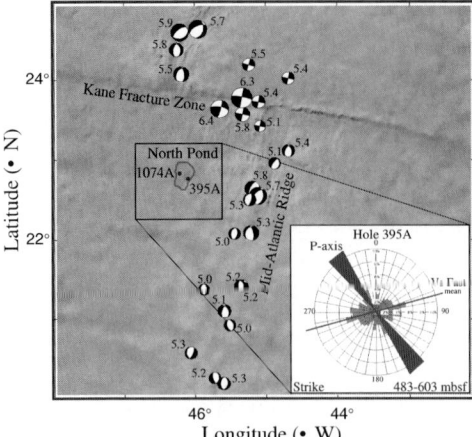

Fig. 8. Bathymetric map of the area in the vicinity of the Kane Fracture Zone and the location of Hole 395A. Focal mechanisms from CMT Harvard solutions are also displayed. Numbers next to focal mechanisms indicate earthquake magnitudes (Mw). The insert shows strike orientations from structural picks obtained from FMS analyses below 483 mbsf. The mean orientation of earthquake P-axis data is aligned 110° from the mean V_S fast-angle orientation.

(ridge-normal) compression is prevalent on the plate scale in the oceanic lithosphere (Wiens & Stein 1984). Focal plane solutions of earthquakes along the Kane fracture zone show a consistent NW–SE compression direction. Earthquakes below the adjacent rift valley are generally normal faults with smaller magnitude that align with the strike of this segment of the ridge. The inset in Figure 8 indicates the measured fast V_S angle, the strike of subvertical features seen in the FMS over the deepest 120 m of Hole 395A, and the range of P-axis orientations for shallow (10–15 km) earthquake focal mechanisms along the Kane Fracture Zone. Although the mean fast angle is parallel to the strike of intermediate and steeply dipping fractures, it falls oblique to the ridge axis and approximately 110° from the direction of regional earthquake compression. We conclude that steeply dipping fractures in the vicinity of the borehole and the fast V_S angle at shallow depths in the crust are aligned, but are not ridge-parallel. These observations may be local for this site, but could also reflect the influence of regional or other borehole-related stresses that contribute to anisotropy in the shallow ocean crust.

In Hole 735B, V_P anisotropy and shear-wave splitting has been observed in laboratory velocity measurements due to preferred mineral

orientations in isolated intervals (Iturrino *et al.* 1991). Most of the deformed gabbros exhibit a well-developed foliation dipping 25–35° with varying degrees of velocity anisotropy that is attributed to preferred plagioclase, amphibole and pyroxene orientations observed in thin sections (Iturrino *et al.* 1991, 2002). These laboratory results, however, do not take into account the effects of large fractures that may contribute to TI-H anisotropy. FMS images from Holes 735B, and from nearby 1105A, show steep (60–90°) and intermediate (30–60°) dipping fractures that strike in a general WNW–ESE direction (Fig. 9). In the upper part of these holes, reliable estimates of the fast V_S angle are generally parallel to the strike of these features. Figure 9 also shows CMT focal mechanism solutions near Hole 735B, and illustrates three significant strike-slip events recorded along the Atlantis II Fracture Zone. For one event on the Atlantis Bank, the regional earthquake compression axis is perpendicular to the fast V_S angle. The steeply dipping fractures in the vicinity of the hole and the observed V_S anisotropy are aligned, but they are oblique to the Atlantis II Fracture Zone and perpendicular to the strike of the Southwest Indian Ridge. Although the proximity of Hole 735B to the SWIR axis and the Atlantis II Fracture Zone may introduce complex tectonic effects that cause fracturing and faulting oblique to the ridge, the contribution of regional and borehole-related stresses to the observed anisotropy in this hole may be equally significant. Further analysis is required to differentiate the causes of the observed anisotropy in this complex crustal environment.

Conclusions

The analyses of dipole sonic log waveforms recorded in Holes 395A and 735B provide new information about V_S anisotropy in the oceanic crust.

- The magnitude of V_S anisotropy in Hole 395A is 5.3%, on average, and 5.8% in Hole 735B. In both cases, anisotropy tends to decrease with depth where the overburden pressure increases and the closure of cracks and infilling of fractures by chemical alteration occurs. Anisotropy decreases more rapidly with depth in the fractured basalt encountered in Hole 395A than in the gabbro in Hole 735B; however, the magnitude of V_S anisotropy correlates with the degree of deformation and compositional changes observed in cores from Hole 735B.

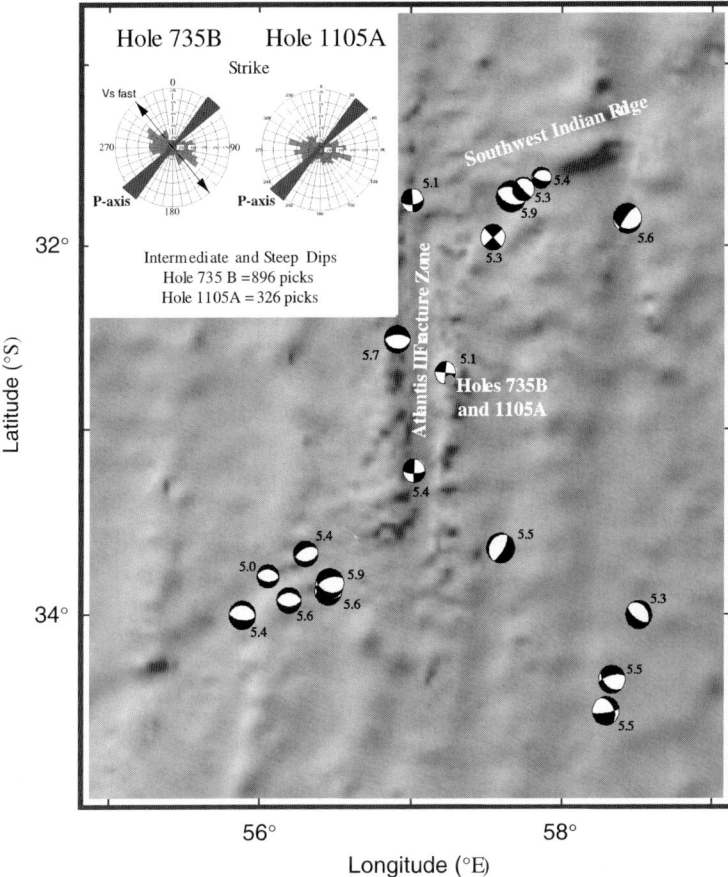

Fig. 9. Bathymetric map of the area in the vicinity of the Atlantis II Fracture Zone, indicating the locations of Holes 735B and 1105A. Focal mechanisms from CMT Harvard solutions are also displayed. Numbers next to focal mechanisms indicate earthquake magnitudes (Mw). The insert shows strike orientations from structural picks obtained from FMS analyses from both holes. The mean orientation of earthquake P-axis data is perpendicular to the fast-angle orientation (arrow).

- Resolving the sources of V_S anisotropy in the ocean crust is complex at both Atlantic Ocean and Indian Ocean sites, caused in part or wholly by intrinsic effects from fracturing, foliation, and/or porosity heterogeneity in the vicinity of the borehole, as well as by stresses potentially inducing changes in the local borehole environment. In Hole 395A, variations in anisotropy can at least be partially attributed to pillow morphologies and fracturing in oceanic Layer 2. In Hole 735B, the primary influence is probably the texture, fabrics and fracturing in these deformed gabbroic rocks, although this is complicated by the proximity of the hole to the Atlantis Fracture Zone and related tectonic stress effects.

- The fast angle of V_S anisotropy determined from this analysis shows somewhat similar results for the two sites. The results from Hole 395A vary considerably about a mean direction of N75°E over the lower portion of the hole, oblique to the strike of the nearest segment of the Mid-Atlantic Ridge and aligned 110° from the earthquake compression axis along the Kane Fracture Zone. In Hole 735B, the principal fast angle varies considerably about a mean direction of S45°E over the upper portion of the hole, perpendicular to the nearest segments of the Southwest Indian Ridge and aligned 90° from the earthquake compression axis along the Atlantis II Fracture Zone.

We thank the crew, technical staff, and the scientists of the *JOIDES Resolution* during ODP Legs 174 and 176 for their efforts at sea that enabled these data to be acquired. The suggestions of two anonymous reviewers and the editors are appreciated, and improved the manuscript. The US Science Support Program, the USSSP Site Augmentation program, and the US National Science Foundation supported this research.

References

ALFORD, R. M. 1986. Shear data in the presence of azimuthal anisotropy. *56th Annual International Meeting, Society of Exploration Geophysics, Expanded Abstracts*, 476–479.

ANDO, M., ISHIKAWA, T. & YAMAZAKI, F. 1983. Shear-wave polarization anisotropy in the upper mantle beneath Honshu, Japan. *Journal of Geophysical Research*, **88**, 5850–5864.

BECKER, K., MALONE, M. J. *et al.* 1998. *Proceedings of ODP, Initial Reports*, **174B**, College Station, TX (Ocean Drilling Program).

BRYAN, W. B., JUTEAU, T. *et al.* 1988. *Proceedings of ODP, Initial Reports*, **109**, College Station, TX (Ocean Drilling Program).

Cornell University GIS group 1998. *Centroid Moment Tensor Seismicity Catalog and Focal Mechanisms*, Cornell University GIS group, Ithaca, New York. World Wide Web Address: (http://atlas.geo.cornell.edu).

CRAMPIN, S. 1981. A review of wave motion in anisotropic and cracked elastic media. *Wave Motion*, **3**, 343–391.

DIANAUT, 1992. The DIANAUT Program. *Geophysical Research Letters*, **19** (5), 493–524.

DICK, H. J. B., NATLAND, J. H., MILLER, D. J. *et al.* 1999. *Proceedings of ODP, Initial Reports*, **176**, College Station, TX (Ocean Drilling Program).

DICK, H. J. B., SCHOUTEN, H. *et al.* 1991. Tectonic evolution of the Atlantis II Fracture Zone. *In*: VON HERZEN, R. P., ROBINSON, P. T., *et al.*, *Proceedings of ODP, Scientific Results*, **118**, Ocean Drilling Program, College Station, TX, 359–398.

DZIEWONSKI, A. & WOODHOUSE, J. 1983. Studies of the seismic source using normal-mode theory. *In*: KANAMORI, H. & BOSHCHI, E. (eds) *Earthquakes: Observation, Theory, and Interpretation: Notes From the International School of Physics, Enrico Fermi*, North-Holland Publishing Company, Amsterdam, 45–137.

FRYER, G. & WILKENS, R. 1990. Fractures, cracks, and large-scale porosity. *In*: PURDY, G. M. & FRYER, G. J. (eds) *Proceedings, JOI/USSSP Workshop, Physical Properties of Volcanic Seafloor*, 121–126.

HYNDMAN, R. D. & SALISBURY, M. H. 1984. The physical nature of young oceanic crust on the Mid-Atlantic Ridge, Deep Sea Drilling Project Hole 395A. *In*: HYNDMAN, R. D., SALISBURY, M. H. *et al. Initial Reports of DSDP*, 78B, US Government Printing Office, Washington, 839–848.

ITURRINO, G. J., CHRISTENSEN, N. I., KIRBY, S. & SALISBURY, M. H. 1991. Seismic velocities and elastic properties of oceanic gabbroic rocks from Hole 735B. *In*: VON HERZEN, R. P., ROBINSON, P. T., ADAMSON, A. *et al.* 1991. *Proceedings of ODP, Scientific Results*, **118**, 227–244.

ITURRINO, G. J., CHRISTENSEN, N. I., BECKER, K., BOLDREEL, L. O., HARVEY, P. K. H. & PEZARD, P. 1995. Physical properties and elastic constants of upper crustal rocks from core-log measurements in Hole 504B. *Proceedings of ODP, Scientific Results*, **140**, Ocean Drilling Program, College Station, TX, 273–291.

ITURRINO, G. J., ILDEFONSE, B. & BOITNOTT, G. 2002. Velocity structure of the lower oceanic Crust. Results from Hole 735B, Atlantis II Fracture Zone. *In*: NATLAND, J. H., DICK, H. J. B., MILLER, D. J. & VON HERZEN, R. P. (eds) *Proceedings of ODP, Scientific Results*, 176, 71 pp.

JOYCE, R., PATTERSON, D. & THOMAS, J. 1998. Advanced interpretation of fractured carbonate reservoirs using four-component cross-dipole analysis. *39th Annual Meeting, Society of Professional Well Log Analysts, Transactions*, Paper R.

KERN, H., LIN, B. & POPP, T. 1997. Relationship between anisotropy of P and S wave velocities and anisotropy of attenuation in serpentinite and amphibolite. *Journal of Geophysical Research*, **102**, 3051–3065.

KIMBALL, C. V. 1998. Shear slowness measurements by dispersive processing of the borehole flexural mode. *Geophysics*, **63**, 337–344.

LANG, S. W., KURKJIAN, A. L., MCCLELLAN, J. H., MORRIS, C. F. & PARKS, T. W. 1987. Estimating slowness dispersion from arrays of sonic logging waveforms. *Geophysics*, **52**, 530–544.

MELSON, W. G., RABINOWITZ, P. D. *et al.* 1978. *Initial Reports of DSDP*, **45**, US Government Printing Office, Washington DC.

MOOS, D. 1990. Petrophysical results from logging in DSDP Hole 395A, ODP Leg 109. *In*: DETRICK, R., HONNOREZ, J., BRYAN, W. B., JUTEAU, T. *et al.* 1990. *Proceedings of ODP, Scientific Results*, **109**, 237–253.

MOOS, D. & MARION, D. 1994. Morphology of extrusive basalts and its relationship to seismic velocities in the shallow oceanic crust. *Journal of Geophysical Research*, **99**, 2985–2994.

MÜLLER, M., BOYD, A. & ESMERSOY, C. 1994. Case studies of the dipole shear anisotropy log. *Transactions, 64th Annual Symposium Society of Exploration Geophysics, Expanded Abstracts*, 1143–1146.

PATTERSON, D. & SHELL, G. 1997. Integration of cross-dipole acoustics for improved formation evaluation. *Transactions, SPWLA, Annual Symposium*, **38**, paper E.

PURDY, G. M. 1987. New observations of the shallow seismic structure of young oceanic crust. *Journal of Geophysical Research*, **92**, 9351–9362.

ROBINSON, P. T., VON HERZEN, R. P., ADAMSON, A. *et al.* 1989. *Proceedings of ODP, Initial Reports*, **118**, Ocean Drilling Program, College Station, TX.

SCHOENBERG, M. & SAYERS, C. 1995. Seismic anisotropy of fractured rock. *Geophysics*, **60**, 204–211.

SHEARER, P. & ORCUTT, J. A. 1985. Anisotropy in the oceanic lithosphere – theory and observations from

Ngendei seismic refraction experiment in the southwest Pacific. *Geophysical Journal of Royal Astronomical Society*, **80**, 493–526.

SHEARER, P. & ORCUTT, J. A. 1986. Compressional- and shear-wave anisotropy in the oceanic lithosphere – the Ngendei seismic refraction experiment. *Geophysical Journal of Royal Astronomical Society*, **87**, 967–1003.

SINHA, B. K., NORRIS, A. N. & CHANG, S. K. 1994. Borehole flexural modes in anisotropic formations. *Geophysics*, **59**, 1037–1052.

SINHA, B. K., KANE, M. & FRIGNET, B. 2000. Dipole dispersion crossover and sonic logs in a limestone reservoir. *Geophysics*, **65**, 390–407.

STAKES, D., MÉVEL, C., CANNAT, M. & CHAPUT, T. 1991. Metamorphic stratigraphy of Hole 735B. *In*: VON HERZEN, R. P., ROBINSON, P. T. *et al. Proceedings of ODP, Scientific Results*, **118**, Ocean Drilling Program, College Station, TX, 153–180.

STEPHEN, R. A. 1981. Seismic anisotropy observed in the oceanic crust. *Geophysical Research Letters*, **8**, 865–868.

TANG, X. & CHUNDURU, R. K. 1999. Simultaneous inversion of formation shear-wave anisotropy parameters from cross-dipole acoustic-array waveform data. *Geophysics*, **64** (5), 1502–1511.

THOMSEN, L. 1977. Weak elastic anisotropy. *Geophysics*, **51**, 1954–1966.

WADE, J. M., HOUGH, G. V. & PEDERSEN, S. H. 1998. Practical methods employed in determining permeability anisotropy for optimization of a planned water flood. *Annual Technology Conference, Society of Petroleum Engineers*, SPE paper **4896**.

WHITE, R. S. & WHITMARSH, R. B. 1984. An investigation of seismic anisotropy due to cracks in the upper oceanic crust at 45°N, Mid-Atlantic Ridge. *Geophysical Journal of the Royal Astronomical Society*, **79**, 439–467.

WIENS, D. & STEIN, S. 1984. Intraplate seismicity and stresses in young oceanic lithosphere. *Journal of Geophysical Research*, **89**, 11 442–11 464.

WINKLER, K. W., SINHA, B. K. & PLONA, T. J. 1998. Effects of borehole stress concentrations on dipole anisotropy measurements. *Geophysics*, **63**, 11–17.

ZOBACK, M., MOOS, D., MASTIN, L. & ANDERSON, R. N. 1985. Well bore breakouts and *in situ* stress. *Journal of Geophysical Research*, **90**, 5523–5530.

Reactive flow and permeability prediction – numerical simulation of complex hydrogeothermal problems

J. BARTELS[1], C. CLAUSER[2], M. KÜHN[3], H. PAPE[2] & W. SCHNEIDER[3]

[1]*Geothermie Neubrandenburg Ltd., Postfach 110120, D-17041 Neubrandenburg, Germany*
[2]*RWTH Aachen University, Applied Geophysics, Lochnerstr. 4-20, D-52056 Aachen, Germany (e-mail: c.clauser@geophysik.rwth-aachen.de)*
[3]*CSIRO–ARRC/Exploration and Mining, 26 Dick Perry Ave, Technology Park, Kensington, Perth, WA 6151, Australia*

Abstract: Simulating complex flow situations in hydrogeothermal reservoirs requires coupling of flow, heat transfer, transport of dissolved species, and heterogeneous geochemistry. We present results of simulations for typical applications using the numerical simulator SHEMAT/Processing SHEMAT. Heat transfer is non-linear, since all thermal fluid and rock properties depend on temperature. Due to the coupling of fluid density with both temperature and concentrations of dissolved species, the model is well suited to simulate density-driven flow.

Dissolution and precipitation of minerals are calculated with an improved version of the geochemical modelling code PHRQPITZ, which accurately calculates geochemical reactions in brines of low to high ionic strength and temperatures of $0-150°C$. Changes in pore space structure and porosity are taken into account by updating permeability with respect to porosity changes due to precipitation and dissolution of minerals. This is based on a novel relationship between porosity and permeability, derived from a fractal model of the pore space structure and its changes due to chemical water–rock interaction.

A selection of model studies performed with SHEMAT completes the review. Examples highlight both density-driven and reactive flow with permeability feedback. With respect to the former, the thermohaline free convection Elder's problem, and density-driven free convection in a coastal aquifer with geothermal exploitation, are considered. Mineral redistribution and associated permeability change during a core flooding experiment; reaction front fingering in reservoir sandstone; and long-term changes in reservoir properties during the operation of a geothermal installation, are all considered in relation to reactive flow with permeability feedback.

This review discusses some of the recent advances in simulating complex flow and transport in porous media using the public-domain and easy-to-use numerical simulation system SHEMAT/Processing SHEMAT (Clauser 2003). The processes considered are characterized by non-linear coupling of flow, heat transfer, transport of dissolved species, and chemical reactions, in particular precipitation and dissolution.

The technical development of SHEMAT/Processing SHEMAT was initiated by the need to predict the long-term behaviour of reservoir properties in productive hydrogeothermal fields. Hydrothermal heat-mining installations usually involve two or more production and injection wells. In order to pay off, operation times of 30 years and more are required, during which sufficiently large flow rates must be maintained at the injection and production wells. Therefore,

permeability evolution, particularly immediately around the wells, is a key parameter for commercially successful heat mining.

Reservoir fluids are often highly saline brines, such as, for instance, in the north German sedimentary basin, where thermal waters are almost saturated with sodium chloride, alkali earth sulphates, and carbonates. When cooled brines are re-injected into a reservoir, the chemical equilibrium between the formation water and the reservoir rocks is disturbed. This could cause temperature-induced dissolution or precipitation of minerals, if not prevented by kinetic barriers. In the worst case this may cause a more or less irreversible sealing of the reservoir.

Additionally, this complex regime is characterized by steep gradients in temperature and concentration of dissolved salts, as well as by comparatively fast processes associated with

From: HARVEY, P. K., BREWER, T. S., PEZARD, P. A. & PETROV, V. A. (eds) 2005. *Petrophysical Properties of Crystalline Rocks*. Geological Society, London, Special Publications, **240**, 133–151.
0305-8719/05/$15.00   © The Geological Society of London 2005.

the high flow rates and temperature differences required for economic heat mining. Such a complex scenario can be analysed by numerical simulation only. However, presently available numerical models (e.g. 3DHYDROGEOCHEM (Cheng & Yeh 1998) and TOUGHREACT (Pham *et al.* 2001)), or a number of chemical reaction modules for the THOUGH2 simulator (White 1995; White & Mroczek 1998) do not meet all of the requirements for a suitable numerical code. In particular, their chemical data-sets are not valid over the entire required temperature range of 0–150°C or for high ionic strength. SHEMAT, in contrast, comprises a chemical speciation code, which is valid for high temperatures and salinities typical for many hot water reservoirs.

The permeability–porosity relationship implemented in current simulation codes is based on a pore-space geometry, which is too simple for most real sandstones. The popular relation in which permeability varies with the cube of porosity is based on a geometrical pore-space model consisting of smooth spherical grains or parallel bundles of capillary tubes. In contrast, data measured on core samples indicate that permeability varies with porosity taken to powers of both smaller and larger than three, depending on the range of porosity and rock type. Therefore, a new relationship between porosity and permeability was derived (Pape *et al.* 1999). This approach was implemented in SHEMAT, a finite difference (FD) code for coupled flow and heat transfer (Clauser & Villinger 1990).

Model initialization requires information or at least assumptions on the permeability distribution in the model domain. In the absence of measurements on core samples, permeability must be estimated from other parameters known from borehole or laboratory measurements. For these purposes, a permeability estimator was developed and integrated in the graphical user interface Processing Shemat (PS). This versatile graphical user interface generally facilitates model set-up and modification, parameter control, and post-processing.

Here we provide information on these novel features and illustrate them by characteristic applications. Details of the physical, chemical, and numerical approaches, the parameters and their dependences, and the governing equations, can be found in the quoted literature.

General model features

SHEMAT (**S**imulator for **HE**at and **MA**ss **T**ransport) is a code with special features for simulating steady-state and transient processes in hydrogeothermal reservoirs in two and three dimensions. It is particularly well suited to predict the long-term behaviour of heat-mining installations in hot aquifers with highly saline brines. It can handle a wide range of timescales, both of technical and of geological processes. Moreover, it is also a truly multipurpose simulator for a wide variety of thermal and hydrogeological problems.

SHEMAT uses a finite difference method to solve the partial differential equations. The flow and transport equations are solved on a Cartesian 2D or 3D grid with co-ordinates x, y, z or, alternatively, and on a 2D vertical, cylindrically symmetrical grid with radius and depth co-ordinates r and z, respectively.

Flow is calculated from the combination of the Darcy equation with mass conservation. Heat transfer is composed of advective and conductive contributions. Heat may be generated by basal heat flow, rock heat production due to the contributions of the constituent minerals, and fluid heat production. The rock heat production rate may vary in space according to variations in the concentration of minerals in the model domain. The fluid heat production rate is assumed to be constant.

Three schemes are available for the spatial discretization of the advection term in the transport equations for heat and dissolved species:

(1) the Il'in flux blending scheme (Clauser & Kiesner 1987);
(2) the common pure upwind scheme; and
(3) the Smolarkiewicz diffusion corrected upwind scheme (Smolarkiewicz 1983). The resulting system can be solved explicitly, implicitly or semi-implicitly. For implicit and semi-implicit time-weighting, the sets of linear equations are solved iteratively.

In SHEMAT, the different flow, transport, and reaction processes can be coupled in various ways. Different kinds of coupling can be invoked individually. Flow is non-linear because of the pressure dependence of the fluid properties: density, viscosity and compressibility. Heat transport is non-linear because of the temperature dependence of rock and fluid thermal conductivity, and volumetric thermal capacity. In this way, flow and heat transport are mutually coupled.

In detail, fluid thermal conductivity is calculated after Kestin (1978). The weighting between fluid and rock thermal conductivity, λ_f and λ_r, is either linear:

$$\lambda(\phi, T, P) = (1 - \phi)\lambda_r(T) + \phi\lambda_f(T, P) \quad (1)$$

or geometrical:

$$\lambda(\phi, T, P) = \lambda_r^{(1-\phi)}(T) + \lambda_f^{\phi}(T, P) \quad (2)$$

with porosity ϕ.

The temperature dependence of rock thermal conductivity is accounted for according to Zoth & Hänel (1988):

$$\lambda_r(T > 400°C) = 770/(350 + T) + 0.7$$

$$\begin{aligned}\lambda_r(T < 400°C) = \lambda_r(T > 400°C) \\ \times [(\lambda_r(T_{ref})/(770/(350 + T_{ref})) \\ + 0.7) - ((\lambda_r(T_{ref}))/ \\ (770/(350 + T_{ref}) + 0.7) - 1) \\ \times (T - T_{ref})/(400 - T_{ref})] \quad (3)\end{aligned}$$

where T and T_{ref} are in °C. In SHEMAT the reference temperature is chosen as $T_{ref} = 20°C$.

The isobaric specific heat capacity of rocks, c_P, can be described by a second-order polynomial:

$$c_p(T) = A_0 + A_1 T + A_2 T^2 \quad (4)$$

where the coefficients depend on rock type, and T is in °C.

Flow depends on heat transport via the temperature dependence of the fluid properties density, viscosity and compressibility. Fluid density depends on temperature and the concentration of dissolved salt. This can give rise to buoyancy-driven free convection. Density coupling is implemented via a linear relation between salt concentration and brine density. Flow and heterogeneous chemical reactions are coupled in such a way that porosity is modified according to the volume consumed or liberated by precipitation or dissolution. The corresponding permeability change can be calculated from a variety of relations discussed in detail below. These permeability changes result in a modification of the flow field, which again influences the transport of the reaction species.

New approaches and modelling tools

Chemical reactions in brine up to 150°C

Equilibrium solubility. SHEMAT's chemical speciation module CHEMEQ (Kühn *et al.* 2002*b*) is a modification of the geochemical modelling code, PHRQPITZ (Plummer *et al.* 1988). Plummer *et al.* (1988) developed PHRQPITZ from the aqueous ion-pairing model PHREEQE (Parkhurst *et al.* 1980) by implementing the Pitzer virial-coefficient approach (Pitzer & Mayorga 1973, 1974; Pitzer & Kim 1974;

Pitzer 1975). This makes it possible to calculate geochemical reactions in brines and other highly concentrated electrolyte solutions using the Pitzer virial-coefficient approach for correcting the activity coefficients. Options for modelling chemical reactions include aqueous speciation and mineral saturation as well as mineral solubility. The Pitzer treatment of the aqueous model is largely based on the equations as presented by Harvie & Weare (1980) and Harvie *et al.* (1984).

We modified this code to include the calculation of temperature-dependent Pitzer coefficients. An expanded database of Pitzer interaction parameters is provided for the system Na–K–Ca–Mg–Sr–Ba–Si–H–Cl–SO_4–OH–(HCO_3–CO_3–CO_2)–H_2O, which is valid for temperatures from 0 to 150°C. The Pitzer treatment is largely based on the data of Greenberg & Moller (1989). Data for the incorporated carbonic acid system (in parentheses in the list above) are valid for temperatures from 0 to 90°C, according to He & Morse (1993). The database is extended with Pitzer interaction parameters for the elements Ba and Sr (Monnin 1999), as well as Si (Azaroual *et al.* 1997).

The temperature dependence of the solubility of several of the minerals within the database is not known, and large errors could result if calculations are performed at temperatures other than 25°C for these solids. The minerals which are tested for the temperature range 0–150°C are the chlorides halite (NaCl) and sylvite (KCl), the sulphates mirabilite ($Na_2SO_4 \cdot 10H_2O$), arcanite (K_2SO_4), glaserite ($NaK_3(SO_4)_2$), anhydrite ($CaSO_4$), gypsum ($CaSO_4 \cdot 2H_2O$), glauberite ($Na_2Ca(SO_4)_2$), labile salt ($Na_4Ca(SO_4)_3 \cdot 2H_2O$), syngenite ($K_2Ca(SO_4)_2 \cdot H_2O$), celestite ($SrSO_4$), and baryte ($BaSO_4$), as well as quartz and silica (SiO_2). The phases tested for the temperatures 0–90°C are calcite ($CaCO_3$) and carbon dioxide (CO_2).

It is an important feature that the data-set for the Pitzer coefficients and the mineral reactions can be adapted or extended by a skilled user without the need to access the model code.

Reaction kinetics. For a *single* mineral species, i, the change in concentration in a time interval Δt due to precipitation or dissolution is

$$\Delta c_{min,i} = \Delta t \times A_{react} \times r_0$$
$$\times \underbrace{\exp{(E_{act}/k_B T)}}_{\text{thermal activation}}$$
$$\times \underbrace{(c_{ion}(t) - c_{eq}(t))/(c_{eq}(t))}_{\text{supersaturation}} \quad (5)$$

where $A_{react} = A_{total} \times x_{min,i}$ is the reactive surface; A_{total} is the total specific internal surface in $m^2\,m^{-3}$; $x_{min,i} = 0\ldots1$ is the fraction of the total surface covered by the mineral i; $\Delta c_{min,i}$ is the change in concentration (mmol L^{-1}) of the mineral i; $c_{ion}(t)$ and c_{eq} are the concentration and the equilibrium concentration (mmol L^{-1}) of the ion species of the ion-pair forming mineral i; r_0 is the reaction rate (mol $s^{-1}\,m^{-2}$); E_{act} is the activation energy for the molecular surface processes associated with dissolution and precipitation, and k_B is Boltzmann's constant, equation (5) reflects the essential functional dependencies of the reaction kinetics. The reaction-specific kinetic parameters of equation (5), i.e. A_{react}, r_0 and E_{act} have to be specified by the modeller via the corresponding input window of the user interface. They can be found in the literature or determined experimentally.

Fractal and other relationships between porosity and permeability, for coupling reaction to flow

The novel relationship between permeability and porosity changes implemented in SHEMAT is based on the assumption that the shape of the internal rock surface follows a self-similar rule. Thus, the theory of fractals can be applied. The fractal relationship between permeability k and porosity ϕ based on the Kozeny–Carman equation was expressed by Pape *et al.* (1999) as a general three-term power series in porosity where the exponents $E_{f,i}(i = 1, 2, 3)$ depend on the fractal dimension of the internal surface of the pore space:

$$k = A\phi^{E_{f,1}} + B\phi^{E_{f,2}} + C\phi^{E_{f,3}}. \qquad (6)$$

The coefficients A, B and C need to be calibrated for each type of sedimentary basin or pore space modification, i.e. porosity change due to chemical reactions. Equation (6) reflects the fact that in different intervals of porosity different processes dominate the changes in porosity and permeability. In SHEMAT this is approximated by equation (7), and the option available to the modeller to define different exponents for three different porosity intervals. In equation (7), k_0 and ϕ_0 denote the initial values, which represent the same information as the coefficients in equation (6):

$$k = k_0(\phi/\phi_0)^{E_f}. \qquad (7)$$

Further, a number of well-established k–ϕ relations from the literature are implemented and

can be selected as an alternative to the fractal relation (equation (7)). They are summarized in Zarrouk & O'Sullivan (2001):

(1) An equation of Weir & White (1996), for the deposition of spheres on a surface in dense, rhombohedral packing; ϕ_c is a critical porosity, below which the permeability vanishes:

$$k = k_0\left\{1 - \left[1 - \left(\frac{\phi - \phi_c}{\phi_0 - \phi_c}\right)^{1.58}\right]^{0.46}\right\}. \qquad (8)$$

(2) The Blake–Kozeny equation (McCume *et al.* 1979) for flow in packed columns and applied permeability changes due to matrix acidizing in hydrocarbon wells:

$$k = k_0(\phi/\phi_0)^3\left(\frac{1 - \phi_0}{1 - \phi}\right)^2. \qquad (9)$$

(3) The Blake–Kozeny equation modified by Lichtner (1996) for the dependence of permeability on porosity in a mixture of K-feldspar, gibbsite, kaolinite and muscovite:

$$k = k_0(\phi/\phi_0)^3\left(\frac{1.001 - \phi_0^2}{1.001 - \phi^2}\right). \qquad (10)$$

(4) The Kozeny–Stein equation for the precipitation of silica in the vicinity of injection wells (Itoi *et al.* 1987):

$$k = k_0(\phi/\phi_0)^3\left(\frac{1 - \phi_0}{1 - \phi}\right)^2\left[\sqrt{\frac{\phi_0 - \phi}{3(1 - \phi_0)} + \frac{1}{4}}\right. $$
$$\left. + \frac{\phi_0 - \phi}{3(1 - \phi_0)} + \frac{1}{2}\right]. \qquad (11)$$

(5) The equation of Schechter & Gidley (1969) for permeability changes due to matrix acidizing in hydrocarbon wells in limestone:

$$k = k_0(\phi/\phi_0)^2 e^{2(\phi - \phi_0)}. \qquad (12)$$

Two case studies in the Application section below, i.e. the core flooding experiment and the heat-mining simulation, demonstrate, among others, the sensitivity of flow and transport behaviour on the chosen porosity–permeability relation and its parameters.

Graphical user interface with permeability estimator

The graphical user interface Processing SHEMAT (PS), based on PMWIN (Chiang & Kinzelbach 2001), supports grid generation and grid refinement as well as input of geological structures, material properties, mineral composition of the reservoir rocks, and geochemical fluid data. It also makes it possible to specify coupling parameters, monitoring points, and numerical and control parameters. It provides a quick first visualization of the simulation results and export options to various input formats of standard visualization software.

Among a number of tools supporting model set-up and result analysis (e.g. field interpolator, digitizer, background map functions, and a log monitoring tool) PS comes with a permeability estimator. Depending on existing data from measurements in the laboratory or in boreholes, different relations are available, which make it possible to estimate the initial permeability of a given domain in the model. A total of seven different methods are classified according to the input parameters required; all of them require a value for porosity.

Fractal exponents and coefficients (fractal Kozeny–Carman)

The fractal relationship for an average sandstone may be applied if no other information is available except for porosity and the fact that no secondary diagenetic cementation occurred:

$$k = A\phi + B\phi^2 + C(10\phi)^{10}. \tag{13}$$

The values of $A = 31$, $B = 7463$, and $C = 191$ are derived from a calibration with a large set of measurements of k and ϕ in the laboratory; the default values of 1, 2 and 10 for the exponents result from a fractal approach assuming a self-similar pore space structure (Pape et al. 1999). The user can modify them.

Capillary-bound-water fraction, tortuosity, and a fractal exponent

Depending on the ratio of the volume fractions of free fluid, ϕ_{ff} and capillary-bound water, ϕ_{cb}, where $\phi = \phi_{cb} + \phi_{ff}$, permeability k is estimated from two different relations:

$$\left.\begin{array}{l} k = C(100\phi_{ff})^{E_{f,3}} \\ k = \dfrac{1}{8T}(1000^2\phi_{ff} + 50^2\phi_{cb}) \end{array}\right\} (\text{nm}^2) \begin{cases} \phi_{ff} > 0.75\phi \\ \phi_{ff} < 0.75\phi. \end{cases} \tag{14}$$

Tortuosity, T, is defined by:

$$T = F\phi, \tag{15}$$

where F is the electrical formation factor. It may be obtained either from the ratio of the electrical resistances R_t and R_f of the bulk rock and the pore fluid, respectively:

$$F = R_t/R_f \tag{16}$$

or from Archie's law:

$$F = \frac{a}{\phi^m} \tag{17}$$

where a is a lithology-dependent factor with values in the range of 0.6–2.0 (default: $a = 0.7$); m is a cementation exponent, with values in the range of 1 to 3 (default: $m = 2.0$).

The ϕ_{cb} can be determined with nuclear magnetic resonance (NMR) measurements in the laboratory or in boreholes. The exponent $D_{f,3}$ and coefficient C are identical to those discussed in method 5 for a series of sandstone cementation types. They are required if the free fluid porosity dominates, i.e. for $\phi_{ff} > 0.75 \, \phi$.

Tortuosity, internal surface, and fractal dimension

If the pore surface area per pore volume, S_{por} (nm^{-1}), the fractal dimension, D_f (—), and the tortuosity, T (—) are known, k can be derived from:

$$k = 0.223\frac{\phi}{T}(S_{por})^{-2/(3-D_f)} \quad (\text{nm}^2). \tag{18}$$

From nitrogen adsorption (BET) measurements, the internal surface related to mass, S_m, is often known. It is related to S_{por} via density, ρ, by $S_{por} = S_m\rho/\phi$. Alternatively, S_{por} can be determined from the capillary-bound-water fraction obtained from NMR measurements: $S_{por} = \phi_{cb}/(50\phi)$ or from the shale volume fraction, V_{sh}: $S_{por} = 0.313 \, V_{sh}$, where S_{por} is in μm^{-1} and V_{sh} is the non-dimensional volume shale fraction with values between 0 and 1.

For fine-grained crystalline matrix cement, S_{por} (μm^{-1}) can be calculated from the cement mineral fraction V_{cem}, and an average grain size of the cement crystallites, $r_{grain,cem}$ (μm):

$$S_{por} = \frac{1-\phi}{\phi}\left\{0.4 + \left(\frac{3}{r_{grain,cem}} - 0.4\right)V_{cem}\right\}. \tag{19}$$

Processing SHEMAT assumes default values of $D_f = 2.36$ in sandstone and $D_f = 2.0$ for purely quartz cemented sandstone with $\phi > 0.1$. In an average sandstone, $S_{por} = 4\ \mu m^{-1}$ can be assumed as a reasonable average value. The shale volume fraction V_{sh} can be determined from measurements of the natural γ-radiation.

Further, the volume fractions of shale or cement can be derived from the density of the sandstone, if the density contrast between quartz and the cement mineral is significant, and if there is only one type of cementation. If ρ is the density of the bulk rock, the density of the pure mineral phase without pores, ρ_{solid}, is:

$$\rho_{solid} = \frac{\rho}{1 - \phi}. \tag{20}$$

In a shaly sandstone V_{sh} and V_{QF} denote the volume fractions of shale and quartz or feldspar relative to the solid volume (i.e. $1-\phi$). Then $V_{QF} + V_{sh} = 1$, and the relative shale volume V_{sh} is given by:

$$V_{sh} = \frac{\rho_{solid} - \rho_{QF}}{\rho_{sh} - \rho_{QF}}. \tag{21}$$

In the same way, if V_{cem} is the cement mineral fraction of a cemented sandstone, $V_{QF} + V_{cem} = 1$, and one obtains the relative cement mineral volume V_{cem} as:

$$V_{cem} = \frac{\rho_{solid} - \rho_{QF}}{\rho_{cem} - \rho_{QF}}. \tag{22}$$

Table 1 lists average densities for some common minerals.

Fraction of clay minerals

If only the volume fraction of shale is known for different ranges of the shale fraction (determined e.g. as described in method 3 above), one obtains the coefficients for equation (13), which are listed in Table 2.

Structure of cement minerals

If the porosity reduction in sandstone is mainly due to cementation and not to compaction, the general relationship of equation (6) reduces to a single exponential term with appropriate combinations of coefficients and fractal exponents for different cementation structures:

(1) Pure quartz sandstone with quartz cementation

$$\left.\begin{array}{l} k = 303(100\phi)^{3.05} \\ k = 0.0275(100\phi)^{7.33} \end{array}\right\} (nm^2)\left\{\begin{array}{l} \phi > 0.08 \\ \phi < 0.08 \end{array}\right. \tag{23}$$

Table 1. *Some mineral densities useful for calculating the shale and matrix cement volume fractions,* V_{sh} *and* V_{cem}

Mineral	Density (kg m^{-3})
Quartz	$\rho_Q = 2650$
Alkali feldspar	$\rho_F = 2570$
Plagioclase	$\rho_P = 2620$
(Quartz + feldspar) in very clean quartz sandstone	$\rho_{QF} = \rho_Q$
(Quartz + feldspar) in average quartz sandstone	$\rho_{QF} = 2630$
Shale minerals	$\rho_{sh} = 2630$
Calcite	$\rho_{cem} = 2720$
Anhydrite	$\rho_{cem} = 2960$

(2) Pure quartz sandstone with coarse-grained anhydrite cementation

$$k = 0.309(100\phi)^{4.84}. \tag{24}$$

(3) Fine-grained ($\phi < 25\ \mu m$) cementation formed on a technical time-scale (re-injection)

$$k = 178(10\phi)^{11.33}; \tag{25}$$

(4) Calcite coating of grains formed on a technical time-scale (re-injection):

$$k = 42073(10\phi)^5. \tag{26}$$

Pore throat radii R25/R50 (from mercury injection)

Required values are tortuosity, T, and the quartile values of the pore throat radius distribution, $R25$ and $R50$ (in nm), corresponding to a mercury saturation of the pore space of 25% and 50%, respectively

$$\left.\begin{array}{l} k = \dfrac{\phi}{8T}(R50)^2 \\[2mm] k = \dfrac{\phi}{32T}(R25)^2 \end{array}\right\} (nm^2)\left\{\begin{array}{l} R50 \geq (R25)/2 \\ R50 < (R25)/2 \end{array}\right. \tag{27}$$

Note that $R25$ is always larger than $R50$, because mercury needs to pass through ever

Table 2. *Coefficients for equation 13 for different ranges of* V_{sh}

V_{sh}	A	B	C
<0.02	31	7463	191
0.02–0.10	155	37 315	630
0.10–0.30	6.2	1493	58
>0.50	0.1	26	1

smaller pore throats with increasing mercury saturation.

If tortuosity is not available then it can be calculated from the effective grain radius, r_{grain}, and the fractal dimension, D_f, using the relation:

$$T^{-1} = 0.1 + 0.5(R50/r_{grain})^{0.67(D_f-2)}. \quad (28)$$

Reasonable default values are $D_f = 2.36$ and $r_{grain} = 200\,000$ nm.

Particle-size distribution

Permeability can be estimated from particle-size distribution in various ways. The Kozeny–Köhler approach (see Langguth & Voigt 1980) requires values for the lower and upper limits of each grain-size class i, Rl_i and Ru_i, and its volume fraction, V_i, for computing an effective grain size, r_{grain}:

$$r_{grain} = \frac{\sum V_i}{\sum V_i/\bar{R}_i}, \quad \text{with:}$$
$$\frac{1}{\bar{R}_i} = \frac{1}{2}\left(\frac{1}{Rl_i} + \frac{1}{Ru_i}\right), \quad (29)$$

from which hydraulic conductivity K follows as $K \propto (r_{grain})^2$. Relating the associated permeability k to that of an average sandstone with an average grain size of 200 000 nm (equation (13), Table 2) yields permeability k as:

$$k = \left(\frac{r_{grain}}{200\,000}\right)^2$$
$$\times (155\phi + 37\,315\phi^2 + 630(10\phi)^{10}).$$

Applications

The importance of density-driven flow and reactive transport with permeability feedback is illustrated for five example simulations.

Density-driven flow

Thermohaline Elder's problem. Thermohaline flow becomes relevant when the density of a fluid varies substantially as a function of temperature and the concentration of dissolved salts. This occurs, for instance, in geothermal energy production or waste disposal in salt formations. Thus, thermohaline effects are important in the production of mineralized thermal water, in thermal water production in coastal aquifers, and in groundwater flow near salt domes.

The original 'Elder's problem' for salt transport was devised by analogy with a laboratory experiment and associated numerical calculations by Elder (1967). Heat convection in a

porous medium was modelled in a Hele–Shaw cell experiment, where a viscous fluid, trapped between two closely spaced vertical plates, was heated from below. This produced a typical, transient 'fingering' pattern.

Based on the analogy between heat- and salt-induced density-driven transport, parameters for a standardized numerical experiment were derived from this experiment (see, for example, Kolditz *et al.* 1998). On a 2D vertical cross-section of 600 m × 150 m, a continuous, constant salt source of normalized concentration $c = 1$ and a length of 300 m is located at the centre of the upper boundary. The bottom of the model is assigned a fixed concentration of $c = 0$. All boundaries of the cross-sections are impervious, except for the two top corner points, where a constant pressure of $p = 0$ is prescribed (by analogy with Fig. 1, but isothermal). Detailed studies of this unstable, non-linear flow problem showed that the numerical solution depends on the level of approximation of the flow equation, the grid resolution, and the numerical method applied.

There is no closed solution to this problem. It is therefore generally accepted to compare the results obtained from one numerical simulation with those obtained with other codes. The evolution of the fingering pattern of the brine is strongly controlled:

(1) by the dispersion of the dissolved salt brought about by the prescribed physical diffusion and dispersion, and
(2) by the numerical dispersion characteristic of the chosen approximation of the advection term in the transport equation.

For a sufficiently fine spatial resolution, all advanced simulation codes for density-driven flow tend to converge towards a typical transient pattern (Kolditz *et al.* 1998, figs. 4 & 5 therein).

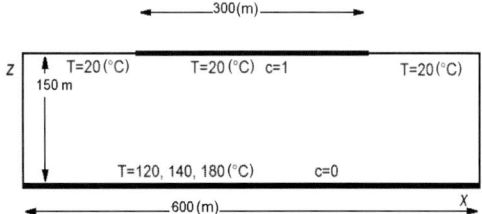

Fig. 1. Definition of the 2D vertical thermohaline Elder's problem (redrawn from Diersch & Kolditz 1998). The boundaries are closed for flow, heat and salt transport. Dissolved salt concentration, c, is fixed at the 300 m section at the upper boundary and at the bottom. Temperature is fixed at the upper and lower boundary.

This was verified also for the SHEMAT simulation code (Bartels *et al.* 2000; Clauser 2003).

The model for the thermohaline Elder's problem presented in this section is an extension of the original 'Elder's problem' for salt transport. Diersch & Kolditz (1998) have discussed thermohaline Elder's problem in detail. They generalized the Elder's problem for salt transport by applying a gradient in temperature opposite to the initial gradient in concentration. This simplification of a thermohaline flow situation made it possible to discuss the general characteristics of thermohaline flow, which is also referred to as double-diffusive convection. A complex, transient flow pattern evolves, which shows significant differences to the purely concentration-driven case, and which depends on the degree of heating at the bottom.

For model verification purposes, we reproduced the simulations of Diersch & Kolditz (1998) using the new SHEMAT code. In contrast to their simulation, however, viscosity varies with temperature in our simulation.

Model geometry and boundary conditions are shown in Figure 1. Simulation results for the concentration and temperature distribution after 10 years and 20 years are presented in Figures 2 and 3, respectively. The influence of thermal buoyancy is systematically increased by raising the fixed bottom temperature from 120°C to 180°C. For comparison, the first row (**a**) of Figures 2 and 3 shows results for a simulation with density and viscosity independent of temperature. This is identical to the result of the Elder's problem for salt transport. In this case, heat is transported like an inert tracer.

If density is temperature dependent in the simulation, diffusion of dissolved salt from the central part of the upper boundary increases fluid density, initiating down-flow. On the other hand, heating from the bottom decreases fluid density, initiating buoyant up-flow. The fingering patterns of salt concentration and temperature shown in Figures 2 & 3 are the result of the interaction of these two opposing driving forces for this type of thermohaline flow.

The difference to the isothermal Elder's problem for salt transport can be clearly seen already in case (**b**), i.e. for a bottom temperature of 120°C, and becomes very pronounced if the fixed bottom temperature is increased further to 140°C and 180°C, as in cases (**c**) and (**d**), respectively. In a similar way to what was reported by Diersch & Kolditz (1998), the change in the flow field is not monotonous, and its geometry may change altogether when the bottom

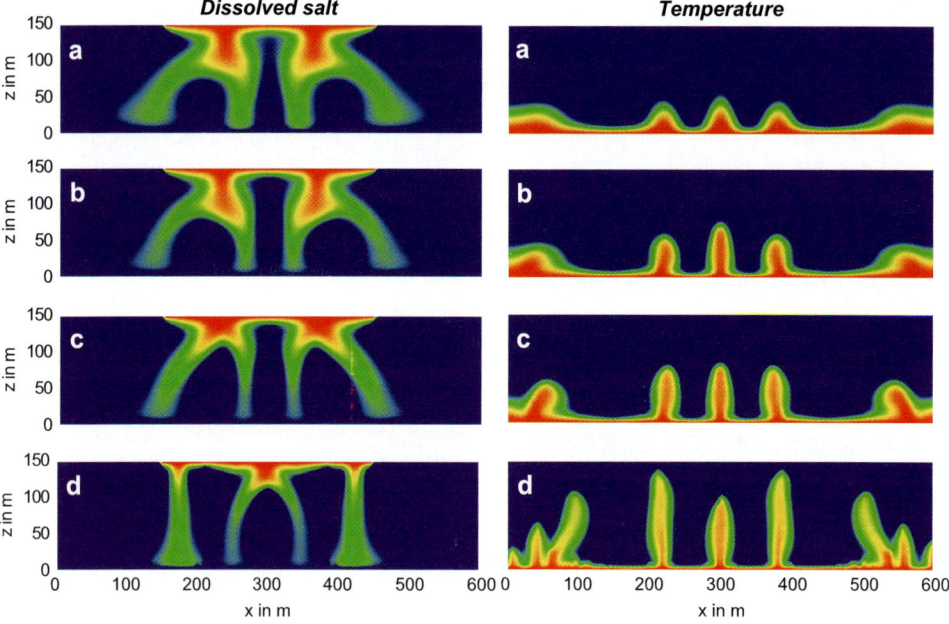

Fig. 2. Normalized distribution of dissolved salt (red: $c = 0$, blue: $c = 1$) and temperature (red: T_{max}, blue: 20°C) after a simulation period of 10 years for the cases: (**a**) $T_{bottom} = 120$°C and density independent of temperature and (**a**), (**b**) $T_{bottom} = 120$°C, (**c**) $T_{bottom} = 140$°C, (**d**) $T_{bottom} = 180$°C.

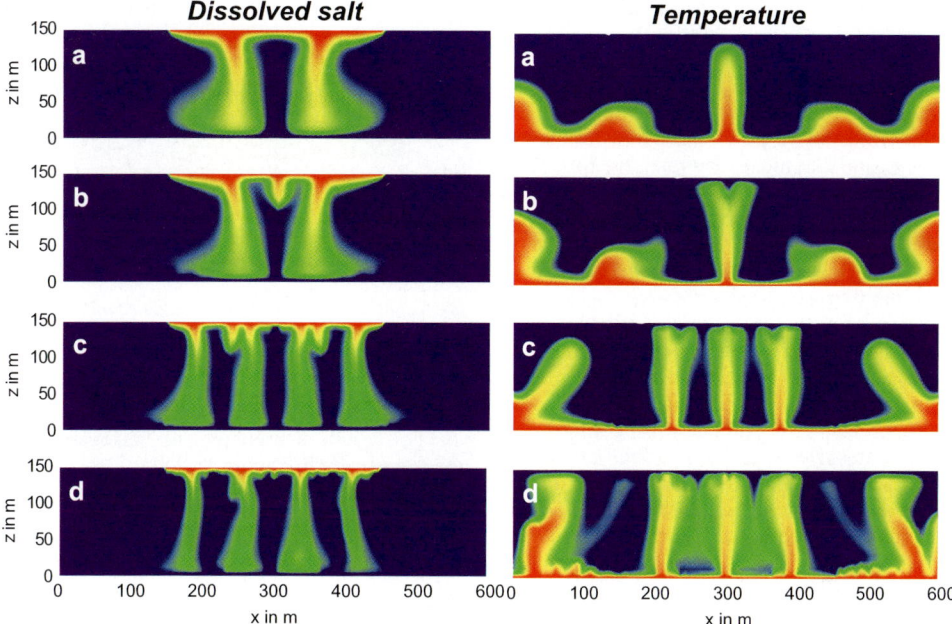

Fig. 3. Normalized distribution of dissolved salt (red: $c = 0$, blue: $c = 1$) and temperature (red: T_{max}, blue: $20°C$) after a simulation period of 20 years for the cases: (**a**) $T_{bottom} = 120°C$ and density independent of temperature and (**a**), (**b**) $T_{bottom} = 120°C$, (**c**) $T_{bottom} = 140°C$, (**d**) $T_{bottom} = 180°C$.

heating is increased. In particular, the flow direction at the centre changes from upward to downward at late stages (panel 'd' in Fig. 2 and panels 'b' and 'd' Fig. 3) and the number of convection cells shown by the salt and temperature distributions increases (panel 'd' in Fig. 2 and panels 'b', 'c' and 'd' in Fig. 3).

For the highest specified bottom temperature (panels 'd' in Figs 2 and 3), the fields show some asymmetry for the first time after 20 years. This is due to the high degree of non-linearity, in which small perturbations may become strongly amplified. While this is confirmed, in principle, by the results of Diersch & Kolditz (1998), existing differences between the flow patterns of their versus our simulations can be attributed to the constant versus temperature-dependent treatment of viscosity, respectively. In summary, these simulations illustrate the importance of non-linear coupling between heat and solute transport in thermohaline systems. They also verify that SHEMAT is a numerical code well suited to model coupled thermohaline flow and transport.

Waiwera coastal geothermal system. Waiwera is a small township north of Auckland, on the east coast of New Zealand's North Island. A low-temperature geothermal reservoir is located underneath the township. The Waiwera geothermal aquifer was first recognized from the presence of hot springs emanating on the local beach front. Utilization of the thermal water began in 1863. At that time, boreholes discharged naturally by artesian flow. The last reported natural artesian flow from boreholes occurred in 1969 (ARWB 1980). Until the early 1970s, the hot water use increased gradually. After the early 1970s this trend was accelerated. In 1975, residents informed the Auckland Regional Water Board (ARWB) (now the Environmental Management Department of the Auckland Regional Council (ARC)), of their concern about declining water levels. The Water Board initiated a study designed to assist in the protection, allocation and management of the resource.

A number of papers describe a numerical 3D coupled flow, heat transfer, and solute transport model of the Waiwera area (Stöfen *et al.* 2000; Kühn *et al.* 2001). This model defines the natural state of the Waiwera reservoir, and is used to simulate varying production regimes and to address the questions:

(1) Is the heat exploitation sustainable?

(2)   Is re-injection desirable and helpful?
(3)   Do chemical reactions alter the reservoir?

The simulations focus on the complex inter-actions of coupled flow, heat transfer, solute transport, and chemical reactions. Results based on a simulation time of 6000 years show that only about 600 years are required to achieve a steady state from the initial conditions. This is due to strong convective heat transfer and advec-tive solute transport at Waiwera.

The source of the geothermal fluid at Waiwera is groundwater, which percolates down to depths in excess of 1200 m into the deep Waiheke Grey-wacke basement rock. There it is heated and acquires its characteristic chemical composition. The hot water rises rapidly via a fault or fracture zone into the well-fractured, 400-m-thick Waite-mata Group Sandstone which forms the geother-mal aquifer. Weathered surface rocks and alluvial deposits confine the aquifer at the top. The centre of the aquifer is 100 m landwards from the beachfront. Temperature profiles suggest an elongated fracture zone in the Waite-mata Group Sandstone. Figure 4 shows the con-ceptual model in cross-section, with the assumed flow paths of the different types of water entering the aquifer. At the western margin of the geother-mal aquifer the geothermal fluid is cooled by conductive heat loss and dilution with cold, non-geothermal groundwater. At the eastern, seaward margin of the aquifer there is a sea-water–freshwater interface.

Figure 5 shows the resulting temperature dis-tribution of the simulated natural state in cross-section through the centre of the aquifer. The corresponding chloride concentrations shown in Figure 6 reflect the mixing of fresh, geothermal, and estuary water. A freshwater/geothermal water boundary develops to the west and a geothermal water/estuary water boundary to

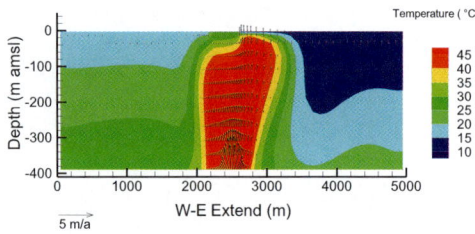

Fig. 5. Temperature distribution in the natural state; west–east cross-section through the centre of the aquifer; arrows show the Darcy velocity (from Kühn *et al.* 2001).

the east. The inflowing geothermal water pre-vents the estuary water from entering the aquifer.

The good agreement of calculated and measured temperature profiles verifies the natural state model. Figure 7 shows one example for a profile at approximately 2575 m from the western margin of the model (cf. Fig. 5).

The results for the natural state yield the initial temperature and salt concentration for pro-duction simulations not shown here. Based on the results of these simulations, it can be con-cluded that:

(1)   the historical exploitation was not sustain-able, but, after modifications of the pro-duction regime, the geothermal system is now recovering again;
(2)   re-injection of the discharged water is of no help for producing, in a sustainable way, a larger amount of geothermal water.

The study of the chemical regime in the reser-voir shows that fresh water, geothermal water, and sea-water are in thermodynamic equili-brium with calcite. In spite of mineral reactions involving some precipitation and dissolution of calcite it can be concluded that precipitation and dissolution of calcite do not alter the

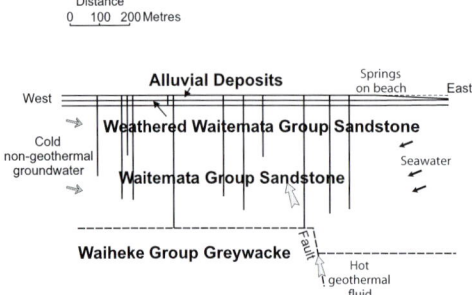

Fig. 4. Conceptual model of the Waiwera geothermal aquifer (after Stöfen & Kühn 2003).

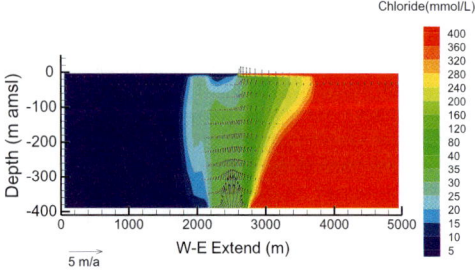

Fig. 6. Chloride concentration in the natural state; west–east cross-section through the centre of the aquifer; arrows show the Darcy velocity (from Kühn *et al.* 2001).

Temperature (°C)

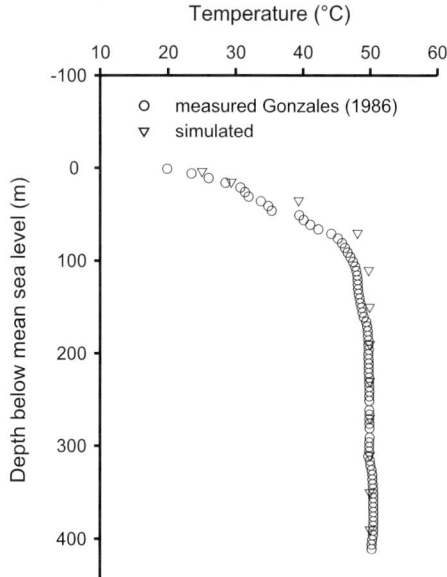

Fig. 7. Measured and simulated temperature profiles at borehole no. 74, situated 2575 m from the western margin of the model (see Fig. 5; after Stöfen & Kühn 2003).

aquifer: the simulations showed no effect on the hydraulic reservoir properties.

Reactive flow with permeability feedback

Core flooding experiment under reservoir conditions with mineral redistribution and associated permeability change. Due to the lack of analytical or experimental benchmarks for testing of the highly non-linear properties of numerical codes for buoyancy driven flow and reactive transport a one-dimensional laboratory core flooding experiment was performed. It simulates the physical and chemical processes at a thermal front propagating through a reservoir by making temperature, pressure, flow rate and salinity in a sandstone core comparable to the conditions in a deep geothermal hot-water reservoir. The model is verified by comparison of simulation results with porosity and permeability observed after the core flooding experiment (Bartels *et al.* 2002).

The experiment was performed on a core of Bentheim sandstone, which is a very clean quartz sandstone, cemented by quartz. Its diameter and length are 29 mm and approximately 500 mm, respectively (Fig. 8). The core was segmented in three parts. The warm outflow segment was heated to a constant temperature of 100°C, while the mantle of the central segment was

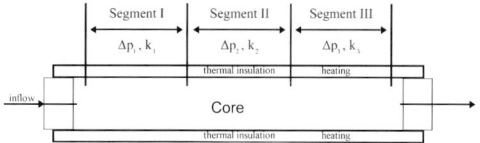

Fig. 8. Core flooding experiment experimental set-up; total length is 0.5 m.

thermally insulated. Fluid flowed from the cold to the warm end, and the outflowing brine was re-injected again at the lower temperature of 80°C. This gave rise to a steady-state temperature gradient of about 1 K cm^{-1} over a distance of 20 cm in the central segment.

Prior to flooding, anhydrite had been deposited in the pore space along the entire length of the core. Since anhydrite cannot be assumed to be precipitated homogeneously in the core, the variation of permeability along the core was measured using a custom-made ring gas-permeameter. Permeability was measured along the length of the core after both the preparation of the core with anhydrite and the flooding experiment (Fig. 9). As permeability is the only continuously measured rock property, all further information on porosity and mineral redistribution had to be inferred from it.

Flooding of the core was maintained at a specific discharge rate of 1.1 metre per day over a period of 20 days. In an operated heat-mining reservoir this corresponds to a distance of 10–100 m from the injection well. The overburden pressure was 10 MPa, and the salinity was 100 g L^{-1} of NaCl. The concentration of the dissolved calcium was analysed at the core outlet. After the end of the flooding experiment, permeability was measured again along the length of the core with the ring gas permeameter (Fig. 9). At positions where significant changes in permeability were observed, samples were taken from the core, and permeability, porosity and anhydrite content were determined.

Due to the special preparation of the core prior to flooding, the only process which may bring about the observed changes in permeability is transport and relocation of anhydrite by dissolution within the low-temperature region, followed by precipitation in the downstream region of elevated temperature. The transient changes in the distribution of the anhydrite along the core are illustrated in three snapshots corresponding to different stages (Fig. 10, bottom panel).

Clearly, the core flooding experiment shows that anhydrite is continuously relocated downstream across the temperature front. In general,

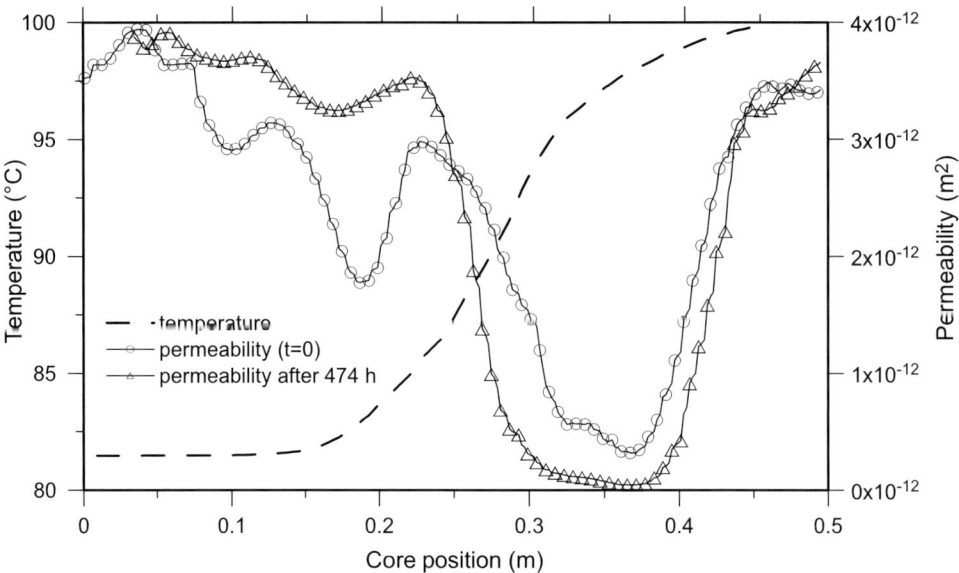

Fig. 9. Variation of permeability with position *x* along the core after preparation with anhydrite measured with a ring permeameter before (circles) and after flooding (triangles). The simulated steady-state temperature field is shown as a dashed line (from Bartels *et al.* 2002).

the final numerical simulation result fits the observed mineral content, porosity and permeability. In detail, both the porosity and permeability reduction in the zone of precipitation and the maximum measured value of anhydrite concentration are reproduced with good accuracy. The simulated and measured outflow concentrations of dissolved anhydrite are 24.7 and 24.5 mmol kg^{-1} H_2O, respectively.

Deviations occur between observation and simulation with respect to the position of the dissolution front, which propagates downstream from left to right. Compared to the data, the simulated reduction of porosity and permeability at the end of the simulation commences 5–8 cm too far downstream (Fig. 10, top panel). On the other hand, the simulated anhydrite content is within the uncertainty range of the laboratory analysis (Fig. 10, bottom panel).

The core flooding experiment shows that increase and reduction of permeability are associated with the dissolution and precipitation of anhydrite, respectively. A dissolution front propagates downstream, which is associated with a permeability increase up to the level of the unprepared core. This strongly suggests that anhydrite is completely removed from the upstream part of the core during the experiment.

Dissolution in this low-temperature region is due to the fact that more anhydrite can be dissolved upstream, at 80°C, than downstream,

at 100°C. There the concentration of dissolved anhydrite (0.0245 mol kg^{-1}) determined in fluid analyses is within thermodynamic equilibrium (Kühn *et al.* 2002*b*). This justifies the assumption of chemical equilibrium in the successful simulation described here. Equilibrium means that the chemical reaction rates are sufficiently rapid, and deviations from equilibrium at a fixed core position caused by transport are equilibrated quasi-instantaneously by dissolution or precipitation. As a consequence, the largest permeability reduction is found where the gradient of temperature is largest.

Finally, this simulation demonstrates that the improved chemical reaction model implemented in SHEMAT and already verified by solubility data (Kühn *et al.* 2002*b*) yields quantitatively correct results for complex flow and transport regimes typical for managed hot water reservoirs, i.e. reactive flow of highly saline brines through a porous medium at varying temperature and high pressure.

The fractal relation between porosity and permeability (equation (7)) is applied twice at sensitive points of our model validation procedure. Therefore, it is an important feature of this simulation that the good fit of the data by the model was achieved without a further fit of E_f in equation (7). Rather, the fractal exponent $E_f = 11.3$ in our experiment, which corresponds to a fractal dimension of $D_f = 2.45$ typical for rough pore walls, was

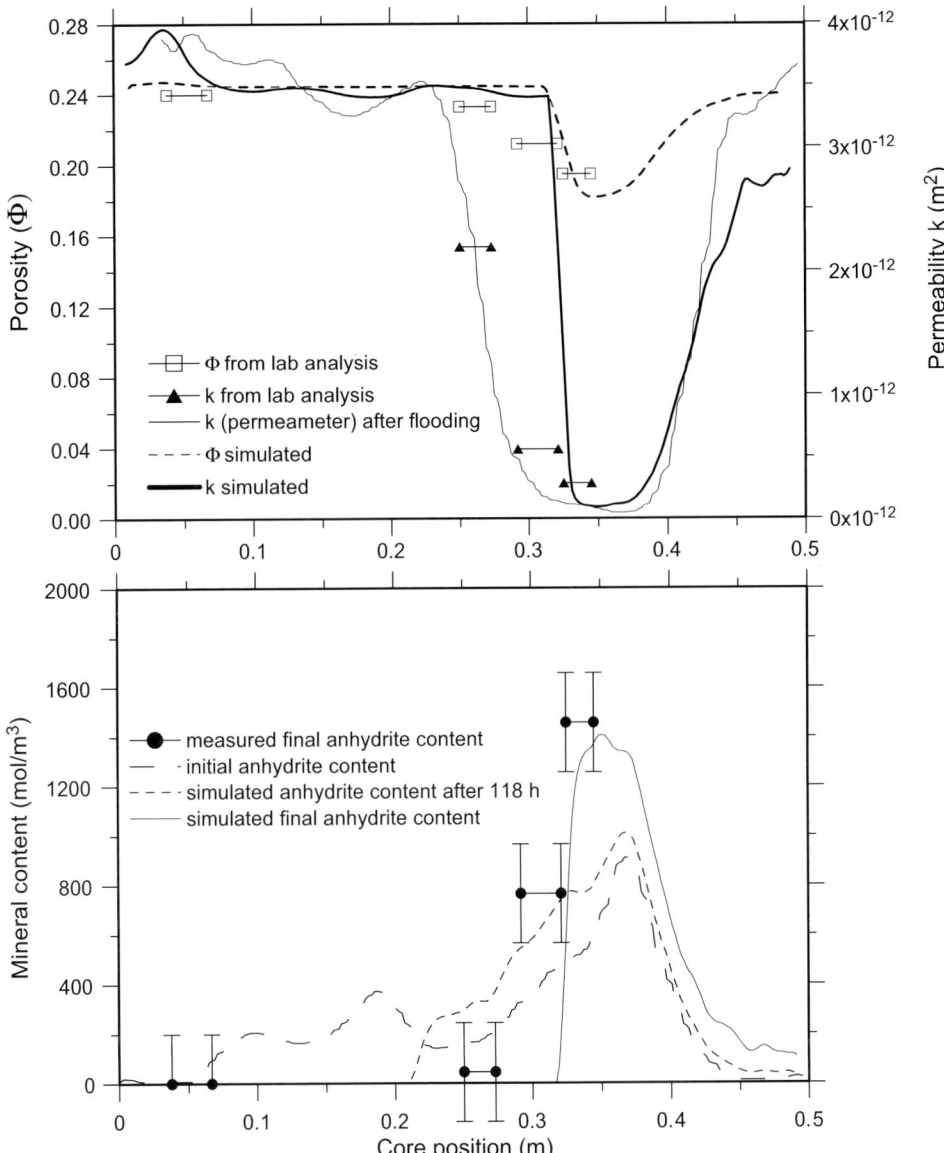

Fig. 10. Comparison of simulation results with data. Top panel: simulated final porosity (bold dashed line), permeability (bold full line), and data from measurements on core specimens (horizontal bars) and with the ring permeameter (fine line); Bottom panel: simulated mineral content before core flooding (long dash), after 118 hours of flooding (short dash), and after 20 days of core flooding (solid line), compared to data measured on core specimens after flooding (horizontal bars with error bars). The length of the horizontal bars in both panels corresponds to the size of the analysed core specimen (from Bartels *et al.* 2002).

obtained purely from a straightforward application of the fractal theory to the data. The results therefore indicate that the frequently made assumption of smooth and spherical grains (corresponding to $E_f = 3$ or $D_f = 2$) is not justified in the case of precipitation of anhydrite in a comparatively rapid

flow regime – typical for managed geothermal reservoirs.

Reaction front fingering in anhydrite-cemented sandstone. A 3300-m-deep borehole (Allermöhe 1, near Hamburg, Germany) taps a 70-m-thick sandstone aquifer with a temperature of 125°C. Temperature and thickness of the aquifer are suitable for hydrogeothermal heat mining. However, it was found that the pore space, with an original porosity of up to 20%, is filled to a large extent by anhydrite. Mineralogical investigations showed that the anhydrite cement either completely fills the pore space or forms isolated, cloud-like or layered structures. A pumping test produced only 3 m³ per hour of formation water: far too little for an economical use of the resource (Baermann *et al.* 2000*b*).

Laboratory core flooding experiments were conducted on anhydrite-cemented sandstone samples from the Allermöhe borehole, to study the feasibility of a gentle stimulation, avoiding the use of chemicals. By such a stimulation the borehole would be connected through the cemented areas immediately around it to more permeable parts of the aquifer. The experiments resulted in a sudden permeability increase after an unexpectedly short time. X-ray tomographic images showed that discrete flow channels had developed within the cores (Baermann *et al.* 2000*a*).

The conditions required for preferential flow path formation are described in the literature. Ormond & Ortoleva (2000) showed that the interaction between mineral reaction and mass transport in rocks might lead to instability of the reaction front and the development of channel-like voids.

The numerical model for the simulation of the laboratory core flooding experiment was set up according to the data published by Baermann *et al.* (2000*a*). Due to a lack of information on the exact initial distribution of anhydrite in the core sample, an initial heterogeneity at the centre of the core inlet is assumed. Starting from there, a preferential flow path develops in the otherwise homogeneous core (Kühn & Stöfen 2001). At first, two channels develop. After some time, one of them ceases to grow further. The remaining flow path increases in width towards the centre of the core, and finally merges with the first one. Thus, only one finger survives, which finally reaches the outlet of the model (Fig. 11).

Simulating the core flooding experiment with SHEMAT yields acceptable results: breakthrough time, mass of water, calcium concentration and amount of dissolved anhydrite agree well with the laboratory experiment (Fig. 12). Hence, the sudden flow breakthrough observed in the core flooding experiments performed on anhydrite-cemented core samples can be adequately explained by the development of preferential flow paths.

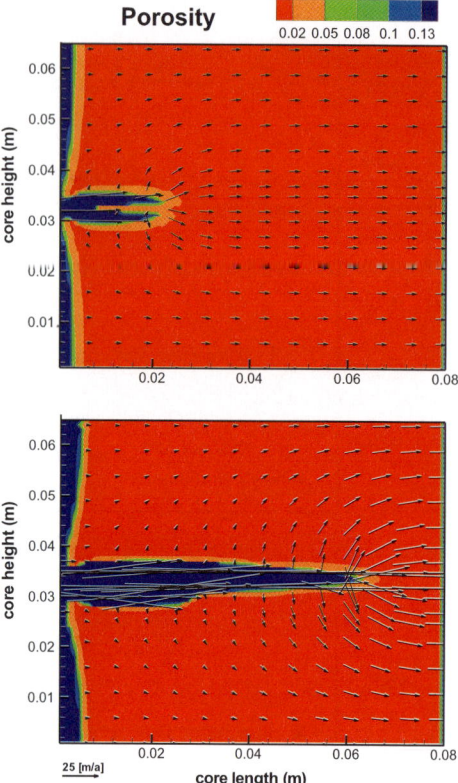

Fig. 11. Porosity distribution in the core after 20 days (top panel) and 45 days (bottom panel). Arrows show the Darcy velocity in m a⁻¹, scaled according to the reference arrow shown (from Kühn & Stöfen 2001).

Long-term reservoir changes induced by heat mining. The variation of the reservoir properties in a sandstone reservoir due to heat mining with re-injection was simulated over the entire assumed operation time. The site studied is located near Stralsund, north-east Germany, and the reservoir rock is a Detfurth sandstone (Buntsandstein), typical for the North German Basin. Two of the boreholes closest to the town are used for production, a third one for re-injection. This way, the distance over which the hot water is conveyed into the district heating distribution network is minimized (Fig. 13). The two production wells yield 50 m³ of hot water per hour. After heat extraction, all of the water is re-injected at a temperature of 20°C. For diagnostic reasons, a conservative tracer is additionally injected in the simulation, in order to illustrate the transport of dissolved species in the model reservoir. Due to the high salinity of 280 g L⁻¹ TDS (total dissolved

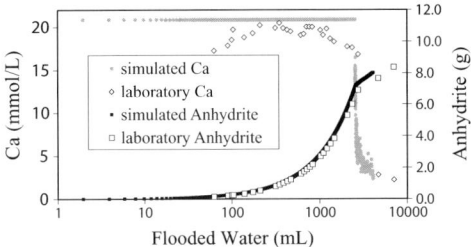

Fig. 12. Calcium concentration (open diamonds: experiment; filled diamonds: simulation) in the solution discharging from the core, and total amount of anhydrite dissolved during the simulation (open squares: experiment; filled squares: simulation) versus the water volume flooded through the core (from Kühn & Stöfen 2001).

solid), all of the produced water is re-injected at 20°C and 0.1 MPa. Figure 14 and the cross-section in Figure 15 show the cold-water front after 50 years of operation in relation to the tracer front, which has already reached the production wells. Propagation of the low-temperature region is retarded because the re-injected cold water also cools the reservoir rock on its way to the production well.

Re-injection of cool water of higher viscosity than the natural reservoir fluid, results in a continuous reduction of hydraulic conductivity. Qualitatively, this effect should be partially balanced by thermally induced mineral reactions.

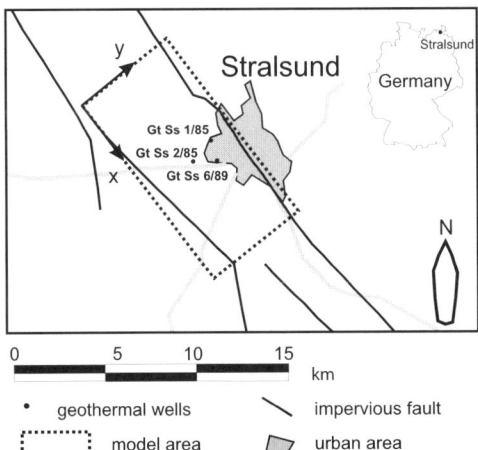

Fig. 13. Stralsund geothermal site with the wells Gt Ss 1/85, Gt Ss 2/85, and Gt Ss 6/89 (black dots). The reservoir is partly delimited by impervious faults (black lines). The model region (dotted rectangle) covers about 12 × 6 km. (from Kühn et al. 2002a).

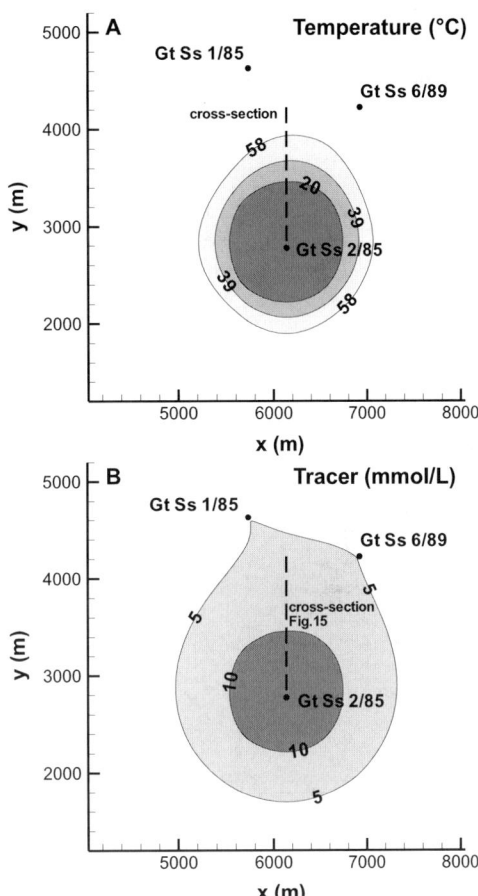

Fig. 14. Thermal (**a**) and tracer front (**b**), propagating from the injection well Gt Ss 2/85 to the production wells Gt Ss 1/85 and Gt Ss 6/89 after 50 years of operation. The fronts are represented by the isotherms in °C, and the concentration isolines in mmol L^{-1} (arithmetic means), respectively. Black dots indicate the wells. At this instant, the tracer just reaches the production wells, while the thermal front still trails behind (from Kühn et al. 2002a).

In detail, the simulations show that the dissolution of anhydrite in the vicinity of the injection well (Fig. 15, bottom panel and Fig. 16a) appears to more than compensate the effect of anhydrite precipitation at the propagating thermal front (Fig. 15, top panel and Fig. 16a). Thus, this redistribution of anhydrite leads to a net increase in injectivity. Moreover, the simulation of reactive flow also shows that calcite is precipitated around the injection well and dissolved at the thermal front (Fig. 15, bottom panel and Fig. 16b). However, this does not alter reservoir properties significantly.

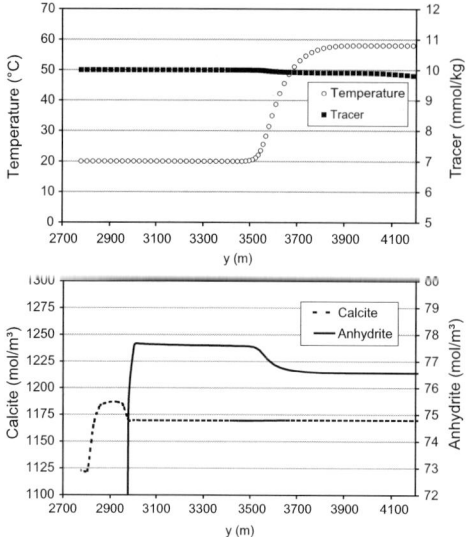

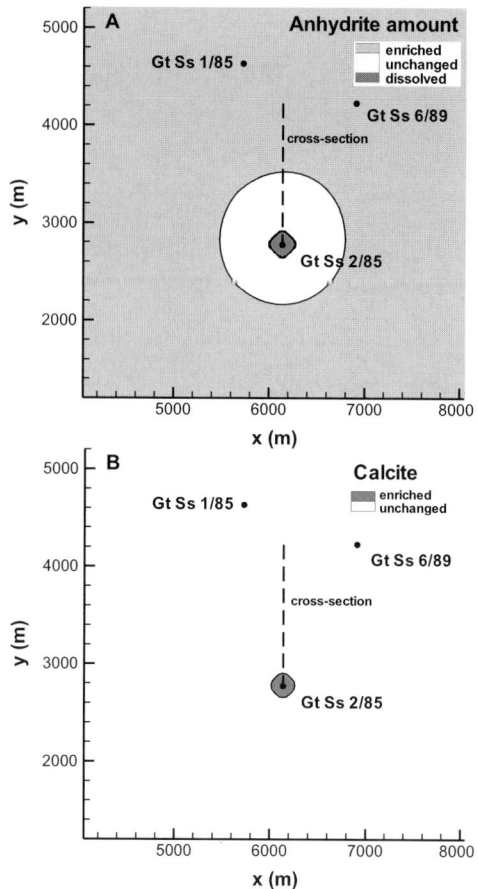

Fig. 15. Reaction fronts after 50 years of brine re-injection along the cross-sections in Figure 14 from the injection well (left) to the production wells (right). Thermal and tracer fronts are shown compared to the mineral phases calcite and anhydrite in Figure 16 (after Kühn *et al.* 2002*a*).

In summary, for complex, mutually coupled processes, a quantitative understanding of the net effect of the counteracting effects of viscosity and mineral redistribution on the overall injectivity requires coupled numerical simulations. In our example, the injectivity of the reservoir is influenced primarily by the viscosity effect, but mineral reactions balance this negative trend to some degree. The importance of the mineral reactions depends on the fractal exponent in the permeability–porosity relation, see equation (7) and Figure 17. The exponent reflects the structure of the pore space, which is formed by the anhydrite cementation. Invoking the fractal relation between porosity and permeability leads to a well-head pressure reduction of 10–20 m (depending on the assumed type of cementation) at the end of the operation period (Fig. 17). For the studied potential heat-mining installation at Stralsund it can be concluded that its operation will not be affected adversely by a long-term reduction of the injectivity of the injection well (Kühn *et al.* 2002*a*).

Conclusions

The novel capabilities of the simulator presented here, and their combination, allow a quantitative prediction of reservoir behaviour in natural as

Fig. 16. Spatial distribution of anhydrite (**a**) and calcite (**b**) in mol m^{-3} in the vicinity of the injection well Gt Ss 2/85, after 50 years of brine re-injection. The region labelled 'dissolved' no longer contains anhydrite. The area labelled 'enriched' refers to anhydrite concentrations between the initial one and the maximum amount of 77.7 mol m^{-3}. The area labelled 'unchanged' corresponds to the initial concentration of 76.5 mol m^{-3}. Calcite varies from the initial concentration of 1170 mol m^{-3} labelled 'unchanged' up to 1187 mol m^{-3} labelled 'enriched'. The regions of anhydrite dissolution and calcite precipitation coincide (from Kühn *et al.* 2002*a*).

well as under exploitation conditions. The highly non-linear physics and chemistry coded in the simulator require a quantitative verification. For lack of closed solutions, this is achieved by comparison of model results with laboratory data and field measurement as documented in this paper. Thus knowledgeable users can employ SHEMAT with confidence.

This simulator makes available to geohydrodynamic modellers the novel and innovative

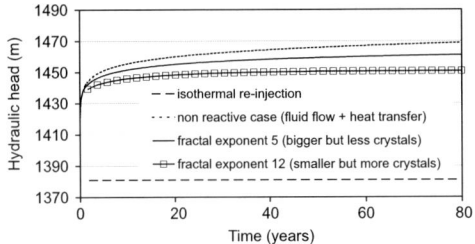

Geothermal Congress 2000, Kyushu–Tohoku, Japan, May 28–June 10, 3997–4002.

Fig. 17. Variation of the pressure head at the injection well (shown as hydraulic head) over 80 years of reservoir exploitation (after Kühn *et al.* 2002a).

porosity–permeability relation derived by Pape *et al.* (1999), which is based on a fractal geometrical model of the pore space, and was calibrated on large petrophysical data-sets. This relationship generally yields an exponent for porosity significantly greater than 3.0 – the generally accepted value in basin studies. This makes a considerable difference in respect to the predicted reservoir behaviour and the match of data and simulation.

The research reported in this paper was supported by the German Federal Ministries for Education, Science, Research, and Technology (BMBF) under grant 032 69 95A-D, and for Economics and Technology (BMWi) under grant 0327095.

References

ARWB, 1980. *Waiwera Water Resource survey – Preliminary Water Allocation/Management Plan.* Auckland Regional Water Board, Technical Publication No. 17.

AZAROUAL, M., FOUILLAC, C. & MATRAY, J. M. 1997. Solubility of silica polymorphs in electrolyte solutions, I. Activity coefficients of aqueous silica from 25° to 250°C, Pitzer's parameterization. *Chemical Geology*, **140**, 155–165.

BAERMANN, A., KRÖGER, J. & ZARTH, M. 2000a. Anhydrite cement in Rhaetian sandstones in Hamburg: X-ray and NMR tomography studies and leaching tests (in German). *Zeitschrift für Angewandte Geologie*, **46** (3), 144–152.

BAERMANN, A., KRÖGER, J., TAUGS, R., WÜSTENHAGEN, K. & ZARTH, M. 2000b. [Anhydrite cement in Rhaetian sandstones in southeastern Hamburg: morphology and structure] (in German). *Zeitschrift für Angewandte Geologie*, **46** (3), 138–144.

BARTELS, J., KÜHN, M., PAPE, H. & CLAUSER, C. 2000. A new aquifer simulation tool for coupled flow, heat transfer, multi-species transport and chemical water–rock interactions. *Proceedings, World*

BARTELS, J., KÜHN, M., SCHNEIDER, W., CLAUSER, C., PAPE, H., MEYN, V. & LAJCZAK, I. 2002. Core flooding laboratory experiment validates numerical simulation of induced permeability change in reservoir sandstone. *Geophysical Research Letters*, **29** (9), 10.1029/2002GL014901.

CHENG, H.-P. & YEH, G.-T. 1998. Development and demonstrative application of a 3-D numerical model of subsurface flow, heat transfer, and reactive chemical transport: 3DHYDROGEOCHEM. *Journal of Contaminant Hydrology*, **34**, 47–83.

CHIANG W. H., KINZELBACH W. 2001. *3D-Groundwater Modelling with PMWIN*. Springer-Verlag, Berlin and Heidelberg.

CLAUSER, C. (ed.) 2003. *SHEMAT and Processing Shemat – Numerical Simulation of Reactive Flow in Hot Aquifers*, Springer-Verlag, Berlin and Heidelberg.

CLAUSER, C. & KIESNER, S. 1987. A conservative, unconditionally stable, second-order three point differencing scheme for the diffusion convection equation. *Geophysical Journal, Royal Astronomical Society*, **91**, 557–568.

CLAUSER, C. & VILLINGER, H. 1990. Analysis of conductive and convective heat transfer in a sedimentary basin, demonstrated for the Rheingraben. *Geophysical Journal International*, **100** (3), 393–414.

DIERSCH, H.-J. & KOLDITZ, O. 1998. Coupled groundwater flow and transport: 2. Thermohaline and 3D convection systems. *Advances in Water Resources*, **21** (5), 401–425.

ELDER, J. W. 1967. Transient convection in a porous medium. *Journal of Fluid Mechanics*, **27** (3), 609–623.

GONZALES, C. N. 1986. *Interpretation of Downhole Temperature Survey at Weiwera Thermal Area.* Report-o. 86.07, Geothermal Institute, University of Auckland.

GREENBERG, J. P. & MOLLER, N. 1989. The prediction of mineral solubilities in natural waters: a chemical equilibrium model for the Na–K–Ca–Cl–SO₄–H₂O system to high concentration from 0 to 250°C. *Geochimica et Cosmochimica Acta*, **53**, 2503–2518.

HARVIE, C. E. & WEARE, J. H. 1980. The prediction of mineral solubilities in natural waters: the Na–K–Mg–Ca–Cl–SO₄–H₂O system from zero to high concentrations at 25°C. *Geochimica et Cosmochimica Acta*, **44**, 981–997.

HARVIE, C. E., MOLLER, N. & WEARE, J. H. 1984. The prediction of mineral solubilities in natural waters: the Na–K–Mg–Ca–H–Cl–SO₄–OH–HCO₃–CO₃–CO₂–H₂O system to high ionic strengths at 25°C. *Geochimica et Cosmochimica Acta*, **48**, 723–751.

HE, S. & MORSE, J. W. 1993. The carbonic acid system and calcite solubility in aqueous Na–K–Ca–Mg–Cl–SO₄ solutions from 0–90°C. *Geochimica et Cosmochimica Acta*, **57**, 3533–3554.

ITOI, R., FUKUDA, M., JINNO, K., SHIMIZU, S. & TOMITA, T. 1987. Numerical analysis of the injectivity of wells in the Otake geothermal field, Japan, *In*: FREESTONE, D. H. & NICHOLSON, K. (eds) *Proceedings of the 9th New Zealand Geothermal Workshop 1987, November 4–6, 1987, Geothermal Institute, Auckland University Press, Auckland, New Zealand*, 103–108.

KESTIN, J. 1978. Thermal conductivity of water and steam. *Mechanical Engineering*, **100** (8), 1255–1258.

KOLDITZ, O., RATKE, R., DIERSCH, H. -J. & ZIELKE, W. 1998. Coupled groundwater flow and transport. 1. Verification of variable density flow and transport models. *Advances in Water Resources*, **21** (1), 27–49.

KÜHN, M. & STÖFEN, H. 2001. Reaction front fingering in anhydrite cemented sandstone. *In*: SIMMONS, S., DUNSTALL, M. G. & MORGAN, O. E. (eds) *Proceedings of the 23rd New Zealand Geothermal Workshop 2001, Auckland University Press, Auckland*, 207–212.

KÜHN, M., BARTELS, J. & IFFLAND, J. 2002*a*. Predicting reservoir property trends under heat exploitation: interaction between flow, heat transfer, transport, and chemical reactions in a deep aquifer at Stralsund. *Geothermics*, **31**, 725–749.

KÜHN, M., BARTELS, J., PAPE, H., SCHNEIDER, W. & CLAUSER, C. 2002*b*. Modelling chemical brine–rock interaction in geothermal reservoirs. *In*: STOBER, I. & BUCHER, K. (eds) *Water–Rock Interaction*, Kluwer Academic Publishers, Dordrecht, 147–169.

KÜHN, M., STÖFEN, H. & SCHNEIDER, W. 2001. Coupled flow, heat transfer, solute transport, and chemical reactions within the thermal water reservoir Waiwera, New Zealand. *In*: SEILER, K.-P. WOHNLICH, S. (eds) *New Approaches Characterizing Groundwater Flow*, Balkema Publishers, Netherlands 1001–1005.

LANGGUTH, H.-R. & VOIGT, R. 1980. *Hydrogeologische Methoden*. Springer-Verlag, Berlin and Heidelberg.

LICHTNER, P. C. 1996. Continuum formulation of multicomponent–multiphase reactive transport. *Reviews in Mineralogy and Geochemistry*, **34** (1), 83–129.

MCCUME, C. C., FORGLER, H. S. & KLINE, W. E. 1979. An experiment technique for obtaining permeability–porosity relationship in acidized porous media. *Industrial and Engineering Chemistry Research Fundamentals*, **18** (2), 188–192.

MONNIN, C. 1999. A thermodynamic model for the solubility of barite and celestite in electrolyte solutions and seawater to 200°C and to 1 kbar. *Chemical Geology*, **127**, 141–159.

ORMOND, A. & ORTOLEVA, P. 2000. Numerical modelling of reaction-induced cavities in a porous rock. *Journal of Geophysical Research*, **105** (B7), 16 737–16 747.

PAPE, H., CLAUSER, C. & IFFLAND, J. 1999. Permeability prediction for reservoir sandstones based on fractal pore space geometry. *Geophysics*, **64** (5), 1447–1460.

PARKHURST, D. L., THORSTENSON, D. C. & PLUMMER, L. N. 1980. *PHREEQE, a Computer Program for Geochemical calculations*. US Geol Survey, Dallas TX.

PHAM, M., KLEIN, C., SANYAL, S., XU, T. & PRUESS, K. 2001. Reducing cost and environmental impact of geothermal power through modeling of chemical processes in the reservoir. *In*: *Proceedings of the 26th Stanford Geothermal Workshop, Stanford CA (LBNL – 47644)*.

PITZER, K. S. 1975. Thermodynamics of electrolytes. V. Effects of higher order electrostatic terms. *Journal of Solution Chemistry*, **4** (3), 249–265.

PITZER, K. S. & KIM, J. J. 1974. Thermodynamics of electrolytes. IV. Activity and osmotic coefficient for mixed electrolytes. *Journal of the American Chemical Society*, **96** (18), 5701–5707.

PITZER, K. S. & MAYORGA, G. 1973. Thermodynamics of electrolytes. II. Activity and osmotic coefficient for strong electrolytes with one or both ions univalent. *Journal of Physical Chemistry*, **77** (19), 2300–2308.

PITZER, K. S. & MAYORGA, G. 1974. Thermodynamics of electrolytes. III. Activity and osmotic coefficients for 2–2 electrolytes. *Journal of Solution Chemistry*, **3** (7), 539–546.

PLUMMER, L. N., PARKHURST, D. L., FLEMING, G. W. & DUNKLE, S. A. 1988. A computer program incorporating Pitzer's equations for calculation of geochemical reactions in brines. *US Geol Survey Water-Resources Investigations, Report* **88–4153**.

SCHECHTER, R. S. & GIDLEY, J. L. 1969. The change in pore size distribution from surface reaction in porous media. *American Institute of Chemical Engineers Journal*, **15** (3), 339–350.

SMOLARKIEWICZ, P. K. 1983. A simple positive definite advection scheme with small implicit diffusion. *Monthly Weather Review*, **111**, 479–486.

STÖFEN, H. KÜHN, M. 2003. Waiwera Coastal geothermal system. *In*: CLAUSER, C. (ed.) 2003. *SHEMAT and Processing Shemat – Numerical Simulation of Reactive Flow in Hot Aquifers*, Springer-Verlag, Berlin and Heidelberg, 297–316.

STÖFEN, H., KÜHN, M. & SCHNEIDER, W. 2000. A heat transfer and solute transport model as a tool for sustainable management of the thermal water resource waiwera, New Zealand. *In*: SIMMONS, S., MORGAN, O. E. & DUNSTALL, M. G. (eds) *Proceedings of the 22nd New Zealand Geothermal Workshop 2000, Auckland University Press, Auckland* 273–278.

WEIR, G. J. & WHITE, S. P. 1996. Surface deposition from fluid flow in a porous medium. *Transport in Porous Media*, **25**, 79–96.

WHITE, S. P. 1995. Multiphase non-isothermal transport of systems of reacting chemicals. *Water Resources Research*, **32** (7), 1761–1772.

WHITE, S. P. & MROCZEK, E. K. 1998. Permeability changes during the evolution of a geothermal field due to dissolution and precipitation of quartz. *Transport in Porous Media*, **33**, 81–101.

XU, T. & PRUESS, K. 2001. Modelling multiphase non-isothermal fluid flow and reactive geochemical transport in variably saturated fractured rocks: 1. Methodology. *American Journal of Science*, **301**, 16–33.

ZARROUK, S. J. & O'SULLIVAN, M. J. 2001. The effect of chemical reaction on the transport properties of porous media. *In*: SIMMONS, S. F., DUNSTALL, M. G. & MORGAN, O. E. (eds) *2001 Proceedings of the 23rd New Zealand Geothermal Workshop 2001, November 7–9, 2001, Auckland*

University Press, Auckland, New Zealand, 231– 236.

ZOTH, G. & HÄNEL, R. 1988. Appendix. *In*: HÄNEL, R., RYBACH, L. & STEGENA, L. (eds) *Handbook of Terrestrial Heat Flow Density Determination*, Kluwer Academic Publishers, Dordrecht, 447–468.

Permeability of rock samples from the Kola and KTB superdeep boreholes at high *P–T* parameters as related to the problem of underground disposal of radioactive waste

A. V. ZHARIKOV[1], V. I. MALKOVSKY[1], V. M. SHMONOV[2] & V. M. VITOVTOVA[2]

[1]*Institute of Geology of Ore Deposits, Petrography, Mineralogy and Geochemistry of the Russian Academy of Sciences, Staromonetny per., 35, 119017, Moscow, Russia*
(e-mail: vil@igem.ru)

[2]*Institute of Experimental Mineralogy of the Russian Academy of Sciences, 142432, Chernogolovka Moscow district, Russia*

Abstract: An experimental study of the samples collected from a depth of 3.8–11.4 km in the Kola and KTB superdeep boreholes, and from the Earth's surface at the Kola drilling site was carried out at temperatures up to 600°C and pressures up to 150 MPa. The study was focused on the estimation of *in situ* permeability of the deep-seated rocks, their palaeo-permeability during metamorphic transformations, and their protective properties for HLW disposal. Permeability dependencies on pressure and temperature were obtained. An increase in confining pressure leads to a decrease in rock sample permeability. The temperature trends obtained are of different types: permeability may decrease within the entire temperature range, or it may firstly decrease, reach its minimum and then decrease. It was found that this permeability behaviour is due to rock microstructure transformations caused by the competing effects of temperature and effective pressure. A possible *in situ* permeability trend for the superdeep section was proposed. A numerical simulation of convective transport was performed in order to determine a safe depth for the HLW well repository. The estimates obtained show that HLW well repositories can be used safely at relatively shallow depths.

One of the major results of investigations in the Russian–Kola (SG-3) and German–KTB–Oberpfaltz superdeep boreholes is the presence of mobile fluids at unexpectedly great depths (Emmerman & Lauterjung 1997; Kozlovsky 1987). Rock permeability is the main parameter controlling fluid heat and mass transfer in the geological media. Therefore, it is very important to estimate the permeability of the deep-seated crustal rocks *in situ*. The temperature at the bottom of the SG-3 hole is about 220°C; in the KTB it is higher – about 265°C. The porosity and permeability values obtained on core samples from SG-3 at ambient temperature and atmospheric pressure, increased with depth. The studies on core samples revealed that this tendency is due to the progressive effect of microcracks that appeared while samples were being cored and brought to the surface, inducing distortion of the rock's physical properties (Gorbatzevich & Medvedev 1986; Wolter & Berckhemer 1989). Therefore, in such a case, it is quite reasonable to perform measurements in a way that compensates for this damage: i.e. at high temperatures and pressures.

The ancient rocks of SG-3 and KTB underwent many changes related to fluids. For the Kola Series rocks, the most important events were progressive metamorphism at temperatures of 500–600°C and retrograde metamorphism at temperatures of about 300°C. The rocks of the Erbendorf–Vohenstrauss Zone also underwent progressive metamorphic transformations under amphibolite-facies conditions (at temperatures up to 700°C). This was one more reason to study the permeability of the samples from superdeep boreholes at high temperature and pressure.

Most countries with a well-developed nuclear industry have accepted the underground disposal of high-level radioactive waste (HLW) as the preferable means for its isolation. In this case, the most hazardous scenario would be the leakage of HLW from the repository and its convective transport to the Earth's surface by groundwater. Since the behaviour and velocity of groundwater flow depend significantly on the spatial distribution of rock permeability, the determination of this parameter is very important for safety assessment of HLW underground

From: HARVEY, P. K., BREWER, T. S., PEZARD, P. A. & PETROV, V. A. (eds) 2005. *Petrophysical Properties of Crystalline Rocks*. Geological Society, London, Special Publications, **240**, 153–164.
0305-8719/05/$15.00 © The Geological Society of London 2005.

disposal. In order to isolate HLW reliably from the biosphere, it is advisable to build a repository at considerable depth. Therefore, the lithostatic pressure in the zone of disposal will be rather high. The temperature of the surrounding rocks in the vicinity of an underground repository should significantly increase due to heat generation by HLW. Hence, the rock permeability has to be estimated at high pressure and temperature, and this is another reason for carrying out measurements at high temperature and pressure.

Therefore, the experimental study of rock sample permeability at temperatures of up to 600°C and pressures of up to 150 MPa has been carried out. This $P-T$ range corresponds:

(1)   to the conditions of deep-seated rocks of the superdeep boreholes *in situ*;
(2)   to the conditions of progressive and regressive metamorphism of the Kola Series rocks;
(3)   to the conditions of the rocks surrounding the deep underground HLW repository.

Samples, experimental set-up and procedure

The SG-3 borehole was drilled in the centre of the synformal structure in the NE part of the Baltic Shield. It penetrates Proterozoic meta-

volcanic rocks of the Pechenga Complex down to a 6842 m, and Archaean faulted basement: gneisses and amphibolites of the Kola Series, in the depth interval from 6842 m down to 12 262 m (Fig. 1). Therefore the opportunity was taken to compare the properties of samples of the same rock types, collected from depths down to 12 km, and from the Earth's surface.

The KTB borehole is drilled in the Erbendorf–Vohenstrauss Zone in the western part of the Bohemian Massif. Its depth is 9101 m. The borehole section is represented by rocks from Late Proterozoic to Palaeozoic age, transformed to amphibolites, metagabbros and associated meta-tuffites, and garnet–mica gneisses (Fig. 1). Thus both of the superdeep borehole sections are represented by the rocks of similar composition and metamorphic grade and, it may be assumed, with similar physical properties. These rocks occur in a very large depth interval: from the Earth's surface down to 12 km. Therefore, in order to reveal the depth effect on rock properties, we collected samples from different depths. Typical rocks from the Kola Series were selected for the experiments: three gneisses and five amphibolite samples. Five core samples were collected from great depths (8.8–11.4 km) and three unweathered samples of the same rock type were collected from the surface, on the Mustatunturi Ridge 50 km to the northeast of SG-3 (Fig. 1).

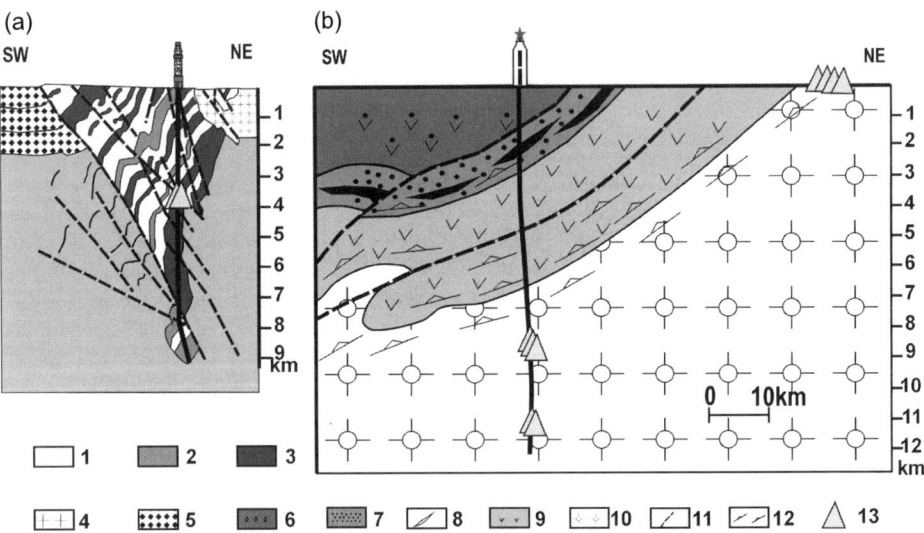

Fig. 1. KTB (**a**) and SG-3 (**b**) borehole sections. Modified after Emmerman & Lauterjung (1997) and Lobanov *et al.* (1999). Key: KTB section: 1, paragneisses; 2, variegated sequences; 3, metabasites; 4, granites; 5, sediments. SG-3 section: Proterozoic Pechenga Complex: 6, metavolcanic rocks; 7, metasedimentary rocks; 8, mafic and ultramafic intrusions; 9, metavolcanic rocks and metaandesites; 10, Archaean faulted basement (gneisses and amphibolites of the Kola Series); 11, faults; 12, shear zones; 13, points of sampling.

The details on sample collection and their properties are presented in Table 1, in Lobanov *et al.* (1999) and Zharikov *et al.* (2003). KTB is represented by metagabbro and amphibolite samples collected from moderate depths of 3.6–3.8 km. It can be seen from Table 1 that core and surface sample compositions are similar and, in contrast, reservoir properties are different: core sample porosity is about twice as high, and core sample permeability is one order higher than in the samples collected from the Earth's surface. We believe that such a massive contrast is caused by the damaging effect of overprinted microcracks. There are many reasons for their initiation: drilling, coring, the relief of strong heterogeneous stresses, and cooling. This notion can be illustrated by Figure 2, showing the porosity and permeability of specimens cut at different distances from the core axis. It can be seen that porosity and permeability increase strongly from the central part of the sample to its peripheral parts. Therefore, one of the reasons for performing high P–T experiments was to find out if reloading and reheating can restore core sample properties.

The experiments were performed on specimens of 9.6 mm in diameter and 15 mm in length, all cut from the central part of the core using a set-up producing hydrostatic pressure (described in Shmonov *et al.*, 1994, 2002 and Zharikov *et al.*, 2003). Flow direction parallel to the core axis simulated filtration along an ascending branch of the thermal convection cell, from a HLW underground repository to the Earth's surface. In order to avoid any fluid–rock interaction, argon was used as a flowing fluid. The data of the applied steady-state method was recalculated using the Klinkenberg technique (Klinkenberg 1941). Klinkenberg established a relationship between rock permeability to water (k_w), to gas (k_g), gas pressure (p), and a constant depending on pore size (b):

$$k_g = k_w \left(1 + \frac{b}{p}\right).$$

Using this technique, several measurements of gas permeability were carried out at different gas

Table 1. *Composition and initial reservoir properties of the studied samples*

	Sample number	Depth of collection (m)	Rock type	Mineral composition (vol. %)	Porosity (%)	Permeability (m²)
1	P 61	Earth's surface	Amphibolite	Hbl–70, Pl–25, Qtz–2, Bt–2, Cl, Ep, Sph–1	0.96	1.4×10^{-18}
2	P 17	Earth's surface	Gneiss	Pl–50, Qtz–40, Bt–9, Ap–1	0.64	4.7×10^{-18}
3	P 19	Earth's surface	Gneiss	Pl–50, Qtz–35, Bt–10, Ms–4, Spn–7, Carb–1	0.49	8.0×10^{-18}
4	Glp	3847	Amphibolite	Pl–44, Hbl–39, Sc–7, Bt–4, Mag–3, Qtz–1	0.99	1.4×10^{-17}
5	D3v	3620	Metagabbro	Cpx–40, Pl–38, Qtz–8, Hbl–4, Bt–6, Grt–3, Ore–2	1.56	2.2×10^{-17}
6	31 421	8812	Gneiss	Pl–40, Qtz–35, Mc–5, Ms–10, Ep–7, Bt–5, Spn, Carb–1	2.01	1.1×10^{-17}
7	31 571	8863	Amphibolite	Hbl–60, Pl–35, Spn–5, Carb–1	1.46	6.10×10^{-17}
8	31 863	8940	Amphibolite	Hbl–65, Pl–25, Spn–5, Carb, Ep, Ore–5	1.56	8.0×10^{-17}
9	43 639	11 400	Amphibolite	Hbl–55, Pl–40, Bt–2, Qtz–2, Ep, Carb, Ore–1	1.80	3.1×10^{-17}

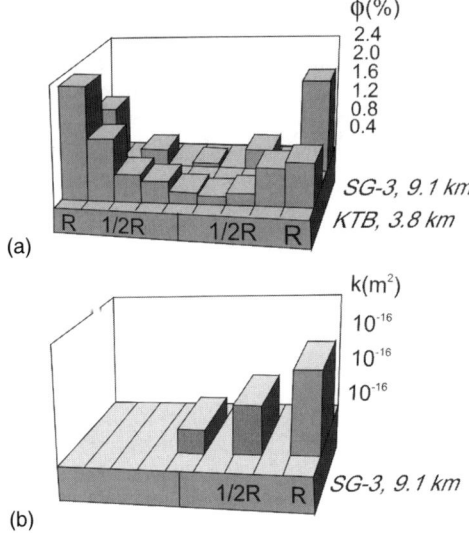

(a)

(b)

Fig. 2. Porosity (**a**) and permeability (**b**) of the specimens cut at a different distance from the core axis.

pressures in order to calculate each value of permeability to water.

Our goal was to make 42 permeability measurements at temperatures of 20, 100, 200, 300, 400, 500 and 600°C and pressures of 30, 50, 80, 100, 120 and 150 MPa on each sample and in each experiment. We tried to reduce the amount of heating–cooling and loading–relief during the experiment run, in order to minimize extra damage to the rock structure that might be related to experimental conditions. Hence, the following pressure and temperature path was used: the given temperatures were reached upon simultaneous heating and loading, and then permeability measurements during pressure cycles at constant temperature were performed.

Experimental results

The data on the studied samples revealed no correlation between rock composition and its permeability behaviour at high $P-T$. The same result was noted earlier, when we studied other tight rocks (see Zharikov *et al.* 2003, where a complete summary, containing all the measurement data on the superdeeps, including the results of anisotropy study, is given, or Shmonov *et al.* 2002, where all the data obtained by our laboratory at high $P-T$ are presented). The samples of different rock types numbered about 50. However, we have revealed some tendencies regarding permeability behaviour at

different $P-T$, and here we will present only the most typical trends.

The pressure–permeability trends obtained on the Kola surface amphibolite, the amphibolite sample collected in KTB from shallow depths (3.8 km) and the Kola core sample – an amphibolite from a depth of 11.4 km – are presented in Figure 3a. An increase of confining pressure always leads to a decrease of rock sample permeability. These data agree with those presented in Huenges (1993) and Morrow *et al.* (1994). The trends in behaviour change with temperature: in Figure 3c it can be seen that the curves obtained at elevated temperatures (400–600°C) are more inclined to the pressure axis.

The temperature trends obtained on the same samples are presented in Figure 3b. To make the graphs more illustrative, the averaged curves are plotted there. The curves present high (100–150 MPa) and low (30–80 MPa) effective pressures. It can be seen that behaviour of temperature trends depends on effective pressure and depth of sampling.

- With high effective pressure, temperature–permeability trends are monotonous for the samples collected from the surface and from moderate depths (3.8 km). Temperature increase leads to a permeability decrease. In contrast, the temperature–permeability trends of the samples collected from great depth (8.8–11.4 km) have inversions: firstly, decreases and then increases.
- With low effective pressure, inversions occur on all the obtained trends for the samples: collected from the great and shallow depths and from the surface as well. The inversions occur at a variety of temperatures: 200, 300 and 500°C.

The sharp, threshold-like form of the pressure and temperature trends of permeability can also be noted.

As mentioned above, 42 permeability measurements in the pressure range from 30 to 150 MPa and the temperature range from 20 to 600°C were carried out on each studied sample. This has allowed us to draw permeability diagrams in $P-T$ co-ordinates, as presented in Figure 3c. From the figure, it can be seen that permeability of the same sample at different $P-T$ can vary by many decimal orders. The upper boundaries of $10^{-17}-10^{-18}$ m^2 are noted in all the samples within the entire pressure range and in the temperature interval of 20–100°C, as well as in samples from depths of 11.4 km at low pressure and temperatures of 500–600°C. Domains of minimum values ($<10^{-21}$ m^2) are shaded grey. Their positions vary for the samples from different depths. A sample from the surface has the widest parameter domain: its boundary begins at 200°C in the zone of high pressure and extends through the entire temperature range at pressures more than 50 MPa. A sample from a depth of 3.8 km has a smaller domain: at temperatures of 400–600°C its boundary shifts to the zone of high pressure. And, finally, the smallest 'non-permeability' domain is noted for the sample from a depth of 11.4 km: it occupies the area of temperatures from 350 to 550°C and pressures from 80 to 150 MPa.

In order to consider and explain the origin of the trends obtained, microstructural investigations were performed in a specially designed cell, which produced temperatures up to 600°C and pressures up to 100 MPa (Shmonov *et al.* 1994). For direct observation of the sample surface at high $P-T$ this cell was placed entirely into a scanning electron microscope. The data on microcrack aperture and length distributions were used for computer simulation of rock permeability. The details of the studies are presented in Shmonov *et al.* (2002) and Zharikov *et al.* (2003).

Summarizing the data obtained, we may draw the conclusion that the observed permeability behaviour is due to the initiation and closure of a variety of microcrack families as a result of the competitive effects of temperature and pressure. When the pressure effect dominates (for example, at temperatures lower than 200°C) a family of low aspect ratio microcracks closes, and permeability decreases (Fig. 3). In contrast, when temperature dominates, high aspect ratio microcracks open, or new ones appear, and permeability increases (for example, at low effective pressures, and temperatures higher than 500°C, Fig. 3). Therefore, inversions may appear within the permeability–temperature trends.

Application of the experimental data

Our experiments, the results of which are presented in this paper, were aimed at estimating rock permeability in the upper continental crust. Our prime focus was rock permeability at typical $P-T$ *in situ*, as well as what would happen if a new source of heat-generation – a HLW underground repository – was added there. No doubt the number of samples studied is insufficient for a detailed assessment, but some general tendencies can be noted.

The dependencies of permeability on simultaneous loading and heating, simulating a depth increase, are presented in Figure 4. The curves were plotted using the following assumptions.

Lithostatic pressure was changing according to the mean density of the rocks in SG-3 (Kozlovsky 1987) is equal to 2.84 *gh*, a fluid pressure we consider to be hydrostatic. We assume water density to be equal to 1.0 g cm^{-3}. Hence, the effective pressure is evaluated as $p = 1.84$ *gh*. The temperature was changing according to the mean values of the temperature gradients in the SG-3 and KTB boreholes: 18°C km^{-1} and 27°C km^{-1}, (Emmerman & Lauterjung 1997; Kozlovsky 1987). The pressures and temperatures that were chosen for the experiments are presented in Table 2 (lines 1 and 2 in Figure 3c also show such $P-T$ changes). The averaged curves for SG-3 core and surface samples and KTB core samples are presented in Figure 4.

At initial $P-T$, the SG-3 core samples show the highest values of permeability and porosity. The values of the parameters in the KTB core samples are lower, and in the surface samples they are lowest: permeability is reduced by a factor of ten and porosity is reduced two- to threefold. We attribute this to the minimal amount of overprinted microcracks there. Hence, the permeability values obtained on superdeep core samples at ambient temperature and under pressure are abnormally high. Heating and loading, simulating depth increases, lead to a permeability decrease with non-constant gradient. With loading and heating up to $P-T$ parameters of 6 km, the difference between core and surface sample permeability values disappears. With further increases in $P-T$, the difference reappears. As a result, at $P-T$ conditions of 10 km the difference between minimum and maximum permeability values reaches two decimal orders. The lowest values (10^{-21} m^2) occur in Kola surface samples and in KTB cores, the highest values (10^{-19} m^2) occur in SG-3 core samples. It is still unknown why such permeability behaviour occurs at $P-T$ conditions of great depths: is it due to the real difference in rock properties, or due to the very strong effect of overprinted microcracks in SG-3 deep cores? Since the curves of surface samples and KTB samples, which were collected at much lesser depth, are similar, we can assume that, in this case, reloading and reheating up to $P-T$ *in situ*, evidently did not lead to the complete closure of overprinted microcracks in SG-3 core samples.

Since the permeability of all the studied samples decreases with depth, we can suppose that, for the conditions of the upper crust – up to a depth of 10 km – the effect of pressure prevails over the temperature effect. In such cases, it is reasonable to compare our results with data

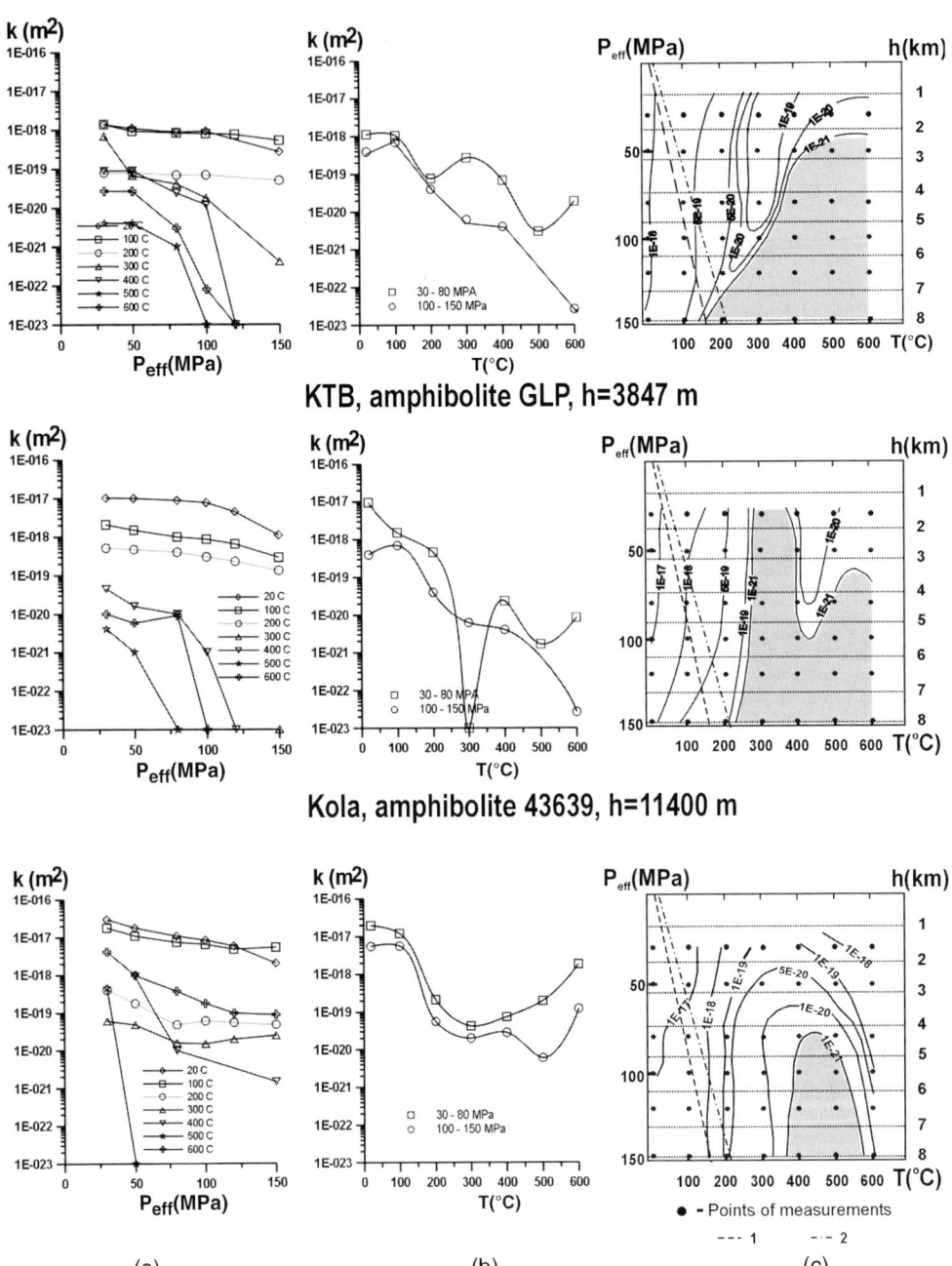

Fig. 3. Permeability dependencies on pressure (**a**) and temperature (**b**), and permeability diagrams of *P–T* co-ordinates (**c**).

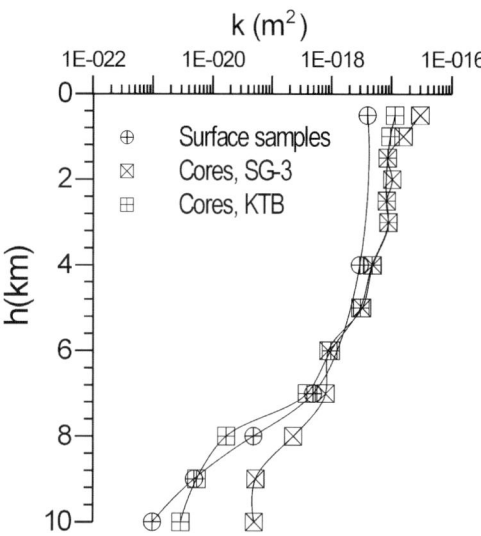

Fig. 4. Permeability dependencies on simultaneous loading and heating simulating *in situ* depth increase.

one can see that at a $P-T$ of 1 km (Table 2) the permeability values are $10^{-18}-10^{-17}$ m^2. If a repository appears there, the pressure *in situ* will not increase, but the temperature in vicinity can rise, because HLW is a source of heat. Let us suppose that the temperature will rise to 200°C. If we move along the T axis to this value, we can see that the permeability will decrease down to 10^{-18} to 10^{-19} m^2.

Progressive metamorphism of the Kola rocks took place at temperatures of 550–600°C under conditions of low effective pressure. At such $P-T$ parameters, the permeability of the studied samples was high enough (up to 10^{-18} m^2 and higher) to permit a fluid flow to penetrate the whole volume of the rock massif. In contrast, at the $P-T$ of retrograde metamorphism (a temperature of about 300°C and high pressure) the permeability values were a few decimal orders lower (10^{-21} m^2 and lower), so the fluid flow could be concentrated only in the large disjunctive zones.

Numerical simulation of thermal convection for safe HLW disposal

Rock permeability is a critical component of the input data for the simulation of radionuclide migration from an underground repository of HLW, the results of which are used as the basis for a safety assessment of HLW disposal. Let us consider the process of convective transport of radionuclides from a HLW repository. HLW are usually represented by solid and chemically stable waste forms, which incorporate radio-active materials.

Mined and well repositories are the principal types of repositories considered for geological HLW disposal (Cooley *et al.* 1990; John 1982; Kedrovsky *et al.* 1991). In mined repositories,

obtained only under pressure and at ambient temperature (Huenges 1993; Morrow *et al.* 1994). There one can see the same permeability behaviour – a decrease – and the same orders of its values.

The experimental curves presented here were obtained for temperature gradients measured recently in the boreholes. Let us consider how rock permeability may change, if a HLW under-ground repository is constructed. A repository must be built at a depth that can ensure long-term and safe waste isolation from the biosphere. An estimation of the appropriate depth will be considered later in this paper. However, we can say that such a depth will be no more than 1 km. From the permeability diagrams (Fig. 3c),

Table 2. P–T-*parameters of the experiments simulating* in situ *depth increase*

Depth (km)	P–T-conditions within the SG-3 borehole		P–T-conditions within the KTB borehole	
	Effective pressure (MPa)	Temperature (°C)	Effective pressure (MPa)	Temperature (°C)
1	18	18	18	27
2	37	36	37	55
3	55	24	55	82
4	74	72	74	109
5	92	90	92	137
6	110	108	110	164
7	129	126	126	191
8	147	144	147	218
9	166	162	166	246
10	184	180		

canisters containing HLW are emplaced in short boreholes drilled in the floor and walls of underground drifts. In well repositories, canisters are stacked along the axis in the lower part of each well (Fig. 5). Well repositories are much easier to design and construct, and are less expensive. These advantages are especially important because of the significant amount of HLW accumulated in temporary storage areas. That is why the process of HLW leakage will be considered in this paper, as applied to the well repository.

The main mechanism of radionuclide migration is their transport by groundwater. The groundwater flow in the vicinity of the repository consists of two components: a forced one – caused by regional flow, and a thermoconvective component caused by heat generation in HLW. It is obvious that the low velocity of the regional groundwater flow is one of the main requirements placed upon selection of a repository site. However, it should be taken into account that HLW generates substantial amount of heat. As a result, a HLW repository will trigger the thermal convection of groundwater. Under conditions of weak regional flow, free thermal convection of groundwater can be one of the primary potential hazards for the long-term safety of a repository site (Wang *et al.* 1989).

Let us consider the simplest model of thermoconvective transport of radionuclides from the well repository shown in Figure 5, in Boussinesq's approximation. In assuming cylindrical symmetry for the problem, the set of governing equations can be written in cylindrical co-ordinates as follows:

$$\frac{\partial u}{\partial z} + \frac{1}{r}\frac{\partial}{\partial r}(rv) = 0;$$

$$u = -\frac{k}{\mu}\left\{\frac{\partial p}{\partial z} - \rho g[1 - \beta(T - T_0)]\right\},$$

$$v = -\frac{k}{\mu}\frac{\partial p}{\partial r};$$

$$\frac{\rho_r c_r}{\rho c}\frac{\partial T}{\partial t} + u\frac{\partial T}{\partial z} + v\frac{\partial T}{\partial r} = \frac{\lambda}{\rho c}\left[\frac{\partial^2 T}{\partial z^2} + \frac{1}{r}\frac{\partial}{\partial r}\left(r\frac{\partial T}{\partial r}\right)\right];$$

$$\frac{\partial C}{\partial t} + \frac{1}{\varphi}\left(u\frac{\partial C}{\partial z} + v\frac{\partial C}{\partial r}\right) = \frac{\partial}{\partial z}\left(d_{zz}\frac{\partial C}{\partial z} + d_{zr}\frac{\partial C}{\partial r}\right)$$

$$+ \frac{1}{r}\frac{\partial}{\partial r}r\left(d_{rr}\frac{\partial C}{\partial r} + d_{rz}\frac{\partial C}{\partial z}\right) - \kappa C \qquad (1)$$

where z and r are cylindrical co-ordinates (the z axis runs along the well axis downwards, with $z = 0$ at the Earth's surface); u and v are the vertical and horizontal components of the groundwater flow velocity field respectively; k is the permeability of enclosing rocks; μ, ρ, β and c are the viscosity, density, coefficient of thermal expansion and specific heat of the groundwater; λ, ρ and c_r, are the thermal conductivity, density and specific heat of enclosing rocks; φ is porosity; p is pore pressure; g is acceleration due to gravity; T is temperature, T_0 is initial temperature; C is the concentration of the considered radionuclide in the groundwater; t is time; d_{zz}, d_{zr}, d_{rz}, d_{rr} are dispersion components, and κ is the radioactive decay constant.

Values of the hydrodynamic dispersion components are defined as:

$$d_{zz} = D + \alpha_L\frac{u^2}{\varphi w} + \alpha_T\frac{v^2}{\varphi w};$$

$$d_{rr} = D + \alpha_T\frac{u^2}{\varphi w} + \alpha_L\frac{v^2}{\varphi w};$$

$$d_{rz} = d_{zr} = (\alpha_L - \alpha_T)\frac{uv}{\varphi w}$$

where D is molecular diffusivity (corrected for tortuosity factor), α_L and α_T are linear parameters of dispersion, and w is absolute value of flow velocity.

Boundary conditions for equation (1) are specified in the form:

$$r = 0, z \in [0, z_1] \cup [z_2, \infty),$$

$$v = 0, \quad \frac{\partial C}{\partial r} = 0,$$

$$\frac{\partial T}{\partial r} = 0;$$

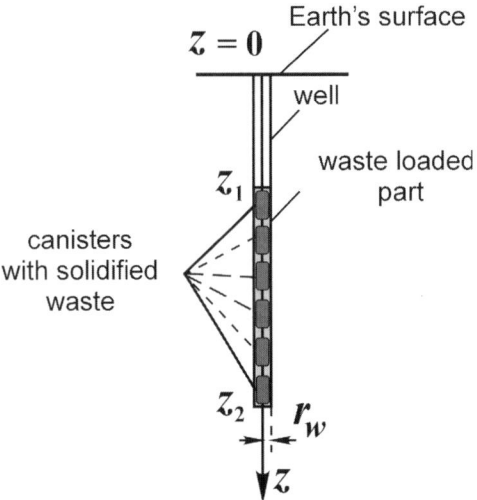

Fig. 5. Schematic diagram of a HLW well repository.

$r = r_w, z \in (z_1, z_2),$

$v = 0, C = \gamma C_{sat}(T);$

$$\frac{\partial T}{\partial r} = -\frac{M\omega}{2\pi\lambda};$$

$$r \to \infty, \quad v \to 0, \frac{\partial C}{\partial r} \to 0, \frac{\partial T}{\partial r} \to 0; \qquad (2)$$

$$z = 0, \quad v = 0, T = T_0, \frac{\partial C}{\partial z} = 0;$$

$$z \to \infty, \quad v \to 0, \frac{\partial C}{\partial r} \to 0, \frac{\partial T}{\partial r} \to 0.$$

Here z_1 and z_2 are the depths of the top and bottom of the waste-loaded interval within the well, respectively; γ is the concentration of the considered radionuclide in the waste form; $C_{sat}(T)$ is the temperature-dependent saturation concentration of matrix components in the aqueous solution at the specified temperature; M is waste mass per unit length of the loaded part of the well; and ω is the heat generation rate per unit of waste mass.

Initial conditions are written as:

$$t = 0, \quad C = 0, T = T_0. \qquad (3)$$

The boundary problem (equations 1–3) was solved numerically by the finite-difference method for different governing parameters. The maximum concentration of a particular radionuclide at the Earth's surface was accepted as a criterion of safety assessment, i.e. it is assumed that the repository site is safe if this maximum concentration is less than the prescribed maximum allowable concentration of the radionuclide. The strictest constraints are imposed upon the concentration of ^{90}Sr, for which the maximum allowable concentration (MAC) is 2×10^{-15} g cm^{-3} (Krauskopf 1988). That is why it is reasonable to select this radionuclide for determining the safety criterion.

Let us consider an example. As mentioned above, permeability is the most important parameter in the equations. We have estimated the matrix permeability of the rocks from SG-3 and KTB. At the probable depths of a well repository, it ranges from 10^{-19} to 10^{-18} m^2. Brace (1980) revealed that the permeability of rock mass is higher than the matrix permeability obtained, due to the occurrence of fractures that are not usually represented in the samples. He estimated a factor of 10^3 for correcting this. The data from the KTB investigations (Huenges 1993; Huenges et al. 1997) agree with Brace's estimations: their matrix permeability ranges from 10^{-20} to 10^{-19} m^2, and the rock-mass permeability is

10^3 higher, i.e. 10^{-16} to 10^{-17} m^2. Unfortunately, the results of pumping tests in SG-3 are not so detailed. However, as we mentioned above, SG-3 and KTB boreholes are drilled in analogous geological structures and the same types of tight rocks occur in the sections. Therefore, we assumed the correction factor value to be the same – 10^3. The maximum matrix permeability values that we estimated in the vicinity of the HLW were in the range 10^{-19} to 10^{-18} m^2. This means that rock-mass permeability values might be in the range of 10^{-16} to 10^{-15} m^2. We will consider the case when $k = 10^{-16}$ m^2. The thermophysical properties of water are specified as:

$$\rho = 1000 \, \text{kg m}^{-3}, c = 4200 \, \text{J/(kg K)},$$

$$\beta = 5 \times 10^{-4} \, 1/\text{K}, \mu = 5 \times 10^{-4} \, \text{Pa s}.$$

An analysis of thermophysical properties of rocks shows that $\rho_r c_r$ and ρc are of the same order of magnitude (Dortman 1984). If we specify that they are equal, then the simulation results are approximately valid for many types of rocks. That is why, in the case being considered, it is taken that $\rho_r c_r \cong \rho c$. Thermal conductivity is specified as $\lambda = 1.26 \, \text{W/(m K)}$ (what is close to lower boundary of the thermal conductivity range for intrusive rocks). Since it is supposed that rocks at repository site do not contain large-scale tectonic disturbances, dispersion parameters are specified as $\alpha_L = \alpha_T = 0.5$ m (Bochever & Oradovskaya 1972). Porosity value is specified as $\varphi = 0.003$.

We assume that $r_w = 1$ m, and the length of the waste-loaded part of the well is 500 m, i.e. $z_2 - z_1 = 500$ m. It is also assumed that each canister occupies 4 m of the well length, and that the heat generation of HLW at $t = 0$ is 2.16 kW per canister (Wang et al. 1989). Therefore, $M\omega_0 = 504$ W m^{-1}.

The dependence $C_{sat}(T)$, according to (Chambre & Pigford 1984), may be given as:

$$\ln C_{sat}(T) = 2.303 \frac{T - T_0}{\Delta T} + B \qquad (4)$$

where ΔT is the temperature range within which C_{sat} changes by 10 times. From the data on solubility of amorphous siliceous compounds, it is specified that $\Delta T \cong 200$ K, and that saturation concentration of the waste form at 20°C is about 5×10^{-5} g cm^{-3} (Chambre & Pigford 1984). The fraction by weight of radioactive materials in the waste form is 10–35% (Hench et al. 1986), and these materials contain about 5% of ^{90}Sr. Since γ is a product of these values, $\gamma = 0.005-0.0165$. It was specified that $\gamma = 0.01$.

Calculations at different values of z_1 showed that the maximum concentration of ^{90}Sr in the groundwater at the Earth's surface is less than MAC, if the depth of waste disposal z_1 is more than 218 m. Therefore, the selected criterion of disposal site safety is satisfied at the disposal depth, which is quite acceptable from both the technical and economic points of view. More sophisticated analysis for the assessment of reciprocal influence between thermoconvective transport processes in neighbouring wells shows that this influence is not significant if the distance between the wells at realistic values of governing parameters is less than 100 m (Malkovsky *et al.* 1995).

As we can see, the boundary problem considered in equations (1) and (2) describes a convective-heat- and mass-transfer process in a uniform porous medium. The availability of such a model is based on a repository safety requirement that a rock massif selected for the HLW repository may not contain large-scale irregularities (and, in particular, tectonic disturbances). Hence, changes in the transport properties of these rocks at elevated pressures and temperatures can be estimated from a limited number of experiments on core samples. Since any real rock massif is not perfectly uniform, the model considered can be used mainly for a preliminary estimation of repository parameters. Refined results should be obtained from calculations on the basis of the extended precision technique (Malkovsky *et al.* 1998), taking into account the rock heterogeneity and topography-driven flow of groundwater, which calls for more detailed studies of the transport properties of these rocks.

The computer simulation results presented show that a HLW well repository site can be safely located at relatively shallow depths. The probable values of *in situ* permeability of a rock massif at the depth of the projected well repository were used in the model. This model does not take into account the changes in permeability due to HLW heat generation, but the results of experiments demonstrate that an increase in pressure and temperature leads to a decrease in permeability for types of rocks being examined, in the range of $P-T$ parameters being considered. Therefore, one can expect that a HLW repository site will remain safe even under temperature increase caused by heat generation in the disposed waste. Secondly, as has been mentioned before, the values used in the simulation are close to the permeability of unweathered rock-mass. The total permeability of the geological structure can be higher.

The Kola SG-3 section is intersected by numerous large fracture systems such as faults and shear zones (Kozlovsky 1987). The permeability of these zones is higher than that which we used for our calculations about 10^{-14} m^2 (Tutubalin *et al.* 1995). However, our main focus was to estimate a safe depth for a HLW well repository, i.e. to find a minimum vertical thickness of a rock-mass containing only microfractures or discrete fractures. The safe depth for a HLW repository, according to our calculations, was estimated to be about few hundred metres. This is the minimal vertical thickness of a well section above the repository that is not intersected by fracture zones. We assumed the length of well-loaded part to be 500 m. Therefore, the minimum vertical dimension for a rock-mass block for a HLW repository is about 700 m. Referring to the data on SG-3 (Kozlovsky 1987) we believe that it is possible to find this type of block in the borehole section. Of course, any repositories for critical wastes are unacceptable in superdeep boreholes. However, similar geological structures that can be considered as potential sites for HLW underground disposal are located in the region, and the estimates obtained show that a well repository site can be safely located at a relatively shallow depth.

Summary

An experimental study of the samples collected from a depth of 3.8–11.4 km in the Kola and KTB superdeep boreholes and from the Earth's surface at the Kola drilling site was carried out at temperatures up to 600°C and pressures up to 150 MPa.

(1) Due to the effect of overprinted microcracks, the porosity and permeability of the core samples are higher than the porosity and permeability of the samples collected from the surface. Therefore, the values obtained on the core samples at ambient temperature and pressure are abnormally high.

(2) An increase in confining pressure leads to a decrease in the permeability of the rock samples. The behavioural trend changes with temperature: the curves obtained at elevated temperatures (400–600°C) are more inclined to the pressure axis.

(3) The behaviour of the temperature–permeability trends depends on the effective pressure and on the depth of sample selection.

- With high effective pressure, the temperature–permeability trends are monotonous for the samples collected from the surface and from moderate depth (3.8 km). The temperature increase leads to a permeability decrease. In contrast, the temperature–permeability trends of samples collected from great depths (8.8–11.4 km) show inversions.
- With low effective pressure, inversions occur on all of the trends obtained for the samples: i.e. those collected from the great and shallow depths and from the surface as well. The inversions occur at a variety of temperatures: 200, 300 and 500°C.

(4) It was found that this permeability behaviour is due to rock microstructural transformations caused by the competing influences of temperature and effective pressure.

(5) Simultaneous heating and loading, simulating an increase in depth, lead to a decrease in sample permeability.

(6) A numerical simulation of convective transport was performed in order to determine a safe depth for a HLW well repository. The estimations obtained show that a well repository site can be safely located at relatively shallow depths. The experimental results obtained demonstrate that increases in pressure and temperature lead to a decrease of permeability for examined types of rocks in the considered range of P–T parameters. Therefore, one can expect that the HLW repository site will remain safe even during the temperature increase caused by heat generation in the disposed waste.

This work was supported by grants from the Russian Foundation for Basic Researches (03-05-641153, 05-05-65136, 05-07-90340, and 05-05-65060) and by Programs 5 and 10 of the Earth Sciences division of the Russian Academy of Sciences.

References

BOCHEVER, F. M. & ORADOVSKAYA, A. E. 1972. [*Hydrogeological substantiation of groundwater and water intakes protection from contaminants.*] Nedra, Moscow (in Russian).

BRACE, W. F. 1980. Permeability of crystalline and argillaceous rocks. *International Journal Rock Mechanics and Mining Science*, **17**, 241–251.

CHAMBRE, P. L. & PIGFORD, P. H. 1984. Prediction of waste performance in a geologic repository. *In*: *Proceedings of the Materials Science Society*,

Scientific Basis for Nuclear Waste Management VII, Boston, Massachusetts, USA, November 1983, Elsevier, 985–1008.

COOLEY, C. R., CANDACE, Y. C. & SHNEIDER, K. J. 1990. High level radwaste plans in nine countries. *Chemical Engineering Progress*, 68–74.

DORTMAN, N. B. (ed.) 1984. [*Physical Properties of Rocks and Minerals.*] Nedra, Moscow (in Russian).

EMMERMAN, R. & LAUTERJUNG, R. 1997. The German Continental Deep Drilling Program KTB: overview and major results. *Journal of Geophysical Research*, **102**, 18 179–18 201.

GORBATZEVICH, F. F. & MEDVEDEV, R. V. 1986. [Mechanisms of rocks decompaction on their stress relief.] *In*: [*Ore Geophysical Investigations at Kola Peninsula.*] Izdatelstvo Kolskogo Filiala AN SSSR, Apatity, 83–89 (in Russian).

HENCH, L. L., CLARK, D. E. & HARKER, A. B. 1986. Nuclear waste solids. *Journal of Material Science*, **21**, 1457–1478.

HUENGES, E. 1993. Profile of permeability and formation pressure down to 7,2 km. *In*: EMMERMANN, R., LAUTERJUNG, J. & UMSONST, T. (eds) KTB report 93-2. *Contributions to the 6th Annual KTB-Colloquium Geoscientific Results, Giessen, 1–2 April, 1993*, 279–285.

HUENGES, E, ERZINGER, J., KUCK, J., ENGESER, B. & KESSELS, W. 1997. The permeable crust: geohydraulic properties down to 9101 m depth. *Journal of Geophysical Research*, **102**, 18 255–18 265.

JOHN, C. M. S. 1982. Repository design. *Underground Space*, **6**, 247–258.

KEDROVSKY, O. L., SHISHITS, I. Yu. & GUPALO, T. A. 1991. [Substantiation of conditions for localization of high-level wastes and spent fuel in geological formations.] *Atomnaya Energiya (Atomic Energy)*, **70**, 294–297 (in Russian).

KLINKENBERG, L. J. 1941. The permeability of porous media to liquids and gases. *American Petroleum Institute Drilling and Productions Practice*, 200–211.

KOZLOVSKY, Ye. A. (ed.) 1987. *The Superdeep Well of the Kola Peninsula*. Springer, Berlin, 558 pp.

KRAUSKOPF, K. B. 1988. Geology of high-level nuclear waste disposal. *Annual Review of Earth and Planet Sciences*, **16**, 173–200.

LOBANOV, K. V., GLAGOLEV, A. A., ZHARIKOV, A. V. & KUZNETSOV, A. V. 1999. [A comparison between Archean rocks from the section of the Kola superdeep drill hole and their surface analogues.] *Geoinformatika*, **4**, 38–50 (in Russian).

MALKOVSKY, V. I., PEK, A. A. & OMELYANENKO, B. I. 1995. Influence of the interwell distance on the thermoconvective transport of radionuclides by groundwater from a two-well high-level nuclear waste repository. *In*: *Proceedings of the 5th International Conference on Radioactive Waste Management and Environmental Remediation 'ICEM'95'* (Berlin, Germany, 3–7 September, 1995). ASME, NY, 1, 819–822.

MALKOVSKY, V. I., PEK, A. A. & TSANG, C. F. 1998. A new formulation of convective transfer for simulation of mass transport with large concentration variations. *In*: *Proceedings of the 4th International*

Conference of the International Association for Mathematical Geology 'IAMG'98' (Napoli, 1998). De Frede Editore, Napoli, 1, 304–308.

MORROW, C., LOCKNER, D., HICKMAN, S., RUSANOV, M. & ROCKEL, T. 1994. Effects of lithology and depth on permeability of core samples from the Kola and KTB boreholes. *Journal of Geophysical Research*, **99**, 7263–7274.

SHMONOV, V. M., VITOVTOVA, V. M. & ZARUBINA, I. V. 1994. Permeability of rocks at elevated temperatures and pressures. *In: Fluids in the Crust. Equilibrium and Transport Properties.* SHMULOVICH, K. I., YARDLEY, B. W. D. & GONCHAR, G. G. (eds) Chapman and Hall, London, 285–313.

SHMONOV, V. M., VITOVTOVA, V. M. & ZHARIKOV, A. V. 2002. [*Fluid Permeability of the Earth's Crust Rocks.*] Nauchniy Mir, Moscow, 216 pp (in Russian).

TUTUBALIN, A. V., GRICHUK, D. V. & MALKOVSKY, V. I. 1995. A complex hydrodynamic and thermo-dynamic model of a convective hydrothermal system. *In: Water–Rock Interaction. Proceedings of the 8th International Symposium.* A. A. Balkema, Rotterdam, 763–766.

WOLTER, K. E. & BERCKHEMER, H. 1989. Time dependent strain recovery of cores from KTB – deep drill hole. *Rock Mechanics and Rock Engineering*, **22**, 273–287.

WANG, J. S. Y., MANGOLD, D. C. & TSANG, C.-F. 1989. Thermal impact of waste implacement and surface cooling associated with geological disposal of high level nuclear waste. *Environmental Geology and Water Sciences*, **11**, 183–239.

ZHARIKOV, A. V., VITOVTOVA, V. M., SHMONOV, V. M. & GRAFCHIKOV, A. A. 2003. Permeability of the rocks from the Kola superdeep borehole at high temperature and pressure: implication to fluid dynamics in the continental crust. *Tectonophysics*, **370**, 177–191.

Integration of electrical and optical images for structural analysis: a case study from ODP Hole 1105A

S. L. HAGGAS[1], T. S. BREWER[2], P. K. HARVEY[2] & C. J. MACLEOD[3]

[1]IHS Energy, Enterprise House, Cirencester Road, Tetbury, Glos, GL8 8RX, UK

[2]Department of Geology, University of Leicester, Leicester, LE1 7RH, UK

(e-mail: tsb5@leicester.ac.uk)

[3]School of Earth, Ocean and Planetary Sciences, Cardiff University,
Main Building, Park Place, Cardiff, CF10 3YE, UK

Abstract: A significant problem with basement core data acquired by the Ocean Drilling Program (ODP) is that it is unorientated with respect to grid north and in those cores with <100% core recovery the data may also be inaccurately located within the borehole. In order for structural data from basement cores to be of any real value, the recovered core must firstly be accurately located and orientated. Here, we develop methods for core reorientation and accurate location within a borehole that integrates wireline borehole electrical images with whole-core digital images. Azimuthal orientation and location of individual core pieces allows for the detailed interpretation of structural data, including fabrics and fractures and reorientation of other spatially anisotropic properties of core, such as palaeomagnetism.

ODP Hole 1105A is one of a pair of holes drilled on the Atlantis Bank in the SW Indian Ocean, which penetrate sections of the lower ocean crust. Hole 1105A was drilled to a depth of 158 metres below sea-floor (mbsf), with core recovery of 83% and the vast majority of the recovered material being represented by gabbroic rocks.

Zones of crystal–plastic deformation occur throughout the core, but overall their frequency increases downhole. In the deformed zones, the gabbros are transformed into equigranular or porphyroclastic gneissose textured rocks, characterized by very prominent planar fabrics. These localized deformation zones relate to the early exhumation history of the lower crust in an environment similar to core-complex formation in continental regions.

Using a combination of digital core and formation microscanner (FMS) images, it is now possible to fully orientate structural features in a borehole and so map the oxide gabbro and crystal–plastic deformation zones in Hole 1105A. From the analysis of the crystal–plastic fabrics, the majority appear to dip southwards, with a few (c.5%) dipping to the north. This would suggest that the major exhumation surfaces in this hole are represented by southward-dipping structures, formed at high temperatures during unroofing of the Atlantis Bank.

The present understanding of ocean crust structure and development has relied largely on data obtained through geophysical, e.g. (Hill 1957; Detrick et al. 1994), and ophiolite studies, e.g. (Kidd 1977; Christenson 1978). Further invaluable information on ocean-crust architecture has been provided by the drilling and recovery of in situ basement sections, allowing more detailed investigation into mid-ocean ridge processes e.g. Holes 504B (Cann et al. 1983; Anderson et al. 1985; Becker et al. 1988, 1992; Dick et al. 1992; Alt et al. 1993) and 735B (Robinson et al. 1989; Dick et al. 1991, 1999). Unfortunately, core recovery in the majority of DSDP and ODP basement boreholes is low (<50%), and much of the recovered core is often biased towards more massive and competent lithologies (Brewer et al. 1998). The curatorial practice of the Ocean Drilling Program, i.e. moving recovered core pieces to the top of the cored interval (Alt et al. 1993), introduces a systematic bias into the curated depths of recovered core pieces, resulting in many individual pieces being placed anything up to several metres above their true position within the cored section. Agrinier & Agrinier (1994) have provided a statistical solution to this problem, but this only provides a probabilistic estimate. The interpretation of the structural setting of a particular borehole is further complicated, as

From: HARVEY, P. K., BREWER, T. S., PEZARD, P. A. & PETROV, V. A. (eds) 2005. *Petrophysical Properties of Crystalline Rocks*. Geological Society, London, Special Publications, **240**, 165–177.

recovered basement cores are unorientated with respect to north. Hence, while the dip of veins, foliations and other structural features can be recorded, the strike of the features can only be measured within the reference frame of the core. Oriented coring devices are extremely rare (MacLeod *et al.* 2002) which places severe restrictions on the ability to perform structural and tectonic studies.

In order to fully interpret core data derived from ocean boreholes, it is necessary to supplement these with wireline logging data, which provides a near-continuous record of the physical and/or chemical properties of the rock surrounding the borehole walls (Brewer *et al.* 1998). Tools such as the routinely deployed Formation Microscanner (FMS) record geophysical data on a scale sufficiently fine that it is possible to map the borehole walls – imaging a variety of different structural features (Pezard *et al.* 1990; MacLeod *et al.* 1994, 1995). The FMS tool also contains a magnetometer, which allows the determination of the *in situ* orientation of imaged features, and hence provides the potential for the reorientation of recovered core (Haggas *et al.* 2001).

The technique discussed by Haggas *et al.* (2001) has been developed further in ODP Hole 1105A; this particular borehole represents a shorter drilled section than the nearby Hole 735B, but has high core recovery, good-quality wireline logging data for the cored interval, and a set of digital core images, which were obtained post-cruise, for 90% of the recovered material.

Geological setting

ODP Hole 1105A is situated approximately 1.3 km ENE of Hole 735B on the Atlantis Platform along the eastern transverse ridge of the Atlantis II Fracture Zone (Pettigrew *et al.* 1999) (Fig. 1). The Atlantis II Fracture Zone is a large-offset, left-lateral transform, which cuts the Southwest Indian Ridge (SWIR) at 31°S 57°E (Dick *et al.* 1991). The SWIR is classified as an ultra-slow spreading centre with a half-rate of approximately 0.8 cm/a. (Fisher & Sclater 1983), and forms a ridge–ridge–ridge triple junction with the Central Indian Ridge and the Southeast Indian Ridge (Dick *et al.* 1991). Extension of the SWIR has resulted from migration of the triple junction to the north-east, creating a number of fracture zones, including the Atlantis II Fracture Zone (Dick *et al.* 1991). Site 1105A is situated in an area expected to have received a relatively low magma supply, as it is located close to where the ridge axis terminates at the transform; this

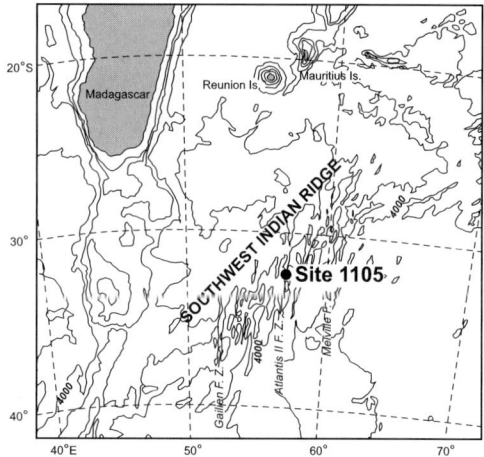

Fig. 1. Bathymetric map showing the location of Site 1105. Contours are at 1000 m intervals. Modified from Robinson *et al.* (1998) and Haggas *et al.* (2001).

suggests mechanical deformation is the dominant accretion process in this section of crust (Pettigrew *et al.* 1999).

Hole 1105A was drilled to a depth of 158 metres below the sea-floor (mbsf); 143 m were cored, recovering 118.43 m of predominantly gabbroic rock, giving an average core recovery of 82.8% (Pettigrew *et al.* 1999). The recovered lithologies were divided into 141 intervals on the basis of variations in mineralogy, mineral mode, texture and grain size; these were grouped into four lithological units (Pettigrew *et al.* 1999) (Fig. 2). The recovered core shows a wide range of brittle to ductile deformational features, including alteration veins, joints, and magmatic and crystal–plastic foliations (Pettigrew *et al.* 1999).

Core data

High-resolution core images were collected of whole core pieces from Hole 1105A, by use of the DMT Colour CoreScan™ system; the method used is discussed in greater detail by Haggas *et al.* (2001). As the Leg 179 core had been split and sampled, it was necessary to reconstruct the whole core using foam spacers and tape; this unfortunately results in the obscuration of some of the fine-scale detail, as a number of structural features have been removed by core sampling. Over 106 m of the Leg 179 core was imaged using this method, representing 90% of the total recovered material; an example of the core images produced is

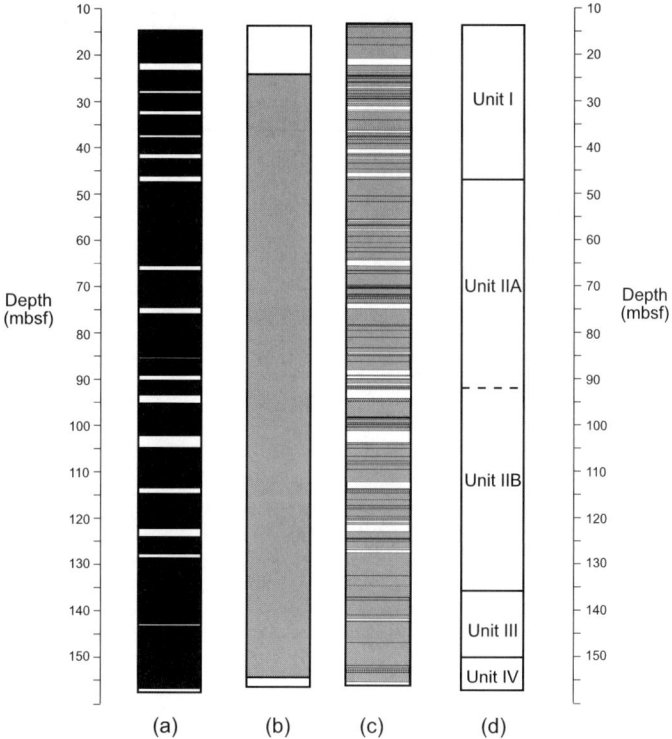

Fig. 2. Summary of available core and log data from ODP Hole 1105A: (**a**) average core recovery with depth; (**b**) availability of FMS data; (**c**) availability of digital core images; (**d**) lithostratigraphy of Hole 1105A, after Pettigrew *et al.* (1999).

shown in Figure 3. The images obtained were then reconstructed into core barrel lengths, using the software DMT Corelog[TM] (Haggas *et al.* 2001). The reconstructed core barrel images were used to identify and measure structural features in the core, including veins, fractures and foliations (Fig. 4).

Logging data

Four down-hole logging tool strings were deployed during Leg 179, and are discussed in detail by the shipboard scientific party (Pettigrew *et al.* 1999). The Formation Microscanner[TM] (FMS) was deployed as part of the second tool string, along with a natural gamma-ray tool (NGT). Two passes of the tools were made, and excellent-quality data were acquired during both passes (Pettigrew *et al.* 1999). The problems encountered in ODP Hole 735B due to excessive resistivity contrasts (Dick *et al.* 1999; Haggas *et al.* 2001) were largely eliminated in Hole 1105A, due to flushing of the borehole with resistive freshwater drilling mud prior to deployment of the tool (Pettigrew *et al.* 1999).

The FMS tool is an electrical imaging device allowing for the visual representation of structural features near to the borehole wall (Brewer *et al.* 1998; Pezard *et al.* 1990). Measurements are made by four pads, which contain two overlapping rows of buttons against a background electrode; these pads are pressed against the borehole walls as the tool is pulled uphole (Brewer *et al.* 1998; Lovell *et al.* 1998). The raw data acquired by the tool consist of a number of microresistivity curves, which are then processed as a high-resolution electrical image, allowing structural features such as bed boundaries, conductive/resistive fractures, and brecciated units to be identified (Rider 1996). It is often possible to distinguish between individual lithological units on the FMS profile; this is particularly true in the case of oxide-rich lithologies, which are easily distinguished from oxide-poor horizons due to their highly conductive nature.

Fig. 3. Unrolled core image of olivine gabbro from the Hole 1105A core. As the core had been split and sampled, foam spacers were used to ensure that the core piece could be rotated evenly during the imaging process.

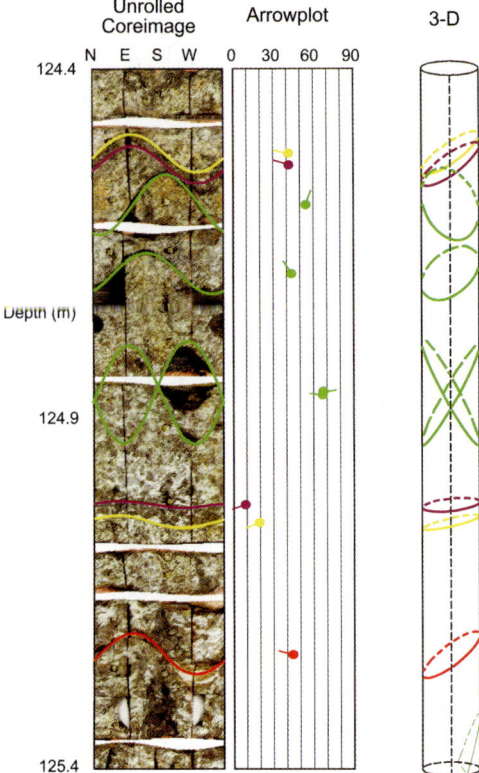

Fig. 4. Reconstructed section using the DMT software CoreLog™. Dips of fractures (green), veins (red), fabrics (purple) and lithological contacts (yellow) are shown. The core pieces have been rotated so that the red marker line, marked on each piece by the structural geology team, is in the same orientation for every piece.

The FMS tool string also includes a general purpose inclination tool (GPIT) and two callipers set 180° apart, which can be used to determine the borehole size and ellipticity (Brewer *et al.* 1998). The orientation of the borehole with respect to local Earth field values, and the orientation of the sensors within the hole, are measured by an accelerometer and a magnetometer contained within the GPIT (Alt *et al.* 1993).

As the FMS images are accurately located and orientated within the borehole, the true dip and direction of a bed boundary or structural feature can be determined. Individual features are picked on the FMS data as sinusoids, from which the amount and direction of dip can be calculated by the Schlumberger software GeoFrame™ (Haggas *et al.* 2001).

Error estimation

Potential errors arise in the logging data, due to inaccurate location of logs, precision and accu-

racy of the tools and the vertical coverage of the logs. The location error of down-hole logs, however, should be ≪ 1 m if shore-based processing has been properly achieved (Brewer *et al.* 1998). The maximum extension of the FMS tool's arms is 15 inches; therefore, blurred FMS images may be generated in those boreholes with a diameter in excess of this, as the electrode pads will not be in contact with the borehole walls (Lovell *et al.* 1998). A further point to note with the FMS data, is that in ODP boreholes only 20% of the borehole walls are imaged; hence, some features may not be seen due to their orientation (Ayadi *et al.* 1998; Brewer *et al.* 1998), however, coverage may be improved with multiple tool passes. The FMS is therefore more likely to image subhorizontal to inclined structures than near-vertical features. Individual tools have different precision and accuracy; however, the worst measurement errors experienced are approximately 10%; this is considerably better than the potential maximum location error associated with core recovery (Brewer *et al.* 1998).

The GPIT used to orientate the FMS logs may be affected by the presence of magnetic minerals (including Fe–Ti oxides) in the formation (Goldberg *et al.* 1992). Hence, in Fe–Ti oxide gabbro horizons, the orientation of the FMS logs may be inaccurate, due to the strong magnetization of oxide-rich layers; this effect of magnetic minerals on the FMS data should be clearly evident on the logs, marked by sharp changes in the azimuth of the tool. However, in Hole 1105A, fluctuations in the orientation of the local magnetic field across oxide-rich intervals appear to be small (Fig. 5), with limited effect on the FMS tool orientation (Pettigrew *et al.* 1999).

Before correlation between core and log data can be achieved, it is necessary to accurately locate individual core pieces within the reference frame of the log data. The first stage in this process is to ascertain the potential location error of each core piece, which is directly related to core recovery. Recovery in Hole 1105A ranges from 44.4% (18R) to 109.0% (20R), and shows no systematic variation down-hole (Fig. 6a). Four of the drilled cores have recovery exceeding 100%; this phenomenon can occur when cored material is left in the borehole after previous coring runs, and is known as 'poling' (Dick *et al.* 2000). A clearer representation of the potential for core location error is presented if the maximum location error for individual core barrels is plotted (Fig. 6b). The highest potential location error occurs in core 18R, in which each individual core piece may be as much as 2.8 m from its true position. In those cores with >100% core recovery (9R, 13R, 17R and 20R), the maximum location error is a negative figure, as some core pieces may be more accurately located in a previously cored interval.

A correlation can be noted between core recovery and the physical properties of the core, as evidenced from the FMS data. Those sections with lower than average core recovery tend to be represented on the FMS images by areas of highly variable conductivity, which can be interpreted as foliated or intensely fractured horizons (Fig. 7). Higher core recovery tends to occur in the more massive sections of

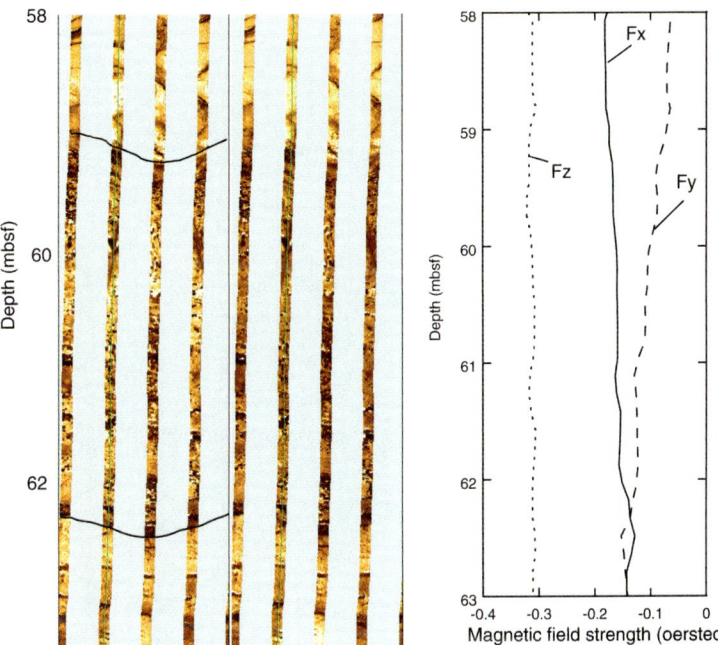

Fig. 5. Magnetic field curves for Hole 1105A across an oxide-gabbro interval (F_x = magnetic field on the *x*-axis, F_y = magnetic field on the *y*-axis, F_z = magnetic field on the *z*-axis). The coarse grain-size of the oxide gabbro results in a distinctive stippled effect on the FMS image.

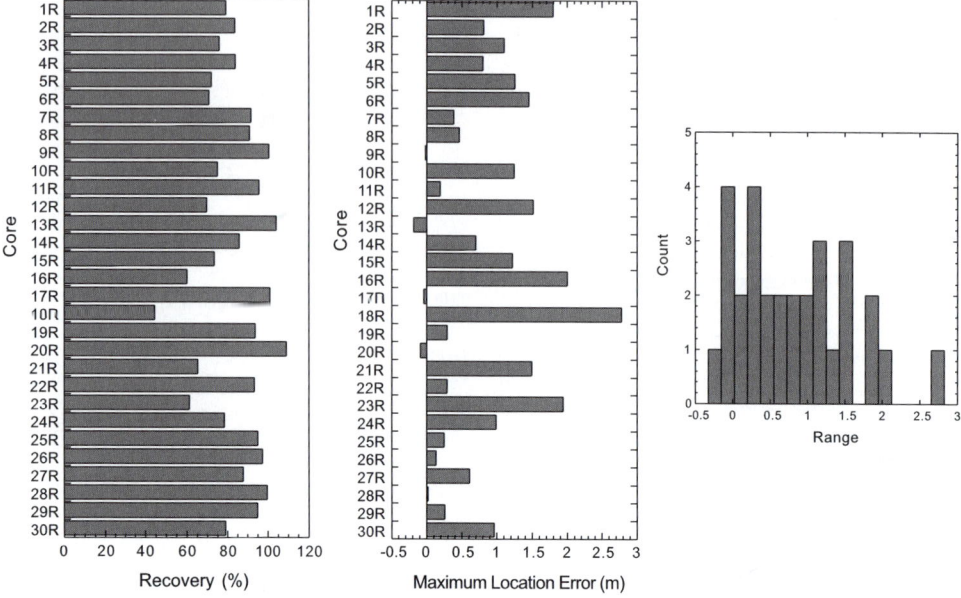

Fig. 6. Percentage core recovery (**a**) and maximum location error (**b**) for core in Hole 1105A. The maximum location error for a core barrel is calculated by subtracting the percentage core recovery from 100% (to give the percentage of the cored interval not recovered), then multiplying this figure by the total cored length of that particular core barrel. (**c**) Histogram of location errors.

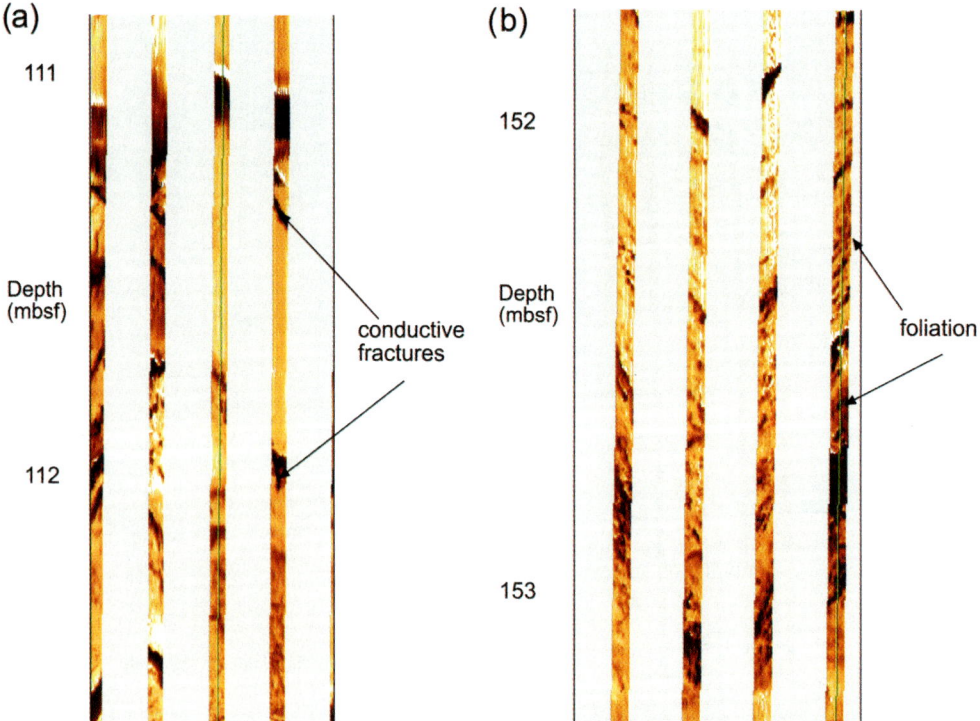

Fig. 7. FMS images of (**a**) fractured and (**b**) foliated gabbroic material in Hole 1105A.

the core with lower fracture densities; this needs to be considered carefully when attempting to interpret the structural core data as a small number of key deformation zones may be missing from the recovered material.

Identification and correlation of lithological intervals

The Shipboard Scientific Party described 141 rock intervals in the Hole 1105A core, divided due to variations in modal proportions, grain size and/or texture (Pettigrew *et al.* 1999). It is not easy to attempt a similar detailed division of the FMS image data, as small variations in modal proportion and grain size are difficult to establish on the electrical images. However, the contrast in log response between oxide-rich and oxide-poor rock intervals allows a basic down-hole stratigraphy to be produced from the down-hole logging data (Fig. 8). Oxide-poor gabbros, such as olivine gabbro, and olivine-bearing gabbro, are imaged on the FMS data as areas of uniformly high resistivity; these are difficult to distinguish from oxide-bearing gabbro and oxide-bearing olivine gabbro, which typically contain ≤1% Fe–Ti oxide. Oxide-rich lithologies (e.g. oxide gabbro, oxide olivine gabbro and olivine-bearing oxide gabbro) typically contain ≥5% Fe–Ti oxides; these are imaged as highly conductive horizons, often with a distinct stippled effect in intervals with coarse grain size (Fig. 5). A number of the highly conductive horizons identifiable on the FMS images are areas of intense crystal–plastic deformation; these sections have a distinctly layered internal structure and are discussed in more detail later.

Oxide-rich lithologies comprise 21.8% of the recovered core from the logged interval; this compares with 23.6% of conductive horizons present on the FMS data. Oxide-rich intervals of variable thickness occur throughout Hole 1105A, and there appears to be very little systematic variation with depth. In many instances, conductive horizons on the FMS data are a few cm thicker than the corresponding oxide gabbro interval in the core; this occurs because the recovered core often comprises small pieces with one or both contacts unrecovered. Conversely, across some rare intervals, the oxide horizon in the recovered core appears to be thicker than the corresponding FMS data. There are two possible reasons for this:

(1) In the case of small core pieces, if these are not obviously contiguous, then a gap of approximately 1 cm is left between each

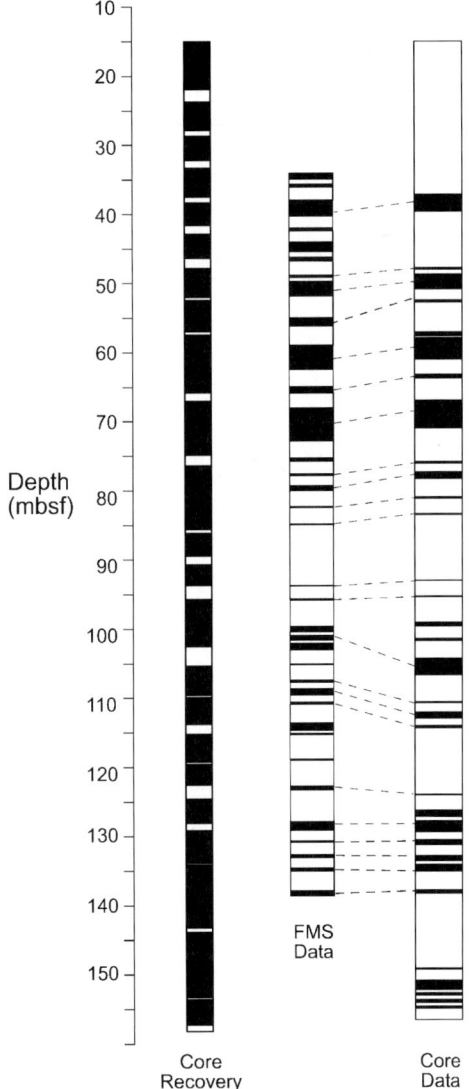

Fig. 8. Log of conductive Fe–Ti horizons from FMS and core data in Hole 1105A, showing the likely correlation between the two data-sets.

piece in the curated core barrel; this may be in excess of the actual missing material.

(2) Contacts between oxide-rich and oxide-poor intervals are commonly gradational, and therefore may have been placed at a slightly different location by the shipboard scientific party than suggested by the FMS data.

The location error of oxide-rich core intervals appears to vary down-hole. Core pieces in the

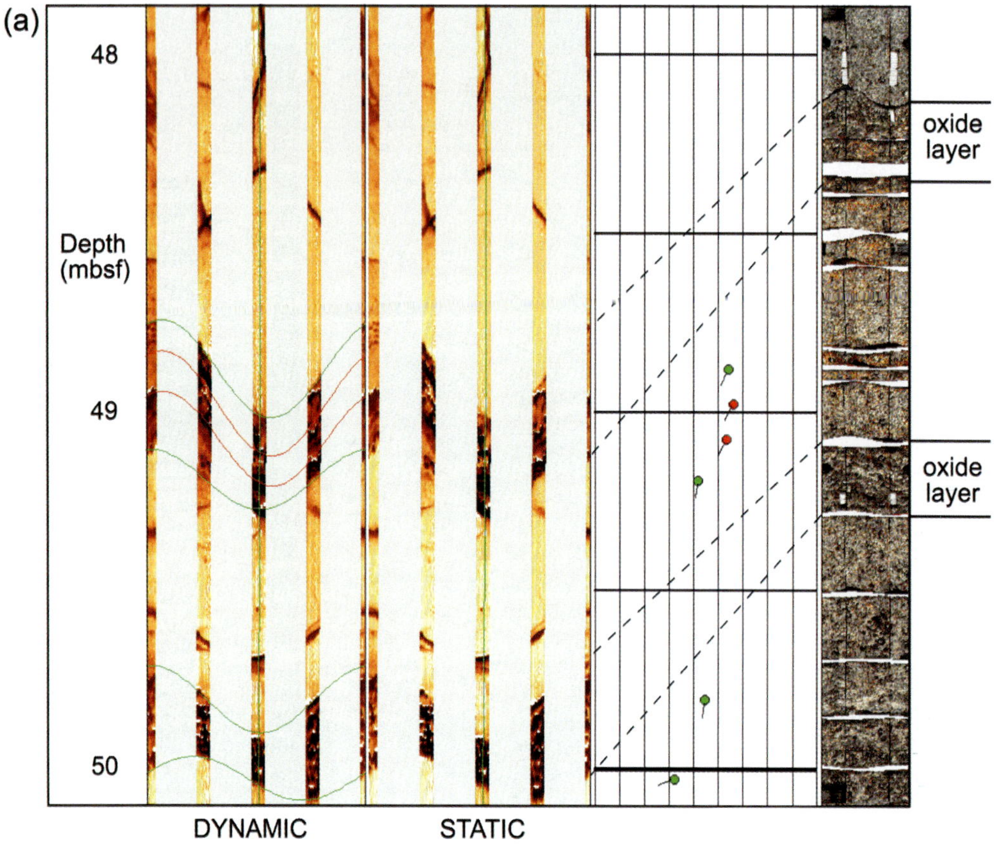

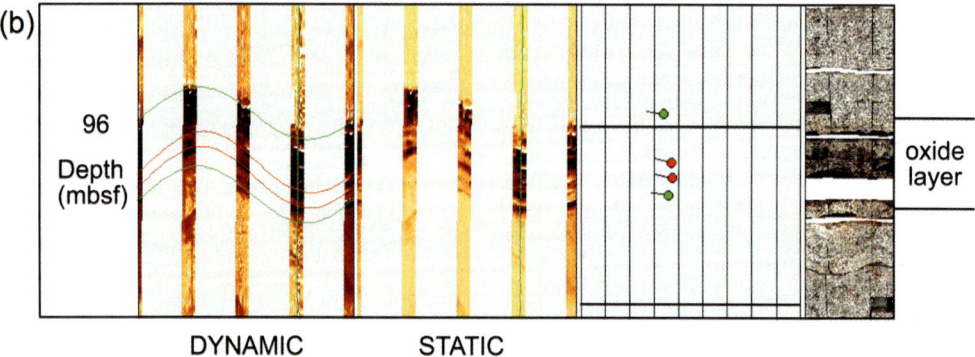

Fig. 9. FMS images showing Fe–Ti oxide gabbro layers within olivine gabbros. Both statically and dynamically processed data are shown. The dynamic data are processed using a sliding window of 2 m for data normalization, whereas the static data are processed using a static window. 'Tadpoles' on the plot indicate the amount and direction of dip of fabrics and layer boundaries. (a) Two Fe–Ti oxide gabbro layers, the upper horizon is dipping at *c*.44° to 193°, the lower horizon is dipping *c*.43° to 185°; (b) Fe–Ti oxide gabbro layer dipping eastwards at *c*.36° to 275° degrees.

upper parts of the core are c.80 cm from their location on the down-hole logs; however, this location error decreases down-hole until it is only a few centimetres at 96 mbsf. Progressing down-hole, between 100 and 115 mbsf, the core data require a depth shift of c.50 cm upwards. Finally, at the base of the hole, the difference between core and log depths are again only a few centimetres. These variations in core location error are predominantly a function of variable core recovery down-hole (Fig. 6).

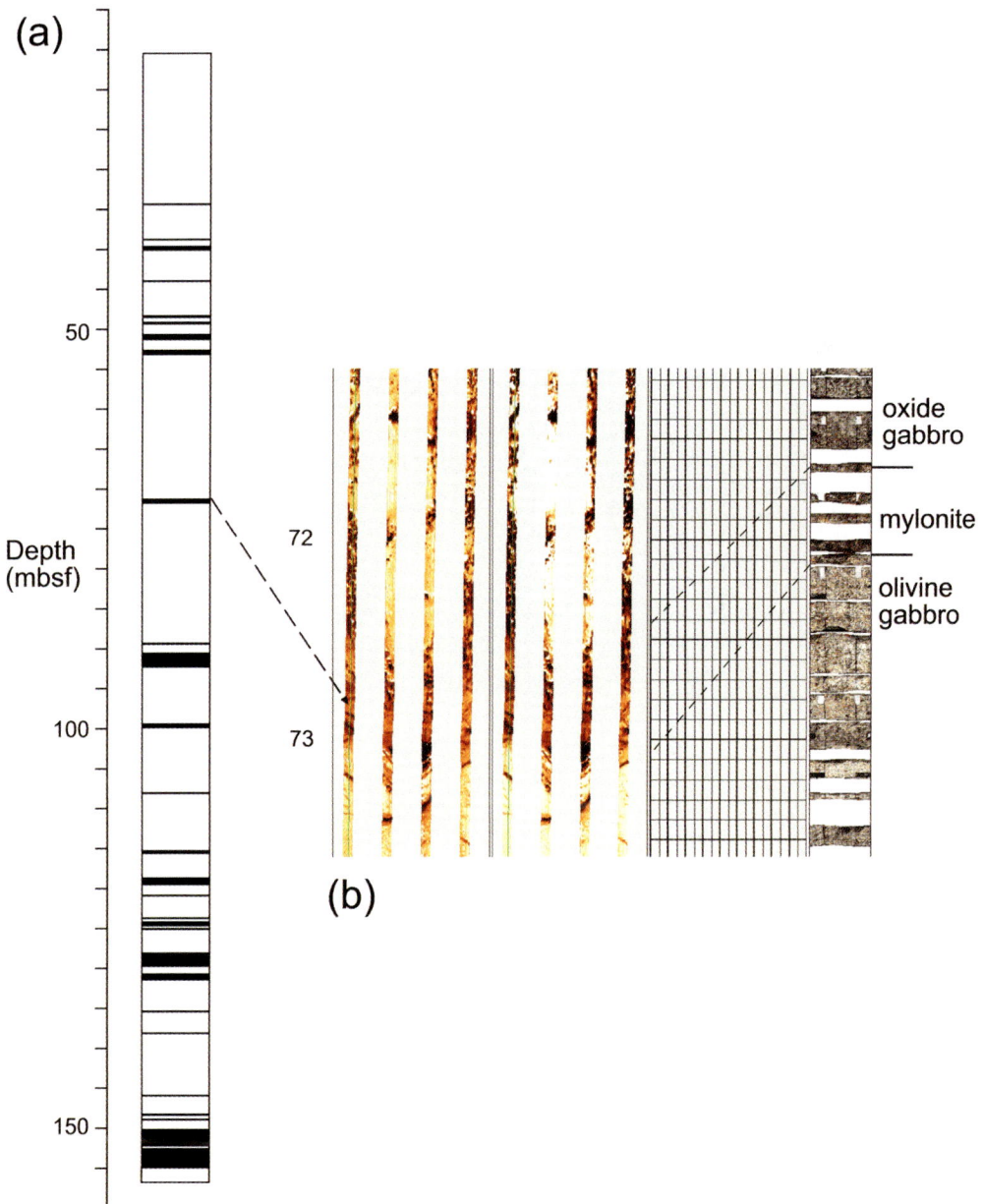

Fig. 10. (**a**) Log of the position and thickness of crystal–plastic fabrics throughout the Hole 1105A core. (**b**) FMS image of a mylonite at the base of a coarse-grained oxide gabbro interval.

Once a number of individual lithological intervals have been identified on the FMS data, these can be relocated and, if contacts have been recovered, rotated to their true orientation in the reference frame of the log data. Three thin intervals of Fe–Ti oxide gabbro were easily identified on both the core and borehole images (intervals 24, 26 and 83), and are used here as examples of the location and reorientation process. The upper and lower contacts of interval 24 with intervals 23 and 25 are easily identified on both the optical and electrical images (Fig. 9a). In the core, interval 24 is a 27-cm-thick Fe–Ti oxide-rich layer; the upper contact is located at 48.14 mbsf and dips 43° with a dip direction of 286° (subsequently, all orientations are expressed as dip/dip direction; e.g. 43°/286°), and the lower contact is at 48.41 mbsf and dips 45°/283°. The same interval is located on the FMS data between 48.9 and 49.2 mbsf, and the upper and lower contacts dip at 46°/195° and 40°/190° respectively. If the FMS data are regarded as the true reference frame, the core data for this interval require a depth shift downwards of *c*.80 cm and an average rotation of 93° anticlockwise. Ninety-five centimetres below interval 24, interval 26 is also easily identified on the FMS data (Fig. 9a). The dip of upper contact of this interval has been measured in the core as 43°; this compares with a dip and strike of 45°/185° calculated on the FMS data.

Interval 83 is an olivine-bearing oxide gabbro located at 96.05 mbsf in the core (Pettigrew *et al.* 1999) (Fig. 9b). The interval is 15 cm thick, with upper and lower contacts dipping at 11°/034° and 10°/037° respectively. This oxide gabbro horizon has been identified on the FMS data at 96.0 to 96.2 mbsf, hence is almost at its correct location, but requires a rotation of 83.5° anticlockwise to correlate with the orientated log data.

Unfortunately the identification, location and orientation of oxide-poor horizons are considerably more complicated than for the oxide-rich gabbro horizons. Location and orientation of massive olivine gabbro horizons are only achievable using their relationships with accurately located oxide gabbros, or if the gabbros contain cross-correlated structural features such as crystal–plastic fabrics or fractures/veins.

Identification and correlation of crystal–plastic fabrics

A number of horizons with crystal–plastic deformation occur throughout the 1105A core and their down-hole distribution is illustrated in Figure 10a. Only those horizons with strong foliation to ultramylonite were logged and compared with the electrical images, as it is unlikely that the FMS data will have imaged very weak foliation and minor deformations such as kinked cleavages. Horizons of crystal–plastic deformation occur throughout the core; however, they increase in frequency and thickness in the lower 80 m. Particularly thick deformed horizons occur in cores 29R and 30R – towards the

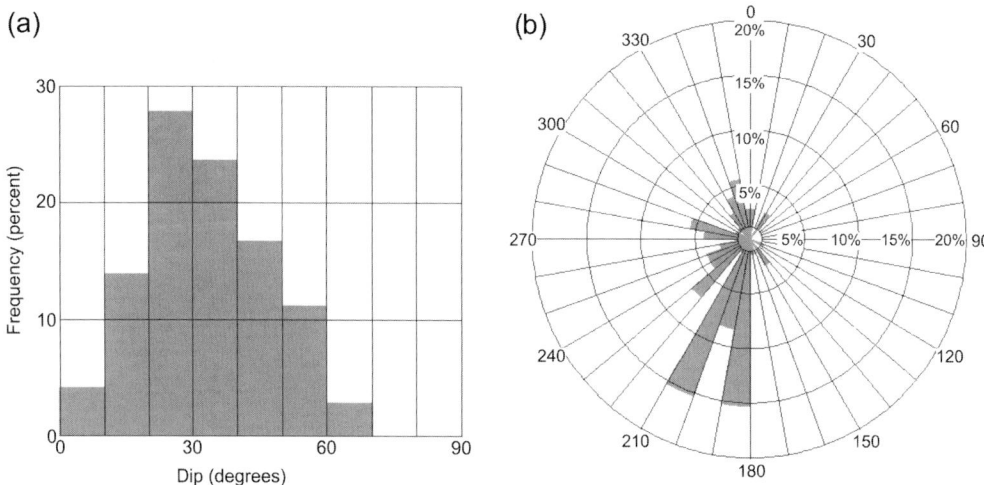

Fig. 11. (a) Range of dips of crystal–plastic fabrics in Hole 1105A, as calculated from the FMS image data. **(b)** Azimuths of crystal–plastic fabrics in Hole 1105A from FMS data.

(a)   (b)

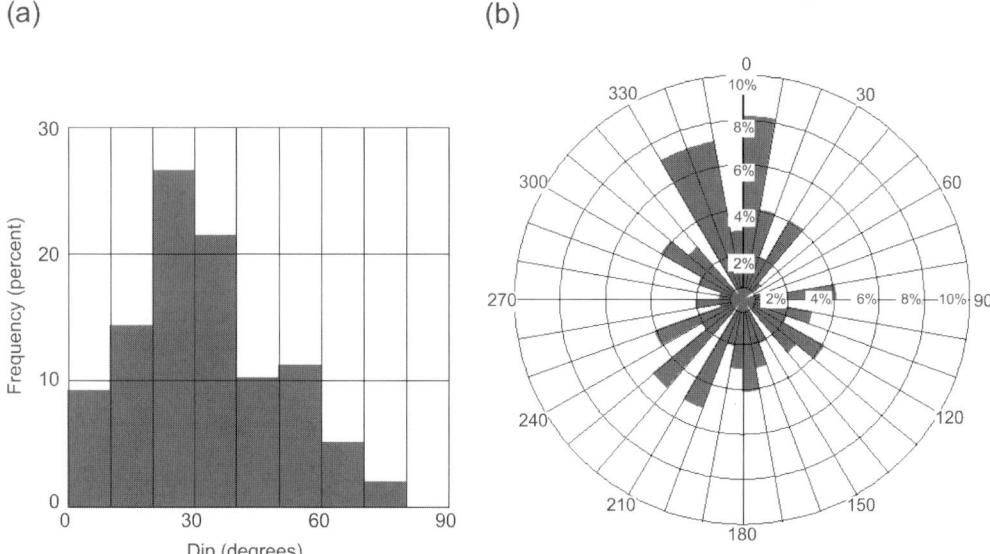

Fig. 12. (**a**) Range of dips of fractures in Hole 1105A, as calculated from the FMS image data; (**b**) azimuths of fractures in Hole 1105A from FMS data.

base of the drilled section. Areas of crystal–plastic deformation tend to be associated with oxide-rich and oxide-bearing intervals, but several instances also occur in olivine-bearing gabbro intervals. A similar association between oxide-rich intervals and moderate to strong crystal–plastic deformation was also noted in the nearby Hole 735B (Dick *et al.* 1999). However, it is also important to note that not all the oxide horizons in the recovered core are deformed; many have very low fabric intensity. The dips of crystal–plastic fabrics in the core range from 0 to 75°, but there appears to be no systematic variation in dip angle with increasing depth down-hole. The down-hole decrease in fabric intensity noted in Hole 735B (Dick *et al.* 1999) is not observed in Hole 1105A; however, the Hole 1105A core represents a much shorter drilled section.

Due to the accurately located and orientated core images, it is possible to correlate foliated sections of the core with the same features on the FMS images. This allows the down-hole variation in dip and azimuth of crystal–plastic fabrics to be determined. Strongly foliated intervals are imaged on the FMS data as sections with alternating bands of high and low conductivity; the example illustrated in Figure 10b is a mylonitized horizon at the base of a oxide gabbro horizon.

Figure 11a illustrates the distribution of dips of crystal–plastic fabrics throughout the logged section. The mean dip value (20–30°) and range of dips (0° to 70°) are consistent with those measured from the core data. It should be noted, however, that any near-vertical fabrics might not be seen on the FMS images, due to incomplete borehole coverage. The fabrics do not show a strong variation in azimuthal directions with depth; however, there does appear to be a slight concentration of azimuthal directions between 180° and 210° distributed throughout the core.

Conclusions

For full location and orientation of core pieces, high-quality log data and comprehensive scanned core images are essential. Oxide-rich horizons are easily identified on the borehole electrical images; hence, the majority of these can be accurately located and, if contacts have been recovered, orientated with respect to grid north. Variations in the magnetic field across oxide gabbro horizons appear to be small, and hence are presumed to have a very small influence on the azimuth of the FMS tool. In ODP Hole 1105A, a large proportion of the oxide-rich horizons in the core appear to have been recovered successfully; most of the unrecovered material in the core is likely to be intensely fractured, suggesting that recovery is influenced more by physical properties than lithology.

Comparison of crystal–plastic fabrics within the core with the oriented logging data shows no systematic variations in dip angle or azimuth with depth; however, there does appear to be a concentration of dip azimuths between 180 and 210°. This would suggest that the major exhumation surfaces in this hole are represented by southward-dipping structures, formed at high temperatures during unroofing of the Atlantis Bank.

This work is supported by NERC small grant GST/02/2691. We thank Dr G. Rafat of Deutsche Montan Technologie GmbH (DMT) for technical assistance and software for use with the DMT CoreScan®, and Schlumberger Ltd for the provision of their log and imaging software, GeoFrame™.

References

AGRINIER, P. & AGRINIER, B. 1994. On the knowledge of the depth of a rock sample from drilled core. *Scientific Drilling* 4, 259–265.

ALT, J. C., KINOSHITA, H., STOKKING, L. B. & the Leg 148 Shipboard Scientific Party. 1993. *Proceedings of the Ocean Drilling Program, Initial Report*, **148**, College Station, TX (Ocean Drilling Program).

ANDERSON, R. N., HONNOREZ, J. & the Leg 83 Shipboard Scientific Party. 1985. *Initial Reports of the Deep Sea Drilling Project*, **83**, 13–118.

AYADI, M., PEZARD, P. A., LAVERNE, C. & BRONNER, G. 1998. Multi-scalar structure at DSDP/ODP Site 504, Costa Rica Rift; I, Stratigraphy of eruptive products and accretion processes. *In*: HARVEY, P. K. & LOVELL, M. A. (eds) *Core-Log Integration*. Geological Society, London, Special Publications, **136**, 297–310.

BECKER, K., SAKAI, H. & the Leg 111 Shipboard Scientific Party. 1988. *Proceedings of the Ocean Drilling Program, Initial Report*, **111**, College Station, TX (Ocean Drilling Program).

BECKER, K., FOSS, G. & the Leg 111 Shipboard Scientific Party. 1992. *Proceedings of the Ocean Drilling Program, Initial Report* **137**, College Station, TX (Ocean Drilling Program).

BREWER, T. S., HARVEY, P. K., LOVELL, M. A., HAGGAS, S. L., WILLIAMSON, G. & PEZARD, P. 1998. Ocean floor volcanism; constraints from the integration of core and downhole logging measurements. *In*: HARVEY, P. K. & LOVELL, M. A. (eds) *Core–Log Integration*. Geological Society, London, Special Publications, **136**, 341–362.

CANN, J. R., LANGSETH, M. G. & the Leg 111 Shipboard Scientific Party 1983. Sites 501 and 504: sediments and ocean crust in an area of high heat flow on the southern flank of the Costa Rica Rift. *In*: CANN, J. R. & LANGSETH, M. G. (eds) *Initial Reports of the Deep Sea Drilling Project*, **69**, 31–174.

CHRISTENSON, N. I. 1978. Ophiolites, seismic velocities and oceanic crustal structure, *Tectonophysics*, **47**, 131–157.

DETRICK, R., COLLINS, J., STEPHEN, R. & SWIFT, S. 1994. In situ evidence for the nature of the seismic layer 2/3 boundary in oceanic crust. *Nature*, **370**, 288–290.

DICK, H. J. B., MEYER, P. S., BLOOMER, S., KIRBY, S., STAKES, D. & MAWER, C. 1991. Lithostratigraphic evolution of an in-situ section of oceanic layer 3. *In*: VON HERZEN, R. P. & ROBINSON, P. T. (eds) *Proceedings of the Ocean Drilling Program, Scientific Results*, **118**, College Station, TX (Ocean Drilling Program), 439–540.

DICK, H. J. B., ERZINGER, J., STOKKING, L. B. & the Leg 140 Shipboard Scientific Party 1992. *Proceedings of the Ocean Drilling Program, Initial Report*, **140**, College Station, TX (Ocean Drilling Program).

DICK, H. J. B., NATLAND, J. H. *et al.* 2000. In-situ section of the lower ocean crust: results of ODP Leg 176 drilling at the Southwest Indian Ridge. *Earth and Planetary Science Letters*, **179**, 31–51.

DICK, H. J. B., NATLAND, J. H., MILLER, D. J. & the Leg 176 Shipboard Scientific Party 1999. *Proceedings of the Ocean Drilling Program, Initial Report*, **176** (CD-ROM), College Station, TX (Ocean Drilling Program).

DICK, H. J. B., SCHOUTEN, H. *et al.* 1991. Tectonic evolution of the Atlantis II Fracture Zone. *In*: VON HERZEN, R. P. & ROBINSON, P. T. (eds) *Proceedings of the Ocean Drilling Program, Scientific Results*, **118**, College Station, TX (Ocean Drilling Program), 359–398.

FISHER, R. L. & SCLATER, J. G. 1983. Tectonic evolution of the Southwest Indian Ocean since the Mid-Cretaceous: plate motions and stability of the pole of Antarctica/Africa for at least 80 Myr. *Geophysical Journal of the Royal Astronomical Society*, **73**, 553–576.

GOLDBERG, D., BROGLIA, C. & BECKER, K. 1992. Fracture permeability and alteration in gabbro from the Atlantis II fracture zone. *In*: HURST, A., WORTHINGTON, P. F. & GRIFFITHS, C. M. (eds) *Geological applications of wireline logs: II. Geological Society, London, Special Publications*, **65**, 199–210.

HAGGAS, S. L., BREWER, T. S., HARVEY, P. K. & ITURRINO, G. I. 2001. Relocating and orientating cores by the integration of electrical and optical images: a case study from Ocean Drilling Program Hole 735B. *Journal of the Geological Society, London*, **158**, 615–623.

HILL, M. N. 1957. Recent geophysical exploration of the ocean floor. *In*: AHRLENS L. H. (ed.) *Physics and Chemistry of the Earth 2*, Pergamon Press, London, 129–163.

KIDD, R. G. W. 1977. A model for the process of formation of the upper oceanic crust. *Geophysical Journal of the Royal Astronomical Society*, **50**, 149–183.

LOVELL, M. A., HARVEY, P. K., BREWER, T. S., WILLIAMS, C., JACKSON, P. D. & WILLIAMSON, G. 1998. Application of FMS images in the Ocean Drilling Program: an overview. *In*: CRAMP, A., MACLEOD, C. J., LEE, S. V. & JONES, E. D. W.

(eds) *Geological Evolution of Ocean Basins: Results from the Ocean Drilling Program.* Geological Society, London, Special Publications, **131**, 287–303.

MACLEOD, C. J., CELERIER, B. & HARVEY, P. K. 1995. Further techniques for core reorientation by core-log integration: application to structural studies of the lower oceanic crust in Hess Deep, Eastern Pacific. *Scientific Drilling*, **5**, 77–86.

MACLEOD, C. J., ESCARTIN, J. *et al.* 2002. Direct geological evidence for oceanic detachment faulting; the Mid-Atlantic Ridge, 15 degrees 45'N. *Geology*, **30**, 879–882.

MACLEOD, C. J., PARSON, L. M. & SAGER, W. W. 1994. Reorientation of cores using the Formation MicroScanner and borehole televiewer; application to structural and paleomagnetic studies with the Ocean Drilling Program: *In:* HAWKINS, J. W., PARSONS, L. M. & ALLAN, J. F. (eds) *Proceedings of the Ocean Drilling Program, Scientific Results,* **135**, College Station, TX (Ocean Drilling Program), 301–311.

PETTIGREW, T. L., CASEY, J. F., MILLER, D. J. & the Leg 179 Shipboard Scientific Party 1999. *Proceedings of the Ocean Drilling Program, Initial Report,* **179** (CD-ROM), College Station, TX (Ocean Drilling Program).

PEZARD, P. A., LOVELL, M. A. & the Ocean Drilling Program Leg 126 Shipboard Scientific Party 1990. Downhole images: electrical scanning reveals the nature of the subsurface oceanic crust. *EOS, Transactions of the American Geophysical Union*, **71**, p. 709.

RIDER, M. 1996. *The Geological Interpretation of Well Logs.* Whittles Publishing, Caithness, 175 pp.

ROBINSON, P. T., VON HERZEN, R., ADAMSON, A. C. & the Leg 118 Shipboard Scientific Party 1989. *Proceedings of the Ocean Drilling Program, Initial Report,* **118** (CD-ROM), College Station, TX (Ocean Drilling Program).

Electrical properties of slow-spreading ridge gabbros from ODP Hole 1105A, SW Indian Ridge

F. EINAUDI[1,2], P. A. PEZARD[3], B. ILDEFONSE[3] & P. GLOVER[4]

[1]*Laboratoire de Géophysique et d'Hydrodynamique en Forage, ODP-NEB, ISTEEM, cc 056, Université de Montpellier II, 34095 Montpellier, Cedex 05, France*
(e-mail: einaudi@dstu.univ-montp2.fr)
[2]*Laboratoire de Pétrologie Magmatique, Université d'Aix Marseille III, Faculté des Sciences de Saint Jérôme, 13397 Marseille Cedex 20, France*
[3]*Laboratoire de Tectonophysique, ISTEEM, CNRS UMR5568, Université de Montpellier II, 34095 Montpellier, Cedex 5, France*
[4]*Département de géologie et de génie géologique, Faculté des sciences et de génie, Université Laval, Sainte-Foy, G1K 7P4, Québec, Canada*

Abstract: Physical properties of gabbroic samples from Ocean Drilling Program Hole 1105A were measured in the laboratory, with a particular emphasis on the analysis of electrical properties. This data-set includes the major lithologies sampled in ODP Hole 1105A: gabbros, olivine gabbros, oxide-rich gabbros, and, for all rock types, different ranges of alteration were sampled: from fresh to highly altered. All these lithologies correspond to the seismic Layer 3 layer of the oceanic crust, and large-scale geophysical data interpretation requires a complete understanding of the physical properties of rocks in this section. Electrical conductivities measured on brine-saturated gabbros reveal strong excess conductivity for samples rich in oxide minerals and, to a lesser extent, for altered samples. However, the classical models do not explain the excess conductivity reported in the oxide-rich samples when saturated with brine. The electrical conduction via electronic processes in metallic minerals has been taken into account in our analysis of the electrical properties. The oxide minerals' contribution has been independently estimated through measuring dry electrical resistivity. These measurements allowed quantification of the electronic conduction, which can reach 80% of the full conductivity for the most oxide-rich gabbros.

Through the last three decades, the Deep Sea Drilling Project (DSDP) and Ocean Drilling Program (ODP) have provided a unique opportunity to study the composition and structure of the oceanic crust. Our knowledge of the *in situ* structure of lower oceanic crust has largely been based on geophysical and ophiolite studies. Deep drilling investigations of lower oceanic crust have been achieved in several locations, taking advantage of tectonic exposure, e.g. Hess Deep (ODP Leg 147; Mével *et al.* 1996), the Mid-Atlantic Ridge (ODP Leg 153 at Mark Area; Karson *et al.* 1997) and the SW Indian Ridge (ODP Legs 118, 176 and 179; Von Herzen *et al.* 1991; Dick *et al.* 1999; Pettigrew *et al.* 1999). Boreholes drilled on the SW Indian Ridge have allowed study of lower-crustal rocks analogous to ophiolitic sequences, and assessment of the seismic nature of oceanic

Layer 3. The analysis of geophysical data requires a complete understanding of the physical properties of the investigated section. For this purpose, laboratory measurements of the physical properties of oceanic samples provide direct insights into the physical structure of the oceanic crust. These measurements can be compared with *in situ* down-hole measurements. In this paper, we present a petrophysical study of gabbroic samples from ODP Hole 1105A, located 1.5 km from the reference ODP Site 735. The physical properties are investigated at room pressure and temperature in order to characterize the penetrated massif, as well as to determine the influence of alteration due to fluid circulation on rock properties.

Porosity, density, electrical properties (formation factor, surface conductivity), bulk magnetic susceptibility and compressional velocity

From: HARVEY, P. K., BREWER, T. S., PEZARD, P. A. & PETROV, V. A. (eds) 2005. *Petrophysical Properties of Crystalline Rocks*. Geological Society, London, Special Publications, **240**, 179–193.
0305-8719/05/$15.00 © The Geological Society of London 2005.

have been measured on a set of 34 samples collected during ODP Leg 179. Electrical properties of gabbros have been investigated in detail by Pezard *et al.* (1991) and Ildefonse & Pezard (2001) in ODP Hole 735B, to identify downhole changes in electrical properties, porosity structure and alteration. An important result is a change in porosity as a function of depth. These works also highlighted the importance of oxide minerals in the measured electrical resistivity. In the oxide-mineral-rich samples, it was pointed out from saturated measurements that oxide ionic conductivity was involved. Unfortunately, the 735B mini-cores were lost, and further investigation of complementary measurements such as oxide content, magnetic susceptibility and dry resistivity were not possible. Our data-set, however, does allow a full investigation of the petrophysical properties in such oxide-rich gabbros.

Due to extreme sensitivity, and in spite of complexity, electrical methods are among the most precise indirect tools for the analysis of rock structures. At low frequencies (<1 kHz), electrical properties of saturated rocks are influenced by the nature of the rock matrix; the chemical composition and salinity of the saturating fluid; the cation-exchange processes along pore surfaces, and/or the movement of fluids in the porous medium (e.g. Waxman & Smits 1968; Olhoeft 1981; Walsh & Brace 1984; Katsube & Hume 1987; Pezard 1990; Pezard *et al.* 1991; Revil *et al.* 1998). Rock-forming minerals are mostly silicates, having a high resistivity (10^6 to 10^{14} ohm m), but, when the rock matrix contains conductive minerals (Ti–Fe oxide in gabbros), the conduction current through the matrix may be appreciable as oxides reach a resistivity of 10^{-6} ohm m (Olhoeft 1981; Guéguen & Palciauskas 1992). The influence of the oxide minerals on the various physical properties are investigated in this paper.

Geological setting: the Atlantis II Bank

ODP Hole 1105A is located on top of the shallow Atlantis Bank (720 m below sea-level), about 18 km east of the active Atlantis transform fault and 93 km south of the present-day ridge axis (Fig. 1). The Atlantis Bank is the shallowest and largest (*c.*35 km^2) of a series of north–south aligned flat-topped platforms (Dick *et al.* 1999). This shallow structure is interpreted as the result of progressive unroofing along a north-dipping low-angle detachment fault (Karson & Dick 1984; Cannat *et al.* 1991; Dick *et al.* 1991), in a tectonic situation similar to that of the inside-corner highs of the Mid-Atlantic Ridge (Tucholke & Lin 1994; Cann *et al.* 1997; Tucholke *et al.* 1998; Ranero & Reston 1999).

The SW Indian Ridge (SWIR) is one of the slowest spreading sections of the mid-ocean ridge system, with a full spreading rate of 16 mm/a (Fisher & Sclater 1983; Dick *et al.* 1991). Slow-spreading ridges are characterized

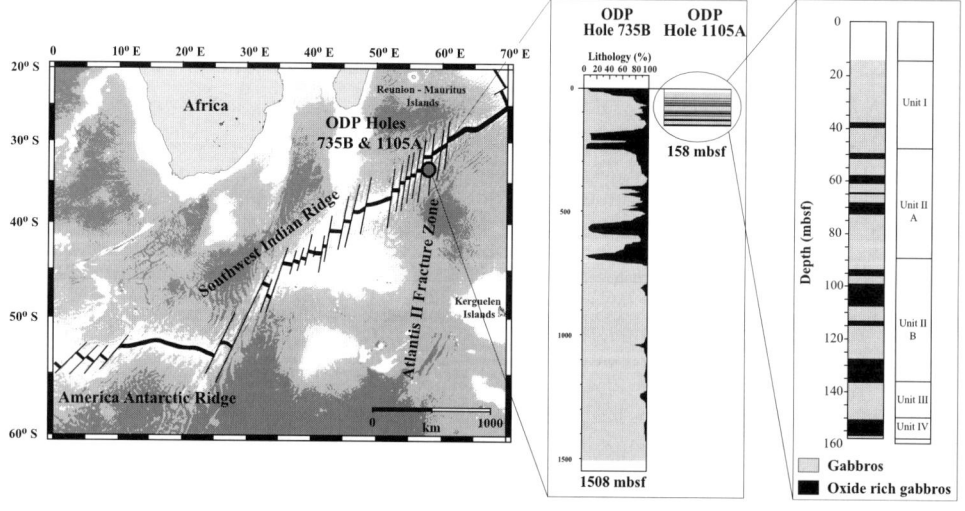

Fig. 1. ODP (Ocean Drilling Program) Hole 1105A and 735B location. ODP Hole 1105A, reaches a depth of 158 mbsf (metres below the sea-floor) and ODP Hole 735B – a total depth of 1507 mbsf. Schematic diagrams of showing the lithostratigraphy of ODP Holes 735B (modified from Dick *et al.* 1999) and 1105A (from Pettigrew *et al.* 1999) Hole 735B as a function of depth. Oxide gabbro intervals are coloured black.

by a low magma budget, and extension appears to be accommodated by a pervasive deformation. About 50% of the dredged rocks along the Atlantis II Fracture zone are peridotites and gabbros (Fisher & Sclater 1983; Dick *et al.* 1991; MacLeod *et al.* 1998). The Atlantis Bank itself mostly consists of outcropping gabbro. Models proposed to explain the exposure of gabbros at the sea-floor involve large-scale normal faulting of the newly created oceanic crust. The gabbroic crust was accreted about 11.5 Ma ago at the SWIR axis (Dick *et al.* 1991).

ODP Hole 1105A is located at 1.3 km from the reference Hole 735B, which penetrates 1508 metres below the sea-floor (mbsf), following two ODP drilling cruises: ODP Leg 118 (Von Herzen *et al.* 1991) and ODP Leg 176 (Dick *et al.* 1999, 2000). The exceptionally high core recovery makes of these two boreholes quasi-continuous sample of the *in situ* gabbroic crust.

Basement stratigraphy

In Hole 1105A, a sequence of gabbroic rocks was drilled to a total depth of 158 mbsf. Core recovery in such an environment is very high and reaches 82.8% for the whole section – similar to core recovery obtained in ODP Hole 735B (Dick *et al.* 1999). The recovered rocks in Hole 1105A are divided into four main rock types, ranging from gabbro (36%), olivine gabbro (43%), oxide gabbro (17%), and olivine oxide gabbro (4%). The full section is presented in detail in Pettigrew *et al.* (1999). It is composed of 54 lithological units on the basis of variation in rock type, mineral abundances and grain size. On a broader scale, four major lithological units have been determined from petrological description (Fig. 1; Pettigrew *et al.* 1999). Hole 1105A lithology consists of:

(1) a gabbroic unit (15–48 mbsf) characterized by more primitive rock types and by a scarcity of oxide gabbro;
(2) a gabbroic unit (48–136 mbsf) characterized by a high abundance of oxide gabbro and oxide-bearing gabbro;
(3) a gabbroic unit (136–150 mbsf) characterized by a lack of oxide gabbros; and finally,
(4) an oxide-rich gabbro and oxide-bearing gabbro unit (150–157 mbsf). In oxide-rich rocks, oxide content reaches up to 20–25% modal Fe–Ti minerals.

Sampling

This study includes a series of 34 samples from ODP Hole 1105A. All minicores were drilled vertically into the core after splitting. The samples include all the major lithologies identified (gabbros, oxide-rich gabbros, olivine gabbros: Pettigrew *et al.* 1999) and different states of secondary alteration, from unaltered to highly altered. The samples are located along the whole section, and have been characterized according to their oxide mineral abundance and degree of alteration, from thin-section analysis. Three distinctive groups have been identified in order to allow petrographical characterization of each gabbro type, related to the objective of this paper: fresh, altered and oxide rich (Fig. 2). Distribution and abundance of oxide minerals in the thin section as shown in Figure 2 were determined by using heightened contrast to enhance the distribution of oxides (black) compared to the silicate matrix (white). Oxide mineral abundances determined by image analysis vary from 0.2 to 31.2%. Microprobe analyses obtained from ODP Hole 735B oxide minerals (Natland *et al.* 1991; Niu *et al.* 2002) show that the oxides are dominated by ilmenite and ilmenite–hematite–magnetite solid solution. For the following discussion, they will be referred to the oxide rich category. In this study, the notion of alteration is related to the surface electrical properties of alteration phases in gabbros, such as smectites, illites or zeolites (Robinson *et al.* 1991; Dick *et al.* 1999; Pettigrew *et al.* 1999).

Furthermore, petrophysical measurements from Pezard *et al.* (1991) and Ildefonse & Pezard (2001) obtained from ODP Hole 735B samples are integrated in the following discussion and compared to the 1105A analyses. Most of these measurements have been made following the same analytical procedure. This dataset does not represent the entire set of lithologies, since the primary criterion for choosing the samples was the freshness of the core. It results in a rather homogeneous sampling, without strongly altered or Fe–Ti oxide-rich samples.

Petrophysical properties from down-hole measurements

Physical properties were measured *in situ* using down-hole tools at Hole 1105A (Pettigrew *et al.* 1999). Wireline logging provides a continuous geophysical characterization of the penetrated formations, and is a good indicator of the lithology distribution. The Triple Combo (resistivity, porosity, density, natural radioactivity and temperature measurements) and the FMS-Sonic tool strings were deployed (Pettigrew *et al.* 1999). The FMS-Sonic tool string provides a measurement of sonic velocity, and includes a

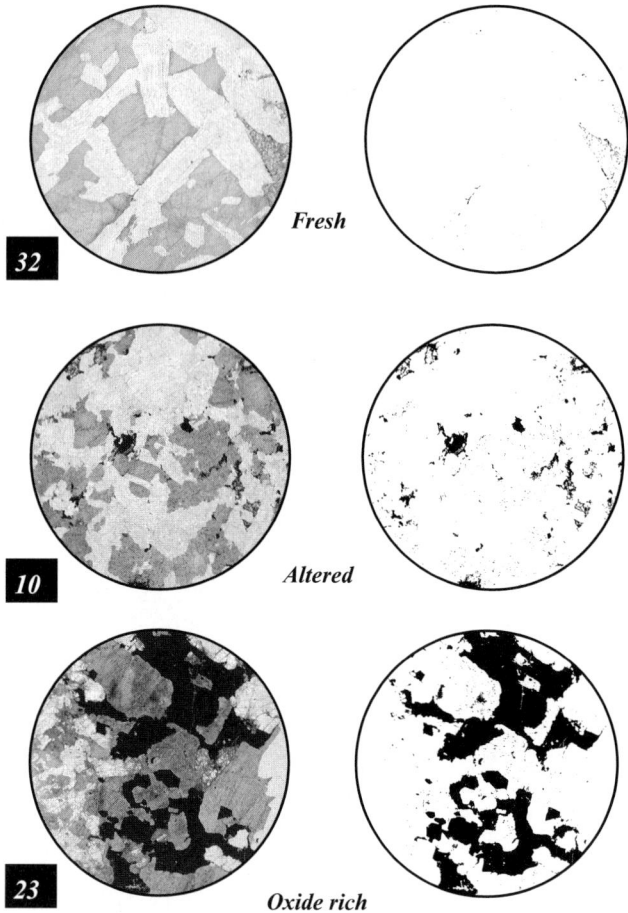

Fig. 2. Standard transmitted micrograph of three representative samples. The sample diameter is about 2.5 cm. Distribution and abundance of oxide minerals in the thin section has been quantified by using heightened contrast to enhance the distribution of oxides (black) compared to the silicate matrix (white). The oxide mineral content for the ODP Hole 1105A samples are reported in Table 1. Sample 32 (140.64 mbsf); sample 10 (61.95) and sample 23 (110.3 mbsf).

micro-electrical imaging device (FMS) together with calipers and gamma-ray measurement to enable depth corrections between tool strings. Full descriptions of the measurement principles of the logging tools used on the ODP program can be found in Goldberg (1997). Figure 3 presents some of these geophysical measurements used in this discussion.

The borehole shape is shown by two orthogonal calipers (Fig. 3a), and appears to be smooth (diameter close to 10 in.), indicating that the borehole conditions required for reliable down-hole measurements are excellent here. Natural gamma-ray measurements (Fig. 3b) exhibit low values along the whole section – coherent in oceanic crust. Slight variations in

gamma-ray measurements are indicative of changes in the degree of alteration. In the lowermost part of the borehole (from 103 mbsf to the bottom), gamma-ray values are slightly higher, and can be attributed to the highest degree of alteration in the gabbros (Fig. 3b). At 103 m, a strong increase in natural gamma-rays is recorded. This increase is correlated with other geophysical parameters such as low density, variation in the borehole shape (caliper), or low recovery in this interval, and may be indicative of the presence of a faulted zone or a fractured interval. Such fractured zones have been previously described in the neighbouring Hole 735 with the borehole televiewer (Goldberg *et al.* 1991; Dick *et al.* 2000). Overall, variations in

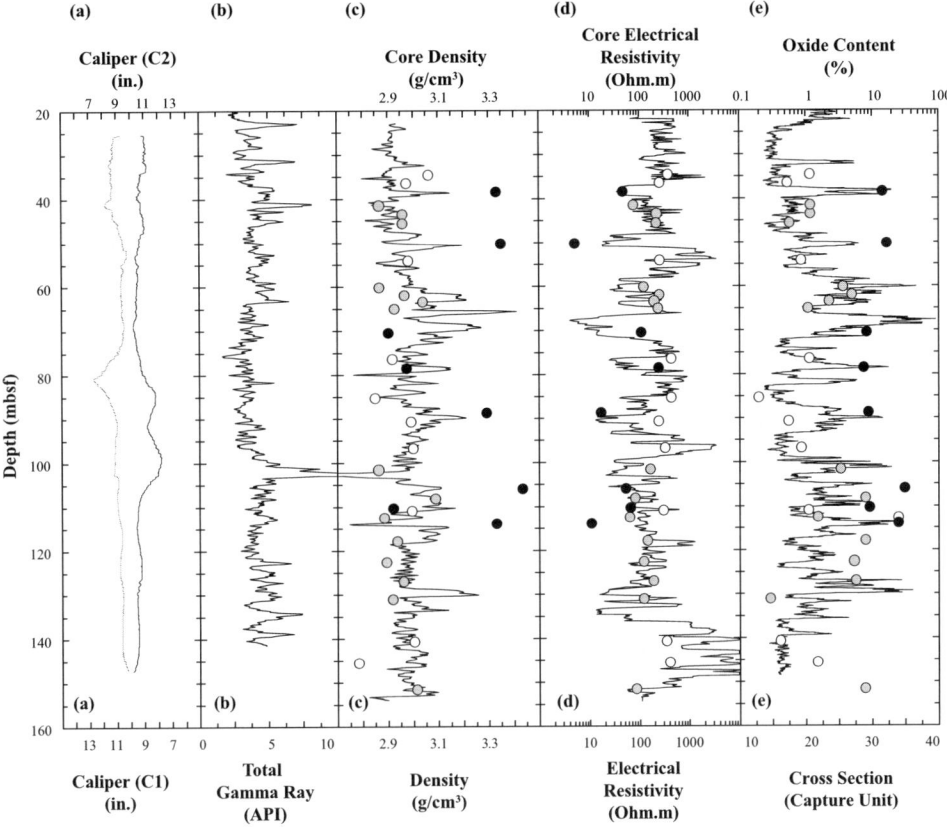

Fig. 3. Down-hole measurements and mini-core measurements along ODP Hole 1105A gabbroic basement. All probes from Schlumberger and laboratory measurements were determined on the same mini-core (1.7 cm long and 2.5 cm in diameter). Mini-core measurements are reported in Table 1. White circles: fresh samples; grey circles: altered gabbros; and black circles: oxide-rich gabbros. (**a**) Caliper measurements (inches) from the Formation Micro Scanner. (**b**) Down-hole total gamma-ray content (API units) from the Hostile Natural Gamma Ray Probe. (**c**) Down-hole bulk density measured by the Hostile Environment Probe. (**d**) Down-hole electrical resistivity measured by the Dual Induction Tool. Mini-cores were saturated with NaCl (30 g L^{-1}, similar to the sea-water) and electrical resistivity measured at 100 Hz. (**e**) Down-hole capture cross-section measured with the accelerator porosity sonde. This diagram gives the oxide abundance determined on thin sections by image analysis.

the density profile mostly correspond to variations in oxide mineral abundances.

Electrical resistivity (Fig. 3d) measured with the dual induction tool (DIT) is the only quantitative assessment of the formation resistivity. Strong variations in resistivity are recorded along the section and are directly linked to the lithology. Typically, values lower than 100 ohm m are indicative of oxide-rich layers. Electrical resistivity measurements appear to be affected by highly resistive gabbros encountered in the lowermost part of the borehole. In this interval (below 140 mbsf), the induction tool reached saturation where formation resistivity values are above 9700 ohm m. Another important parameter used to distinguish the different

gabbroic units is the capture cross-section (Fig. 3e). The thermal neutron absorption is a measure of the formation's ability to absorb neutrons. This measure is useful, as it is a linear function of the formation geochemistry. In ODP Hole 1105A, the extreme range in the composition of the gabbros makes this measure a good proxy to identify oxide-rich layers. Harvey *et al.* (1996) showed that some rare earth elements (REE: Gd, Sm, Eu, Dy) and B, Cl and H have the highest cross-section values. All these elements are abundant in oxide-rich gabbros, due to their higher degree of differentiation (high REE content as observed in the neighbouring ODP Hole 735B, Coogan *et al.* 2001) and also the higher degree of alteration (corresponding to the highest H content due to hydrous minerals). High values of the capture cross-section are well correlated with oxide mineral abundance measured in thin sections.

All down-hole measurements highlight the fact that oxide-rich layers are not negligible along the whole section. Good agreement is observed between the down-hole measurements and laboratory measurements presented in the following section.

Physical properties of minicores

Measurements of physical properties have been performed on 25-mm-diameter mini-cores sampled along the drilled section. For each sample, porosity, grain density, compressional velocity, magnetic susceptibility and electrical resistivity at different salinities were measured at laboratory temperature and atmospheric pressure. The results are presented in Table 1. A particular emphasis has been placed on electrical measurements, in order to investigate changes in pore-space morphology and alteration, as well as the influence of oxide minerals on electrical properties.

Porosity and grain density

The porosity and density were determined by the standard immersion method corresponding to a triple weighing (with dry, saturated and immersed samples). This method allows direct computation of porosity and grain density, independently of sample size and shape, with:

$$\Phi = 100 \times \frac{(W_{\text{sat}} - W_{\text{dry}})}{(W_{\text{sat}} - W_{\text{i}})} \qquad (1)$$

where W_{dry} is the dry weight, W_{sat} is the saturated weight; W_{i} is the immersed weight, and Φ the porosity.

The grain density (ρ_g) is calculated as the following:

$$\rho_g = \frac{W_{\text{dry}}}{(W_{\text{sat}} - W_{\text{i}})} \times \rho_w \qquad (2)$$

where ρ_w is the saturating fluid density.

Porosity, bulk density, and grain density are summarized in Table 1. For rocks with porosities lower than 1%, as in most of the samples in this study, the absolute computed error is lower than 2% for densities (i.e. about 0.06 g cm^{-3}), and about 15% for porosities (i.e. $\leq 0.1\%$). Values of porosity are generally low in most samples (Table 1). The highest porosities (Fig. 4c) correspond to the altered and oxide-rich samples

(respectively 1.31% and 1.94% on average). Bulk density (Fig. 4a) is more variable, with an average of 2.95 g cm^{-3} in fresh and altered samples, although it increases to 3.43 g cm^{-3} in oxide-rich samples.

Acoustic compressional velocity

Compressional velocity (P-wave velocities: V_P) measurements were performed for each samples at 2.25 MHz, using a Panametrics Epoch III-2300 device. The Teflon jacket used for the electrical measurements was kept in place, in order to reduce the influence of desaturation during the measurements. V_P was measured on saturated samples with a 30 g l^{-1} NaCl brine. The precision of these measurements was determined as better than 1%, i.e. of the order of 50 ms^{-1}, determined by repeat measurements and standard measurements. The measured velocities (V_P) values range from 6.0 to 7.1 km s^{-1} (Fig. 4b) and are comparable to the values for gabbroic rocks measured on Hole 735B (Leg 118: $6.713 \pm 0.383 \text{ km s}^{-1}$ (Von Herzen *et al.* 1991); Leg 176: $6.777 \pm 0.292 \text{ km s}^{-1}$, Dick *et al.* 1999), and are then representative of seismic Layer 3. The highest values were measured in fresh samples (with a mean V_P of 6.871 km s^{-1}), and a slight decrease is observed in the altered samples (with a mean V_P of 6.427 km s^{-1}). Oxide-rich samples have markedly lower V_P values with a large game of value (from 4.530 to 4.5 km s^{-1}). These lowest velocity values can be attributed to the abundance of Fe–Ti oxide minerals, and the higher variability is likely to be caused by the variable degree of alteration and porosity (Fig. 4c) and variation in modal oxide abundance in these samples.

Magnetic susceptibility

A Kappabridge was used under the scalar mode for measuring the bulk magnetic susceptibility (Table 1). This measurement proved to be particularly useful on board to detect the presence of thin metallic oxide seams, as the magnetic susceptibility was measured on continuous cores recovered (Pettigrew *et al.* 1999; Natland & Dick 2001). The magnetic susceptibility is an intrinsic material property related to the induced magnetization intensity that may be measured on a sample. The magnetic susceptibility is a function of the nature, concentration and grain size of magnetic minerals within the sample. As a consequence, magnetic susceptibility data may be used as a proxy to clearly

Table 1. *Physical properties of the ODP Hole 1105A gabbroic samples*

Depth (mbsf)	M_g (g cm^{-3})	M_b (g cm^{-3})	ϕ (%)	χ (10^{-3} SI)	F	$Cs + C_{mineral}$ (mS/m)	τ	m	V_P (km s^{-1})	Oxide-content (%)	C_{100} (mS m^{-1})	C_{dry} (mS m^{-1})	$C_{dry}/C_s + C_{mineral}$ (%)	q
Fresh														
34.57	3.06	3.05	0.58	10.33	1744.8	0.191	10.03	1.45	6.695	1.16	7.49	0.039	20.4	5.38
36.48	2.97	2.97	0.36	11.42	1145.1	0.007	4.17	1.25	7.059	0.54	11.08	0.002	32.1	5.12
54.13	2.98	2.97	0.63	2.71	1217.6	0.183	7.67	1.40	6.751	0.85	10.28	0.083	45.4	4.87
76.41	2.92	2.91	0.49	12.51	2051.1	0.054	10.14	1.44	6.65	1.14	5.91	0.008	14.4	5.71
85.28	2.85	2.84	0.44	6.68	1758.2	0.030	7.69	1.38	7.041	0.19	7.87	0.009	30.5	4.06
90.7	2.99	2.98	0.68	1.61	1165.7	0.086	7.90	1.41	6.558	0.55	10.77	0.043	50.0	4.59
96.7	3.00	2.99	0.44	6.98	1520.2	0.010	6.66	1.35	7.053	0.85	9.74	0.002	21.8	5.62
110.89	2.99	2.98	0.51	6.11	1409.0	0.027	7.20	1.37	6.965	1.09	8.84	0.006	22.5	5.72
140.64	3.00	2.99	0.53	6.48	1500.8	0.091	7.98	1.40	7.147	0.41	8.17	0.016	17.3	4.53
145.57	2.78	2.77	0.50	3.82	1767.4	0.111	8.87	1.41	6.789	1.49	7.67	0.016	14.7	5.90
Mean	2.95	2.94	0.52	6.86	1528.0	0.079	7.83	1.39	6.871	0.83	8.78	0.02		5.15
Standard deviation	0.08	0.08	0.09	3.64	301.87	0.067	1.72	0.05	0.21	0.40	1.66	0.03		0.62
Altered														
43.48	2.96	2.93	1.39	4.54	1548.1	0.847	20.92	1.71	6.576	1.19	8.95	0.362	42.7	4.90
41.52	2.87	2.85	1.07	2.30	364.8	0.455	3.89	1.30	6.500	1.82	34.35	0.156	34.2	5.64
45.55	2.96	2.94	1.05	4.53	1127.9	0.435	11.84	1.54	6.777	0.58	11.81	0.070	16.1	4.54
60.19	2.87	2.83	1.94	13.12	844.0	1.405	16.41	1.71	6.233	3.65	15.94	2.145	>100	6.03
61.95	2.97	2.95	0.73	16.55	1540.0	0.524	11.24	1.49	6.487	4.98	8.24	0.096	18.4	7.69
63.39	3.04	3.02	0.74	28.39	1165.3	0.651	8.66	1.44	6.568	2.24	11.44	0.102	15.7	6.05
64.97	2.93	2.91	0.64	23.71	1372.6	0.567	8.83	1.43	6.703	1.08	9.36	0.091	16.1	5.10
101.58	2.86	2.83	1.88	6.47	1096.1	1.059	20.58	1.76	6.01	3.33	12.43	0.410	38.7	6.36
108.11	3.09	3.05	1.66	22.42	565.2	2.076	9.40	1.55	6.739	7.93	23.66	0.439	21.1	8.50
112.51	2.88	2.83	2.67	0.79	432.4	2.408	11.56	1.68	5.583	1.51	30.33	1.756	72.9	4.81
117.82	2.94	2.90	1.59	38.78	1024.8	1.397	16.28	1.67	6.236	7.97	14.11	0.352	25.2	8.61
122.64	2.89	2.87	1.34	24.89	912.5	1.832	12.21	1.58	6.779	5.36	15.39	0.312	17.0	7.48
151.58	3.01	3.00	0.55	43.37	681.0	2.202	3.75	1.25	6.609	7.78	19.79	3.007	>100	7.69
131.06	2.92	2.89	1.49	2.14	771.6	1.304	11.50	1.58	6.459	0.29	16.97	0.242	18.5	3.79
126.95	2.96	2.94	0.97	22.75	1292.0	0.936	12.51	1.54	6.147	5.68	10.59	0.256	27.3	7.70
Mean	2.94	2.92	1.31	16.98	982.6	1.207	11.97	1.55	6.427	3.69	16.23	0.65		6.33
Standard deviation	0.07	0.07	0.58	13.64	374.70	0.667	5.03	0.15	0.33	2.76	7.81	0.90		1.53
Oxide rich														
38.39	2.94	2.92	1.31	101.32	1078.0	7.013	6.14	1.42	6.436	19.69	17.45	8.552	>100	11.43
50.21	3.33	3.31	1.07	28.62	529.4	4.476	3.74	1.29	6.947	14.48	24.47	7.773	>100	9.66
70.46	3.35	3.33	0.70	249.51	257.2	18.314	0.52	0.87	6.530	16.69	45.49	—	—	—
78.5	2.90	2.88	1.08	64.41	893.6	1.446	9.62	1.50	6.365	8.28	14.93	2.880	>100	7.90
88.76	2.97	2.96	0.59	14.42	1223.3	0.095	7.26	1.39	6.706	7.49	7.91	0.027	28.0	9.39
106	3.29	3.26	1.37	86.76	569.3	13.038	2.59	1.22	6.251	8.81	29.81	—	—	—
110.3	3.43	3.39	1.87	195.10	824.8	6.326	15.44	1.69	6.066	31.23	23.91	2.469	39.0	17.03
113.88	2.92	2.89	1.49	16.01	450.0	1.933	6.70	1.45	6.210	9.28	28.29	1.331	68.8	8.59
135.63	3.33	3.15	7.95	205.66	134.8	10.043	9.99	1.91	4.531	25.11	107.02	10.943	>100	13.27
Mean	3.16	3.12	1.94	106.87	662.3	6.965	6.89	1.41	6.227	15.67	33.25	4.85		11.04
Standard deviation	0.22	0.21	2.29	88.78	367.63	5.966	4.47	0.29	0.69	8.36	29.62	4.17		3.20

All petrophysical measurements were performed on mini-cores at Cerege, except for dry conductivity determined at the University of Montpellier II, ISTEEM.
Key: M_g, grain density (g cm^{-3}) and M_b, bulk density (g cm^{-3}); ϕ, porosity (%); χ, magnetic susceptibility (10^{-3} SI); F, electrical formation factor; and C_x, surface conductivity (mS m^{-1}) calculated from the Revil & Glover (1998) formulation. τ, tortuosity; m, cementation factor; V_P, acoustic compressional velocity (km s^{-1}). Oxide content (%) was determined from thin sections. Conductivity (mS m^{-1}) was measured on saturated samples (100 g l^{-1}, conductivity of the brine: 4.2 S m^{-1}). Conductivity was measured on dry samples (S m^{-1}); C_{dry}/C_X ratio, connectivity exponent q.

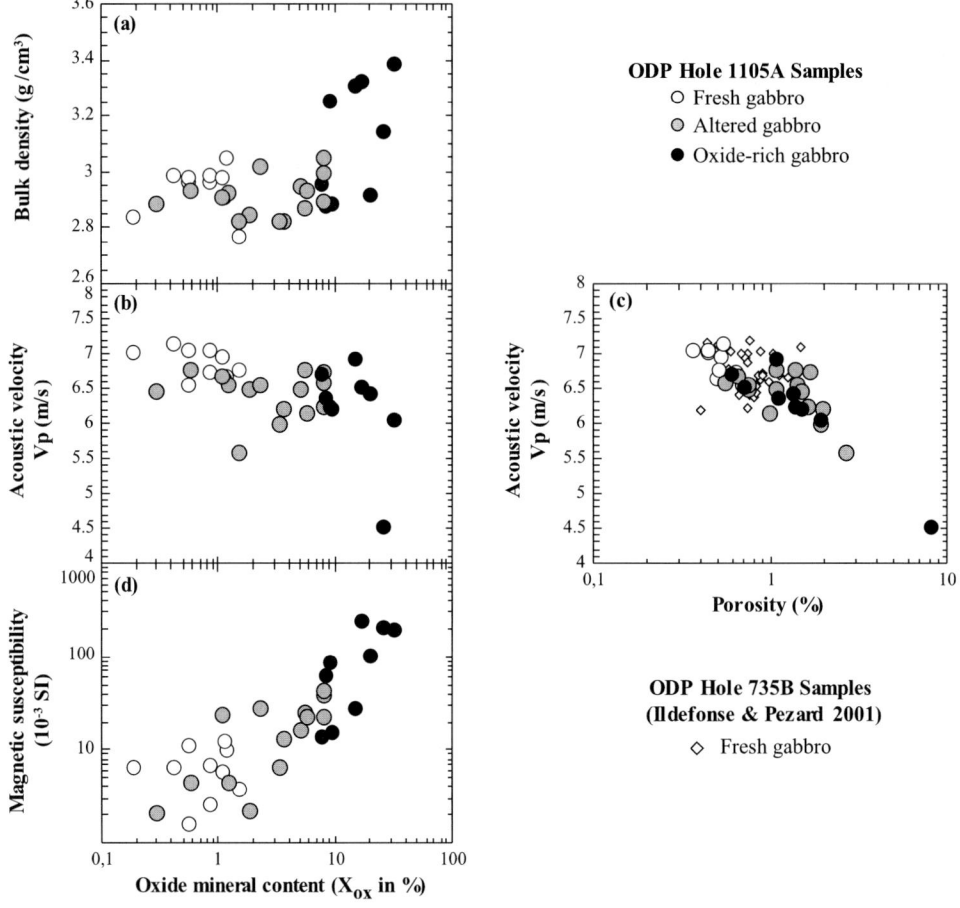

Fig. 4. (a) Grain density (g cm^{-3}), and (b) compressional velocity V_P (km s^{-1}), versus oxide mineral content (%) determined from thin-section analysis; (c) V_P (km s^{-1}) versus porosity (%); (d) Magnetic susceptibility (10^{-3} SI) versus oxide mineral content.

identify the occurrence of oxide minerals. As expected, a good positive correlation between the estimated oxide content is observed from thin section (2D) and magnetic susceptibility (3D) (Fig. 4d). The scatter that does occur could be explained by the highly variable content of oxide in the gabbros, or by the difference of investigation methods (2D versus 3D).

Electrical resistivity

Background. The sensitivity of the electrical properties of porous media to fluid content and alteration provides a powerful means to detect conductive horizons such as fractures, either open or mineralized, in a resistive matrix (e.g. Walsh & Brace 1984; Katsube & Hume 1987; Pezard & Luthi 1988). At the metre-scale,

in situ measurements of electrical resistivity are often analysed to derive porosity (Archie 1942; Brace *et al.* 1965; Becker 1985). The different modes by which the current is transported may be identified in the laboratory, providing insights into the degree of alteration of the rock, with the analysis of surface electrical conduction (Waxman & Smits 1968; Pezard 1990; Revil & Glover 1998). The method used was initially derived for porous media such as sandstones and clays (Waxman & Smits 1968; Revil & Glover 1998). It has also been used successfully in various low-porosity igneous rocks, such as basalts (Pezard 1990; Bernard 1999; Einaudi *et al.* 2000), gabbros (Pezard *et al.* 1991; Ildefonse & Pezard 2001), and granites (Pape *et al.* 1985). For the oxide-free samples, the electrical conductivity in the samples depends upon a combination of pore volume and pore surface conduction.

In a porous media comprising a matrix considered as infinitely resistive and a connected pore space saturated with a conductive electrolyte, two electrical conductivities are involved: an electrolytic conduction mechanism within the pore volumes, and a surface conduction mechanism at the interface between the electrolyte and minerals. While the electrolytic conduction is directly related to the nature and salinity of the pore fluid, the surface conduction is due to the presence along pore surfaces of charges which appear to guarantee the electroneutrality of the medium. In most geological settings, the mineral surfaces are charged negatively. The cations present in the fluid along pore surfaces migrate through a diffuse layer to contribute to the overall conduction (Waxman & Smits 1968; Clavier et al. 1977).

In the simplest case, i.e. when the surface conduction component is negligible with respect to the electrolytic component (in the case of a pure quartz sand, for instance), the total conductivity of the pore space (C_o) can be considered as directly proportional to that of the saturating fluid (C_w):

$$C_o = \frac{C_w}{F} \qquad (3)$$

The dimensionless F coefficient is called the 'electrical formation factor'. The factor F describes the contribution of pore-space topology in the matrix to the electrical resistivity of a fluid-saturated medium. The electrical formation factor (F) has been reported to be dependent upon rock texture; distribution of pores and pore throat sizes, connectivity between pores; and flow-path tortuosity, all of which illustrate the internal 3D topology of the pore space of the analysed rocks.

When surface processes cannot be neglected with respect to electrolytic conduction, which is generally the case in the presence of alteration phases, the previous equation does not hold. The full expression for the electrical conductivity of the porous media is the sum of two contributions: a bulk electrolytic conductivity and surface conductivity processes (Waxman & Smits 1968; Clavier et al. 1977). Waxman & Smits (1968) have proposed an empirical model to take into account the excess conductivity (C_s) due to surface processes, with:

$$C_o = \frac{C_w}{F} + C_s \qquad (4)$$

where C_s is the surface conductivity resulting from the presence of the electrical double layer (Clavier et al. 1977), and F is the electrical formation factor.

A new and more complete approach based on the pore space microgeometry has been proposed by Revil & Glover (1998), with:

$$C_o = \left(\frac{C_w}{F}\right)\left[1 - t^f_{(+)} + F\xi + f(F,\xi)\right] \qquad (5)$$

where $t^f_{(+)}$ is the Hittorf number of cations in the electrolyte (Revil & Glover 1998), and ξ is a dimensionless parameter defined by Kan & Sen (1987). The $f(F, \xi)$ is a complex function of F and ξ detailed by Revil & Glover (1998). The predictive capacity of this model covers the adequate salinity range. From the single approach of (5), the RG model is more precise, yielding the following high- and low-salinity end-members:

$$C_o = \left(\frac{C_w}{F}\right)[1 + 2\xi F(F - 1)](\text{HS})$$
$$\qquad (6)$$
$$C_o = \xi C_w\left[1 - \frac{\xi - 1}{\xi F}\right] \qquad (\text{LS})$$

At high fluid salinity, the dominant path for current transport depends on the pore volume topology, which is represented by the electrical formation factor (F). At low salinity, adsorbed cations are transported along the fluid–grain interface, and the electrical conduction is dominated by surface electrical conductivity (C_s). Further details on this model can be found in Revil & Glover (1998), Revil et al. (1998) and Ildefonse & Pezard (2001). For the following discussion, we used the Waxman & Smits (1968) formulation.

Experimental design

In order to distinguish between the electrolytic and the surface conduction mechanisms, samples were analysed by measuring the electrical resistivity with variable saturating fluid salinity. The measurements were performed at seven different saturating fluid salinities (from 0.2 to 100 g l^{-1}; i.e. conductivity varying from 0.04 to 11 S m^{-1}) with a two-electrode Wayne–Kerr bridge. The samples were saturated with NaCl solution at near-vacuum conditions for the first measurement. In the following, the bath fluid was regularly checked and brines were changed when changes in the electrical response of the system had ceased. The device used a spring to keep electrodes in contact with each mini-core. The sides of the cores were wrapped in insulating Teflon tape to restrict the current from passing down the sides of the mini-core. A 10 mV signal was applied to perform each of the measurements at seven frequencies. Only measurements realized at 100 Hz are presented in this study. This frequency value was chosen to be similar to that used for down-hole measurements.

From these measurements (Fig. 5), we applied the detailed relationship of Revil & Glover (1998) to compute the electrical formation factor (F) and surface conductivity (C_s) of each sample.

Porosity structure from electrical measurements

In the absence of clay or alteration phases, the electrical formation factor F depends primarily on the porosity value and structure. In the absence of clay or alteration phases, F depends primarily on the porosity value and structure. The surface conductivity C_s is then negligible when compared to the fluid conductivity (C_w). The F is measured or modelled to increases as the porosity decreases or becomes more tortuous. An empirical relationship between the electrical formation factor (F) and porosity (Φ) was proposed by Sundberg (1932), later followed by Archie (1942) for sedimentary rocks, yielding Archie's law (Fig. 6a). This relationship is still widely used in oil industry. Thus the formation factor (F) was related to porosity by:

$$F = \Phi^{-m} \qquad (7)$$

where m refers to the cementation factor of the sedimentary rock, and is related to the degree of cementation between individual grains. This relationship may also be expressed in terms of degree of connectivity of the inner pore space, or the inverse with electrical tortuosity τ (Walsh & Brace 1984; Katsube & Hume 1987; Pezard 1990; Pezard *et al.* 1991; Guéguen & Palciauskas 1992), yielding:

$$F = \frac{\tau}{\Phi} \qquad (8)$$

While the cementation factor describes the non-uniformity of the section of the conductive channels, the tortuosity relates to the complexity of the path followed by the electrical current (e.g. Guéguen & Palciauskas 1992) or, in a more general sense, the efficiency of electrical flow processes (Clennell 1997). Many different analytical and numerical methods were developed to derive transport properties of porous

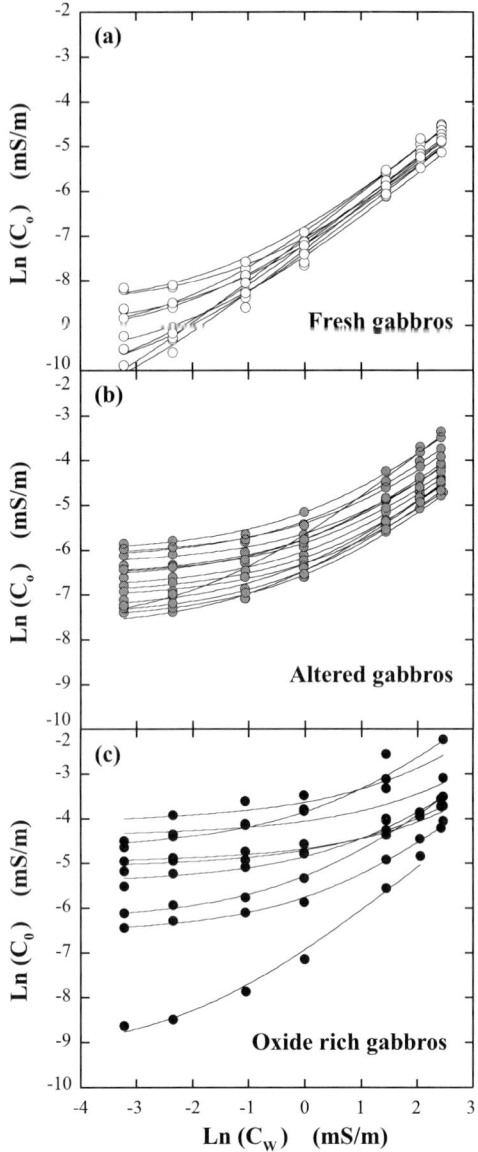

Fig. 5. Plots of the core conductivities as a function of the saturating fluid conductivities. Best-fit curves are calculated using the Revil & Glover (1998) model. Based on thin sections, samples have been divided into three types: (**a**) fresh samples, (**b**) altered samples and (**c**) oxide- rich samples.

media from microstructural parameters. Clennell (1997) provides a useful review of the different approaches and results that have been obtained.

The electrical formation factor F characterizes the ratio between the bulk electrical conductivity and the electrical conductivity of the saturating

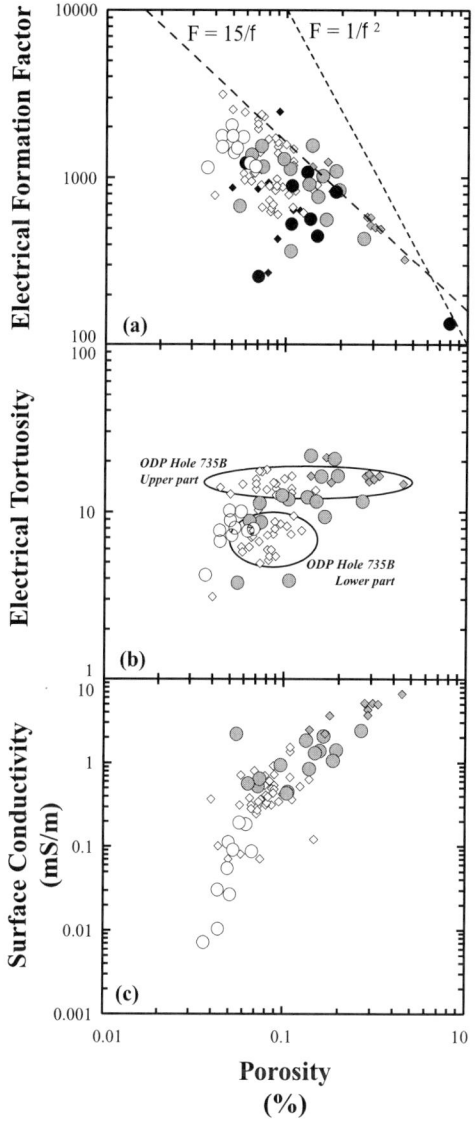

ODP Hole 1105A Samples
○ Fresh gabbro
◔ Altered gabbro
● Oxide-rich gabbro

ODP Hole 735B Samples
(Ildefonse & Pezard 2001)
◇ Fresh gabbro
◈ Altered gabbro
◆ Oxide-rich gabbro

Fig. 6. Plots of the electrical formation factor
(**a**), electrical tortuosity (**b**), and surface conductivity
(**c**) from electrical measurements as a function of porosity.

fluid; the saturated rock becomes more conductive as F decreases. Figure 6a shows that the freshest samples exhibit the highest electrical formation factor, with a restricted range. The presence of altered phases or oxides induces lower F than the ones for fresh samples (Fig. 6a). No relationship appears between the electrical resistivity and the porosity. Therefore it appears to be difficult to derive a reliable estimate of the porosity from the electrical resistivity in such rocks. The power-law dependence of F on Φ has been observed over a wide range of porosity in both real rocks (Ruffet *et al.* 1991) and artificial porous media (Sen *et al.* 1981; Mattisson & Knackstedt 1997), but the problem of the physical basis for Archie's law is still debatable.

Ildefonse & Pezard (2001) have shown that there is a partitioning in the porosity structure as a function of depth. In ODP Hole 735B, the cementation factor and the electrical tortuosity clearly exhibit a bimodal distribution as a function of depth (Fig. 6b, transition at about 825 mbsf). In ODP Hole 1105A gabbros, such partitioning is not observed, and can be explained by the fact that the hole is much shallower and also probably due to our highly heterogeneous sampling.

The surface conductivity (C_s) presents a strong decoupling of the gabbroic samples into two groups (Fig. 6c). The most altered samples, as expected, have the highest surface conductivities, with an average value of 1.207 mS m^{-1}. C_s decreases to 0.079 mS m^{-1} on average in the freshest samples. Furthermore, C_s values calculated for oxide-rich gabbros are not reported in this diagram. It appears that the high C_s values (reaching 18.314 mS m^{-1}) incorporate the oxide mineral contribution of the bulk conductivity. These 'pseudo C_s' values calculated in oxide-rich gabbros integrate both cations transported along the fluid–grain interface and the conduction via electronic processes in metallic minerals. These 'pseudo-C_s' values for oxide-rich gabbros are reported in Table 1, and show that 'pseudo C_s' in oxide-rich samples reaches 7 mS m^{-1} on average, i.e. about 80 times higher than C_s in the freshest samples. A new model should be proposed to include the effect of a conductive phase such as the Fe–Ti oxides in the studied gabbros, and a new constitutive law for electrical conduction is hence required here to account for the conductive phase in the mineral matrix.

Electrical properties in the oxide-rich gabbros

The presence of numerous Fe–Ti oxide-rich levels in ODP Hole 1105A (Pettigrew *et al.* 1999) leads to a more complex conduction mechanism for the transport of electrical charges in the rock. Matrix conduction mechanisms via electronic processes in metallic oxide grains may then become significant (Drury & Hyndman 1979; Pezard 1990; Pezard *et al.* 1991). As pointed out above, the conduction via electronic processes in metallic minerals is not taken into account in these formulations of the electrical properties. No model includes the effect of a conductive phase such as the oxides present in the studied gabbros, and a new constitutive law for electrical conduction is hence required to account for the conductive phase in the mineral matrix. Pezard *et al.* (1991), from ODP Hole 735B gabbros, proposed:

$$C_o = \frac{C_w}{F} + [C_s + C_{mineral}] \quad (9)$$

The $C_{mineral}$ term is a constant related to the oxide content for a given sample, which does not depend on the salinity of the saturating fluid. Thus, the different contributions to the electrical conduction are identified, and estimated. The oxide contribution to the measured electrical resistivity is achieved through dry resistivity measurements.

Dry measurements

The measurements of dry sample conductivity were made in a two-electrode cell with an experimental device which allows multi-frequency measurements. The signal measured on a dry sample is a complex impedance (*Z*) characterized by two components: an in-phase *R* (or resistive) signal, and an out-of-phase *X* (or reactive) signal. The complex impedance can be expressed as follows:

$$Z = R + iX \quad (10)$$

where *i* is the complex operator ($i^2 = -1$). The complex electrical resistivity is calculated from the *Z* measurements by:

$$\rho = Z^*(A/L) = \rho' + i\rho'' \quad (11)$$

where *A* is the cross-sectional area of the mini-core and *L* is its length. The ρ' and ρ'' are respectively the real and imaginary parts of the complex resistivity. The electrical resistivity is the value corresponding to the measured real part of *z* at the frequency where ρ'' equals zero.

The dry conductivity values present a large distribution range (Fig. 7). However, as expected, oxide-rich gabbros present by far the highest conductivities. The dry conductivities

show a large distribution, with five-order magnitudes from 2.3×10^{-6} to 1.9×10^{-1} S m^{-1}. For comparison, saturated measurements at 100 g l^{-1} vary over only one order of magnitude (from 5.9×10^{-3} to 1.1×10^{-1} S m^{-1}, Table 1), since electrical conductivity is here dominated by electrolytic processes. In a general sense, the difference between these two measurements represents the conductivity of surface and minerals. In the range 0–2% of oxide mineral abundance (fresh gabbros), their contribution to measured conductivity is about 26%, between 2 and 15% (altered gabbros), the contribution is about 37%; and for oxide contents above 15% (oxide-rich gabbros), the oxide mineral contribution reaches 82%. In some cases, particularly for oxide-rich samples, the ratio between the dry conductivity and the saturated conductivity is higher than one (Table 1). Several explanations may be involved, such as the accuracy of the electrical measurements; or the accuracy of oxide content determination (2D versus 3D); or the quality of desaturation of each sample.

In the case of multi-phase conducting porous systems, a mixing model such as:

$$C_o = \sum_{i=1}^{N} C_i(X_i) \quad (12)$$

can be used (Guéguen & Palciauskas 1992; Glover *et al.* 2000). This applies to the case of a mineral, solid-phase conducting current is parallel to the more usual fluid-related phases described by Waxman & Smits (1968) or Revil & Glover (1998). Guéguen & Palciauskas (1992) and Glover *et al.* (2000) have proposed a modified relationship for a two conducting phases medium:

$$C_o = C_1(1 - X_2)^n + C_2(X_2)^p \quad (13)$$

with C_o corresponding to the measured conductivity, C_1 and C_2 respectively to the phase 1 and phase 2 conductivities, and X_2 being the volume fraction of the conducting phase 2 (with $X_1 + X_2 = 1$). This modified model has two exponents (*n* and *p*) that describe the connectivity of each of the two phases. As a consequence, we may propose a more analytical expression for the mineral-phase contribution by combining this approach and equation (9); the total electrical conductivity (C_o) can be expressed as:

$$C_o = \left(\frac{C_w}{F} + C_s\right) + C_{ox}(X_{ox})^q \quad (14)$$

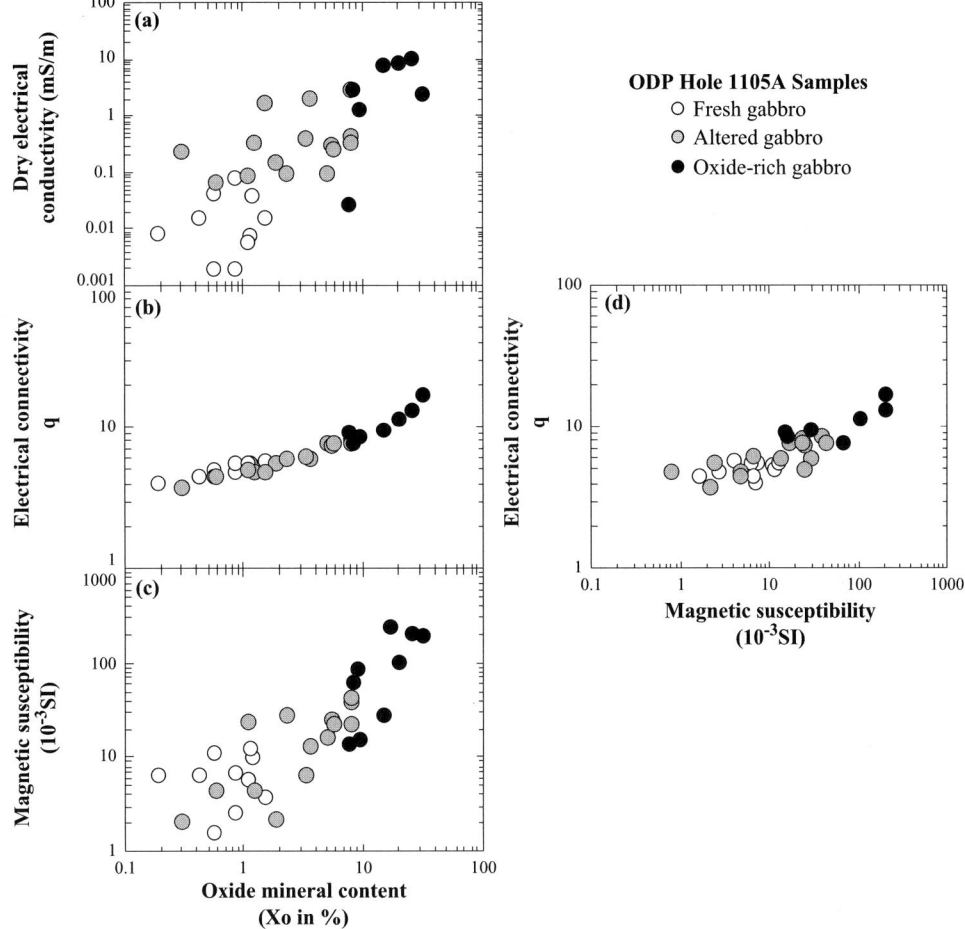

Fig. 7. Plots of (**a**) dry electrical conductivity; (**b**) electrical connectivity exponent q, and (**c**) magnetic susceptibility versus oxide mineral content (X_o). (**d**) Electrical connectivity exponent versus magnetic susceptibility (10^{-3} SI).

with C_{ox} being the intrinsic electrical conductivity of the oxide minerals, and X_{ox} their volume fraction determined from thin section; q corresponds to the connectivity exponent of the oxide phase. In the case of dry conductivity measurements, the measured electrical conductivity (C_{dry}) is only related to the conductivity of oxide minerals, and can be expressed as:

$$C_{dry} = C_{ox}(X_{ox})^q \qquad (15)$$

with C_{ox} equal to 10^6 S m^{-1} (order of magnitude for magnetite after Olhoeft 1981; Guéguen & Palciauskas 1992). From equation 14, we can calculate the solid-phase q exponent referred to as the 'solid connectivity exponent'. The q exponent is a function of the fractional volumes of each phase, because low phase connectivities and

high phase connectivities are associated with large volume fractions.

Conclusions

The physical properties of a set of 34 gabbroic samples from ODP Hole 1105A (SWIR), have been measured in the laboratory, with a particular emphasis on the analysis of electrical properties. The electrical 'formation factor' F of each sample, as well as a pseudo-surface conductivity (C_s) have been extracted from the gabbros; there is evidence of three conducting phases (matrix, surface conductivity and fluid conductivity). Dry measurements clearly show that most of the excess conductivity evident in brine-saturated gabbros can be attributed to ionic conduction for the oxide-rich samples. Our investigations of electrical properties of ODP Hole 1105A samples enhanced the oxide mineral contribution on electrical conductivity. The conductivity measured on dry samples is correlated to the oxide mineral content, and their contribution can reach up to 80% of the measured conductivity on saturated samples.

This research used samples and data that were provided by the Ocean Drilling Program (ODP). ODP is sponsored by the US National Science Foundation (NSF) and participating countries under management of Joint Oceanographic Institutions (JOI), Inc. We wish to thank D. Hermitte for his help in the laboratory. The authors thank the reviewers and editors for their helpful comments, which substantially improved this manuscript.

References

ARCHIE, G. E. 1942. The electrical resistivity log as an aid in determining some reservoir characteristics. *Journal of Petroleum Technology*, **5**, 1–8.

BECKER, K. 1985. Large-scale electrical resistivity and bulk density of the oceanic crust, DSDP Hole 504B, Costa Rica Rift. *In*: ANDERSON, R. N., HONNOREZ, J. and BECKER, K. *et al.* (eds) *Initial Reports of the Deep Sea Drilling Project*, **83**, Washington (US Government Printing Office), 419–427.

BERNARD, M.-L. 1999. *Etude expérimentale des propriétés physiques des roches pyroclastiques de la montagne pelée*. PhD thesis, Université de Paris IV, 260 p.

BRACE, W. F., ORANGE, A. S. & MADDEN, T. R. 1965. The effect of pressure on the electrical resistivity of water-saturated crystalline rocks. *Journal of Geophysical Research*, **70**, 5669–5678.

CANN, J. R., BLACKMAN, D. K. *et al.* 1997. Corrugated slip surfaces formed at ridge–transform intersections on the Mid-Atlantic Ridge. *Nature*, **385**, 329–332.

CANNAT, M., MÉVEL, C. & STAKES, D. 1991. Stretching of the deep crust at the slow-spreading Southwest Indian Ridge. *Tectonophysics*, **190**, 73–94.

CLAVIER, C., COATES, G. & DUMANOIR, J. 1977. The theoretical and experimental bases for the dual-water model for the interpretation of shaly sands. *Society of Petroleum Engineers, 52nd Annual Technical Conference*. Paper 6859.

CLENNELL, M. B. 1997. Tortuosity: a guide through the maze. *In*: LOVELL, M. A. & HARVEY, P. K. (eds) *Developments in Petrophysics*. Geological Society, London, Special Publications, **122**, 299–344.

COOGAN, L. A., MACLEOD, C. J. *et al.* 2001. Whole-rock geochemistry of gabbros from the Southwest Indian Ridge: constraints on geochemical fractionations between the upper and lower oceanic crust and magma chamber processes at (very) slow-spreading ridges. *Chemical Geology*, **178**, 1–22.

DICK, H. J. B., MEYER, P. S., BLOOMER, S., KIRBY, S., STAKES, D. & MAWER, C. 1991. Lithostratigraphic evolution of an in-situ section of oceanic layer 3. *In*: VON HERZEN, R. P., ROBINSON, P. T. *et al. Proceedings of the Ocean Drilling Program, Scientific Results*, **118**, 439–540, College Station, TX.

DICK, H. J.B., NATLAND, J. H. *et al.* 2000. A long in-situ section of the lower ocean crust: results of ODP Leg 176 drilling at the Southwest Indian Ridge. *Earth and Planetary Science Letters*, **179**, 31–51.

DICK, H. J. B., NATLAND, J. H. *et al.* 1999. *Proceedings of the Ocean Drilling Program, Initial Reports*, **176** [Online]. Available from World Wide Web Address: http://www-odp.tamu.edu/publications/176_IR/176TOC.HTM.

DRURY, M. J. & HYNDMAN, R. D. 1979. The electrical resistivity of oceanic basalts. *Journal of Geophysical Research*, **84**, 4537–4546.

EINAUDI, F., PEZARD, P., COCHEMÉ, J.-J., COULON, C., LAVERNE C. & GODARD, M. 2000. Petrography, geochemistry and physical properties of a continuous extrusive section from the Hilti massif, Oman ophiolite. *Marine Geophysical Researches*, **21**, 387–407.

FISHER, R. L. & SCLATER, J. G. 1983. Tectonic evolution of the Southwest Indian Ocean since the mid-Cretaceous: plate motions and stability of the pole of Antarctica/Africa for at least 80 Myr. *Geophysical Journal of the Royal Astronomical Society*, **73**, 553–576.

GOLDBERG, D. 1997. The role of downhole measurements in marine geology and geophysics. *Reviews of Geophysics*, **35**, 315–342.

GLOVER, P. W. J., HOLE, M. J. & POUS, J. 2000. A modified Archie's law for two conducting phases. *Earth and Planetary Science Letters*, **180**, 369–383.

GOLDBERG, D., BROGLIA, C. & BECKER, K. 1991. Fracturing, alteration, and permeability: in-situ properties in Hole 735B, *In*: VON HERZEN, R. P., ROBINSON, P. T. *et al. Proceedings of the Ocean Drilling Program, Scientific Results*, **118**, 261–269, College Station, TX.

GUÉGUEN, Y. & PALCIAUSKAS, V. 1992. *Introduction à la physique des roches*. Hermann, Paris, 392 pp.

HARVEY, P. K., LOVELL, M. A., BREWER, T. S., LOCKE, J. & MANSLEY, E. 1996. Measurement of thermal neutron absorption cross section in selected geochemical reference materials. *Geostandards Newsletter*, **20**, 79–85.

ILDEFONSE, B. & PEZARD, P. 2001. Electrical properties of slow-spreading ridge gabbros from ODP Site 735, Southwest Indian Ridge. *Tectonophysics*, **330**, 69–92.

KAN, R. & SEN, P. N. 1987. Electrolytic conduction in periodic arrays of insulators with charges. *Journal of Chemical Physics*, **86**, 5748–5756.

KARSON, J. A. & DICK, H. J. B. 1984. Deformed and metamorphosed oceanic crust on the Mid-Atlantic Ridge, *Ofioliti*, **9**, 279–302.

KARSON, J. A., CANNAT, M., MILLER, D.J. & ELTHON, D. E. 1997. *Proceedings of the Ocean*

Drilling Program, Scientific Results, **153**, College Station, TX.

KATSUBE, T. J. & HUME, J. P. 1987. Permeability determination in crystalline rocks by standard geophysical logs. *Geophysics*, **52**, 342–352.

MACLEOD, C. J., DICK, H. J. B., *et al.* 1998. Geological mapping of slow-spread lower ocean crust: a deep-towed video and wireline rock drilling survey of Atlantis Bank (ODP Site 735, South West Indian Ridge). *Interridge News*, **7**, 39–43.

MATTISSON, C. & KNACKSTEDT, M. A. 1997. Transport in fractured porous solids. *Geophysical Research Letters*, **24**, 495–498.

MÉVEL, C., GILLIS, K. M., ALLAN, J. F. & MEYER, P. S. E. 1996. *Proceedings of the Ocean Drilling Program, Scientific Results*, **147**, College Station, TX.

NATLAND, J. H. & DICK, H. J. B. 2001. Formation of the lower oceanic crust and the crystallization of gabbroic cumulates at a very slowly spreading ridge. *Journal of Volcanology and Geothermal Research Letters*, **110**, 191–233.

NATLAND, J. H., MEYER, P. S., DICK, H. J. B. & BLOOMER, S. H. 1991. Magmatic oxides and sulfides in gabbroic rocks from Hole 735B and the later development of the liquid line of descent. *In*: VON HERZEN, R. P., ROBINSON, P. T. *et al. Proceedings of the Ocean Drilling Program, Scientific Results*, **118**, 75–111, College Station, TX.

NIU, Y., GILMORE, T., MACKIE, S., GREIG, A. & BACH, W. 2002. Mineral chemistry, whole-rock compositions, and petrogenesis of Leg 176 gabbros: data and discussion. *In*: NATLAND, J. H., DICK, H. J. B., MILLER, D. J. & VON HERZEN, R. P. (eds) *Proceedings of the Ocean Drilling Program, Scientific Results*, **176** [Online]. World Wide Web Address: *http://www-odp.tamu.edu/ publications/176_SR/chap_08/chap_08.htm.*

OLHOEFT, G. R. 1981. Electrical properties of rocks. *In*: TOULOUKIAN, Y. S., JUDD, W. R. & ROY, R. F. (eds) *Physical Properties of Rocks and Minerals*. McGraw-Hill, New York, 257–330.

PAPE, H., RIEPE, L. & SCHOPPER, J. R. 1985. Petrophysical detection of microfissures in granite. *Transaction SPWLA, 26th Annual Logging Symposium*, Paper P.

PETTIGREW, T. L., CASEY, J. F., MILLER, D. J., *et al.* 1999. *Proceedings of the Ocean Drilling Program Initial Reports*, **179**, [Online]. World Wide Web Address: *http://www-odp.tamu.edu/ publications /179_IR/179TOC.HTML.*

PEZARD, P. A. 1990. Electrical properties of Mid-Ocean Ridge basalt and implications for the structure of the upper oceanic crust in Hole 504B. *Journal of Geophysical Research*, **95**, 9237–9264.

PEZARD, P. A. & Luthi, S.M. 1988. Borehole electrical images in the basement of the Cajon Pass Scientific Drillhole, California; fracture identification and tectonic implications. *Geophysical Research Letters*, **15**, 1017–1020.

PEZARD, P. A., HOWARD, J. J. & GOLDBERG, D. 1991. Electrical conduction in oceanic gabbros, Hole 735B, Southwest Indian Ridge. *In*: VON HERZEN, R. P., ROBINSON, P. T. *et al. Proceedings of the Ocean Drilling Program, Scientific Results*, **118**, 323–331.

RANERO, C. R. & RESTON, T. J. 1999. Detachment faulting at ocean core complexes. *Geology*, **2**, 983–986.

REVIL, A. & GLOVER, P. W. J. 1998. Nature of surface electrical conductivity in natural sands, sandstones, and clays. *Geophysical Research Letters*, **25**, 691–694.

REVIL, A., CATHLES, L. M., LOSH, S. & NUNN, J. A. 1998. Electrical conductivity in shaly sands with geophysical applications. *Journal of Geophysical Research*, **103**, 23 925–23 936.

ROBINSON, P. T., DICK, H. J. B. & VON HERZEN, R. P. 1991. Metamorphism and alteration in oceanic layer 3: Hole 735B. *In*: VON HERZEN, R. P., ROBINSON, P. T. *et al. Proceedings of the Ocean Drilling Program, Scientific Results*, **118**, 541–552. College Station, TX.

RUFFET, C., GUEGUEN, Y. & DAROT, M. 1991. Complex conductivity and fractal microstructures, *Geophysics*, **56**, 758–768.

SEN, P. N., SCALA, C. & COHEN, M. H. 1981. A self-similar model for sedimentary rocks with application to the dielectric constant of fused glass beads. *Geophysics*, **46**, 781–795.

SUNDBERG, K., 1932. Effect on impregnating waters on electrical conductivity of soils and rocks. *Transactions of the American Institute of Mining and Metallurgical Engineers*, **79**, 367–391.

TUCHOLKE, B. E. & LIN, J. 1994. A geological model for the structure of ridge segments in slow spreading ocean crust. *Journal of Geophysical Research*, **99**, 11 937–11 958.

TUCHOLKE, B. E., LIN, J. & KLEINROCK, M. C. 1998. Megamullions and mullion structure defining oceanic metamorphic core complexes on the mid-Atlantic ridge. *Journal of Geophysical Research*, **103**, 9857–9866.

VON HERZEN, R. P., ROBINSON, P. T. *et al.* 1991. *Proceedings of the Ocean Drilling Program, Scientific Results*, **118**, College Station, TX.

WALSH, J. B. & BRACE, W. F. 1984. The effect of pressure on porosity and the transport properties of rocks. *Journal of Geophysical Research*, **89**, 9425–9431.

WAXMAN, M. H. & SMITS, L. J. M. 1968. Electrical conductivities in oil-bearing shaly sands. *Society of Petroleum Engineers Journal*, **8**, 107–122.

Non-invasive characterization of fractured crystalline rocks using a combined multicomponent transient electromagnetic, resistivity and seismic approach

M. A. MEJU

Department of Environmental Science, Lancaster University, Lancaster LA1 4YQ, UK
(e-mail: m.meju@Lancaster.ac.uk)

Abstract: DC resistivity and seismic-refraction methods are well established in basement studies, and the multicomponent transient electromagnetic (TEM) method is also emerging as an effective tool for locating electrically conductive zones in crystalline rock underneath thick overburden. It can be expected that combining these three techniques will furnish a powerful non-invasive approach for characterizing fractured crystalline rocks. The aims of this paper are twofold, namely: (1) to establish the *in situ* geophysical signature of fracture-zones in weathered crystalline rock masses using the TEM profiling method, and (2) determine the resistivity and seismic-velocity relationship in the zones of fractured crystalline rock evinced by TEM profiling. Multicomponent TEM data from terrains with varying degrees of weathering are examined in this paper. The TEM response over a deeply weathered fracture zone in granite is found to be band-limited, with a consistent pattern of vertical voltage response (V_z) at early times enabling accurate location of the wet fracture zone which manifests as a steep conductive feature in 2D DC resistivity imaging. Depending on the thickness of the weathered mantle and the measurement bandwidth, the composite V_z signature consists of three parts: a high amplitude response with a single peak centred on the target fracture zone at very early times; a near invariant or slowly decreasing amplitude response across the fracture zone at some intermediate time; and a marked low-amplitude response at the centre of the fracture zone with higher amplitudes near the flanks (i.e. twin peaks) at later times. The across-strike voltage response (V_x) is diagnostic of conductive fracture zones and exhibits a marked sign reversal. The result of 2D inversion of resistivity, magnetotelluric and seismic-refraction recordings over an intensively fractured granodiorite suggests that resistivity (ρ) and compressional-wave velocity (V_P) can be related in the form $\log_{10} \rho = m \log_{10} V_P + c$, where m and c are constants.

Useful information about the structure or physical property distribution of concealed crystalline rocks can be deduced from non-invasive electrical resistivity and seismic-velocity imaging (e.g. Meju *et al.* 2003; Gallardo & Meju 2003, 2004), but extraneous data are often needed to ascertain that such concealed rocks are significantly fractured. Accurate identification and characterization of zones of fractured rock underneath thick overburden in crystalline terrains have implications for improved natural resource evaluation and fluid transport modelling studies of hard rocks. Electrical and electromagnetic geophysical techniques are widely used for resistivity characterization of crystalline terrains, mostly in connection with geotechnical, hydrogeological and mineral resource investigations (e.g. Palacky *et al.* 1981; Lindqvist 1987; McNeill 1987; Beeson & Jones 1988; Hazell *et al.* 1988; Edet 1990; Frohlich *et al.* 1996;

Meju *et al.* 1999). The DC resistivity and horizontal loop electromagnetic (HLEM) methods are the most popular tools for mapping potential water-bearing fracture zones at shallow depths in crystalline basement terrains. It is beginning to emerge that the time-domain or transient electromagnetic (TEM) method is an attractive alternative portable tool for fracture-zone detection and characterization in the depth range 10–500 m, especially in deeply weathered crystalline terrains (e.g. Peters & de Angelis 1987; Meju *et al.* 2001). The TEM method has the best resolution for subsurface conductivity targets (less than 0.5 ohm m) (e.g. Merrick 1997) and offers a high rate of productivity.

It can be expected that combining conventional resistivity and seismic imaging (e.g. Meju *et al.* 2003; Gallardo & Meju 2003, 2004) with TEM profiling will furnish a more diagnostic tool for characterizing fractured crystalline rocks.

From: HARVEY, P. K., BREWER, T. S., PEZARD, P. A. & PETROV, V. A. (eds) 2005. *Petrophysical Properties of Crystalline Rocks*. Geological Society, London, Special Publications, **240**, 195–206.
0305-8719/05/$15.00   © The Geological Society of London 2005.

First, it is instructive to examine TEM signatures across fractured crystalline rock masses at sites where DC resistivity, HLEM, magnetotelluric (MT) or seismic-refraction data are available for comparison, to ascertain whether a consistent response pattern obtains, and to decipher the TEM criteria for identifying fracture zones underneath thick overburden. Multidimensional inversion of the DC resistivity or MT and seismic data from an area identified as consisting of fractured crystalline rock may then furnish the *in situ* relationship between electrical resistivity and seismic P-wave velocity in the fractured rock mass. This is the main thrust of this paper, and we will focus on the field response characteristics of fracture zones in weathered granitic and dioritic rocks.

The multicomponent TEM method

In the TEM method, electromagnetic energy is applied to the ground by artificial transient pulses. A steady current in a wire loop or long grounded wire (Fig. 1) is suddenly switched off, and the associated magnetic flux falls from its initial steady value to zero. However, during a short interval it is obviously varying with time, and currents will therefore be induced in the conductive earth (e.g. Nabighian & Macnae 1991). Their secondary field decays with time as the currents gradually dissipate on account of the electrical resistance in the conductive earth. This time-dependent field will, in turn, induce a transient electromotive force in a receiver circuit on which it may be acting. The measurable transient response of the ground may be the induced current, voltage (V), or magnetic and electric fields. The transient signal is sampled at discrete (preselected) time intervals

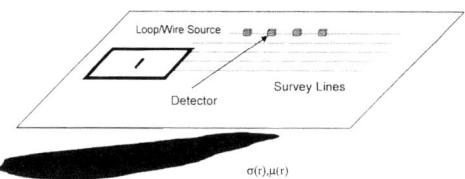

Fig. 1. Generalized layout for multicomponent TEM surveying. The method uses a mobile or fixed transmitter (loop or grounded wire source) and a roving detector (three-component magnetic receiver coil or electric dipole receiver). Survey lines are staked out over prospective conductive targets (e.g. dark body representing a hypothetical zone of fractured rock) in the subsurface, characterized by electrical conductivity $\sigma(r)$ and magnetic permeability $\mu(r)$, which vary with position, r.

called delay times, and the data may be converted to apparent resistivity versus transient time (usually in milliseconds). An important feature of TEM measurements is that extensive signal averaging (stacking of transients) can take place within the few minutes of field measurements, thus reducing the effects of incoherent noise. A single broadband (i.e. multi-spectral) TEM measurement allows the derivation of information from different depths. There are a number of different configurations as regards the position of the receiver (Rx) relative to the transmitter (Tx). In the single loop configuration, the same loop serves as the Tx and Rx. In the central-loop and short offset-loop (or separated-loop) configurations, the receiver is separate and may be a small multi-turn coil or a relatively large wire loop placed at the centre of the Tx loop in the former case or located outside the Tx loop in the latter. The TEM method pulses a localized patch of ground, and accurate measurement of the three spatial components (z, x, y) of the electromagnetic fields induced in the ground allows the reconstruction of the full signature of heterogeneous subsurface conductive targets.

While most standard applications use single-channel (i.e. vertical component) measurements (e.g. Fitterman & Stewart 1986; Buselli *et al.* 1988; Goldman *et al.* 1991, 1994; Meju *et al.* 1999, 2000, 2001), three-component or multi-channel measurements are used in specialized cases (e.g. Fokin 1971; Fouques *et al.* 1986; Smith & Keating 1986; Peters & de Angelis 1987) and provide important extra information as well as constraints in modelling (see Newman *et al.* 1986, 1987). With commercially available portable TEM equipment, it is possible to image fracture zones in crystalline basement rocks underneath several tens of metres of conductive overburden (Meju *et al.* 2001; Meju 2002). The TEM field surveys described in this paper involved single-component and three-component recordings with contiguous small-size (10–50 m-sided) Tx loops using various moving Tx–Rx (dubbed *slingram*) arrangements. For brevity, the use of multiple Tx–Rx configurations and transient recordings in three-co-ordinate directions along a survey line will collectively be called multicomponent profiling in this paper.

Exploration model for fracture zones in weathered crystalline rocks

A working model of a weathered fracture-zone in crystalline rocks is shown in Figure 2. In deeply

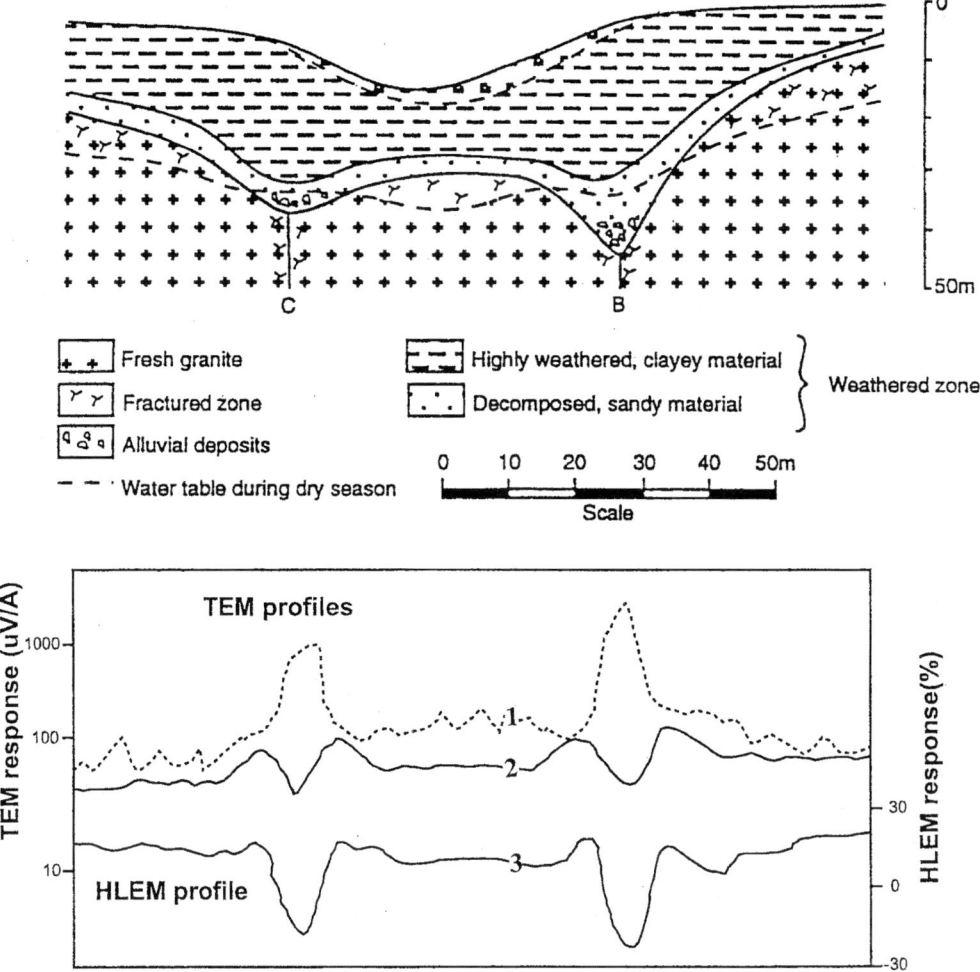

Fig. 2. Exploration model for fracture zones in granite in deeply weathered terrains (after Meju *et al.* 2001). The upper diagram shows the typical geological model of a fractured granitic basement (Jones 1985) and two hypothetical aquiferous zones (B and C) that will be the targets of geophysical exploration. The lower diagram shows the hypothetical vertical-component *slingram* TEM (1,2) and HLEM (3) signatures across these geological targets. The TEM voltage response shows single-peak anomalies at early times (1) and twin-peak anomalies at late times, corresponding respectively to the signatures of the thickened weathered materials and the fracture zones proper.

weathered, intensively fractured, basement complexes (e.g. Palacky *et al.* 1981; Jones 1985), we expect the basal part of the weathering profile to consist of a zone of maximum intergranular porosity and permeability (typically 5 to 10 m thick, and formed by a process of selective mineral decomposition and drainage). This zone, whose position may vary from 15 to 70 m below the surface, could serve as an important aquifer, and may overlie (and serve to recharge) a steep fracture zone which itself may be a few metres

to tens of metres across (Palacky *et al.* 1981; Jones 1985). The weathered layer may contain a water-saturated chloritization zone (or saprolite) and an upper leached zone depending on the type of lithology (Palacky 1987). The saprolite layer is a very important marker zone, identified in many regions of the world as the most electrically conductive layer in a weathered section (e.g. Peric 1981; Palacky 1987). It can be expected that the conductive weathered mantle will be best developed over zones of intense fracturing in a given rock mass (around zones B and C in Fig. 2), which will be more susceptible to weathering than the surrounding relatively solid rock.

A weathered fracture zone in a crystalline basement terrain may thus be characterized by the presence of tabular or elongate preferentially weathered surficial zones (consisting of a leached or mottled upper zone and an underlying highly conductive saprolite zone), and a discordant transition zone of partial weathering above fresh bedrock. For the horizontal component TEM *slingram* profiles over such a structure, we expect the across-strike transient response (V_x) to peak on the flanks of the conductive fracture

zone and be depressed over its centre, such that a prospective target may be defined by a weak response flanked by two large responses of opposite polarity in dual-loop measurements (cf. Newman *et al.* 1987); this is supported by the field observations across steep mineralized and aquiferous prospects by several workers (e.g. Peters & de Angelis 1987; Meju & Fontes 1999), as in the example shown in Figure 3a from measurements across an outcropping fracture-zone in quartz-monzonite in Coed-y-Brenin Forest in northern Wales, which is weathered and possibly mineralized (e.g. Rice & Sharp 1976). For single-loop vertical component measurements across a fracture zone, a single-peak anomaly would be expected over zones of thickened weathered mantle, while twin-peak anomalies are to be expected over the underlying discordant segment, which may contain clay alteration products as known from mining and hydrogeological applications (e.g. Peters & de Angelis 1987; Meju *et al.* 2001), as in the example shown in Figure 3a from Coed-y-Brenin Forest. The along-strike horizontal component (V_y) of the transient response is typically subdued, but may become prominent where there are other off-line conductive fracture zones oriented at oblique angles to the TEM survey line.

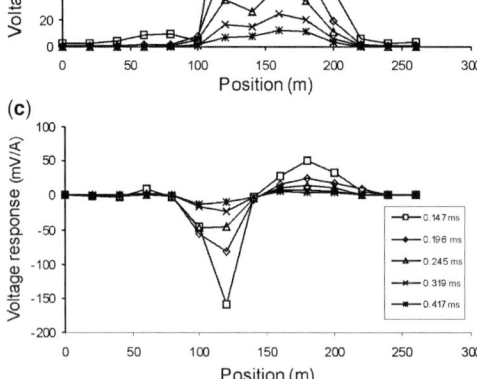

Fig. 3. Multicomponent TEM profiles across an outcropping weathered fracture zone in diorite in Coed-y-Brenin Forest, North Wales. The fracture zone is located at profile position 140 m. (**a**) Single-loop V_z data; (**b**) central-loop V_z data; (**c**) central-loop V_x data.

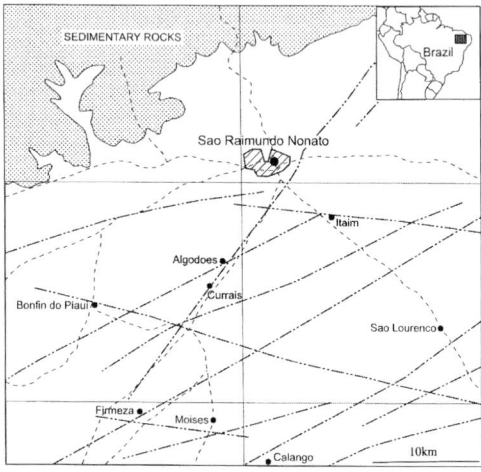

Fig. 4. Map showing the location of the Currais granitic investigation site south of the city of Sao Raimundo Nonato in NE Brazil. Shown are the main linear structural features evinced from aerial photographs and the aeromagnetic map of the surrounding region (Meju *et al.* 2001).

Table 1. *Summary of predicted depth of weathering (DOW) in metres from inversion (Meju 1992) of TEM data, total drilled depth and water yield at some sites of fractured granitic basement rocks south of São Raimundo Nonato*

Site	Predicted DOW	Drilled depth (m)	Water yield (L/h)
Algodoes	55	60	2000
Calango	70	77	380
Currais	64	62	21 000
Firmeza	60	54	2000
Moises	60	64	12 000

Response characteristics of fracture zones in deeply weathered granitic terrains

The combined vertical-component TEM and HLEM methods were used by Meju *et al.* (2001) to accurately locate fracture zones in granitic rocks at 22 sites situated south of the city of Sao Raimundo Nonato in NE Brazil (Fig. 4). Initial analysis of aeromagnetic and aerial photographic data helped to determine the approximate locations of possible lineaments (Fig. 4), while follow-up collocated TEM-HLEM profiling and data inversion

Table 2. *Extracts from driller's lithological logs for the granitic fracture-zone sites at Algodoes, Calango and Firmeza (re-interpreted after Meju et al. 2001)*

Site	Depth interval (m)	Lithological description
Algodoes	0–13	Brown clay
	13–33	Fractured igneous rock with visible quartz grains
	33–60	Fractured granite
Calango	0–6	Clay with sandy top section
	6–16	Altered and broken igneous rock
	16–77	Compact igneous rock (biotite granite)
Firmeza	0–14	Yellowish-brown and red clay (laterite)
	14–16	Compact micaceous igneous rock
	16–54	Fractured grey igneous rock (granite)

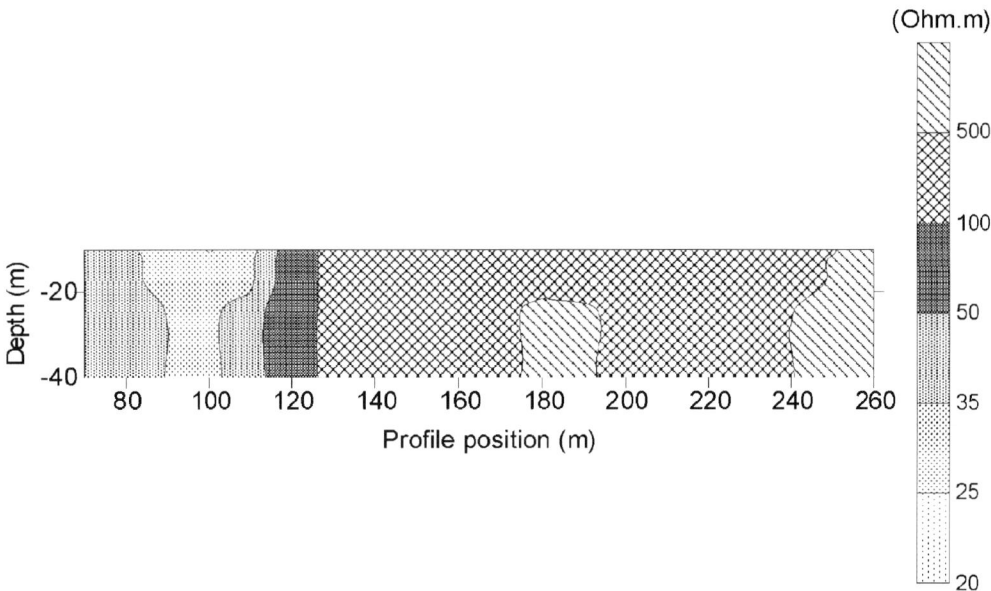

Fig. 5. A 2D resistivity inversion model derived from the Schlumberger resistivity profile across the Currais fracture zone.

pinpointed the targets for drilling. The thickness of the weathered cover ranged from 50 to 75 m at these sites, as shown in Table 1. The anomalous HLEM and TEM responses across the prospective sites were consistent with that shown in Figure 2, and subsequent drilling showed that these sites contain fractured granitic rocks (see sample lithological logs in Table 2).

Detailed multicomponent TEM and HLEM experiments were also conducted across a known fracture zone in the village of Currais (Fig. 4). The Currais site is underlain by strongly foliated granite, and has a thick lateritic cover (Meju & Fontes 1999). Schlumberger DC resistivity profiling, with a constant current electrode separation of 140 m and a station interval of 10 m, helped to locate the fracture-zone; a 62-m-deep borehole found water within fractured granite and gave a yield of 21 000 L/h. The model constructed by inversion of the observed data using a two-dimensional (2D) inverse modelling approach and an initial half-space model of 100 ohm m is presented in Figure 5. The productive borehole is located at a profile position of 100 m. The model lacks vertical resolution due to insufficient depth probing data; the adopted current electrode spacing would effectively probe a depth of about 30 m

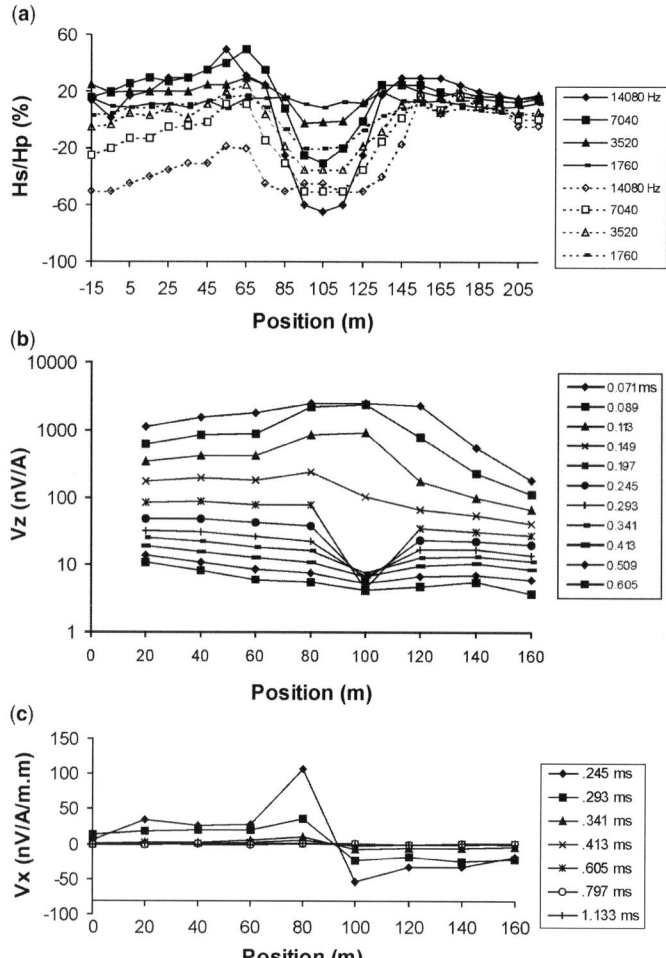

Fig. 6. Collocated HLEM and TEM profiles across the Currais fracture zone (Meju & Fontes 1999). The centre of the fracture zone is at position 100 m on this transect. (**a**) Multi-frequency HLEM profiles. The in-phase component (solid lines) and quadrature component (dashed lines) are shown. (**b**) Single-loop TEM vertical voltage response (V_z) profiles for different delay times. The data are presented in log scale to emphasize the time evolution of the anomalous voltage response. (**c**) Central-loop across-strike voltage response (V_x) profiles.

(based on Meju 1995) and, as such, only the resistivity distribution in the 25–35 m depth range in this model may be justified by the field data. The model suggests the presence of a steep conductive (27–33 ohm m) zone at profile positions 100 to 110 m, coincident with the known zone of intensively fractured aquiferous rock.

Sample conventional multi-frequency HLEM profiles across the Currais fracture zone are presented in Figure 6. The HLEM data (Figure 6a) suggest the presence of a steeply dipping relatively conductive zone at the borehole site. The anomaly is well defined at all the frequencies shown (1760 to 14 080 Hz) in both the quadrature and in-phase profiles. The single-loop induced-voltage (V_z) profiles across the site (Meju & Fontes 1999) are presented in Figure 6b for eleven logarithmically equispaced delay times (0.059 to 0.605 ms). Note the anomalous zone stretching from profile position 80 to 120 m in this figure. There is a single-peak anomaly at early delay times, suggesting the presence of thickened conductive overburden across the fracture zone. At later times the fracture-zone position is characterized by a low-amplitude response with flanking zones of high-amplitude response; the prominent twin-peak anomaly is indicative of the presence of an underlying steep conductive feature at the fracture-zone location. The twin-peak anomaly is asymmetrical, suggesting that it is a dipping fracture zone, the larger peak response being on the down-dip side (cf. Weidelt 1983).

The across-strike horizontal component (V_x) profile data from central-loop profiling (Fig. 6c) exhibit maximum values near the position of the fracture-zone, and show consistent sign reversals possibly over the middle of the fracture-zone in the time bandwidth 0.245 to 0.5 ms. The along-strike horizontal component (V_y) response is of very small magnitude across the known position of the aquiferous fracture zone, and may not be used independently for target identification on a single survey line. Consistent TEM signatures were also obtained across fracture zones in recent successful exploration campaigns in carbonate, metamorphic and volcanic terrains (Meju 2002). It is thus clear that the TEM method is a potent tool for identifying fracture zones in crystalline rocks.

Resistivity–velocity relationship in fractured granodiorite bedrock

Coincident TEM, DC resistivity, MT and seismic surveys were conducted over part of the Mountsorrel granodiorite (MG) that forms the bedrock in Quorn and surrounding areas in England (Fig. 7). This body was unroofed, deeply weathered and then eroded (resulting in a highly irregular surface with deep gullies or wadis) during Permo-Triassic times, and was subsequently overlain by the Mercian Mudstone (MM) deposits. MG outcrops in the southern margin of the site, and is believed to descend

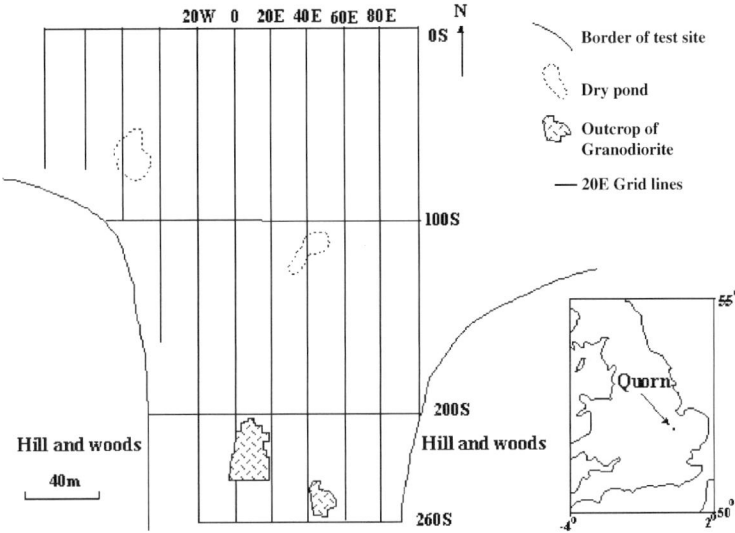

Fig. 7. Map showing the geophysical survey grid at Quorn (Meju *et al.* 2003). The lines run north–south and are spaced 20 m apart. Inset shows the location of Quorn in England.

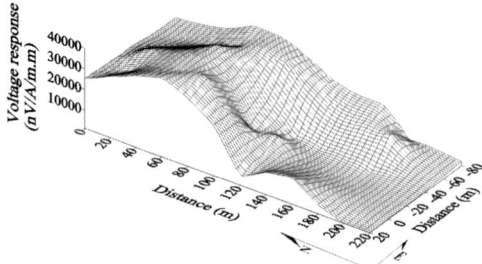

Fig. 8. 3D map of TEM voltage response at 0.016 ms for the geophysical survey grid at Quorn. Note the depressed response at position 120 m on line 20W.

c.400 m south of the survey grid shown in Fig. 7). The geophysical data were recorded along survey lines extending from the subcrop area in the north to the outcrop region in the south (see the north–south lines in Fig. 7). Contiguous 20-m-sided loops were used for the TEM profiling of the entire survey grid. The TEM response map of the survey grid for one delay time is presented in Figure 8. The zone of depressed (trough-like) responses on this map possibly corresponds to the position of a concealed fracture zone or buried wadi (Meju *et al.* 2003).

Line 20W was selected for detailed analysis, since it has a clear twin-peak anomaly at position 120 m in the TEM response profiles, where a shallow fault or fracture zone might be present (this could also be the feather-edge of the Mercia mudstone–granodiorite contact). The absence of a single-peak early-time signature on this profile indicates the thin nature of

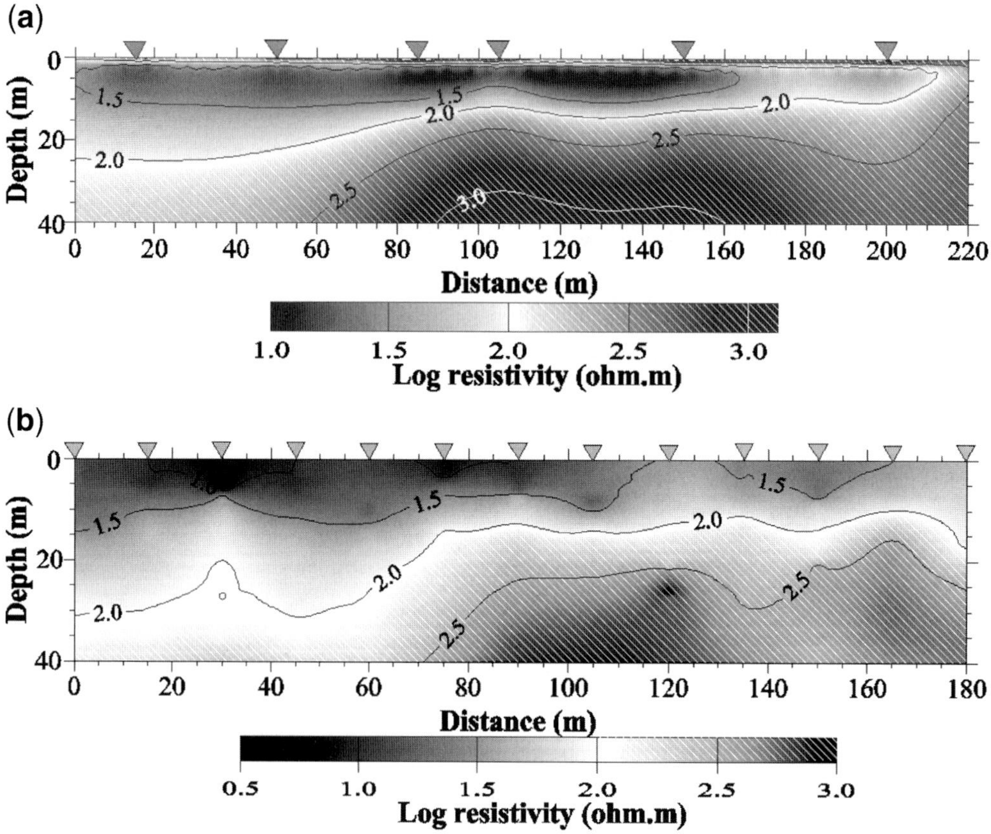

by a series of major steps to the north under sedimentary cover (Meju *et al.* 2003). It is heavily fractured at outcrop and presumably at depth (based on field observations at the largest hardrock quarry in Western Europe, located

(a)

(b)

Fig. 9. 2D resistivity model for line 20W (Meju *et al.* 2003). (**a**) DC resistivity model derived by 2D inversion of field data from six sounding locations (shown by the inverted triangles). (**b**) AMT resistivity model. The 13 sounding positions (15 m apart) are indicated at the top.

the overburden. DC resistivity audiofrequency magnetotelluric (AMT), and seismic-refraction data were recorded on line 20W. The resistivity models furnished by separate 2D inversion of the DC resistivity and AMT data for line 20W (Meju *et al.* 2003) are shown in Figure 9. These models are in excellent agreement and reveal the presence of a steep boundary at profile position 120 m (i.e. 120S in Fig. 7) coincident with the zone of anomalous TEM response. This is interpreted as a fault or fracture zone. The seismic P-wave velocity model from 2D inversion (Fig. 10) shows subsurface structural features similar to the resistivity models, suggesting that there may be a geological basis for correlating these models, The configuration of the boundary between the fracture granodiorite bedrock and its cover materials can be discerned in these models, and is taken to be approximately marked by the 100 ohm m and 3000 m s^{-1} contours.

For the Quorn study site, Meju *et al.* (2003) noted that the resistivity (ρ in ohm m) and P-wave velocity (V_{P} in m s^{-1}) distributions (sampled at coincident grid positions or pixels in the 2D models) are related in the form (see Fig. 11):

$$\log_{10} \rho = m \log_{10} V_{\mathrm{P}} + c \qquad (1)$$

where the constants m and c respectively have values of 3.88 and -11 for the consolidated rocks (>3 m deep) consisting of fractured granodiorite at this site (see trend B in Fig. 11). An inverse relation was found to hold for the unconsolidated soil/drift deposits (i.e. top 3 m), for which $m = -3.88$ and $c = 13$ (see trend A in Fig. 11). It may be noted that Rudman *et al.* (1975, eq. 10) interrelated ρ_{a} and velocity logs from deep wells, using an equation derived assuming ρ_{a} and V_{P} to be functions of porosity. It was suggested that Rudman's equation simplifies to $\log_{10} \rho_{\mathrm{a}} = (m \log_{10} V_{\mathrm{P}} - m \log_{10} B)$, where m and B are empirical constants, and is thus identical to equation (1) determined empirically for the consolidated rocks at Quorn. This

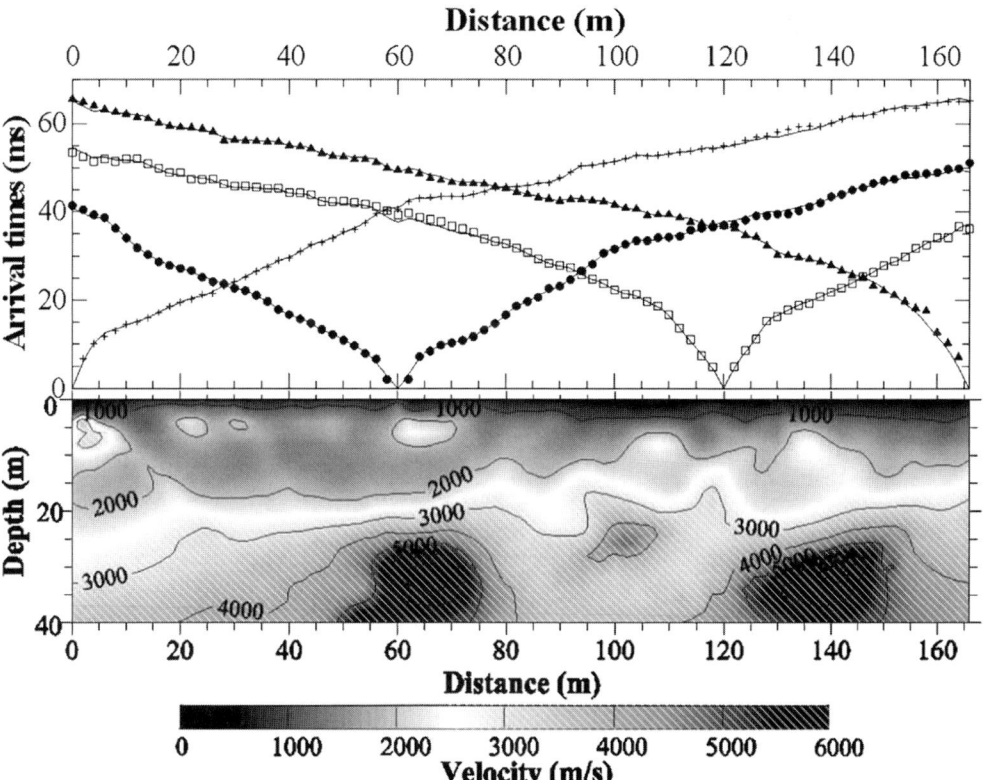

Fig. 10. 2D velocity model for line 20W (Meju *et al.* 2003). The model is shown in the bottom diagram. The fit to the field recordings for different shot points (differentiated by symbols) is shown in the top plot.

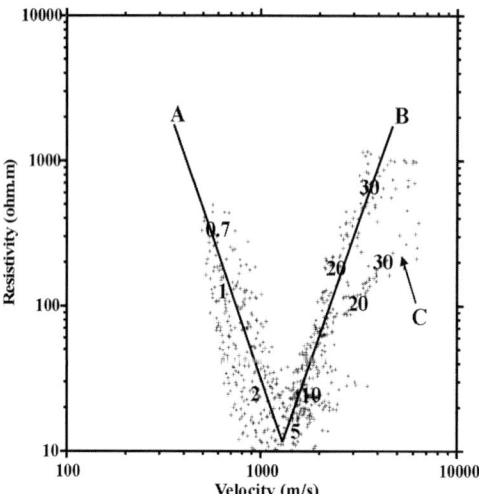

Fig. 11. Relationship between logarithmic resistivity and seismic P-wave velocity on line 20W (Meju *et al.* 2003). The depth of sampling (in metres) is shown for selected points (pixels). Note the identified trends A and B of inverse slope. Trend B was constrained to pass through well-estimated points, thus giving less emphasis to contributions (e.g. zone C) from unresolved deep features in the seismic model.

would suggest that porosity is also a connecting factor for resistivity and velocity in the fractured granodiorite.

Conclusions

An attempt has been made to gauge the usefulness of broadband single-component and three-component TEM *slingram* measurements for characterizing fractured crystalline rocks. In the main, it was demonstrated that portable single-loop and central-loop TEM techniques can identify prospective fracture zones in deeply weathered hard rocks. The multicomponent TEM data presented here seem to suggest that the preferentially weathered zone of high conductance can give rise to anomalous vertical component TEM signatures that are detectable at very early measurement times. At later times, the V_z responses generally appear to exhibit high amplitudes near the flanks of the fracture zone, with a 'subtle low' or rapid amplitude decrease across the middle of the prospective fracture zone, but the latter feature may be overprinted in fracture zones of small depth extent. The results of the present field studies support what is known previously from numerical and physical-scale EM modelling studies (e.g. Newman *et al.* 1987; Wilt & Williams 1989)

i.e. that the horizontal components of the TEM response can pinpoint the location of anomalous conductivity or geological contacts much more accurately than the vertical component. The central-loop across-strike (V_x) data from granitic terrain show peaked transient responses of opposite polarity on either flank, and depressed responses over the central part of the various conductive fracture zone targets.

The integrated geophysical study at Quorn has shown that 2D imaging of collocated high-resolution seismic and geo-electromagnetic surveys can map the complex topography of the Mountsorrel granodiorite bedrock. The structure of the overlying sedimentary materials was also resolved by the 2D imaging surveys. A remarkable correlation was achieved between resistivity and P-wave velocity models after a co-operative inversion of seismic data incorporating the information furnished by resistivity models on lateral variations in the subsurface. A useful relationship is found to exist between the electrical resistivity and seismic-compressional-wave velocity of the fractured granodiorite.

Overall, it has been shown that the TEM method can be combined with resistivity and seismic imaging to locate zones of fractured rock underneath a thick weathered mantle in crystalline terrains. However, there is an increasing need for accurate prediction of the presence of fluids in these potential subsurface reservoirs, as well as the potential yield before drilling. These are important problems for future research, and require 3D (grid-type) surveying and realistic three-dimensional data inversion.

The author wishes to thank S. L. Fontes, P. J. Fenning, D. Groom, L. Gallardo and E. Ulugergerli for their field and technical support.

References

BEESON, S. & JONES, C. R. C. 1988. The combined EMT/VES geophysical method for siting boreholes. *Ground Water*, **26**, 54–63.

BUSELLI, G., BARBER, C. & ZERILLI, A. 1988. The mapping of groundwater contamination with TEM and DC methods. *Exploration Geophysics*, **19**, 240–248.

EDET, A. E. 1990. Application of photogeologic and electromagnetic techniques to groundwater exploration in northwestern Nigeria. *Journal of African Earth Sciences*, **11**, 321–328.

FITTERMAN, D. V. & STEWART, M. T. 1986. Transient electromagnetic sounding for groundwater. *Geophysics*, **51**, 995–1005.

FOKIN, A. F. 1971. *Method Perekhodnikh Protsessov Pri Poickakh Mestorozdeniy Sulfidnikh Rud*. Nedra, Leningrad.

FOUQUES, J. P., FOWLER, M., KNIPPING, H. D. & SCHIMANN, K. 1986. The Cigar Lake uranium deposit: discovery and general characteristics. *Canadian Institute of Mining and Metallurgy Bulletin*, **79** (886), 70–86.

FROHLICH, R. K., FISHER, J. J. & SUMMERLY, E. 1996. Electric–hydraulic conductivity correlation in fractured crystalline bedrock: Central Landfill, Rhode Island, USA. *Journal of Applied Geophysics*, **35**, 249–259.

GALLARDO, L. A. & MEJU, M. A. 2003. Characterisation of heterogeneous near-surface materials by joint 2D inversion of dc resistivity and seismic data. *Geophysical Research Letters*, **30**, 1658–1661.

GALLARDO, L. A. & MEJU, M. A. 2004. Joint 2D inversion of dc resistivity and seismic data using cross-gradients constraints. *Journal of Geophysical Research*, **19** (B3), B03311.

GOLDMAN, M., DU PLOOY, D. & ECKARD, M. 1994. On reducing ambiguity in the interpretation of transient electromagnetic sounding data. *Geophysical Prospecting*, **42**, 3–25.

GOLDMAN, M., GILAD, D., RONEN, A. & MELLOUL, A. 1991. Mapping of seawater intrusion into the coastal aquifer of Isreal by the transient electromagnetic method. *Geoexploration*, **28**, 153–174.

HAZELL, J. R. T., CRATCHLEY, C. R. & PRESTON, A. M. 1988. The location of aquifers in crystalline rocks and alluvium in northern Nigeria using combined electromagnetic and resistivity techniques. *Quarterly Journal of Engineering Geology*, **21**, 159–175.

JONES, M. 1985. The weathered zone aquifers of the Basement Complex areas of Africa. *Quarterly Journal of Engineering Geology*, **18**, 35–46.

LINDQVIST, J. G. 1987. Use of electromagnetic techniques for groundwater exploration in Africa. *Geophysics*, **52**, 456–458.

MACNAE, J. C. 1984. Survey design for multicomponent electromagnetic systems. *Geophysics*, **49**, 265–273.

McNEILL, J. D. 1987. Advances in electromagnetic methods for groundwater studies. Proceedings of Exploration '87: *Applications of Geophysics and Geochemistry*, Geological Survey of Canada Special Publications, 678–702.

MEJU, M. A. 1992. An effective ridge regression procedure for resistivity data inversion. *Computers & Geosciences*, **18**, 99–118.

MEJU, M. A. 1995. Simple resistivity-depth transformations for infield or real-time data processing. *Computers & Geosciences*, **21**, 985–992.

MEJU, M. A. 2002. Geoelectromagnetic exploration for natural resources: models, case studies and challenges. *Surveys in Geophysics*, **23**, 133–205.

MEJU, M. A. & FONTES, S. L. F. 1999. Three-component TEM method for improved location of fracture-zone aquifers in deeply-weathered basement terrains: first field test in NE Brazil. *6th International Congress of the Brazilian Geophysical Society, Expanded Abstracts, Vol. 1. August 1999, Rio de Janeiro*.

MEJU, M. A., FENNING, P. J. & HAWKINS, T. R. W. 2000. Evaluation of small-loop transient electromagnetic soundings to locate the Sherwood Sandstone aquifer and confining formations at well sites in the Vale of York, England. *Journal of Applied Geophysics*, **44**, 217–236.

MEJU, M. A., FONTES, S. L., OLIVEIRA, M. F. B., LIMA, J. P. R., ULUGERGERLI, E. U. & CARRASQUILLA, A. A. 1999. Regional aquifer mapping using combined VES–TEM–AMT/EMAP methods in the semiarid eastern margin of Parnaiba Basin, Brazil. *Geophysics*, **64**, 337–356.

MEJU, M. A., FONTES, S. L., ULUGERGERLI, E. U., LA TERRA, E. F., GERMANO, C. R. & CARVALHO, R. M. 2001. A joint TEM-HLEM geophysical approach to borehole siting in deeply-weathered granitic terrains. *Ground Water*, **39**, 554–567.

MEJU, M. A., GALLARDO, L. A. & MOHAMED, A. K. 2003. Evidence for correlation of electrical resistivity and seismic velocity in heterogeneous near-surface materials. *Geophysical Research Letters*, **30**, 1373–1376.

MERRICK, N. P. 1997. An experiment in geophysical monitoring of a contaminated site. *In*: CHILTON, J. (ed.), *Groundwater in the Urban Environment: Problems, Processes and Management*, A. A. Balkema, Rotterdam, 487–490.

NABIGHIAN, M. N. & MACNAE, J. C. 1991. Time-domain electromagnetic prospecting methods. *In*: NABIGHIAN, M. N. (ed.) *Electromagnetic Methods in Applied Geophysics*, vol. 2A, Society of Exploration Geophysicists, Tulsa, OK. 427–520.

NEWMAN, G. A., ANDERSON, W. L. & HOHMANN, G. W. 1987. Interpretation of transient electromagnetic soundings over three-dimensional structures for the central-loop configuration. *Geophysical Journal of the Royal Astronomical Society*, **89**, 889–914.

NEWMAN, G. A., HOHMANN, G. W. & ANDERSON, W. L. 1986. Transient electromagnetic response of a three-dimensional body in a layered-earth. *Geophysics*, **51**, 1608–1627.

PALACKY, G. J. 1987. Resistivity characteristics of geological targets. *In*: NABIGHIAN, M. N. (ed.) *Electromagnetic Methods in Applied Geophysics–Theory*, vol. 1, Society of Exploration Geophysicists, Investigations in Geophysics Series, no. 3, 53–129.

PALACKY, G. J., RITSEMA, I. L. & DE JONG, S. J. 1981. Electromagnetic prospecting for groundwater in Precambrian terrains in the Republic of Upper Volta. *Geophysical Prospecting*, **29**, 932–955.

PERIC, M. 1981. Exploration of Burundi nickeliferous laterites by electrical methods. *Geophysical Prospecting*, **29**, 274–287.

PETERS, W.S. & DE ANGELIS, M. 1987. The Radio Hill Ni–Cu massive sulphide deposit: a geophysical case history. *Exploration Geophysics*, **18**, 160–166.

RICE, R. & SHARP, G. J. 1976. Copper mineralization in the forest of Coed-y-Brennin, North Wales.

Transactions of Institution of Mining and Metallurgy, **76** (B), 1–13.

RUDMAN, A. J., WHALEY, J. F., BLAKE, R. F. & BIGGS, M. E. 1975, Transformation of resistivity to pseudovelocity logs. *AAPG Bulletin*, **59**, 1151–1165.

SMITH, R. S. & KEATING, P. B. 1996. The usefulness of multicomponent, time-domain airborne electromagnetic measurements. *Geophysics*, **61**, 74–81.

THOMAS, L. & LEE, T. 1988. Time-domain electromagnetic responses of a polarizable Target. *Exploration Geophysics*, **19**, 365–367.

WEIDELT, P. 1983. The harmonic and transient electromagnetic response of a thin dike. *Geophysics*, **48**, 934–952.

WILT, M. J. & WILLIAMS, J. P. 1989. Layered model inversion of central-loop EM soundings near a geological contact. *Exploration Geophysics*, **20**, 71–73.

On the neutron absorption properties of basic and ultrabasic rocks: the significance of minor and trace elements

P. K. HARVEY & T. S. BREWER

Department of Geology, University of Leicester, Leicester, LE1 7RH, UK
(e-mail: pkh@leicester.ac.uk)

Abstract: The neutron absorption macroscopic cross-section, Σ, is measured routinely by neutron porosity tools and, although rarely presented as a logging curve in its own right, is used indirectly for the estimation of (neutron) porosity. One of the reasons that this primary measurement is not often employed directly in petrophysical analysis is the difficulty of interpretation. In particular, little is known about the range of Σ values for common lithologies, or exactly what information the measurement is providing.

In this contribution we demonstrate that excellent estimates of Σ can be calculated, provided that the chemistry of a sample is known in sufficient detail. When applied to a range of geochemical reference materials, it becomes apparent that the minor and trace elements present may have a profound effect on the Σ value of a sample, and, in turn, on the interpretation of neutron porosity measurements. Using this approach we present Σ data for basaltic and ultrabasic rocks, and model the change in Σ with alteration.

Alteration is considered in these models as an increase in alteration minerals (which are mainly clays, but also carbonates and zeolites in basic rock alteration) and changes in the trace-element chemistry of the rocks. Of the trace elements, boron and some of the rare-earth elements are of particular importance. Modelling the variation in Σ with these mineralogical and compositional changes indicates that increases in boron are the most important of these factors in increasing Σ; this is enhanced by the alteration, particularly to clay phases, which generally accompanies an increase in boron.

These models suggest that a Σ log should be able to act as a proxy for alteration trends in basic and ultrabasic crystalline rocks, and a quantitative model for such alteration is described.

The nuclear absorption macroscopic cross-section of a sample (Σ) can be defined as:

$$\Sigma = \sigma_m \rho$$

where ρ = sample density (g cm^{-3}), and σ_m = mass-normalized (thermal absorption) cross-section.

Expressed this way, Σ is in units of cm^{-1}, but for logging purposes is quoted in capture units (cu) = 1000 times Σ, as is σ, for the purposes of this contribution.

The σ_m is a function of the bulk composition of the sample and any included fluids. Assuming that the fluids fill the available pore space, then Σ can be related to porosity as:

$$\Sigma = (1 - \phi)\Sigma_m + \phi\Sigma_f$$

where Σ_m is the nuclear cross-section of the matrix, and Σ_f is the nuclear cross-section of the interstitial fluids.

The macroscopic cross-section, Σ, is measured routinely by neutron porosity tools and, although rarely presented as a logging curve in its own right, is used indirectly for the estimation of (neutron) porosity.

In principle, σ_m can be measured directly in a nuclear reactor, a very precise and expensive procedure. It can, however, also be calculated from a full geochemical analysis, and σ_f can likewise be estimated for a known or assumed fluid composition. In much of the work described later, we have used samples of the oceanic crust obtained by the Ocean Drilling Program; the drilling fluid in this instance was sea-water. If σ_m and σ_f can be calculated, and appropriate densities (ρ_m and ρ_f, respectively) assumed, then Σ, for modelling purposes, can be calculated and compared with the corresponding borehole measurements. For sea-water, whose composition is known in some detail: $\sigma_f = 32.34$ cu, and $\rho_f = 1.017$ g cm^{-3}, giving $\Sigma_f = 32.87$ cu.

From: HARVEY, P. K., BREWER, T. S., PEZARD, P. A. & PETROV, V. A. (eds) 2005. *Petrophysical Properties of Crystalline Rocks*. Geological Society, London, Special Publications, **240**, 207–217.
0305-8719/05/$15.00 © The Geological Society of London 2005.

The calculation of σ_m is given simply by the summation:

$$\sigma_m = \Sigma p_i \sigma_{mi}$$

where p_i is the proportion of the ith element in the sample and σ_{mi} is the mass-normalized (thermal absorption) cross-section for that element.

We have validated this relationship elsewhere (Harvey *et al.* 1996) by comparing direct measurements of σ_m made in the Nessus facility at the Atomic Energy facility at Winfrith in Dorset, UK, with calculated values for a number of geochemical reference standards. For geochemical purposes, the important factor is the magnitude of σ_{mi} for the major elements and those minor and trace elements which have high mass-normalized (thermal absorption) cross-sections. The values of these coefficients are summarized in Figure 1; these values are tabulated, and with appropriate references, in Harvey *et al.* (1996).

From the geochemical viewpoint, the important features about Figure 1 are the importance of several of the rare-earth elements, in particular gadolinium, which has the highest mass-normalized cross-section of any element, and the minor elements boron and lithium. Other elements with high mass-normalized cross-sections include Cd and In, but these elements typically occur in such low levels as to be insignificant in their contribution to σ_m. Of the elements treated in Figure 1 as 'major elements', chlorine has the highest mass-normalized cross-section, followed by hydrogen. For this reason, saline liquids, or sea-water in the case of the Ocean Drilling Program, as formation fluids, make up a significant part of the measured Σ, but one which can be estimated if the porosity is known.

Even though Σ is generated as a curve by a number of nuclear porosity tools, it is not often employed directly in petrophysical analysis, as there is some difficulty of interpretation. In particular, little is known about the range of Σ values for common lithologies, or exactly what information the measurement is providing. For instance, Ellis (1987) quotes the problem as 'a common misconception ... that the error in the

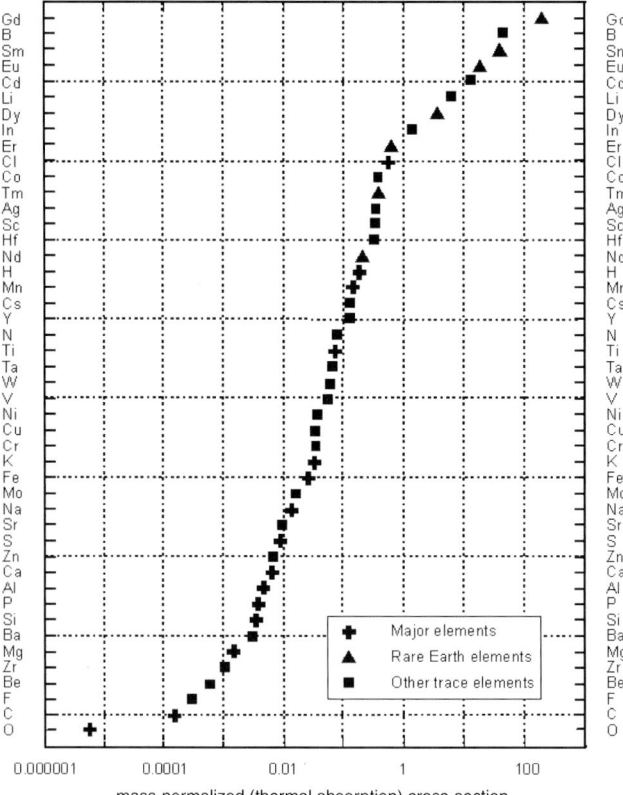

Fig. 1. Variation in cross-section by element. Note the logarithmic scale, and importance, in absolute terms, of B, Li and the REE elements amongst the minor and trace elements.

thermal porosity readings in shales is caused by associated trace elements with large thermal capture cross sections' (p. 254), and, 'the large apparent porosity values are due primarily to the hydrogen concentration associated with the shale matrix.' Broglia & Ellis (1990) reach the same conclusion in an analysis of thermal neutron porosities in gabbros and basalts from Deep Sea Drilling Project and Ocean Drilling Program sites. In their paper they recognize that some minor elements (Cl, Gd, B, Li and Sm are noted) can affect the thermal neutron absorption, but still regard hydrogen in bound water in the pore space and in the alteration phases as the primary source of error in thermal neutron porosities.

The purpose of this contribution is to evaluate these factors, and their potential geological usage, in basic and ultrabasic rocks. Such rocks occur widely, but in particular are the predominant lithologies which make up the oceanic crust.

Mass-normalized cross-section in basic and ultrabasic igneous rocks

To calculate a range of Σ estimates for any rock type it is necessary to know the full composition of the sample, at least with respect to the elements in Figure 1 which occur in other than a minute concentration. There are very few published samples for which there is a sufficiently complete chemical analysis, including the important cross-section elements. In particular, elements like boron and lithium are rarely measured, although other element groups like the rare earths are frequently analysed. Even within the latter, Gd, with by far the highest mass-normalized cross-section, is often not determined. While few data are available, one source is the geochemical reference standards (GRSs); such standards are analysed for comparative purposes for a wide range of elements. Despite the available range of such standards, few have been sufficiently well analysed to reliably calculate Σ. The main problem is the lack of data on the abundance of boron, although gadolinium and lithium are also rarely determined even in these important reference materials.

Mass-normalized cross-section in Geochemical Reference Standards

Of the available Geochemical Reference Standards of basic or ultrabasic composition, 25 have been identified as having sufficiently complete chemical analyses for the reliable estimation of their mass-normalized cross-sections. The main problem with other potentially suitable reference materials is generally the absence of boron abundance data. The data used here are

from Govindaraju (1994), Croudace & Randle (1993), Hallett & Kyle (1993) and Terashima (1993). Of the 25 standards, 15 are of basic composition (12 basalts, three gabbros) and six are ultrabasic rocks, one being a serpentinite.

The cross-section data for these samples are summarized in Table 1 & Figure 2, with the variation in mass-normalized cross-section for basic rocks, shown above, ultrabasic rocks, below, and the one serpentinite, UB-N, is shown at the bottom. In the right-hand panel of Table 1 the main element groups are shown as percentages of their contribution to the mass-normalized cross-section. The groups shown are the 'major elements', the rare-earth elements ('REE elements'), B ('boron') and 'all other' (remaining trace) 'elements'. Major elements here include: O, Si, Al, Ti, Fe, Mg, Mn, Ca, Na, K, P, H, C and Cl.

Over the 15 basalts and gabbros, the mean value for σ_m is 6.95 cu with a standard deviation of 1.13 cu. Table 1 shows that while the major elements make the greatest contribution to σ_m, around 80%, the rare-earth elements are also important, while the contribution of boron is very variable. The latter is an expression of the range of tectonic settings represented by the chance choice of the samples as reference materials. The contribution of all other elements is notable fairly constant and averaging around 3%.

For the five essentially 'fresh' ultrabasic rocks, the mean value for σ_m is 4.49 cu with a standard deviation of 0.92 cu, which is lower than that for the basalts and gabbros. While these samples are barely representative, the overall pattern of elemental contributions to σ_m is predictable: the contribution made by major elements is some 90%, while that of the rare-earth elements and boron is very small (together less than 2%). What is noticeable is the quite different cross-section response for the serpentinite, UB-N. For this reference material, the computed value of σ_m (9.92 cu) is over twice that of the (average) unaltered ultrabasic rock, with only 29.9% of the value being due to the major elements. The high value is due mainly to bound water (6.8%) following serpentinization, and a high boron concentration (147 ppm), which accounts for almost two-thirds (63.3%) of the σ_m value.

Mass-normalized cross-section in samples of the oceanic basement

For comparison with the geochemical reference standards, direct measurements of σ_m were made on 12 basalts (10 from Ocean Drilling Program Leg 148; two from Leg 149), and three serpentinized peridotites (from Leg 149). These measurements were described by Brewer

Table 1. *Distribution of the major components (major elements, REE elements, boron and all other elements) that make up the mass-normalized cross-section in basic and ultrabasic geochemical reference standards*

	σ_m (cu)	Percentage of the mass-normalized (thermal absorption) cross-section due to:			
		Major elements	REE elements	Boron	All other elements
Basic rocks					
JB-1	7.04	74.4	16.5	5.7	3.5
JB-1a	6.88	75.5	16.2	4.9	3.4
JB-2	8.07	71.3	9.1	16.0	3.6
JB-3	7.66	72.4	14.1	10.1	3.4
BCR-1	7.98	79.3	19.5	0.0	1.2
BHVO-1	7.72	78.0	18.9	1.4	1.8
BIR-1	5.53	90.0	7.4	0.3	2.3
BR	7.48	85.0	6.4	4.6	4.1
BE-N	7.09	89.5	6.7	0.0	3.8
DNC-1	5.12	88.3	8.9	0.8	2.1
W-2	5.86	86.7	2.2	8.6	2.5
W-1	8.50	60.3	10.5	27.2	2.1
MRG-1	7.65	94.0	2.6	0.8	2.7
JGb-2	4.76	90.0	2.6	4.4	3.0
JGb-1	6.92	89.5	5.4	2.5	2.6
Mean	6.95	81.6	9.8	5.8	2.8
Std. dev.	1.13	9.5	5.9	7.4	0.8
Ultrabasic rocks					
NIM-P	5.22	80.8	0.0	0.0	19.2
PCC-1	3.25	90.0	0.0	0.0	10.0
NIM-D	5.45	86.5	0.0	3.9	9.6
NIM-N	4.67	93.8	0.9	2.0	3.3
JP-1	3.87	90.2	0.1	1.5	8.2
Mean	4.49	88.3	0.2	1.5	10.0
Std. dev.	0.92	4.9	0.4	1.6	5.8
Serpentinite					
UB-N	9.92	29.9	0.0	63.3	6.8

et al. (1996) and Lofts *et al.* (1996), and are summarized in Figure 3.

The basalts are all of the mid-ocean ridge type (MORB), and, with the exception of one of the Leg 149 samples, show a near-constant value for σ_m of 6.00 cu (average) with a standard deviation of 0.25 cu. The one Leg 149 basalt with a higher σ_m value (7.76 cu) is described as 'altered'. The three serpentinized peridotites from Leg 149 (average σ_m: 12.2 cu) can only be compared with the GRS serpentinite, UB-N, which has a very similar mass-normalized cross-section.

While the number of reliable estimates or direct measurements of σ_m are small for basic and ultrabasic rocks there is a consistency that in summary indicates that:

- for fresh basic rocks the range of σ_m is relatively small; this is particularly true for mid-ocean ridge basalts;

- increases in σ_m can be expected to result mainly from increases in boron, together with bound water, as a result of alteration processes;
- this effect is taken to the limit with the serpentinization of peridotites where elements involved in alteration dominate the cross-section;
- the cross-section (macroscopic or mass normalized) is hence a potential proxy for alteration.

A model for basalt alteration

From the preceding discussion, the measured macroscopic cross-section in basalts can partitioned into:

- a contribution from 'fresh' MOR basalt;
- a contribution due to mineralogical changes accompanying alteration;

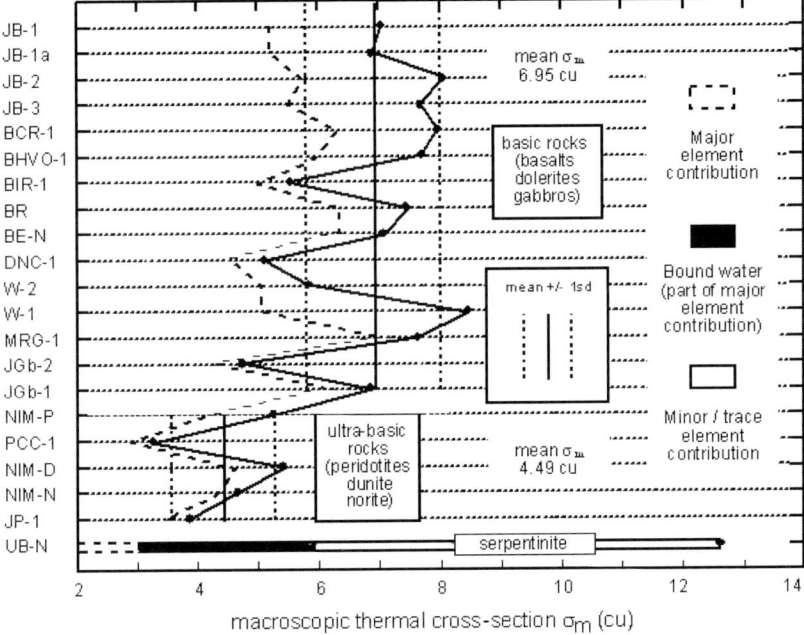

Fig. 2. Variation in the mass-normalized cross-section in basic and ultrabasic geochemical reference standards. For the identification and origin of the individual GRSs see Govindaraju (1994).

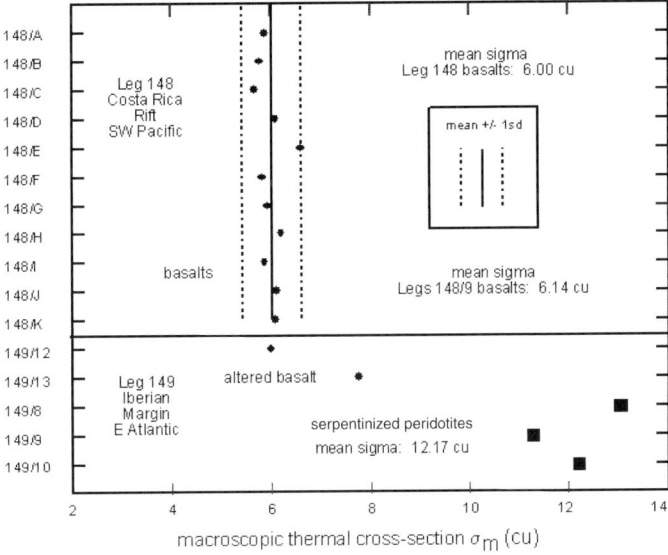

Fig. 3. Variation in the mass-normalized cross-section of mid-ocean ridge basalts and serpentinized peridotites collected during Legs 148 and 149 of the Ocean Drilling Program.

- a contribution due to key elements, like boron or some rare-earth elements, which have significant mass-normalized thermal cross-sections, which are absorbed on to, rather than being part of the lattice of, alteration phases, and
- the contribution due to porosity and the associated pore-filling fluids.

Ignoring the effect of porosity in the first instance and considering the altered basalt to consist of a component of 'fresh' basalt together with a component of alteration phases, then measurements of the macroscopic cross-section in basalts may be described by the relationship:

$$\Sigma_F = w_b \rho_b \sigma_b + (1 - w_b) \rho_a \sigma_a$$

where Σ_F is the macroscopic nuclear cross-section of the formation; σ_b is the mass-normalized cross-section of (fresh) MOR basalt; ρ_b is the density of (fresh) MOR basalt; σ_a is the mass-normalized cross-section of alteration products; ρ_a is the density of the alteration products; w_b is the weight fraction of 'fresh' basalt in the rock fraction.

The Σ_F is available as a log measurement; close estimates of σ_b were described in the previous section, ρ_b and ρ_a can be measured directly or estimated from core data, and modelling (discussed below) enables estimates of σ_a to be obtained. Hence, for an altered basalt with 'zero' porosity, rearrangement of the equation above should enable a estimate of the proportion of 'fresh' (or altered) basalt to be obtained as an 'Index of Alteration':

$$w_b = \frac{\Sigma_F - \rho_a \sigma_a}{\rho_b \sigma_b - \rho_a \sigma_a}$$

If porosity, and associated pore fluids, are included in the model, we get:

$$\Sigma_F = (1 - \phi) w_b \rho_b \sigma_b + (1 - \phi)(1 - w_b)\rho_a \sigma_a$$
$$+ \phi \rho_f \sigma_f$$

where σ_f is the mass-normalized cross-section of the pore fluid, ρ_f is the density of the pore fluid, and ϕ is the formation porosity.

In an attempt to model changes in the macroscopic cross-section with alteration in the following sections, it is assumed that the pore fluid is sea-water (where σ_f and ρ_f are known or easily calculated), and that the petrophysical parameters for 'fresh' basalt and its alteration products are known. Again, rearrangement of the equation above enables an estimate of w_b to be obtained:

$$w_b = \frac{((\Sigma_F - \phi \rho_f \sigma_f)/(1 - \phi)) - \rho_a \sigma_a}{(\rho_b \sigma_b - \rho_a \sigma_a)}$$

In the following sections, the effects of alteration, boron and rare-earth element abundance, and porosity, are modelled for mid-ocean ridge basalts, as represented by the Leg 148/149 basalts. These models are intended to provide the background and estimates of the petrophysical parameters needed to apply the equations above directly to log measurements in order to obtain the degree of alteration in the basaltic sequence.

Modelling changes due to alteration, composition and porosity in MOR basalts

Change in Σ_F with increasing porosity

Assuming sea-water, at least locally in a borehole, as the interstitial fluid in the formation, then the change in cross-section with porosity can be summarized in Figure 4.

There is an initial increase in the macroscopic cross-section to a porosity of about 65%, after which Σ_F decreases in response to the increasing effect of decreasing density. While porosities of 65% or more may occur in sediments, porosities in oceanic basement rocks would not normally be expected to exceed half that figure. So, at less than 5% porosity an increase of 1% would give an increase of about 0.7 cu in Σ_F, decreasing to

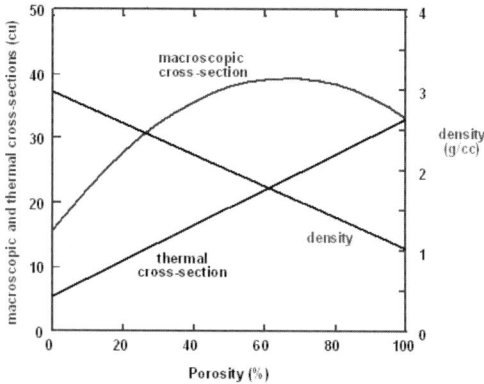

Fig. 4. Variation in the macroscopic and thermal cross-sections of 'fresh' mid-ocean ridge basalt with increasing porosity and with sea-water as the formation fluid.

about 0.4 cu per per cent increase in porosity at 30% porosity.

In the models that follow, porosity is modelled on one (X) axis, and other chemical and mineralogical parameters on the other axis. Replacement and alteration are considered first, followed by the effects of varying boron and rare-earth abundances. Sea-water has been assumed, for modelling purposes, as the interstitial fluid in the formation. This is a reasonable assumption, as sea-water is used as the drilling 'fluid' for most Ocean Drilling Program boreholes, and also penetrates and fills the cracks and fractures in the formation which, in basaltic rocks, provide most of the porosity.

Change in Σ_F with increasing porosity and carbonate replacement

Mineralogically, alteration is mainly to clay and zeolite assemblages, or to carbonates: often one assemblage following the other (Teagle *et al.* 1998). To evaluate the likely changes in macroscopic cross-section, two models are briefly described; the first showing the effect of increasing carbonate alteration of the MOR basalt, and the second involving possible clay and zeolite assemblages. The carbonate here is modelled as pure calcite. Assuming sea-water, at least locally in a borehole, as the interstitial fluid in the formation, then the effect on the cross-section with an increase in calcite (diluting the 'fresh' basalt) and porosity is summarized in Figure 5.

The effect of carbonate replacement is comparatively small, but, in contrast to the addition of the trace elements discussed, the effect of increasing calcite is to decrease Σ_F. The area to the left of the dotted line in Figure 5 marks the region where Σ_F is lower than would be expected for 'fresh' MOR basalt.

Change in Σ_F with increasing porosity and clay alteration

The low-temperature alteration of oceanic basalts typically yields clay, zeolite and carbonate veins infilling fractures, and clay infilling of vesicles and other natural pore spaces. The cross-sectional characteristics of carbonates are described above. Common clay species include saponite, celadonite, and mixed-layer clays such as chlorite–saponite and chlorite–nontronite, while the important zeolites are phillipsite and chabazite. The brown alteration product of olivine, iddingsite, is also a common alteration product.

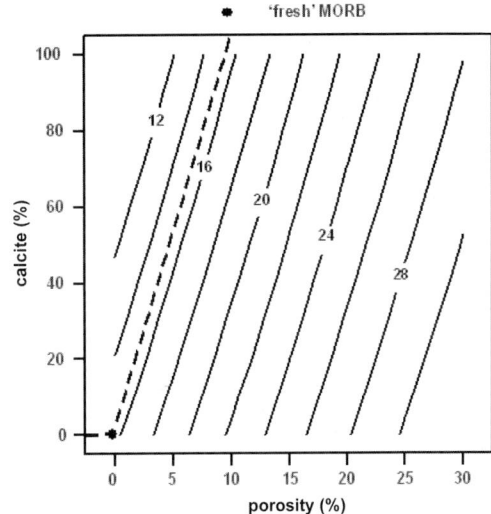

Fig. 5. Variation in the macroscopic cross-section of 'fresh' mid-ocean ridge basalt with dilution by calcite, representing carbonate 'alteration and replacement' and porosity, with sea-water as the formation fluid. The contours are in capture units; a density of 2.71 g cm^{-3} was assumed for the calcite.

There is a wealth of reliable major-element data on all these minerals in precisely the basalts of interest, and, for the purposes of this contribution, representative mineral data from ODP Hole 896A have been employed (Alt *et al.* 1993). Unfortunately, in all cases only major-element data are available, and Σ_F values computed for modelling purposes in the following section would inevitably be low. For some clays, essentially full analyses (major and trace elements including rare earths and boron) are available and described by Herron & Matteson (1993) in an evaluation of elemental composition and nuclear parameters, including the macroscopic cross-section. Clays included in this study were kaolinite, illite and smectite, none of which unfortunately are significant in altered basalt. The data of Herron & Matteson (1993) may, however, give some idea of the extent to which computed values of Σ_F may be low. For the average smectite given by Herron & Matteson (1993), 35% of the Σ_F value is due to rare earths (20.0%), boron (9.7%) and other trace elements (4.5%). Computed without the trace elements, this average smectite has a Σ_F value of 6.12 cu. With the trace elements included, this figure rises to 9.31 cu.

Table 2 summarizes the mineral compositions of the basalt alteration products, together with the average smectite of Herron & Matteson

Table 2. Σ_F *values for some common alteration products of oceanic basalts*

	SiO$_2$	Al$_2$O$_3$	TiO$_2$	Fe$_2$O$_3$	MgO	CaO	Na$_2$O	K$_2$O	MnO	LOI	Σ_F	n
MORB*	49.30	16.74	0.78	9.39	7.55	13.11	1.94	0.10	0.18	0.86	6.00	
Smectite[†]	56.50	17.20	0.20	2.80	3.60	1.90	0.90	0.70	0.05	16.10	9.31	?
Saponite[‡]	45.06	6.03	0.03	10.71	22.80	1.44	0.11	0.12	0.13	13.58	6.52	13
Saponite[§]	45.37	6.78	0.05	10.89	20.75	1.33	0.56	0.59	0.07	13.61	6.68	18
Saponite[§]	44.67	7.13	0.00	11.67	20.29	0.79	0.57	0.18	0.01	14.69	6.83	30
Celadonite[‡]	47.32	3.98	0.08	31.38	5.86	0.78	0.12	6.06	0.07	4.35	9.84	13
Celadonite–nont[‡]	47.32	3.98	0.08	31.38	5.86	0.78	0.12	6.06	0.07	4.35	9.84	35
Iddingsite	33.42	2.74	0.01	41.12	7.66	0.90	0.39	2.55	0.04	11.16	11.98	20
Chlorite–saponite	36.25	7.67	0.01	24.35	20.10	0.38	0.16	0.10	0.21	10.76	8.42	7
Phillipsite	59.70	18.58	0.00	0.72	0.69	1.66	4.69	5.81	0.00	8.16	5.66	6
Chabazite	53.45	20.41	0.00	0.34	0.11	4.03	6.38	2.00	0.00	13.28	5.90	5

*Average composition of MORB samples from ODP Hole 896A on which Σ_F was directly measured Brewer *et al.* (1996).
[†]Average saponite: Herron & Matteson, 1993; [‡]data from Teagle *et al.* (1996); [§]data from Laverne *et al.* (1996).
n: number of analyses averaged.
Nont.: nontronite.
The Σ_F value for MORB is from direct measurements; that for smectite includes a full chemical analysis; all other computed Σ_F values are based on major-element data alone.

(1993), an average MORB composition, and their Σ_F values. The Σ_F value for MORB is from direct measurements; that for the smectite is computed from a full chemical analysis; all other computed Σ_F values are computed from major-element data alone. All the alteration minerals, except the zeolites, have higher macroscopic cross-sections than the 'fresh' basalt, and those which have been computed from their major-element chemistry alone are likely to be lower by 25 to 45%.

As an example of the possible effect of alteration, and an increase in pore fluid, on the formation cross-section a model based on the average smectite of Herron & Matteson (1993) is summarized in Figure 6. The effect of the smectite alteration is comparatively small, increasing Σ_F approximately 1.0 cu per 10% increase (5–6%) in the amount of smectite, up to about 40% smectite.

Change in Σ_F with increasing porosity and boron content

It is well known that clay minerals in marine sediments adsorb boron dissolved in sea-water, and, as a result, many sediments have high boron concentrations (Wedepohl 1978; Zeibig *et al.* 1989). Smectite and serpentinite formed by low-temperature interaction between sea-water and the oceanic basalts or ultramafics also contain large amounts of boron. Some low-temperature altered MORB and serpentine recovered from the sea-floor and from drill holes can have high boron concentrations in excess of 100 ppm (Thompson & Melson 1970; Donnelly *et al.* 1980; Bonatti *et al.* 1984).

Smith *et al.* (1995), in a study involving a number of DSDP/ODP boreholes together with samples from the Oman and Cyprus ophiolites, concluded that there was a good correlation between the concentrations of boron, potassium and $\delta^{18}O$ and the extent of low-temperature alteration. For DSDP Holes 417A, 417D and 418A, drilled into the pillow basalt sequence, the range of boron values encountered was 7.2

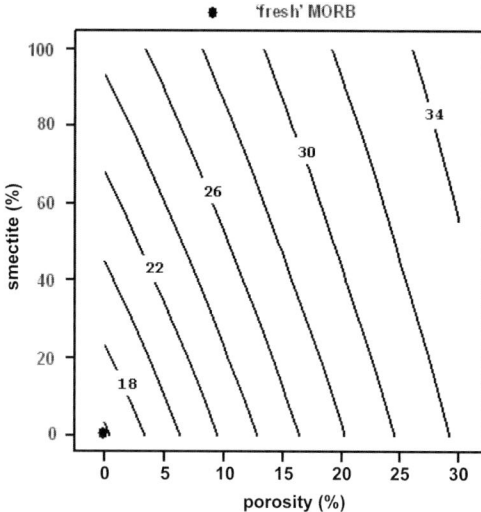

Fig. 6. Variation in the macroscopic cross-section of 'fresh' mid-ocean ridge basalt with dilution by smectite, representing smectite alteration, and porosity, with sea-water as the formation fluid. The contours are in capture units; a density of 2.63 g cm^{-3} (Herron & Matteson 1993) was assumed for the smectite.

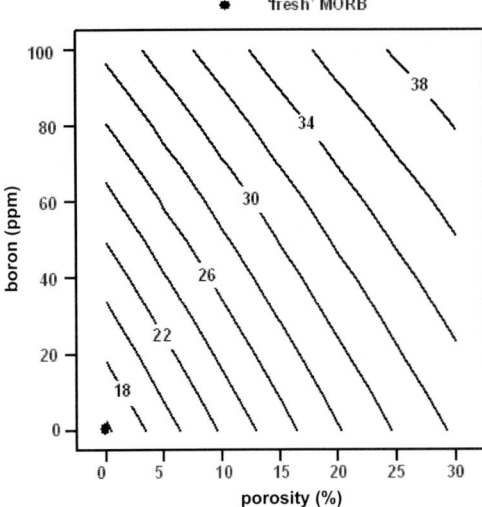

*fresh' MORB

Fig. 7. Variation in the macroscopic cross-section of 'fresh' mid-ocean ridge basalt with increasing boron content and porosity, with sea-water as the formation fluid. The contours are in capture units; a density of 2.99 g cm^{-3} was assumed for the basalt.

to 104 ppm. Because of its high mass-normalized cross-section at thermal energies, the role of boron in alteration is very important. To evaluate this we have modelled the effect of increasing boron on the macroscopic cross-section for 'fresh' basalts; the results are summarized in Figure 7.

The effect of boron, at zero porosity, is to increase Σ_F at a rate of 0.139 cu per ppm boron, decreasing nearly linearly to 0.080 cu per ppm boron at 30% porosity. A doubling of the macroscopic cross-section for 'fresh' MORB (15.7 cu) could hence be achieved by the addition of, say, 30 ppm boron and 30% porosity, or 75 ppm boron and a 10% increase in porosity.

Change in Σ_F with increasing porosity and REE content

Only seven rare-earth elements have mass-normalized thermal cross-sections greater than 0.000 001 g cm^{-3}, and appear on Figure 1. These do, however, include gadolinium, which has by far the largest elemental cross-section, with another four (Sm, Eu, Dy and Er) amongst the top ten largest cross-sectional elements. Small variations here, particularly in gadolinium, may potentially make real changes to the measured macroscopic cross-section. For 'fresh' MOR basalts, the rare-earth abundances are relatively

constant, low and show little variation. Figure 7 shows the variation in Σ_F with increasing rare-earth abundance, and increasing porosity. The 'average' rare-earth content was taken as the average profile of eight MOR basalts described by Sun *et al.* (1979), and is shown in Figure 8 with a REE factor of 1.0; other REE factors are multiples of this abundance profile. Together, the range of rare-earth elements represented in Figure 8 includes most 'fresh' and altered MOR basalts, and covers the average compositions for E- and N-MORB given by Sun & Donough (1989). The abundance ranges covered are: Dy: 0–11.7 Er: 0–7.7, Eu: 0–2.53, Gd: 0–11.1, Nd: 0–18.2, Sm: 0–6.3 and Tm: 0–0.78 ppm.

For comparison the 'average' ocean island basalt of Sun & Donough (1989) is also plotted; the higher rare-earth content of the latter gives a Σ_F value over 20% higher than the corresponding 'average' MORB. This suggests the possibility of using this approach to discriminate between basalt types, although the differences are small and unlikely to be successful unless the rocks are unaltered and with very low porosities.

Discussion and conclusions

The mass-normalized nuclear absorption cross-section of unaltered basic rocks is generally between 6 and 8 cu, with 80% of that value

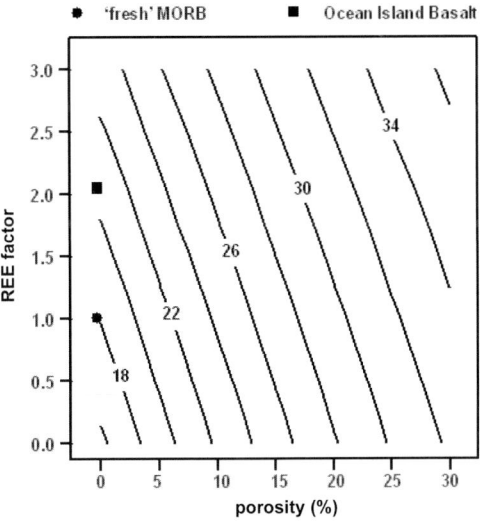

Fig. 8. Variation in the macroscopic cross-section of 'fresh' mid-ocean ridge basalt with increasing rare-earth content and porosity, with sea-water as the formation fluid. The contours are in capture units.

being due to the major-element chemistry, and some 20% to the abundance of trace and minor elements. Of particular significance amongst the latter, because of their very high elemental cross sections, are boron and the rare-earth elements. Fresh mid-ocean ridge basalt has a remarkably constant nuclear absorption macroscopic cross-section (6.0 ± 0.25 cu).

With an increase in porosity and, or, alteration the mass-normalized cross-section generally increases, and can easily double. The Σ_F is measured by a number of nuclear wireline logging tools and, while this curve is rarely used directly in petrophysical interpretation, combined with an independently determined porosity curve, it offers the potential to evaluate and quantify the degree of alteration in basic rock sequences. Given the simple model: $\Sigma_F =$ 'fresh' MORB + alteration + fluid-filled pore space, then, with some assumptions, the alteration signal can potentially be separated. In the Ocean Drilling Program, in studies of the oceanic crust, for example, the pore fluids can be assumed to be sea-water; the absorption properties of MORB are as described above.

The alteration signal is effectively that part of the mass-normalized cross-section which is over and above that required for the fluid-filled pore space and 'fresh' MORB, and is due to the abundance of alteration minerals and their respective densities and cross-sections, and to the abundance of trace elements, of which boron and some of the rare-earth elements (gadolinium in particular) are the most important. To evaluate these factors, a series of simple models were described above, in which the effects of increasing the abundance of different alteration minerals and trace elements vary with 'fresh' MORB and changes in porosity. A number of conclusions arise from these models:

(1) The mass-normalized cross-section does not always have to increase; alteration and replacement by calcite decreases Σ_F as the amount of calcite increases; the effect is significant: 50% replacement of the basalt results in a c.30% decrease in Σ_F.

(2) A lack of full chemical compositions for the clay minerals developed in basic rock by alteration prevent precise modelling of these changes, but such alteration will lead to increases in Σ_F; for the smectite-based model described, 50% replacement of the basalt by smectite at zero porosity would result in an increase in Σ_F of about 35%.

(3) Of the trace elements boron is particular importance; it is normally a few ppm in

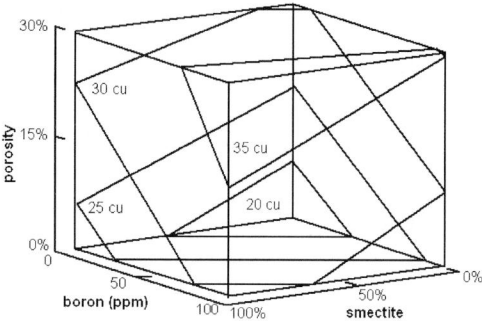

Fig. 9. Variation in the macroscopic cross-section of 'fresh' mid-ocean ridge basalt with increasing boron content, smectite alteration and porosity, with sea-water as the formation fluid. Contour planes for 20, 25, 30 and 35 capture units (cu) are shown.

abundance at the most, in fresh basalts, but can increase rapidly to a few tens of ppm as alteration progresses; the addition of 70 ppm to a 'fresh' basalt results in an increase in Σ_F of about 50%.

(4) Several of the rare-earth elements can make significant contributions to Σ_F, gadolinium being the most important; however, there is relatively little movement of rare-earth elements associated with alteration, so, relatively, far less fluctuation in abundance values than there is associated with boron; a doubling of the 'average' rare-earth abundance for MORB only leads to an increase in Σ_F of about 10%.

These models show that with the alteration of basic rocks the most important factors involved in increasing the macroscopic cross section are the amount of clay present, and the boron content. Of these, boron is the most significant, although both factors tend to operate together. A combined model showing the effects of increasing boron and smectite is summarized in Figure 9, to illustrate these effects.

The Natural Environment Research Council is acknowledged for funding the direct measurement of σ_m in a number of Ocean Drilling Program samples through grant number: GST/02/694.

References

ALT, J. C., KINOSHITA, H., STOKKING, L. B. & THE LEG 148 SHIPBOARD SCIENTIFIC PARTY 1993. *Proceedings of the Ocean Drilling Program, Initial Report*, **148**, College Station, TX (Ocean Drilling Program).

BONATTI, E., LAWRENCE, J. R. & MORANDI, N. 1984. Serpentinization of oceanic peridotites: temperature dependence of mineralogy and boron content. *Earth & Planetary Science Letters*, **70**, 88–94.

BREWER, T. S., HARVEY, P. K., LOCKE, J. & LOVELL, M. A. 1996. The neutron absorption cross-section (Σ) of basaltic basement samples from Hole 896A, Costa Rica Rift. *Proceedings of the Ocean Drilling Program, Scientific Results*, **148**, 389–394.

BROGLIA, C. & ELLIS, D. 1990. Effect of alteration, formation absorption, and standoff on the response of the thermal neutron porosity log in gabbros and basalts: examples from Deep Sea Drilling Project–Ocean Drilling Program sites. *Journal of Geophysical Research*, **95** (B6), 9171–9188.

CROUDACE, I. W. & RANDLE, K. 1993. Fluorine abundances in twenty-nine geological and other reference samples using fast-neutron activation analysis. *Geostandards Newsletter*, **XVII**, 217–218.

DONNELLY, T. W., THOMPSON, G. & SALISBURY, M. H. 1980. The chemistry of altered basalts at Site 417, Deep sea Drilling Project Leg 51. *In*: *Initial Reports of the Deep Sea Drilling Project*, **51**, 1319–1330 US Government Printing Office Washington.

ELLIS, D.V. 1987. *Well Logging for Earth Scientists*. Elsevier, New York, 532 pp.

GOVINDARAJU, K. 1994. 1994 compilation of working values and sample description for 383 geostandards. *Geostandards Newsletter*, **18**, Special Issue, 1–158.

HALLETT, R. B. & KYLE, P. R. 1993. XRF and INAA determinations of major and trace elements in Geological Survey of Japan igneous and sedimentary rock standards. *Geostandards Newsletter*, **XVII**, 127–133.

HARVEY, P. K., LOVELL, M. A., BREWER, T. S., LOCKE, J. & MANSLEY, E. 1996. Measurement of thermal neutron absorption cross section in selected geochemical reference materials. *Geostandards Newsletter*, **20**, 79–85.

HERRON, M. M. & MATTESON, A. 1993. Elemental composition and nuclear parameters of some common sedimentary minerals, 1993. *Nuclear Geophysics*, **7**, 383–406.

LAVERNE, C., BELAROCHI, A. & HONNOREZ, J. 1996. Alteration mineralogy and chemistry of the upper oceanic crust from Hole 896A, Costa Rica Rift. *In*: ALT, J. C., KINOSHITA, H., STOKKING, L. B. & MICHAEL, P. J. (eds) *Proceedings of the Ocean Drilling Program, Scientific Results*, **148**, 151–170.

LOFTS, J. C., HARVEY, P. K., LOVELL, M. A. & LOCKE, J. 1996. The relationship between lithology and the neutron absorption cross-section (Σ) of samples from Ocean Drilling Program Leg 149. *In*: SAWYER, D. S., WHITMARSH, R. B. & KLAUS, A. (eds) *Proceedings of the Ocean Drilling Program, Scientific Results* **149**, 595–599.

SMITH, H. J., SPIVAC, A. J., STRADIGEL, H. & HART, S. R. 1995. The boron isotropic composition of altered ocean crust. *Chemical Geology*, **126**, 119–135.

SUN, S.-S. & MCDONOUGH, W. F. 1989. Chemical and isotopic systematics of oceanic basalts: implications for mantle composition and processes. *In*: SAUNDERS, A. D. & NORRY, M. J. (eds) *Magmatism in the Ocean Basin*. Geological Society, London, Special Publications, **42**, 313–345.

SUN, S.-S., NESBITT, R. W. & SHARASKIN, A. YA. 1979. Geochemical characteristics of mid-ocean ridge basalts. *Earth and Planetary Science Letters*, **44**, 119–138.

TEAGLE, D. A. H., ALT, J. C. & HALLIDAY, A. N. 1998. Tracing the evolution of hydrothermal fluids in the upper oceanic crust: Sr isotope constraints from DSDP/ODP Hole 504B and 896A. *In*: MILLS, R. A. & HARRISON, K. (eds) *Modern ocean floor processes and the geological record*. Geological Society, London, Special Publications, **148**, 81–97.

TEAGLE, D. A. H., ALT, J. C., BACH, W., HALLIDAY, A. N. & ERZINGER, J. 1996. Alteration of upper ocean crust in a ridge-flank hydrothermal upflow zone: mineral, chemistry and isotopic constraints from Hole 896A. *In*: ALT, J. C., KINOSHITA, H., STOKKING, L. B. & MICHAEL, P. J. (eds) *Proceedings of the Ocean Drilling Program, Scientific Results*, **148**, 119–150.

TERASHIMA, S. 1993. Determination of total nitrogen and carbon in twenty-two sedimentary rock reference samples by combustion elemental analyser. *Geostandards Newsletter*, **XVII**, 123–125.

THOMPSON, G. & MELSON, W. G. 1970. Boron contents of serpentinites and metabasalts in the oceanic crust: implications for the boron cycle in the oceans. *Earth & Planetary Science Letters*, **8**, 61–65.

WEDEPOHL, K. H. 1978 (ed.) *Handbook of Geochemistry*, **II/1**, Springer-Verlag.

ZEIBIG, G., KUBANEK, F. & LUCK, J. 1989. Pressure leaching of boron from argillaceous sediments for facies analysis. *Chemical Geology*, **74**, 343–349.

The interpretation of thermal neutron properties in ocean floor volcanics

T. S. BREWER[1], P. K. HARVEY[1], S. R. BARR[1], S. L. HAGGAS[1,2] & H. DELIUS[1]

[1]*Department of Geology, University of Leicester, Leicester, LE1 7RH, UK*
(e-mail: tsb5@leicester.ac.uk)
[2]*IHS Energy, Enterprise House, Cirencester Road, Tetbury, Glos, GL8 8RX, UK*

Abstract: Using near-complete chemical analyses, computed thermal absorption cross-section values (σ) are calculated for ocean-floor basalts from ODP Holes 896a and 1179D. Comparison with nuclear measurement of σ, demonstrates that this computational method is valid and clarifies the interpretation of σ in ocean-floor basalts. Boron, lithium and the rare-earth elements are important controls on the σ-values, and, of these, the distribution of the rare-earth elements is controlled by primary magmatic processes, whereas the distribution of boron and lithium is strongly influenced by secondary low-temperature alteration processes. Consequently, computed σ-values can be used to discriminate between various basalt types and to identify areas of secondary alteration.

Following calibration of the 'fresh basalt' signature, it is possible to interpret log derived neutron absorption measurements of a sample (Σ-values), which, when integrated with other log responses, allow the distribution of alteration within a drilled section to be mapped. Examples from ODP Hole 801C demonstrate the potential of this technique.

For nearly 20 years the Ocean Drilling Program (ODP) has been exploring the oceans through drilling of blanketing sediments and basement rocks. Oceanic basement is typically described in terms of the classic layered seismic model, with layer 1 representing the sedimentary blanket, layer 2a the extrusive volcanics, and layer 2b the sheeted dyke complex that acted as the conduit for the volcanics. An essential and ubiquitous feature of the basement is the large volume of sea-water that circulates through the crust during formation at the mid-ocean ridges. These circulating hydrothermal fluids exert significant control on heat loss, chemical alteration and fluxes within the crust at ridge systems. Although the affects of alteration have been well documented in the majority of basement holes, one of the major unknowns is the distribution and degree of alteration within the drilled section. Interpretation of both these factors is strongly biased by the amount and location of both fresh and altered material recovered. Brewer *et al.* (1998) demonstrated that the non-riser drilling technique used by the ODP favours the recovery of 'fresh' massive basalt, and, that as alteration and veining increase, core recovery decreases in quantity and quality.

This bias in recovery may lead to:

(1) inaccurate interpretation of the lithological architecture and style of alteration;

(2) underestimation of the proportions of the altered material; and

(3) biased estimates of the bulk chemical composition.

Errors in estimating the bulk crustal chemical composition compound the errors in estimates of the chemical fluxes between the lithosphere and hydrosphere during the evolution of the crustal segment. Previously it has been demonstrated how core–log integration can be used to effectively reconstruct the lithological structure in ODP basement holes (Brewer *et al.* 1995; Brewer *et al.* 1998; Brewer *et al.* 1999; Barr *et al.* 2002; Haggas *et al.* 2002; Revillon *et al.* 2002); however, a methodology is developed here for extending the core–log integration into evaluation of the intensity and chemistry of alteration within ocean-floor volcanics.

Chemical data routinely acquired by wireline logging are derived from gamma-ray spectrographic tools, which yield estimates of K, Th and U concentrations. Potassium is preferentially affected by low-temperature alteration, such that altered ocean-floor volcanics have very variable and often-enriched K concentrations.

Other more specialist tools (e.g. the geochemical logging tool, GLT) have been deployed both by industry and the ODP (Harvey & Lovell 1989; Anderson *et al.* 1990), but the quality of such

From: HARVEY, P. K., BREWER, T. S., PEZARD, P. A. & PETROV, V. A. (eds) 2005. *Petrophysical Properties of Crystalline Rocks.* Geological Society, London, Special Publications, **240**, 219–235.
0305-8719/05/$15.00 © The Geological Society of London 2005.

data in ODP boreholes is problematic, due to both environmental (i.e. oversized holes) and technical problems (Brewer *et al.* 1996).

The neutron absorption macroscopic cross-section (Σ), is routinely collected during deployment of the accelerator porosity sonde (APS) in ODP logging operations, but is infrequently used subsequently for formation analysis. Measurements are typically made every 5 cm, and Σ can be defined as:

$$\Sigma = \sigma\rho$$

where ρ = sample density (g cm^{-3}) and σ = mass-normalized (thermal absorption) cross-section.

Expressed this way, Σ is in units of cm^{-1}, but, for logging purposes, it is usually quoted in capture units (cu) = 1000 times Σ, as is σ, for the purposes of this contribution.

In this paper we:

(1) Compare direct measurements of Σ with computed Σ-values from basalts from Hole 896A to validate the computational method;
(2) calculate Σ-values from whole-rock chemical analyses on basalts from ODP Hole 1179D. For each sample detailed petrography is available which allows mineralogical calibration of the calculated Σ-value.
(3) By use of log derived Σ-values we evaluate the alteration within selected intervals from ODP Hole 801C.

Chemical controls on sigma

In natural rocks, the σ, and hence Σ, values are strongly influenced by the concentrations of elements such as Sm, Eu, Gd, Dy, B, Li, Cd and In, which have large capture cross-sections (Brewer *et al.* 1996; Harvey *et al.* 1996; Lofts *et al.* 1996; Harvey & Brewer 2005). Of these elements the concentrations of Cd (119 $\pm$ 22 ppb, Yi *et al.* 2000) and In (61 $\pm$ 7 ppb, Yi *et al.* 2000) are very low in fresh MORB, and little is known about the mobility of both elements during low-temperature alteration. However, from subaerial weathering studies it is unlikely that either element will be strongly enriched during low-temperature alteration. Given the very low concentrations of both Cd and In, their effects on the σ-values are assumed to be extremely small and, as such, are not discussed further.

In typical mid-ocean ridge basalts (MORB), the rare-earth elements (REE) are in general present in low concentrations, <10 ppm; are largely unaffected by low-temperature alteration; and Sm, Eu, Gd and Dy have limited ranges

(e.g. Sun & McDonough 1989). Although Gd has the greatest influence on σ, the net effect of the low and limited concentrations suggests that significant variations in the computed σ are not related to REE. Lithium and B both have low concentrations in fresh MORB (B 0.34 to 0.74 ppm, Chaussidon & Jambon (1994); Li 3 to 6 ppm, Chan *et al.* (1992), Ryan & Langmuir 1987), but both elements, and in particular B, are enriched during low-temperature alteration of MORB (Smith *et al.* 1995).

Potentially, *in situ* measured Σ-values for ocean-floor volcanics may be largely controlled by the secondary distribution of B and/or Li, and, as such, provide a proxy for the intensity, distribution and styles of low-temperature alteration within the formation.

Geological settings

Hole 1179D

Hole 1179D was drilled during ODP Leg 191, and provides samples of Early Cretaceous ocean basement (Kanazawa *et al.* 2001). Site 1179 is located on the abyssal sea-floor, to the east of Japan and to the north of the Shatsky Rise in the NW Pacific Ocean. Hole 1179D penetrates 377 m of sediment and *c.*100 m of basaltic basement. The basement was first encountered at a depth of 375 metres below the sea-floor (mbsf) and consists predominantly of relatively fresh aphyric basalts (both pillow lavas and massive flows), intercalated with a variety of breccias and a small amount of inter-pillow sediment (Kanazawa *et al.* 2001). Basement material was divided into 48 units, which were then subsequently assigned to one of three petrographic groups. Samples reported in this paper span the cored interval, and are representative of 18 of the 48 units.

The 27 core samples cover the range of lithological types recovered from Hole 1179D, and include breccias, pillow basalts and one intercalated sedimentary unit. Breccias are composed of angular to subangular basalt clasts cemented in either a calcite or zeolite matrix. The clasts encompass a variety of basaltic types (e.g. glass, aphyric and phyric hypocrystalline and/or holocrystalline basalts), which are variably altered. In some samples, angular very pale-brown 'fresh' glass containing sparse plagioclase microlites and flow banding is cemented by a sparry calcite (Fig. 1). In contrast, other breccias contain variably altered glassy and/or hypocrystalline basalt fragments which have thin Fe-hydroxide rims and a fringe of euhedral zeolites set in a sparry calcite matrix. Individual basalt clasts are variably altered, many display a con-

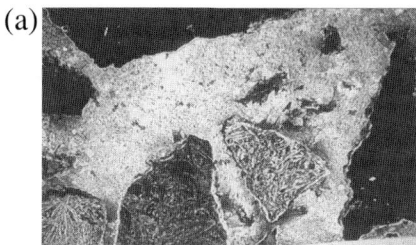

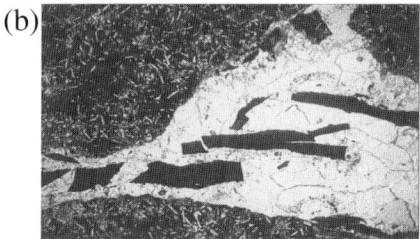

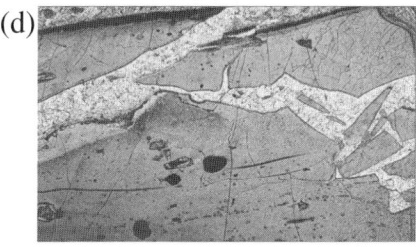

1mm

Fig. 1. Representative petrology of breccias from Hole 1179D. In all examples the cement is composed of calcite. (**a**) Clasts are composed of holocrystalline and hypocrystalline basalts, together with glass. (**b**) Holocrystalline basaltic clasts and small glass clasts which probably represent spalled pillow rims. (**c**) Brown basaltic glass clast showing flow banding. Note the thin rim of zeolite surrounding clasts. (**d**) Brown basaltic glass clasts; note the dark rims to the clast and discoloration of glass at margins, both of which are the result of secondary low-temperature alteration.

centric zonation, with relatively unaltered cores and highly altered rims; many of the rims are composed of pale-yellow palagonitized basalt. Elsewhere, some basalt clasts are completely

replaced by mixtures of saponite, Fe-hydroxides and celadonite. These very different styles of alteration reflect the different origins for the breccias, with some representing volcanic debris (hyaloclastite) cemented *in situ*, and others representing 'sedimentary-style' breccias produced by the accumulation of fragmentary material, probably by mass wastage.

The pillow basalts are composed of a range of textural types from hypocrystalline to holocrystalline. Hypocrystalline basalts comprise a granular framework of lath-shaped plagioclase and intergranular anhedral clinopyroxene with variable proportions of intergranular glass. In samples containing abundant glass (>40%) plagioclase and clinopyroxene form discrete phenocrysts or glomerocrystic aggregates. Holocrystalline pillow basalts are composed of fine to medium (1–3 mm) granular frameworks composed of euhedral to subhedral plagioclase laths, granular anhedral clinopyroxene and rare small pseudomorphs of olivine and minor anhedral opaques. All of the pillow basalts show some evidence of secondary low-temperature alteration, the effects of which can be divided into three broad types:

(1) Vesicle infilling,
(2) veining and/or rinds, and
(3) replacement of igneous minerals.

Several basalts are vesicular (vesicles <10%), and the vesicles are either partially or completely infilled by secondary minerals. Individual vesicles frequently show concentric infills, with an early thin dark-red brown Fe-hydroxide and saponite layer coated by a thin rim of saponite, followed by relatively coarse sparry calcite. Veins are common in the basalts, with the majority forming narrow anastomosing forms. The most abundant vein types in order of decreasing abundance are, pale-brown saponite + celadonite + calcite ± Fe-hydroxides, green celadonite + calcite, calcite and Fe-hydroxides alone.

A number of the pillow basalts contain thin hyaloclastite rinds, which are composed of angular basaltic fragments cemented by calcite. The hyaloclastite debris always shows alteration, from minor partial replacement of igneous minerals by saponite and/or celadonite, to completely altered fragments composed of saponite + celadonite + Fe-hydroxides.

In the pillow basalts, replacement of igneous minerals is ubiquitous, ranging from *c*.5 to 20% (modal), with commonly developed secondary minerals being saponite, celadonite, Fe-hydroxides and carbonates. In the majority of cases, the alteration is heterogeneously distributed, with interstitial glass variably replaced by pale-brown, red-brown or green and pale-green assemblages composed of varying proportions of saponite, Fe-hydroxides, celadonite and minor carbonate. In individual specimens, it appears that original small olivines are replaced by saponite + Fe-hydroxides.

Hole 896A

Hole 896A is located in the equatorial east Pacific, in *c.*5.9-Ma ocean crust that was formed at the Costa Rica Rift (Brewer *et al.*1998). The drill site is located on a bathymetric high overlying a basement high. Basement was first encountered in Hole 896A at 179 mbsf, and the hole was cored from 195.1 to 469 mbsf; and, in the drilled section, core recovery averaged 26.9% (Alt *et al.* 1993). Basalts from Hole 896A are sparsely to highly aphyric tholeiites with plagioclase and olivine present as phenocrysts, although, in one short interval, clinopyroxene phenocrysts occur (Alt *et al.* 1993). The samples analysed in this study are those described by Brewer *et al.* (1996), and are fresh basalts with only very limited evidence of low-temperature alteration (i.e. minor alteration of glass and olivine phenocrysts).

Hole 801C

ODP Hole 801C lies in the Pigafetta Basin, to the east of the Mariana Trench in the West Pacific. During the combined drilling efforts of ODP Legs 129 and 185, Hole 801C was cored to 935.7 mbsf. It penetrates 462 m of Cretaceous sediment and 474 m of Jurassic fast spread (160 km/m.y) oceanic crust (Lancelot *et al.* 1990; Plank *et al.* 2000). Based on descriptions of the recovered core in the basement section, the material consists of 60.6% massive basalt flows, 27.8% pillows, 8.1% breccia and hyaloclastite, and 3.5% inter-pillow sediments and hydrothermal deposits. However, recovery averages only 47%, and more accurate estimates of the lithological proportions, based on core–log integration results, indicate 27.4% massive basalt flows, 33% pillows, 31% breccia and hyaloclastite, and 2.7% inter-pillow sediments and hydrothermal deposits, with 5.9% unclassified (Barr *et al.* 2002).

The basalts are aphyric or slightly phyric, and include both intrusive alkaline and extrusive tholeiitic, normal MORB varieties. Intense oxidative alteration occurs in four discrete zones, at the basement sediment interface, adjacent to two hydrothermal deposits and in the vicinity of a thick breccia unit deep in the hole. Otherwise, with the exception of flow margins and pillow rims, the basalts are only slightly altered

under anoxic conditions. The dominant alteration minerals are calcite, smectite, saponite, pyrite, silica, celadonite and Fe-oxyhydroxides (Lancelot *et al.* 1990; Plank *et al.* 2000). Different mixtures of these minerals define the colour of the core, which varies from grey-black to green-grey to light brown, depending on the degree of alteration. Veins are abundant throughout the core, ranging from <1 mm to over 10 mm thick, and are composed variably of saponite, carbonate, celadonite, pyrite and Fe-oxyhydroxides.

Computation of sigma

All samples were analysed for major and trace elements by X-ray fluorescence spectrometry at the University of Leicester, using standard procedures described by Harvey *et al.* (1973) and Harvey (1989). The REE, B, Li and Be were determined by inductively coupled plasma optical emission spectrometry, using a JY-Ultima-2 spectrometer at the University of Leicester, following the procedures outlined in Appendix 1. Carbon and S determinations were carried out using a LECO CS-125 analyser and the procedures documented in Appendix 2.

Summaries of the data-set are presented in Tables 1 & 2, and from these near-complete whole-rock analyses it is possible to calculate the macroscopic thermal cross-section (Σ) and the thermal absorption cross-section (σ), following the methodology described by Harvey & Brewer (2005). To allow comparison between computed σ-values and directly measured σ-values, REE and B analyses have been made on the set of basalts previously analysed from ODP Hole 896A (Leg 148) by Brewer *et al.* (1996).

Direct measurements of σ on the Hole 896A basalt samples were previously obtained by using the NESSUS facility at the centre of the NESTOR reactor at Winfrith in Dorset, UK (Brewer *et al.* 1996). Comparison of the measured (Brewer *et al.* 1996) with the computed σ-values of the basalts from Hole 896A give very similar results (σ-measured 5.99 ± 0.27, σ-computed 5.62 ± 0.18 cu, Fig. 2a), which validates the computational method, and demonstrates that the computed σ-values are representative of the σ for the basalts.

Computed σ-values for the Hole 1179D basalts range from 6.93 to 8.79 (mean 7.92 ± 0.61), which is greater than that previously reported for unaltered Ocean Floor Basalts (Hole 896A basalts, 5.99 ± 0.27 cu; Brewer *et al.* 1996; Harvey & Brewer 2005). The difference in the computed σ-values for the basalts from Holes 1179D and 896A, correlates with

Table 1. *Whole-rock geochemical analyses for basalts from Hole 896A used in this study.* Major and trace elements analysed by X-ray spectrometry, carbon and sulphur by LECO, rare-earth elements, B, Be and Li by inductively coupled plasma spectrometry. The value 'na' indicates not analysed. Major elements are quoted in weight per cent; trace elements in parts per million and σ in capture units. Depth is expressed as metres below the sea-floor.

Sample	148-A	148-B	148-C	148-D	148-E	148-F	148-G	148-H	148-I	148-J
Depth	317.69	195.39	201.2	289.22	348.43	325.24	286.98	349.03	325.52	354.52
core	14r	1r	2r	11r	17r	15r	11r	17r	15r	18r
section	3	1	1	3	4	1	1	4	1	1
interval	021–024	029–036	030–032	030–035	053–058	094–097	098–109	113–116	122–126	142–145
peice	4	7	7	5	3b	14	8b	11	18	12b
SiO_2	48.66	49.42	49.4	48.99	49.98	49.04	49.2	49.63	49.23	49.42
Al_2O_3	16.62	17.68	17.53	15.92	15.92	17.03	17.24	16.24	17.44	15.81
TiO_2	0.73	0.71	0.7	0.8	0.85	0.76	0.74	0.88	0.78	0.89
Fe_2O_3	9.48	8.33	8.58	9.93	9.95	9.43	9.4	9.8	8.99	9.98
MgO	7.69	7.21	7.19	8.07	7.93	7.24	7.17	7.38	7.05	8.6
CaO	13.04	13.4	13.57	13.04	12.82	13.39	13.29	12.98	13.19	12.41
Na_2O	1.9	2.06	2.01	1.91	1.84	1.91	1.88	2.06	1.95	1.88
K_2O	0.07	0.07	0.07	0.08	0.23	0.07	0.05	0.17	0.05	0.1
MnO	0.19	0.16	0.16	0.18	0.18	0.18	0.17	0.18	0.18	0.19
P_2O_5	0.05	0.05	0.06	0.06	0.07	0.05	0.05	0.07	0.05	0.08
LOI	1.19	1.06	0.93	0.93	0.28	0.91	1.03	0.58	1.01	0.7
Total	99.62	100.15	100.2	99.91	100.05	100.01	100.22	99.97	99.92	100.06
Ba	13	<8	13	10	25	12	11	29	<8	27
Cr	332	395	379	343	283	328	361	282	362	328
Cu	62	81	87	78	61	79	80	48	78	49
Ga	15	15	14	17	15	15	15	17	16	16
Nb	<1	1	<1	2	<1	<1	<1	<1	<1	<1
Ni	148	189	170	136	103	147	155	105	149	100
Pb	<1	<1	<1	<1	<1	<1	<1	<1	<1	<1
Rb	2	2	1	3	5	2	2	4	2	3
S	309	1931	1672	166	433	1185	1568	110	1988	1500
Sc	40	45	41	36	46	42	41	46	46	47
Sr	65	70	68	61	56	64	70	61	66	57
Th	<1	1	<1	<1	<1	<1	<1	<1	<1	2
U	<1	2	2	2	<1	<1	<1	1	2	<1
V	221	227	215	235	254	222	226	256	243	259
Y	23	23	22	24	27	22	21	27	26	28
Zn	52	55	55	49	58	51	55	60	63	57
Zr	39	41	41	42	45	42	46	50	43	48

(continued)

Table 1. *Continued*

Sample	148-A	148-B	148-C	148-D	148-E	148-F	148-G	148-H	148-I	148-J
La	0.89	0.92	0.83	0.97	0.97	0.84	0.72	0.82	0.85	0.87
Ce	3.18	3.39	3.05	3.34	3.43	3.13	2.67	2.62	2.61	2.92
Pr	0.50	0.53	0.49	0.55	0.52	0.47	0.43	0.53	0.50	0.51
Nd	3.36	3.49	2.99	3.37	3.74	3.21	3.09	3.81	3.45	3.71
Sm	1.39	1.60	1.32	1.57	1.68	1.48	1.35	1.71	1.44	1.77
Eu	0.54	0.59	0.54	0.57	0.61	0.54	0.54	0.62	0.59	0.62
Gd	2.33	2.44	2.22	2.45	2.68	2.31	2.23	2.50	2.08	2.69
Dy	2.91	2.99	2.74	2.97	3.22	2.91	2.76	3.34	2.65	3.11
Er	2.09	2.16	2.03	2.22	2.21	2.04	2.03	2.31	1.73	2.24
Yb	2.10	2.19	2.08	2.21	2.28	2.08	2.08	2.26	1.69	2.24
Lu	0.31	0.32	0.30	0.33	0.33	0.32	0.30	0.32	0.26	0.33
Be	0.14	0.13	0.15	0.13	0.16	0.13	0.15	0.16	0.15	0.14
Li	8.71	4.97	4.71	6.75	4.75	5.23	4.49	5.68	4.14	4.75
B	5.45	2.9	2.32	5.05	6.53	5.23	4.04	6.04	4.47	6.97
σ	5.68	5.38	5.31	5.73	5.75	5.62	5.55	5.77	5.50	5.86

the marked compositional difference between basalts in the two holes (Fig. 3). The compositional differences between the basalts in Holes 896A and Hole 1179D, reflects differences in both the primary magmatic processes and the degree of secondary low-temperature alteration.

The REE, Ti and Zr are largely unaffected by low-temperature alteration, so the differences in these elements is consistent with Hole 896A basalts having been derived from a more depleted source than the Hole 1179D basalts, and probably involving different degrees of melting.

In contrast, B is an important element in controlling σ, and Hole 1179D basalts are enriched in this element as compared to Hole 896A basalts (Fig. 3). B in fresh MORB glass has very low concentrations (<1 ppm), but as sea-water contains *c.*4 ppm, this element is significantly enriched during secondary low-temperature alteration (e.g. Ryan & Langmuir 1993; Smith *et al.* 1995). It is therefore important to establish the contribution that the elevated B values are making to the 1179D basalts, and to establish if this is responsible for the difference in σ-values between the two boreholes. In Table 3, we have computed the percentage contribution of various elements to the σ-value. What is noticeable is that in 1179D B accounts for between 7 and 20% of the σ value (Table 3), suggesting that the elevated B values are not controlling the difference. As confirmation of this, the 1179D computed σ-values were recalculated with a B value of 4.9 ppm, which is the average B content of the 896A basalts. The result is to slightly reduce the mean σ value from 7.92 to 7.2, indicating that secondary alteration is not responsible for the difference in σ, rather the original (igneous) chemistry being the control. Therefore, it appears that σ (or Σ from a log measurement) can be used to distinguish between enriched (high σ/Σ) and depleted (low σ/Σ) MOR basalts.

Mineralogical controls on σ-values: Hole 1179D case study

In Hole 1179D, basalts have the lowest and the breccias the highest magnitude and range in σ-values (Fig. 4). The basalts have a narrow range of σ-values, and it is not possible to discriminate between the three petrographic types identified by Kanazawa *et al.* (2001). The inability to distinguish the different petrographic types reflects the limited chemical and mineralogical compositions for each group (Table 2, Fig. 5).

Table 2. *Whole-rock geochemical analyses of samples from Hole 1179D. Major and trace elements analysed by X-ray spectrometry, carbon and sulphur by LECO, rare-earth elements, B, Be and Li by inductively coupled plasma spectrometry. The value 'na' indicates not analysed. Major elements are quoted in weight per cent; trace elements in parts per million and σ in capture units. Depth is measured in metres below sea-floor, and Type refers to the rock types; for the basalts the numbers correspond to the threefold subdivision of Hole 1179D basalts by Kanazawa et al. (2001).*

Sample	SH2	SH3	SH6	SH7	SH10	SH15	SH16	SH17	SH19	SH27	SH11
Depth	377.91	378.33	391.67	393.42	420.13	428.53	430.07	444.97	455.28	470.60	421.07
Type	Group 1	Group 1	Group 1	Group 1	Group 2	Group 2	Group 2	Group 2	Group 3	Group 3	Breccia
SiO_2	49.67	47.37	48.32	48.11	47.84	46.65	50.12	47.39	44.25	46.03	24.74
TiO_2	1.41	1.24	1.22	1.35	1.84	1.93	1.82	1.62	1.52	1.54	0.82
Al_2O_3	14.83	14.04	14.08	14.33	13.64	13.92	15.13	16.03	15.16	15.43	6.35
Fe_2O_3	10.05	11.70	11.37	11.38	12.16	15.29	10.27	11.36	11.28	10.53	8.18
MnO	0.15	0.15	0.19	0.16	0.20	0.19	0.18	0.19	0.21	0.18	0.12
MgO	7.85	7.25	7.46	7.09	6.71	6.13	7.46	5.88	6.24	6.28	4.91
CaO	11.94	12.83	13.34	12.18	12.46	9.80	11.25	12.80	14.26	13.98	29.37
Na_2O	2.65	2.34	2.16	2.59	2.66	2.77	2.75	2.70	2.61	2.68	1.22
K_2O	0.39	0.71	0.51	0.76	0.49	0.76	0.41	0.43	0.34	0.27	0.88
P_2O_5	0.13	0.12	0.12	0.14	0.17	0.28	0.16	0.19	0.17	0.17	0.08
LOI	1.06	2.04	1.43	1.94	2.23	1.75	0.67	1.59	3.95	3.09	23.77
Total	100.14	99.80	100.19	100.03	100.41	99.45	100.22	100.20	100.00	100.17	100.44
C%	0.4520	0.6180	0.5910	0.6670	0.6590	0.4940	0.4070	0.6200	1.0200	0.8850	5.1500
S%	0.0050	0.0042	0.0454	0.0061	0.0077	0.0045	0.0112	0.0027	0.0202	0.0218	0.0555
Ba	19.0	27.2	17.4	40.3	11.5	19.4	12.0	28.6	47.2	27.9	23.9
Co	42.5	39.2	42.0	41.9	45.3	49.9	40.3	45.1	42.9	45.1	24.7
Cr	300.9	237.5	300.5	230.1	122.7	96.6	184.4	332.8	385.2	387.5	56.6
Cu	129.7	80.6	83.4	54.6	79.6	87.4	165.6	79.5	33.8	43.0	39.6
Ga	16.6	16.4	14.8	17.1	19.5	20.2	21.9	20.2	18.4	17.0	9.8
Mo	1.3	1.3	1.4	1.3	1.5	1.5	1.4	1.1	1.4	1.7	1.1
Nb	3.0	2.1	2.6	2.6	2.9	3.6	2.7	3.8	4.4	3.7	1.0
Ni	85.9	70.0	81.6	77.4	71.4	53.6	73.9	107.9	107.6	154.4	35.0
Pb	3.5	3.0	2.4	<1.0	1.2	3.1	2.3	5.1	4.9	8.6	6.5
Rb	8.3	17.6	11.9	17.1	9.4	13.0	7.7	6.2	7.6	6.5	14.4
Sc	51.2	39.6	41.2	51.0	48.5	51.7	54.8	43.8	42.6	41.5	41.2
Sr	102.3	91.0	83.0	96.0	102.5	106.0	102.9	155.0	166.7	169.1	113.3
Th	2.2	<1.0	1.1	2.3	3.3	2.7	4.2	2.7	6.0	4.5	3.9
U	<0.8	4.1	2.4	<0.8	2.9	2.4	3.3	5.6	1.4	0.9	1.6
V	354.1	324.7	345.8	380.9	465.8	469.3	465.1	348.4	332.6	324.5	177.4
Y	33.4	28.2	33.0	34.3	46.8	54.0	41.7	36.9	31.1	32.9	19.5
Zn	92.6	74.3	83.0	84.8	114.4	129.3	108.1	98.7	100.5	94.6	49.8

(continued)

Table 2. *Continued*

Sample	SH2	SH3	SH6	SH7	SH10	SH15	SH16	SH17	SH19	SH27	SH11
Zr	83.5	71.1	73.2	81.4	117.7	119.6	108.9	114.4	107.9	115.7	60.3
La	2.67	2.48	1.99	2.43	3.80	4.91	3.19	3.99	4.00	4.11	1.76
Ce	8.46	7.62	5.96	7.71	10.47	12.66	8.79	11.10	12.30	10.76	5.00
Pr	1.41	1.27	1.07	1.24	1.67	2.01	1.36	1.69	1.84	1.75	0.76
Nd	8.46	7.48	6.75	8.11	11.14	12.37	10.00	9.27	10.97	10.73	4.30
Sm	3.30	2.92	2.33	2.67	4.13	3.99	3.13	3.41	3.39	3.42	1.34
Eu	1.22	1.04	0.95	1.10	1.47	1.66	1.36	1.23	1.30	1.29	0.55
Gd	3.94	3.75	3.54	4.24	6.36	6.19	5.32	4.22	4.29	4.44	2.00
Dy	5.55	4.97	4.37	4.90	7.00	7.71	6.21	5.12	4.52	4.85	2.29
Er	3.39	3.26	2.91	3.33	4.56	5.27	4.12	3.04	2.94	2.86	1.82
Yb	3.55	3.16	2.78	2.70	3.99	4.26	3.46	3.05	2.70	2.94	1.46
Lu	0.51	0.44	0.41	0.45	0.50	0.59	0.62	0.47	0.38	0.53	0.19
Be	0.41	0.20	0.17	0.19	0.23	0.29	0.22	0.29	0.32	0.29	0.09
Li	9.31	7.55	8.46	7.03	13.06	11.50	12.29	22.02	18.91	23.89	10.73
B	14.09	15.66	28.35	40.11	14.25	14.40	32.40	21.59	18.17	19.01	27.30
σ	7.77	6.93	7.70	8.54	8.35	8.79	8.17	7.14	8.21	7.96	11.49

Sample	SH12	SH13	SH14	SH18	SH22	SH23	SH24	SH25	SH9	SH20	SH21
Depth	421.41	426.39	427.35	446.83	465.20	465.31	466.08	466.63	400.59	455.54	458.00
Type	Breccia	Breccia	Breccia	Breccia	Breccia	Breccia	Breccia	Breccia	Altered	Altered	Altered
SiO_2	36.73	34.95	50.09	41.67	30.85	47.13	38.04	33.90	44.15	40.46	44.89
TiO_2	1.25	1.26	1.95	1.22	0.95	1.85	1.17	1.14	1.24	1.21	1.45
Al_2O_3	9.70	9.85	12.36	12.83	9.47	15.89	12.72	12.66	13.25	13.92	14.99
Fe_2O_3	10.99	10.76	14.40	7.93	7.89	10.48	8.06	9.09	11.57	9.66	10.00
MnO	0.11	0.18	0.08	0.14	0.13	0.15	0.12	0.12	0.16	0.22	0.23
MgO	5.91	5.26	7.74	7.02	5.86	8.43	5.54	4.19	5.72	5.01	6.73
CaO	18.36	21.08	2.59	16.07	23.82	8.46	19.15	21.93	14.44	18.44	12.12
Na_2O	1.54	1.91	2.47	2.19	1.41	1.89	1.94	2.08	2.50	2.30	3.24
K_2O	1.33	0.78	2.93	0.94	1.17	1.14	0.83	0.63	0.76	0.49	0.86
P_2O_5	0.10	0.15	0.08	0.13	0.09	0.14	0.15	0.28	0.13	0.25	0.21
LOI	14.16	13.59	4.80	9.56	18.42	4.47	12.37	13.90	5.00	8.28	5.28
Total	100.19	99.76	99.49	99.69	100.06	100.04	100.08	99.92	99.91	100.25	99.98

C%	2.8500	3.1400	0.8400	2.0000	3.9200	0.5310	2.5700	3.0700	1.2900	2.0200	1.1300
S%	0.0783	0.0357	0.0046	0.0512	0.0341	0.0131	0.0171	0.0160	0.0301	0.0089	0.0107
Ba	15.4	44.5	56.9	127.2	64.0	90.1	35.0	25.3	15.4	17.8	36.8
Co	36.5	36.9	46.0	33.6	26.9	40.7	32.5	31.7	41.3	33.3	36.4
Cr	95.0	69.8	139.6	306.0	227.0	380.2	242.4	247.1	198.8	273.9	333.8
Cu	67.6	48.7	68.3	60.2	39.9	69.2	86.1	43.7	45.9	46.0	98.3
Ga	11.5	12.7	14.9	12.5	10.7	13.5	16.6	16.3	17.1	16.2	18.1
Mo	1.3	1.5	1.4	1.3	1.2	1.1	1.5	1.2	1.5	1.0	1.7
Nb	1.9	1.9	3.0	2.8	2.7	5.6	3.1	2.7	1.8	3.5	4.1
Ni	57.9	40.6	63.0	123.9	89.7	139.2	97.5	76.8	60.9	78.9	142.0
Pb	2.4	5.8	<1.0	2.9	4.1	1.8	<1.0	3.1	3.6	3.8	2.2
Rb	23.7	12.6	39.8	11.0	16.7	16.2	14.8	11.8	20.8	10.5	28.5
Sc	41.5	41.6	42.0	41.8	33.5	45.9	36.6	46.3	46.1	37.5	37.7
Sr	108.6	88.6	73.3	116.8	102.9	84.2	134.9	152.6	98.1	169.2	150.4
Th	2.3	2.0	4.5	1.6	3.2	5.9	1.7	5.0	5.1	5.5	2.3
U	1.4	<0.8	<0.8	2.5	<0.8	<0.8	0.9	2.6	4.0	1.8	3.8
V	240.8	291.2	226.5	200.1	156.2	277.1	237.0	315.5	354.6	275.2	288.6
Y	28.9	33.9	23.4	24.9	25.8	31.2	27.4	37.1	31.6	37.2	33.4
Zn	67.3	85.8	99.0	67.9	55.2	72.8	78.7	78.3	91.1	85.2	91.0
Zr	89.1	83.7	107.4	86.8	72.2	123.2	88.7	85.7	78.3	91.7	105.8
La	2.24	3.30	2.00	3.31	3.85	3.40	4.08	5.05	2.57	4.23	4.98
Ce	6.83	8.30	7.16	9.30	8.65	10.05	12.45	12.79	7.75	11.48	12.50
Pr	1.28	1.28	1.20	1.36	1.21	1.55	1.98	2.18	1.27	1.94	1.91
Nd	7.42	9.62	6.36	8.56	7.79	8.84	10.99	12.82	7.90	10.94	11.02
Sm	2.59	2.99	2.38	2.72	2.22	3.26	3.67	4.28	2.91	3.17	3.36
Eu	0.93	1.26	0.83	1.03	0.88	1.12	1.37	1.59	1.12	1.28	1.33
Gd	3.37	5.10	2.94	3.80	3.59	4.01	4.42	5.23	3.48	4.29	4.83
Dy	4.58	5.67	3.89	4.16	3.64	4.35	5.35	6.69	5.47	4.80	5.00
Er	2.77	3.82	2.30	2.54	2.21	2.71	3.04	4.02	3.21	3.22	2.96
Yb	2.92	3.07	2.48	2.27	1.95	2.58	3.06	3.84	3.59	3.10	2.74
Lu	0.43	0.43	0.34	0.38	0.30	0.38	0.36	0.54	0.51	0.43	0.43
Be	0.19	0.19	0.25	0.28	0.19	0.37	0.29	0.24	0.21	0.33	0.41
Li	23.51	13.79	24.67	23.51	19.32	44.62	23.12	12.23	9.71	14.46	18.43
B	73.56	10.26	29.97	19.40	na	25.72	25.24	24.10	41.83	29.54	56.17
σ	12.43	9.91	8.18	8.64	9.59	7.73	9.65	10.27	9.15	9.26	10.08

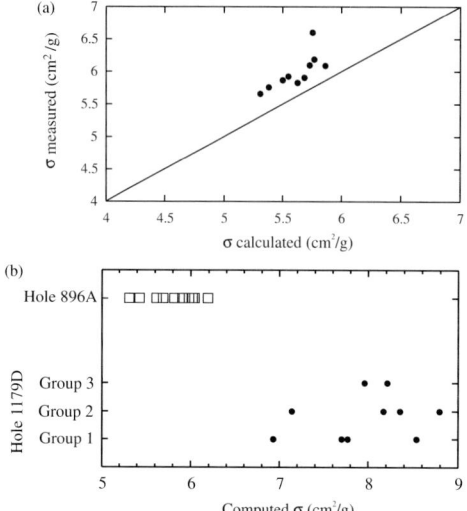

Fig. 2. (a) Comparison of directly measured and computed σ values for Hole 896A basalts, computation based on values in Table 1. **(b)**. Comparison of computed σ-values for basalts from Holes 896A and 1179D.

Altered basalts

More highly altered basalts are characterized by a greater proportion of alteration (>15% modal abundance), which is manifested by either a greater abundance of palagonitized glass or celadonite, saponite and/or Fe-hydroxides. Chemically, the more highly altered basalts are characterized by relatively high loss-on-ignition (LOI > 5%) and B values (B $\geq$ 30 ppm) and elevated σ-values (σ > 9.0 cu) compared to the 'fresh' basalts (Fig. 5). The REE elements are unaffected by the alteration, and have concentrations similar to the basalts (Fig. 5). Alteration proceeded by a variety of hydration reactions involving interactions with sea-water; the breakdown of glass and igneous minerals; and the development of secondary low-temperature minerals. As part of this process B was partitioned/absorbed by the alteration phases, which resulted in variable enrichment. The net result of such alteration has been to increase σ above the normal range of the basalts (Fig. 4).

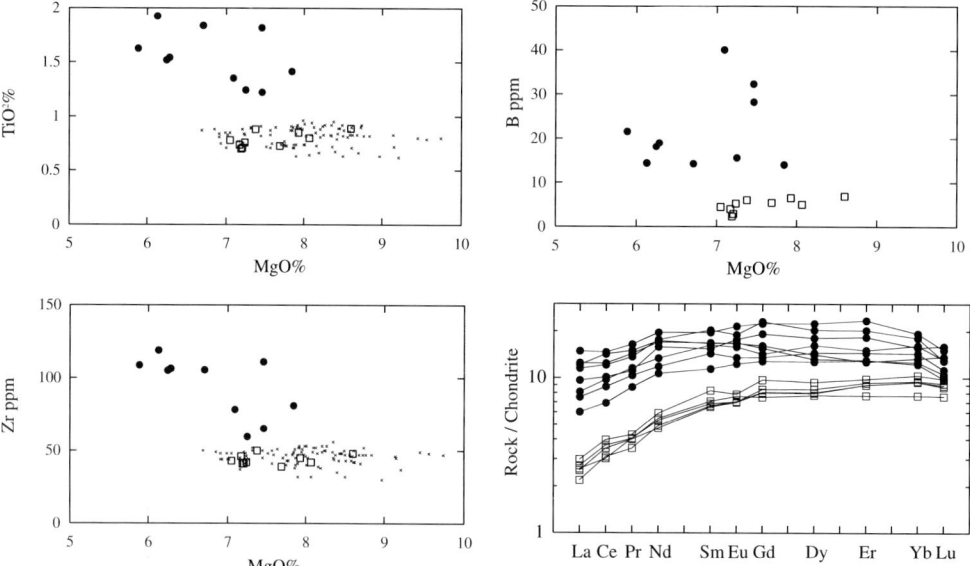

Fig. 3. Comparison of basalts from Hole 896A and Hole 1179D using selected chemical parameters. Squares represent Hole 896A samples analysed in this study; crosses are published Hole 896A analyses (Brewer *et al.* 1996; Teagle *et al.* 1996), and filled circles are Hole 1179D basalts. The rare-earth elements are normalized with respect to the values of Nakamura (1977).

Table 3. *Contribution of various chemical elements to the computed σ value. The modelled σ value was calculated assuming 4.9 ppm boron in all samples. See text for details. Group refers to the classification scheme employed by Kanazawa* et al. *(2001) for Hole 1179D basalts.*

Sample	SH2	SH3	SH6	SH7	SH10	SH15	SH16	SH17	SH19	SH27
Group	1	1	1	1	2	2	2	2	3	3
Computed σ	7.77	6.93	7.70	8.54	8.35	8.79	8.17	7.14	8.21	7.96
% major element contribution	76.2	77.9	72.3	67.6	74.0	75.6	66.7	81.6	76.3	74.3
% REE contribution	13.1	11.5	10.2	10.9	16.8	15.6	14.3	2.4	11.9	12.6
% Boron contribution	8.7	9.0	15.7	20.1	7.3	7.0	17.0	12.9	9.5	10.2
% other traces contribution	2.0	1.5	1.7	1.4	1.9	1.8	2.0	3.0	2.4	2.8
Modelled σ	6.54	6.94	6.7	7.03	7.95	8.39	6.99	6.42	7.64	7.36

Breccias

In Hole 1179D, breccias have the highest magnitude and range of σ-values, and calcite-cemented breccias have the highest and most dispersed values, whereas zeolite-cemented breccias in general have more restricted σ-values (Fig. 4). In the calcite breccias, σ correlates well with the CaO content of the sample: where CaO is a proxy for the proportion of calcite in the sample (Fig. 6). The one exception to this is sample SH12, but in this sample the basaltic clasts are highly altered, suggesting that the chemistry of the altered material is an important control on sigma.

From this mineralogical and chemical study the key conclusions are:

(1) Sigma-values can be used to distinguish between enriched and depleted MOR basalts, as demonstrated by the difference between σ-values from Hole 896A (depleted) and Hole 1179D (enriched) basalts.

(2) The least-altered basalts in Hole 1179D have σ in the range 6.93 to 8.79 (mean 7.92 ± 0.61).

(3) Breccias have the greatest range in σ (8.64 to 12.43 cu), and calcite-cemented breccias

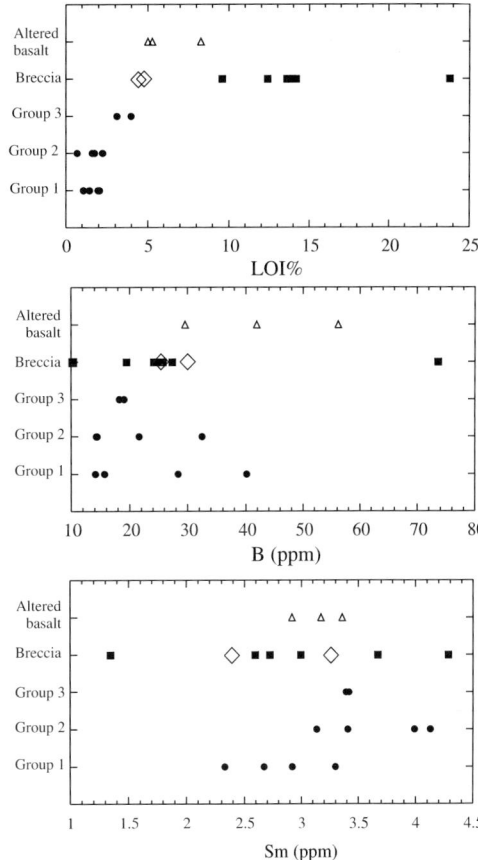

Fig. 5. Variation in selected chemical parameters between basalts, altered basalts and breccias in Hole 1179D. Symbols as in Figure 4.

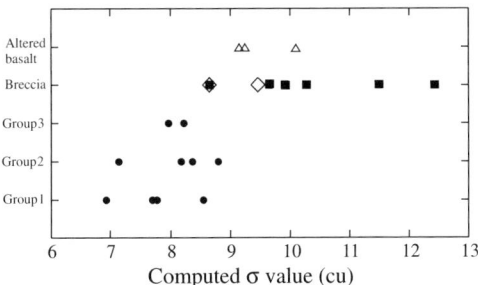

Fig. 4. Computed σ-values for fresh and altered basalts and breccias from Hole 1179D. Basalts (filled circles) are divided into three groups based on the criteria of Kanazawa *et al.* (2001). Breccias are subdivided based on the composition of the cement: calcite-cemented breccias are represented by filled squares, and zeolite-cemented ones are represented by diamonds. Altered basalts are represented by triangles.

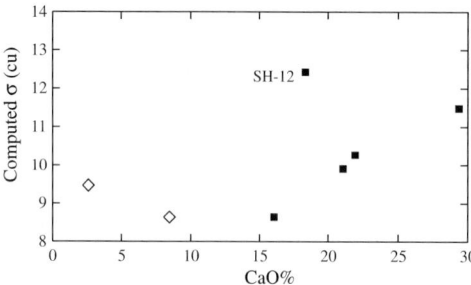

Fig. 6. Correlation between σ and CaO content in Hole 1179D breccias. Symbols as in Figure 4.

appear to have a higher magnitude and range than the zeolite-cemented breccias.

(4)   Alteration of basalts from the same lava sequence causes a significant increase in σ and, in turn, Σ.

Therefore, if Σ is measured continuously down-hole, it should provide a means to estimate the proportion of alteration with the borehole. However, it is important to:

(1)   use whole-rock data to calibrate the unaltered basalt signal;
(2)   integrate the Σ down-hole log with other logging measurements to effectively discriminate between various breccia types and highly altered basalts.

Modelling the variation in alteration across basalt flows in Hole 801C: down-hole logging case study

This case study uses data from Hole 801C, because there are no down-hole logging data available from Hole 1179D. Two basalt flows from Hole 801C (Leg 185) have been chosen to illustrate the way in which alteration varies across the flows, which are both bounded above and below by volcanic breccias. The upper flow (UF) is 8.5 m thick and lies between 755.5 and 764.0 mbsf. The shallow (LLS) and deep (LLD) laterolog resistivity traces across the flow are shown in Figure 7a; the flow is picked out by its high resistivity ($>$100 ohm m), while the breccias have lower resistivity values, below about 80 ohm m. In Figure 7b the curves for neutron porosity (HALC) and the macroscopic thermal cross-section (HSIG), derived from Schlumberger's high-resolution APS, are shown, together with a per cent alteration curve. The latter was generated using the method described by Harvey & Brewer (2005). Using this approach the weight fraction of altered basalt in oceanic basement can be expressed as:

$$w_a = 1 - \left(\dfrac{\dfrac{(\Sigma_F - \phi\rho_f\sigma_f)}{(1-\phi)} - \rho_a\sigma_a}{(\rho_b\sigma_b - \rho_a\sigma_a)} \right)$$

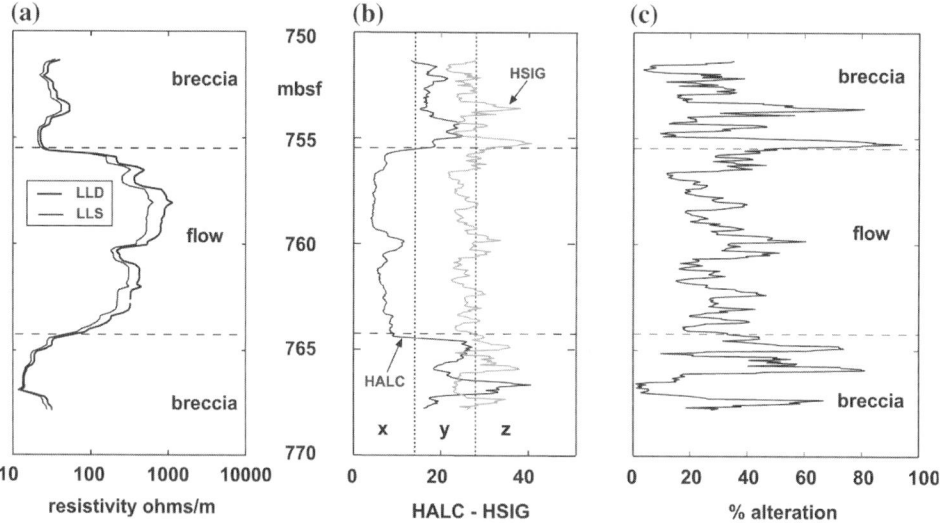

Fig. 7. (a) Laterolog resistivity curves across the upper flow in Hole 801C. (b) HALC (neutron porosity, in per cent), HSIG (macroscopic thermal-cross section, in capture units) and (c) modelled per cent alteration, across the same flow. See text for details.

where: w_a is the weight fraction of altered basalt in the rock fraction; Σ_F is the macroscopic nuclear cross-section of the formation; ϕ is the porosity of the formation; ρ_f is the density of the pore fluid; σ_f is the mass-normalized cross-section of the pore fluid; ρ_b is the density of (fresh) MOR basalt; σ_b is the mass-normalized cross-section of (fresh) MOR basalt; ρ_a is the density of the alteration products; σ_a is the mass-normalized cross-section of alteration products.

For modelling of the basalt flows in Hole 801C, the interstitial pore fluid was assumed to be sea-water, with $\sigma_f = 32.34$ cu and $\rho_f = 1.017$ g cm^{-3}, while, for fresh MORB, the following physical properties were taken: $\sigma_b = 6.00$ cu (± 0.25) and $\rho_f = 3.000$ g cm^{-3} (Harvey & Brewer 2005). Log measurements from the APS provided down-hole data for Σ_F (SIGF curve) and formation porosity (APLC curve). The remaining physical properties, for the alteration products (ρ_a and σ_a), are less easy to define precisely (Harvey & Brewer 2005), but, for the purposes of the models presented here, the following constants have been used: $\sigma_a = 9.31$ cu and $\rho_a = 2.63$ g cm^{-3}. The mass-normalized cross-section of the alteration products (σ_a) is computed from the smectite data of Herron

& Matteson (1993), and is considered to have nuclear properties similar to the clay alteration products found in the altered basalts (Harvey & Brewer 2005). The resulting per cent alteration curve, modelled using these parameters, and shown in Figure 7c, has values between zero and about 95% alteration. While the modelled range is acceptable, the accuracy of these figures is highly dependent on the constants chosen for the alteration products (ρ_a and σ_a). Further work is currently being undertaken to validate and/or modify these constants. At this stage the model shows an average alteration across the flow of about 30%, with predictably higher zones of alteration both above and below the flow in the volcanic breccias.

Figures 7b & 7c show a close relationship between the neutron porosity (HALC) and modelled alteration curves, indicating that porosity is the dominant control on the degree of alteration. This is demonstrated in Figure 8, where these two measurements are cross-plotted. The data fall on to three lines, which pick out three distinct porosity ranges (see Fig. 8b, a histogram of porosity values for this section).

The porosity ranges are a direct expression of the lithologies, with the lowest porosities ($<14\%$; line X in Fig. 8a) occurring within the

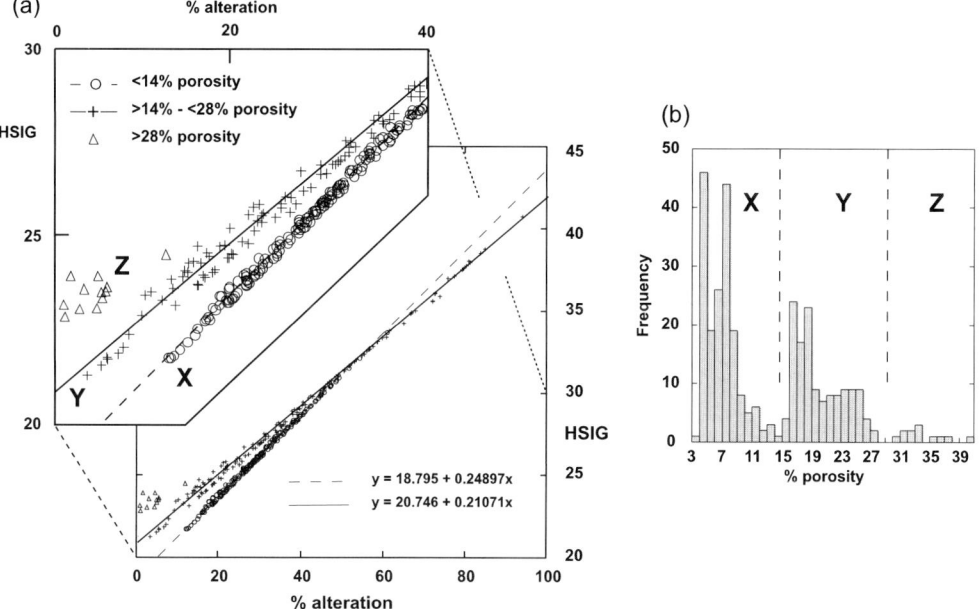

Fig. 8. Relationship between neutron porosity, the modelled amount of alteration and lithology across the upper flow in Hole 801C. (**a**) Cross-plot of HSIG and per cent alteration, divided into three categories depending on porosity. (**b**) Histogram of porosity values showing the essentially natural divisions between low-porosity values in the flow (A) and intermediate range of porosities in the breccias.

flow, while the bulk of the intermediate porosities (14–28%; line Y) are associated with the
breccias. A small additional group with anomalous porosities (>28%; group Z) occurs within
the lower breccia. Here the high porosities are
associated with relatively low levels of alteration, possibly indicating the position of primary
fluid pathways. Regression lines are given in
Figure 8a for lines X and Y the slope increasing
with decreasing porosity. Line X, through the
lowest porosities, gives an intercept of
18.795 cu, only a little above the expected
value of Σ for 'fresh' MORB with zero porosity
(Harvey & Brewer 2005).

Figure 9 shows the down-section distribution
of the three porosity groups. The low-porosity
group (line X) defines the lava flow, while the
higher porosities define the adjacent breccias.
The clustered 'very high' porosity group
(line Z) occurs in the lower breccia, and is
characterized by the lowest resistivity in the
section, and low computed levels of alteration.

The second basalt flow modelled is 11.0 m
thick and lies between 812.0 and 823.0 mbsf.
The LLS and LLD laterolog resistivity traces
across the flow are shown in Figure 10a; the
flow is again picked out by its high resistivity,
higher in the lower part of the flow and decreasing
towards the top. Breccias and pillow lavas with
resistivity values of below about 80 ohm m
lie respectively above and below, the basalt

flow. Figure 10b shows the curves for HALC
and HSIG, as above, derived from Schlumberger's high-resolution APS, while Figure 10c
gives the modelled per cent alteration curve. The
pattern across this flow is more complex than
that of the upper flow described above, with the
neutron porosity falling irregularly from over
20% at the top of the flow to below 5% in the
lower one-third, while the macroscopic thermal
cross-section remains between about 25 and
45 cu, without any apparent systematic trend.
Combining the two curves through the alteration
model (Fig. 10c), however, shows a high level
of alteration in the upper half of pillows below
the flow, and an irregular pattern of alteration
through the flow, which is characterized by
increasing frequency towards the top of the
flow.

Figure 11 shows a cross-plot of HALC and
HSIG, together with a histogram of the neutron
porosity (HALC) measurements. The spread of
neutron porosity values can again be divided
into three groups for descriptive purposes. The
samples in these groups have been identified in
Figure 11b, and each regressed separately
(Fig. 11a). The three groups: 'high' (>28),
'medium' (14–28) and 'low' (<7.5) per cent
porosity, less indicative of the lithology. The
difference in Σ in this example is in response to
the generally higher levels of alteration towards
the top of the flow.

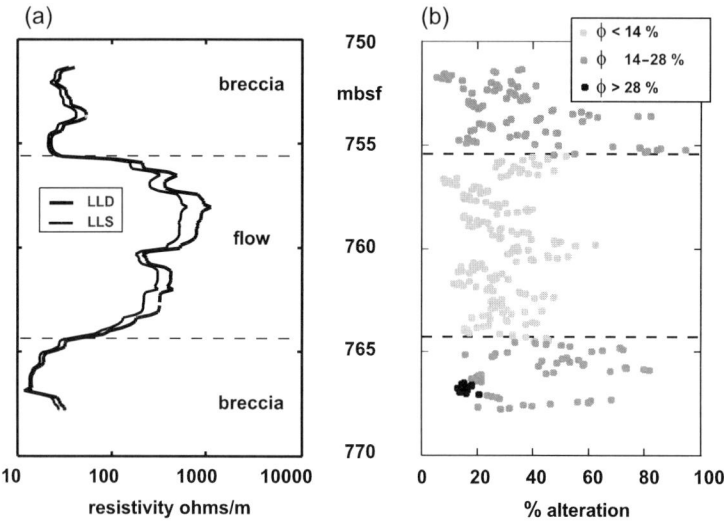

Fig. 9. (**a**) Laterolog resistivity curves across the upper flow in Hole 801C. (**b**) modelled per cent alteration, across
the same flow coded as 'high' (>28), 'medium' (14–28) and 'low' (<14) per cent porosity following the divisions
in the porosity histogram in Figure 8.

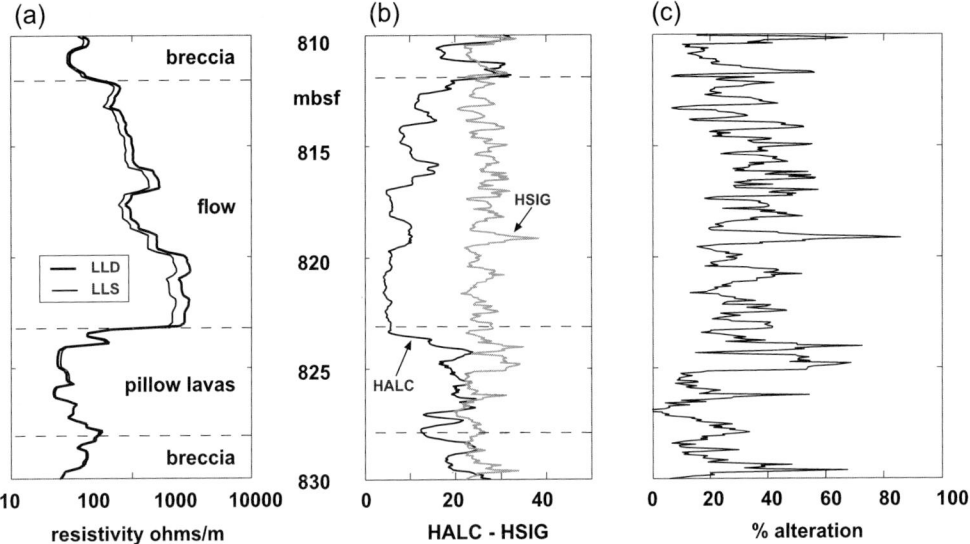

Fig. 10. (**a**) Laterolog resistivity curves across the lower flow in Hole 801C. (**b**) HALC (neutron porosity, in per cent), HSIG (macroscopic thermal cross-section, in capture units) and (**c**) modelled per cent alteration, across the same flow. See text for details.

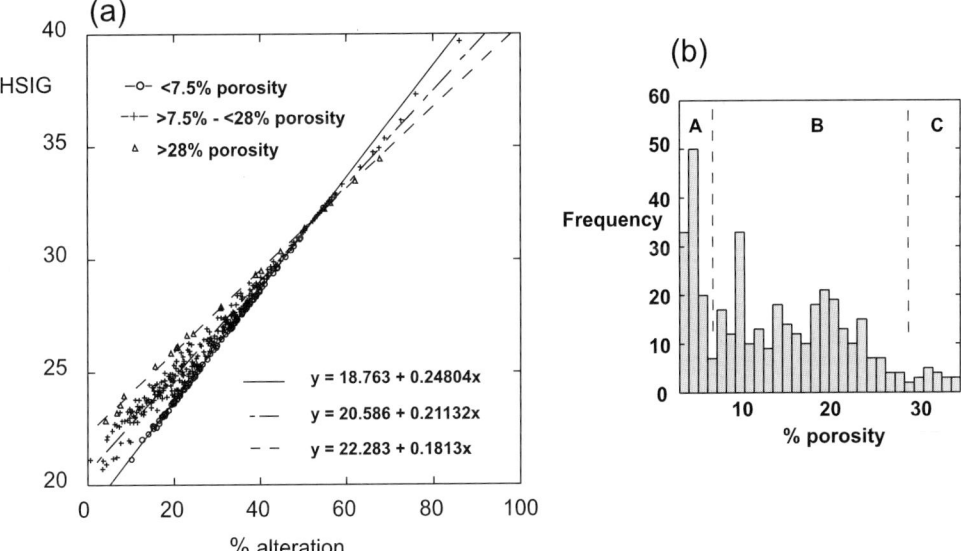

Fig. 11. Relationship between neutron porosity, the modelled amount of alteration and lithology cross the lower flow in Hole 801C. (**a**) Cross-plot of HSIG and per cent alteration, divided into three categories depending on porosity. (**b**) Histogram of porosity values showing the essentially natural divisions between low porosity values in the flow (**A**) and intermediate range of porosities in the breccias.

Conclusions

Computed σ-values derived from near-complete chemical analyses compare favourably with estimates of measured σ-values, and clearly demonstrate that the concentrations of B, Li and the REE strongly influence this value in ocean-floor basalts. The B and Li in ocean-floor basalts are strongly fractionated and frequently enriched by low-temperature alteration. The REE elements are largely unaffected by such processes, and better reflect variations in primary magmatic processes. Consequently, σ-values in ocean floor basalts can be used to:

(1) distinguish between enriched (High σ/Σ) and depleted (Low σ/Σ) MOR basalts, and
(2) to quantify alteration.

Following, calibration of the 'fresh basalt' signature in a drill hole, the Σ-log can be used with other wireline logging data (e.g. resistivity, spectral gamma-ray) to refine the distribution of lithologies and intensity of alteration within a borehole. This is demonstrated with logs through two lava flows in ODP Hole 801C. Furthermore, it is evident that neutron porosity values calculated from wireline logs can give erroneous values, which are strongly biased by the distribution of B, Li and REE. Consequently, such porosity estimates must be treated with caution until the distribution of these large capture cross-section elements are evaluated.

E. Mansley and K. Sharkey supported the analytical work at the University of Leicester, UK.

References

ALT, J. C., KINOSHITA, H., STOKKING, L. B. and Shipboard Scientific Party 1993. *Proceedings of the Ocean Drilling Program, Initial Reports,* **148**, 1–352. College Station, TX (Ocean Drilling Program).

ANDERSON, R. N., ALT, J. C., MALPAS, J., LOVELL, M. A., HARVEY, P. K. & PRATSON, E. L. 1990. Geochemical well logging in basalts: the Palisades sill and the oceanic crust of Hole 504B. *Journal of Geophysical Research,* **95** (B6), 9265–9292.

BARR, S. R., REVILLON, S., BREWER, T. S., HARVEY, P. K. & TARNEY, J. 2002. Determining the inputs to the Mariana Subduction Factory: using core–log integration to reconstruct basement lithology at ODP Hole 801C. *Geochemistry, Geophysics, Geosystems,* **3** (11), doi: 10.1029/2001GC000255.

BREWER, T. S., HARVEY, P. K., HAGGAS, S., PEZARD, P. A. & GOLDBERG, D. 1999. The role of borehole images in constraining the structure

of the ocean crust: case histories from the Ocean Drilling Program. *In*: LOVELL, M. A., WILLIAMSON, G. & HARVEY, P. K. (eds) *Borehole Imaging: Case Histories,* Geological Society, London, Special Publications, 283–294.

BREWER, T. S., HARVEY, P. K., LOCKE, J. & LOVELL, M. A. 1996. The neutron absorption cross-section (Σ) of basaltic basement samples from Hole 896A, Costa Rica Rift. *Proceedings of the Ocean Drilling Program, Scientific Results,* **148**, 389–394.

BREWER, T., LOVELL, M., HARVEY, P. & WILLIAMSON, G. 1995. Stratigraphy of the ocean crust in ODP Hole 896A from FMS images. *Scientific Drilling,* **5** (2), 87–92.

BREWER, T. S., LOVELL, M. A., HARVEY, P. K., WILLIAMSON, G. & HAGGAS, S. 1998. Ocean floor volcanism: constraints from core–log data. *In*: HARVEY, P. K. & LOVELL, M. A. (eds) *Core–log Integration.* Geological Society, London, Special Publications, **136**, 341–362.

CHAN, L. H., EDMOND, J. M., THOMPSON, G. & GILLIS, K. 1992. Lithium isotopic composition of submarine basalts: implications for the lithium cycle in the oceans. *Earth and Planetary Science Letters,* **108**, 151–160.

CHAUSSIDON, M. & JAMBON, A. 1994. Boron content and isotopic composition of oceanic basalts: geochemical and cosmochemical implications. *Earth and Planetary Science Letters,* **121**, 277–291.

HAGGAS, S., BREWER, T. S. & HARVEY, P. K. 2002. Architecture of the volcanic layer from the Costa Rica Rift, constraints from core–log integration. *Journal of Geophysical Research,* **107** (No. B2), ECV 2–1,14.

HARVEY, P.K. 1989. Automated X-ray fluorescence in geochemical exploration. *In*: AHMEDALI, S. T. (ed.) *X-ray Fluorescence Analysis in the Geological Sciences. Advances in Methodology.* Geological Association of Canada, Short Course **7**, 221–257.

HARVEY, P. K. & BREWER, T. S. (2005) On the neutron absorption properties of basic and ultra-basic rocks: the significance of minor and trace elements. *In*: HARVEY, P. K., BREWER, T. S., PEZARD, P. A. & PETROV, V. (eds) *Petrophysical Properties of Crystalline Rocks.* Geological Society, London, Special Publications, **240**, 207–218.

HARVEY, P. K. & LOVELL, M. A. 1989. Basaltic lithostratigraphy in Ocean Drilling Program Hole 504B. *Nuclear Geophysics,* **3** (2), 87–96.

HARVEY, P. K., LOVELL, M. A., BREWER, T. S., LOCKE, J. & MANSLEY, E. 1996. Measurement of thermal neutron absorption cross section in selected geochemical reference materials. *Geostandards Newsletter,* **20**, 79–85.

HARVEY, P. K., TAYLOR, D. M., HENDRY, R. D. & BANCROFT, F. 1973. An accurate fusion method for the analysis of rocks and chemically related materials by X-ray fluorescence spectrometry. *X-ray Spectrometry,* **2**, 33–44.

HERRON, M. M. & MATTESON, A. 1993. Elemental composition and nuclear parameters of some common sedimentary minerals. *Nuclear Geophysics*, **7**, 383–406.

KANAZAWA, T., SAGER, W. W., ESCUTIA, C. & the Leg 191 Shipboard Scientific Party. 2001. *Proceedings of the Ocean Drilling Program, Initial Reports*, **191 [CD-ROM]**. College Station TX (Ocean Drilling Program).

LANCELOT, Y., LARSON, R. L. and Ocean Drilling Program Leg 129 Shipboard Scientific Party, 1990. *Proceedings of the Ocean Drilling Program, Initial Reports*, **129**: College Station, TX (Ocean Drilling Program).

LOFTS, J. C., HARVEY, P. K., LOVELL, M. A. & LOCKE, J. 1996. The relationship between lithology and the neutron absorption cross-section (Σ) of samples from Ocean Drilling Program Leg 149. *Proceedings of the Ocean Drilling Program, Scientific Results*, **149**, 595–599.

NAKAMURA, N. 1974. Determination of REE, Ba, Fe, Mg, Na and K in carbonaceous and ordinary chondrites. *Geochimica et Cosmochimica Acta*, **44**, 287–308.

PLANK, T., LUDDEN, J. N., ESCUTIA, C. & Ocean Drilling Program Leg 185, 2000. Shipboard Scientific Party, *Proceedings ODP, Initial Reports*, **185** (online and CD-ROM). Ocean Drilling Program, Texas A&M University, College Station TX.

REVILLON, S., BARR, S. R., BREWER, T. S., HARVEY, P. K. & TARNEY, J. 2002. A new approach using gamma-ray and geochemical data integration to estimate the inputs to subduction zones, ODP Leg 185, Site 801C. *Geochemistry, Geophysics, Geosystems*, **3** (12), doi: 10.1029/2002GC000344 22p.

RYAN, J. G. & LANGMUIR, C. H. 1987. The systematics of lithium abundances in young volcanic rocks. *Geochimica et Cosmochimica Acta*, **51**, 1727–1741.

SMITH, H. J., SPIVACK, A. J., STAUDIGEL, H., & HART, S. R. 1995. The boron isotopic composition of altered oceanic crust. *Chemical Geology*, **126**, 119–135.

SUN, S.-S. & MCDONOUGH, W. F. 1989. Chemical and isotopic systematics of oceanic basalts: implications for mantle composition and processes. *In*: SAUNDERS, A. D. & NORRY, M. J. (eds) *Magmatism in the Ocean Basins*, Geological Society of London Special Publications, **42**, 313–345.

TEAGLE, D. A. H., ALT, J. C., BACH, W., HALLIDAY, A. & ERZINGER, J. 1996. Alteration of upper ocean crust in a ridge-flank hydrothermal upflow zone: mineral, chemistry and isotopic constraints from Hole 896A. *In*: ALT, J. C., KINOSHITA, H., STOKKING, L. B. & MICHAEL, P. J. (eds) *Proceedings of the Ocean Drilling Program*, **148**, 119–150.

YI, W., HALLIDAY, A. N., ALT, J. C., LEE, D.-C. & REHKÄMPER, M. 2000. Cadium, indium, tin, tellurium and sulphur in oceanic basalts: implications for chalcophile element fractionation in the Earth. *Journal of Geophysical Research*, **105** B8, 18 927–18 948.

Appendix 1 Rare-earth element analysis

All samples were digested in a CEM-MDS2000 microwave using a mixture of 10 ml HF and 3ml $HClO_4$. The rare-earth elements were then separated from the digested solution using ion chromatography columns employing a Dowex resin. The REE were collected in 6N HCl and then converted into nitrates prior to analysis. These solutions were then analysed using a JY-ULTIMA2 sequential ICP; individual analytical lines for each of the REE were selected to optimize sensitivity. The lower limit of detection for the REE is 0.8 ppm with a precision of $\pm 10\%$. A number of international geochemical reference materials were analysed in the same manner as controls on precision and accuracy.

Appendix 2 Carbon and sulphur analysis

Individual samples were weighed into crucibles, with addition of iron and tungsten chip accelerants prior to loading into a radio-frequency induction furnace. The system was initially purged with oxygen, which then continued to stream throughout the combustion process. Power is applied until the accelerants are molten, and carbon is given off as CO_2 and CO and the sulphur is given off as SO_2. A dust filter prevents silicates (a potential infra-red wavelength overlap) from entering the infra-red cells. The combustion gases are passed through a drying tube of magnesium perchlorate, and then pass to the SO_2 infra-red cell. Once determined, the gases pass through a platinized silica gel catalyst to convert any CO to CO_2, any SO_2 is trapped out as SO_3. The CO_2 content is then determined in the CO_2 infra-red cell; gases then vent from the system.

The infra-red absorption cells use a tungsten filament as the source, heated to approximately 850°C. The infra-red is then chopped c.85 Hz and filtered to achieve a monochromatic infra-red wavelength corresponding to the energy of the CO_2/SO_2 adsorption wavelengths respectively. The output from the cells is monitored at 4 Hz, converted from an analogue signal to a digital signal, and the areas of the peaks are integrated. These values are then corrected for sample weight, blank value and calibration factors to give a total carbon/sulphur result.

The lower limit of detection for carbon is 10 ppm, with a precision of $\pm 5\%$ and for S the lower limit is 10 ppm with a precision of $\pm 8\%$.

Microstructure, filtration, elastic and thermal properties of granite rock samples: implications for HLW disposal

V. A. PETROV[1], V. V. POLUEKTOV[1], A. V. ZHARIKOV[1], R. M. NASIMOV[2],
N. I. DIAUR[2], V. A. TERENTIEV[2], A. A. BURMISTROV[3], G. I. PETRUNIN[3],
V. G. POPOV[3], V. G. SIBGATULIN[4], E. N. LIND[4],
A. A. GRAFCHIKOV[5], V. M. SHMONOV[5]

[1]*Institute of Geology of Ore Deposits, Petrography, Mineralogy and
Geochemistry (IGEM), Russian Academy of Sciences, 35 Staromonetny per.,
Moscow 119017, Russia (e-mail: vlad@igem.ru)*
[2]*United Institute of Physics of the Earth (OIFZ), Russian Academy of Sciences,
10 B. Gruzinskaya St, Moscow 123810, Russia*
[3]*Moscow State University (MGU), Vorobievy Gory, Moscow 119899, Russia*
[4]*Krasnoyarsk Scientific Research Institute of Geology and Mineral Resources
(KNIIGiMS), 55 Pr. Mira, Krasnoyarsk 660049, Russia*
[5]*Institute of Experimental Mineralogy (IEM), Russian Academy of Sciences,
Chernogolovka, Moscow district 142432, Russia*

Abstract: This contribution presents preliminary data on the petrographic and mineral–chemical composition of the granitoid and dioritoid core samples of the Itatsky and Kamenny sites in the Nizhnekansky Massif of the Krasnoyarsk Region, in combination with data on structure of their pore space, filtration, elastic, mechanical and thermophysical properties, data necessary to select a site for an underground research laboratory. This laboratory would be a precursor to an underground radioactive waste facility. It was found that the variation of petrophysical properties depends on the degree of metamorphic, hydrothermal–metasomatic and deformational transformations, and differs markedly for two groups of rocks, namely, 1 – granites and granite–gneisses, and 2 – quartz diorites and diorites. Comparative analysis of the data obtained shows that the country rocks of the Kamenny site have a relatively higher degree of heterogeneity in their petrophysical characteristics, than those of the Itatsky site. These results could be applied both to select the preferred site and to make a preliminary estimate of the thermal, hydraulic and mechanical behaviour of country rocks in the vicinity of a prospective underground heat-generating waste repository.

At the Mining and Chemical Combine (MCC) in Zheleznogorsk, Krasnoyarsk Region, there is a radiochemical facility where, for many years, spent nuclear fuel (SNF) has been reprocessed and high-level radioactive waste (HLW) has accumulated (Laverov *et al.* 2000). In 1992, work was initiated to identify suitable sites for HLW underground disposal. After reconnaissance studies, comparative assessment and preliminary geological–geophysical investigation of the Nizhnekansky granitoid massif, which is located very close to the zone for restricting contaminant release of the MCC, the Itatsky (I) and Kamenny (K) sites were chosen (Anderson *et al.* 1998). It was recommended that investigation of

these sites should proceed with the aim of identifying a place for the establishment of an underground laboratory and the eventual construction of an HLW long-term monitored retrievable storage or underground disposal repository.

The Nizhnekansky granitoid massif is located in the southern part of the Yeniseisky mountain ridge, within the boundaries of the Angaro-Kansky block, which is a projection of the Baikalsky basement (Datsenko 1995). The massif is a pluton elongated to the NW. Its total area, including the zone covered by the Jurassic sediments, amounts to 3500 km^2. Gravimetric survey data indicate that the thickness of the massif in the area of the central

From: HARVEY, P. K., BREWER, T. S., PEZARD, P. A. & PETROV, V. A. (eds) 2005. *Petrophysical Properties of Crystalline Rocks*. Geological Society, London, Special Publications, **240**, 237–253.
0305-8719/05/$15.00 © The Geological Society of London 2005.

magma intake channel is at least 5–6 km. The massif is intruded into intensively metamorphosed rocks of Archaean and Early Proterozoic age.

The massif is made up of granites and to a lesser extent leucogranites, plagiogranites, granodiorites, quartz diorites, diorites and their subalkaline analogues. Diorites and quartz diorites belong to the first intrusive phase, while biotite granites, adamellites, leucogranites and their subalkaline varieties belong to the second phase of the massif's formation. The vein rocks of the first phase occur as aplite, pegmatite and rarer granite porphyry dykes, while those of the second phase include dioritic porphyrites and lamprophyres. The absolute age of granites is estimated at 920 ± 50 Ma, based on the work of Volobuev & Zykov (1961) using K–Ar methods (on micas and feldspars) and Pb–U methods (on accessories).

In 1999, at the I and K sites, 500 m and 700 m boreholes were drilled (1I-500 and 1K-700 respectively). The core material of these boreholes is the subject of this research, in the following study areas:

- unusual features of the texture and mineral–chemical composition of rocks;
- filtration, elastic, mechanical and thermophysical characteristics of rocks;
- the effect of superimposed hydrothermal–metasomatic and deformational transformations of rocks on the anisotropy of their physical properties;
- comparative analysis of the prospective sites, aimed at assessing their homogeneity/heterogeneity using the investigated parameters, and choosing the most favourable site for the construction of the underground SNF long-term monitored retrievable storage or HLW disposal facility.

All investigations were performed with the same rock samples. Figure 1 shows the patterns of core sample marking and the arrangement of sensors for measuring elastic-wave velocities in cores. In all subsequent studies for core samples (1), with a zero mark (2), the velocity of P-wave and surface Rayleigh wave propagation is along and perpendicular (3) to the core axis. Furthermore, the core was cut to prepare samples for uniaxial compression resistance tests (4), heat measurements (5), identification of gas permeability (6), volume weight, density and parameters of water saturation (7), as well as petrographic and petrochemical research (8).

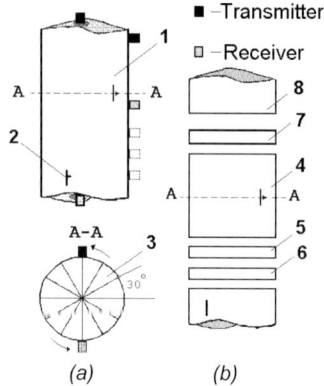

Fig. 1. The patterns of core sample marking (**a**) and preparation for tests (**b**).

Petrography and mineral–chemical composition

Petrographic and mineral–chemical studies have shown that in the sections of the I and K sites, there is a wide range of granitoids and dioritoids, such as leucogranites, granites, adamellites, granodiorites, tonalites, quartz diorites, diorites and quartz monzodiorites, as well as vein formations (spessartites, dioritic porphyrites, aplites, pegmatites, microgranites and granite porphyries). Data on the mineralogy of the main varieties of rocks, including the altered varieties, are summarized in Table 1. The country rocks have been deformed and metamorphosed, and have suffered some local hydrothermal–metasomatic alteration.

During the regional and amphibolite-facies metamorphism (quartz, feldspars, biotite, amphiboles) granite–gneiss, plagiogneiss and crystalline schists were formed. Closely associated with these were some high-temperature postmagmatic metasomatic changes. With subsequent greenschist metamorphism, a variety of schists were formed, including chlorite–epidote–actinolite schists. Deformation occurring locally as zones of mylonitization, cataclasis and blastesis (mylonites-1 and cataclasites-1) up to 50 m in thickness, together with the widespread development of foliated, gneissic and blastocataclastic textures, is also associated with these metamorphic and metasomatic changes.

Low-temperature hydrothermal–metasomatic alteration is observed around zones of cracking in borehole sections, as broad (up to 100 m) zones of chloritization, sericitization and argillization of rocks, which are accompanied by zones

Table 1. *Mineralogical compositions of fresh and altered varieties of granitoids, dioritoids and lamprophyres (volume %)*

Rock	Quartz	K-feldspar	Plagioclase	Biotite	Hornblende	Sec.	Acc.
Leucogranite	25–30	45–50	20–25	2–5	—	1–3	3–5
Granite	30	30–35	25–30	5	—	5–7	5–7
Adamellite	25–30	25–30	30–35	5–7	—	5–7	5–7
Granodiorite	25–30	10–15	45–50	7–10	—	1–3	3–5
Tonalite	15–20	5–10	55–60	10–15	5	1–3	2–3
Quartz diorite	15–20	5	55–60	5–10	10–15	3–5	2–3
Quartz monzodiorite	15	5	55–60	5–10	5–10	3–5	2–3
Diorite	5	5	50–55	5–10	20–25	1–3	2–3
Altered spessartite	—	—	50–55	—	25–30	15–20	3–5
Altered granite–gneiss	25–30	30–35	25–30	—	—	10–15	1

Note: Sec., secondary minerals (muscovite, sericite, chlorite, pyrite, carbonate, clay minerals, leucoxene, pyrite); Acc., accessory minerals (zircon, apatite, sphene, rutile, magnetite).

(up to 15 m) of intensive veinlet development, cataclasis and brecciation (mylonites-2 and cataclasites-2). Brown metasomatic feldspars, carbonate, chlorite, sericite, clay minerals (illite, mixed-layered illite–smectite, kaolinite), leucoxene and hematite are the main minerals in these altered rocks.

Evaluation of the different lithologies in the sections of the I and K sites allows the identification of the following trends. In the upper part of the I site section to a 170-m depth, frequent alternation of granites and granodiorites with tonalites and diorites is observed. The central part of the section is an extensive interval of quartz diorites, diorites and tonalites. Below the 310-m mark and to the very bottom of the borehole section, the alternation of granites, adamellites, quartz diorites and diorites of relatively persistent thickness (up to 50 m) is observed. The quartz diorites and diorites are connected by transition varieties. Vein and dyke formations occur in the lower part of the section, at a depth of 400 m. The distribution of the different lithologies within the boundaries of the section under consideration is as follows (%): granites, including granite proper and adamellites (27), granodiorites (10), tonalites (10), quartz diorites (5), diorites (16), lamprophyres and dioritic porphyrites (2).

In the section of the K site, down to a 500 m depth, leucocratic granitoids (granites, adamellites, leucogranites and alaskites) clearly predominate. At 100–220 m depth, granodiorite interlayers occur. The 220 m to 460 m depth interval includes predominantly leucogranites and alaskites, while below the 500 m mark and down to the very bottom of the borehole section there is a clear predominance of granodiorites, tonalites and quartz diorites. Leucogranites make up 28%, alaskites 7%, granites,

including granites proper and adamellites 21%, granodiorites 18%, tonalites 16%, quartz diorites 5%, and lamprophyres and dioritic porphyrites 5% of the section total, respectively.

In comparing the two sections it is evident that, for the K site, granites and leucogranites predominate (over 50%) and have a persistent thickness, whereas at the I site this is true for quartz diorites and diorites. At the same time, diorites do not occur in the K site section, whereas in the I site section, leucogranites and alaskites are not observed.

With respect to the chemical properties of the rocks, the most notable geochemical characteristics are as follows. Quartz diorites are high-AL potassium–sodium rocks with a high agpaitic coefficient. They are characterized by a normal content of silica and alkalis. From their alkalinity type, the granodiorites and granites belong to the K–Na series. Only the porphyritic blastocataclastic granites of the K site, which do not contain alkaline dark-coloured minerals, occur partially in the field of subalkaline granites. For granodiorites, the Na_2O/K_2O ratio is estimated at 1.0; for granites it is 0.9–1.1; while for porphyritic granites it is only 0.5–0.8. The compositions of the main rock varieties are summarized in Table 2.

With subsequent deformation and metamorphism some changes in chemical composition have been noted. For the mylonites-1 and cataclasites-1, the content of SiO_2 in granites remains constant, although higher concentrations of MgO, CaO, K_2O and LOI, and somewhat lower concentrations of Na_2O, are observed. During transformation of the quartz diorites, slightly higher concentrations of $Na_2O + K_2O$ are observed, while the content of Al_2O_3 and TiO_2 remains unchanged. In the process of the development of deformational and low-temperature

Table 2. *Chemical compositions of fresh and altered rock varieties (wt%)*

	1	2	3	4	5	6	7
SiO_2	76.26	72.05	66.52	62.00	54.83	71.00	51.22
Al_2O_3	13.12	14.09	14.46	18.24	15.50	13.94	16.19
$Fe_2O_3{}^*$	0.75	2.01	4.77	4.50	7.27	4.07	12.39
MnO	0.02	0.07	0.08	0.08	0.12	0.06	0.09
MgO	0.14	0.53	2.20	2.02	8.08	0.86	5.00
CaO	0.49	1.33	3.75	3.76	5.77	1.88	8.01
Na_2O	3.76	3.62	3.31	4.93	3.34	3.46	1.97
K_2O	4.93	5.11	3.44	2.80	1.68	4.64	0.80
TiO_2	0.11	0.24	0.57	0.66	1.05	0.36	0.95
P_2O_5	0.01	0.07	0.15	0.23	0.44	0.09	0.21
LOI	0.36	0.84	0.51	0.58	1.81	1.36	7.15
Total	99.95	99.96	99.76	99.80	99.89	101.72	103.98

Note: 1, leucogranite; 2, granite; 3, granodiorite; 4, quartz diorite; 5, diorite; 6, altered granite–gneiss; 7, altered spessartite. LOI, loss of ignition at 105–1200 degrees centigrade. *Total iron content is expressed as Fe_2O_3.

hydrothermal–metasomatic transformations of the mylonites-2 and cataclasites-2, distinctly higher CaO and LOI contents are observed, while concentrations of elements such as SiO_2, Al_2O_3 and $Na_2O + K_2O$ are lower.

Density and filtration properties

The principal objective of research into the properties of these rocks is the quantitative characterization of the process of their free (capillary) saturation and the identification of the factors governing this process. Using these parameters, a comparative assessment has been undertaken, of both the main varieties of granitoids of the massif, and of some intervals in the boreholes.

The density (g cm^{-3}), volume weight (g cm^{-3}) and effective porosity (Φ, %) of 56 samples were assessed by the method of buoyancy weighing of the preliminary dried samples and those fully saturated with fresh water.

The volume weight corresponds to the mass of 1 cm^3 of an absolutely dry sample, while the density corresponds to 1 cm^3 of its mineral phase. The applied methods of free saturation are not suitable for the investigation of closed porosity. Therefore, the rock density was determined with a some degree of approximation.

Parameters of rock saturation dynamics, such as the arbitrary-instantaneous volume saturation (A, %), average intensity of saturation (B_{mean}, hour^{-1}) and saturation half-time ($T_{1/2}$, hour) were determined using sample weight increases over 0 and 30 s; 1, 3, 5 and 20 min; 1 and 3 h; and 1, 3, 5, 9 and 12 days. In cases of saturation stoppage the measurements were performed for a further three to four days to confirm the state of the sample, although this period was not taken into account in any calculations.

There is a time (t) exponential dependence of the value of pore-space volume saturation (m_i):

$$m_i(t) = A + m_o e^{B_i(t_i - t_A)} \qquad (1)$$

where A (%) is arbitrary-instantaneous saturation, m_o (%) is a percentage of effective porosity filled in the exponential mode, t_i and t_A (hour) are the moments of time of the ith observation and completion of arbitrary–instantaneous saturation accordingly, B_i (hour^{-1}) is the saturation intensity index within the interval between t_i and t_A.

$$B_i = \{ \ln[(\Phi - m_i)/m_0] \}/(t_i - t_A) \qquad (2)$$

where Φ (%) is effective porosity ($\Phi = A + m_0$).

The B_{mean} is approximately equal to a simple average of B_i values, and corresponds to the mean value of the saturation intensity index for the sample under consideration. For rocks in which pores with a size of several microns or several dozen microns determine the effective porosity, the mean value of B is usually 0.3 hour^{-1} at least.

Saturation half-time ($T_{1/2}$) corresponds to the time required for the sample saturation, which is equal to half of the top value of saturation in the mode of exponential dependence, and it follows the completion of the arbitrary-instantaneous saturation. Saturation half-time is calculated by the following formula:

$$T_{1/2} = 0.693/B_{mean}. \qquad (3)$$

For permeability measurements of 24 samples, a hydrostatic setup was used, which was developed by A. A. Grafchikov and V. M. Shmonov and made in the Institute of Experimental

Mineralogy, Russian Academy of Science (RAS). It allows the measurement of permeability in the range from $n \cdot \mathrm{E}\text{-}15$ to $n \cdot \mathrm{E}\text{-}22 \ \mathrm{m}^2$, at temperatures up to 300°C and effective pressures up to 50 MPa. The mini-cores used for these permeability measurements were about 14 mm in diameter, and about 10 mm in length. Measurements were performed at room temperature and under confining pressures up to 3 MPa. For the recalculation of permeability to water, the pulse decay method was used. It was modified to account for the variation in the thermodynamic properties of the flowing gas.

A study of the characteristics of the pore space was performed using modified structural–petrophysical analysis (Starostin 1979). This method combines the identification of a P-wave velocity anisotropy factor in dry (a_d) and fresh-water-saturated (a_s) samples, anisotropy of crack-pore space (a_{st}), calculation of the absolute value of crack–pore voidage (CPV) and intensity of microcrack network development (I), as well as the coefficient of fluid conductivity of rocks (K_{fl}).

The anisotropy of the elastic properties of 56 rock samples was studied using an ultrasonic sounding of discs of 10 mm thickness. Velocities of P-waves (V_P) at a nearly 2 MHz frequency were measured along 12 radial directions (every 30° of the turn) at 90° (in the centre of the sample) to 30° angles of inclination to the plane of the disc under investigation. As a result, the measurements covered the disc volume within a 120° angle.

The measurements were made on both absolutely dry and water-saturated samples. In dry samples, the velocity minima were identified, the most contrasting of which were associated with microcracks and their nets. Velocity maxima are determined by the orientation of mineral grains. In water-saturated samples, anomalies in velocities are determined by grain orientation only.

The data obtained were used to construct diagrams of spatial anisotropy on a Schmidt grid. Thus, for every sample, three diagrams were made: diagrams for dry and water-saturated conditions, and difference diagrams. Difference diagrams were constructed by subtraction of the velocity values for the dry sample from the corresponding values for the water-saturated sample. The maxima of the velocity increases in this case correspond to microcrack networks in the sample. Anisotropy factors for dry (a_d) and saturated (a_s) samples were determined as the percentage ratio of the difference between the maximal and minimal velocities to the average value for the sample. The anisotropy factor of crack–pore space (a_{ps}) was determined by the difference diagram as the ratio of difference between the maximum and the minimum of the velocity increase to the average value for the sample.

The anisotropy of a rock, in general, is characterized by a_d, while the anisotropy associated only with an oriented texture or structure is characterized by a_s. A such factor as a_{ps} indicates the presence of nets of oriented crack–pore channels, which are due to substantial extension in the zones of microcracks formed by such channels, and can distinctly increase the rate of saturation and the distance of component transfer in aqueous solutions. It is clear that, along with the above, it is very important to take into account the volume of the transferred components, which is directly dependent on the value of opening (voidage) of every network. With the pressure gradient in crack zones, which is determined by hydrogeological processes, the overall permeability coefficient is proportional to the square of the crack aperture and to the value of the crack conductivity of the rock. Thus, the real process of the formation of contamination aureoles might be even more intensive than is actually the case with the modelling of capillary saturation.

The absolute value of crack–pore voidage (CPV) was determined by two methods. In the first case it corresponds to the average increase of V_P on difference diagrams. A more physically correct evaluation is based on the theoretical work of O. P. Yakobashvili (1992). According to his findings, the absolute value of fracturing (d/l) is the quotient of the division of the average aperture of the microcrack (d) by the average distance between cracks in the net (l), and it actually corresponds to the value of the crack–pore voidage of the rocks. The absolute value of fracturing is related to the level of anisotropy, as follows:

$$d/l = w[(V_o/V_i)^2 - 1] \qquad (4)$$

where d/l is the absolute value of fracturing $w = 0.05$ for water-saturated and $w = 0.0001$ for dry samples, and V_o and V_i represent the P-wave velocity in a crack-free sample, which was measured along any direction accordingly.

Based on equation (4) and the above-mentioned w values, for the direction of wave propagation perpendicular to a given net of microcracks, the following relation may be used:

$$500[(V_o/V_{sat})^2 - 1] = [(V_o/V_d)^2 - 1] \qquad (5)$$

where V_{sat} and V_d are wave velocities in water-saturated and dry samples.

From equation (5) it follows that $V_o \approx V_{sat}$.

Taking due account of equation (5), equation (4) may be rewritten as follows:

$$d/l \approx 10^{-4}[\Delta V(2V_d + \Delta V)/(V_d)^2] \quad (6)$$

where $\Delta V = V_{sat} - V_d$ are the values of the velocity increase with saturation.

There is a formula derived by O. P. Yakobashvili (1992) for the volumetric absolute value of fracturing, which is a total of partial moduli for three mutually perpendicular directions in a sample. Theoretically, this value is a constant for a sample under consideration. However, such calculations were not performed, because, in this case, it was necessary to study the wave velocities for all directions. From equation (6) it may be shown that the volumetric absolute value of fracturing is approximately proportional to the average increase of wave velocities in the sample.

Thus, the absolute value of fracturing for the samples may be calculated as two options: partial (CPV_{part}) and volumetric or general (CPV_{gen}). The CPV_{part} characterizes the degree of opening for the main network of microcracks, which is usually presented on difference diagrams as contrast and great maxima. The volumetric absolute value was approximately calculated as the value of the P-wave velocity average increase with saturation. An anisotropy diagram for one of the samples is shown in Figure 2.

The intensity of the microcrack network development (I) was calculated as the total of three values: the number of the minima on the

diagrams for dry and saturated samples, and the maxima (corresponding to these minima by orientation) in the difference diagram. This additional analysis allowed more precise determination of the intensity of the microcrack network's development, as with the increase in intensity the maxima (minima) of the diagrams were shown more clearly.

To characterize the intensity of the fluid conductivity of rocks, an integrated additive petrophysical index (K_{fl}) was used:

$$K_{fl} = B_{mean} + 10A + \Phi + a_{st}/100$$
$$+ CPV_{gen}/100 + CPV_{part} + I/10 \quad (7)$$

The applied parameters characterize the rate and volume of saturation of different classes of pores. The coefficients for some parameters were selected so that the values of all characteristics were of the same order. In this case, use is made of both direct (experimental) and indirect parameters obtained from the data of ultrasonic sounding. Saturation half-time is not taken into account in calculations, as it is the reciprocal of B_{mean}. The anisotropy factor for the dry sample (in addition to the anisotropy of the crack space) includes the anisotropy of the solid phases in the rocks, and therefore it is also excluded from the calculations.

Results

The density and volume weight increase as the rocks become more basic: from leucogranites and granites (2.61–2.63 g cm^{-3}) to quartz diorites and tonalites (2.74–2.79 g cm^{-3}). Granodiorites, granite–gneisses and other gneisses demonstrate intermediate values (2.65–

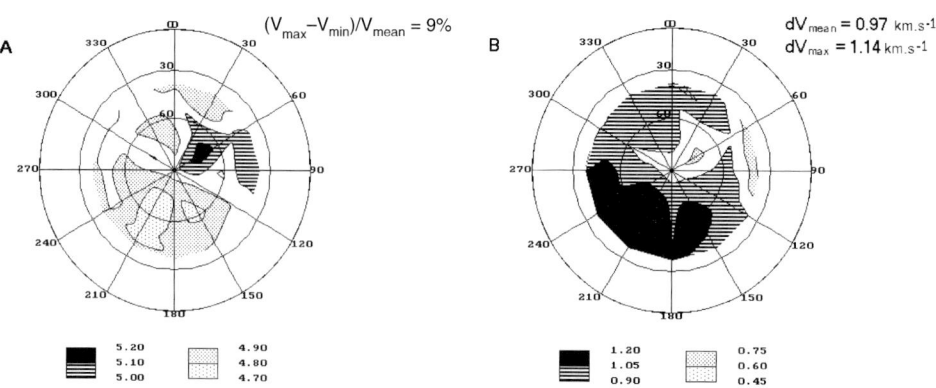

Fig. 2. Diagrams of elastic anisotropy for the core sample (I 459.2) under dry conditions (**a**) and a difference diagram (**b**). Velocity scales (km s^{-1}) and velocity increase (dV) of P-waves are shown below the related diagrams.

2.75 g cm^{-3}). The density and volume weight of the rocks increase with depth at the K site, whereas at the I site the maximum values are typical of the central part of the section.

The values of effective porosity of rocks vary from 0.14 to 0.95%. The samples of granites showing hypergene alteration (up to 1.31%) and spessartite dykes with metasomatic alterations (5.6%) are the exception. Rocks of the I site are less porous, while at the K site the effective porosity of rocks declines with depth.

The research into permeability included the modelling of the upflow, i.e. the filtration occurred in parallel with the axis of the core samples. Table 3 presents the measurements obtained.

The data in Table 3 show that the values of gas permeability vary from n·E-20 to n·E-16 m^2. Maximal values of permeability are typical for cataclastic varieties and rocks with metasomatic alteration. Comparing data on the pore space structure, effective porosity, water saturation dynamics and permeability suggests that in the studied samples secondary porosity prevails, and microcracks are the principal channels for fluid filtration.

Data on the nature of the fluid conductivity coefficient for rocks at the I and K sites are summarized in Figure 3. In general the values of filtration parameters for the K site substantially

exceed those for the I site. Furthermore, the average values of these properties for all rock varieties of the K site are also higher, and this is particularly typical for quartzdiorites, tonalites and intensively metasomatically altered rocks, including spessartite. This suggests that the unusual nature of the pore space structure is the main reason for such a difference.

With approximately equal Φ values for quartz diorites–tonalites (0.42% for K and 0.53% for I) higher values of CPV_{gen} and CPV_{part} (107 and 0.83, 77 and 0.42 accordingly) are more typical of the K site. It is also affects the growth of B_{mean} (7.24 for K and 1.24 for I). Along with this, for metasomatically altered rocks, the dramatic growth of the effective porosity is of great importance, as is the percentage of large pores (A – up to 0.53%). The rate of micropore saturation (B_{mean}) for these rocks at both sites is nearly equal (1.61 hour^{-1} and 1.28 hour^{-1}). However, the opening of the main network of microcracks of metasomatic rocks for the K site substantially exceeds that for the I site ($CPV_{part} = 2.06$ and 0.42 accordingly). It is possible that, with such large values of pore volumes of different types, the rate of micropore saturation decreases due to the complicated structure of the pore channels. This is demonstrated by

Table 3. *Permeability of rock samples from the Nizhnekansky Massif*

Sample no.	Rock	Permeability (m^2)	Determination error (%)
I 88.3	Tonalitic quartz diorite	9.147·E-20	15.81
I 124.7	Altered granite–gneiss	1.181·E-19	5.44
I 142.6	Altered granite–gneiss	3.712·E-20	4.04
I 189.9	Quartz diorite–monzodiorite	2.617·E-19	7.50
I 247.6	Quartz diorite	2.432·E-18	6.23
I 304.2	Quartz diorite	1.858·E-18	2.25
I 330.8	Adamellitic granito-gneiss	6.305·E-19	9.16
I 323.1	Adamellitic gneiss	9.045·E-20	15.46
I 357.2	Quartz diorite	3.092·E-19	5.37
I 459.2	Granite	3.949·E-19	14.09
I 491.7	Adamellitic gneiss	8.201·E-19	6.27
I 504.8	Quartz diorite	9.595·E-19	5.66
K 120.5	Porphyritic adamellite	6.377·E-20	8.25
K 301.6	Leucogranite	4.844·E-19	5.43
K 336.2	Leucogranite	1.064·E-19	1.97
K 420.5	Leucogranite	1.541·E-18	5.47
K 465.1	Leucogranite	3.528·E-18	7.38
K 546.7	Leucogranite	2.537·E-19	3.53
K 560.8	Porphyritic granodiorite	1.488·E-18	4.45
K 613.1	Porphyritic adamellite	2.307·E-18	6.24
K 627.6	Porphyritic cataclastic granite	1.639·E-18	6.61
K 647.4	Quartz diorite	8.589·E-18	3.04
K 674.0	Altered spessartite	5.503·E-16	10.92
K 702.9	Quartz diorite	1.829·E-18	6.09

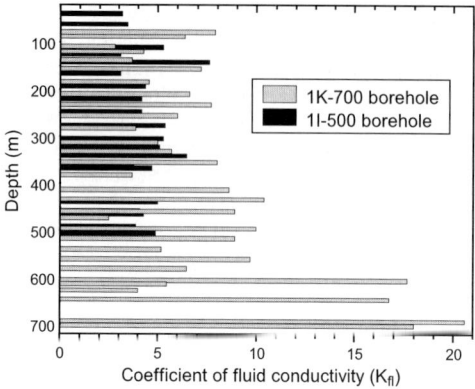

Fig. 3. The variation of the coefficient of fluid conductivity of rocks by depth in the I and K boreholes.

the lower anisotropy factor of the pore space (66% and 89% respectively).

Minimal values of K_{fl} are typical of granodiorites and adamellites, and slightly higher values are more characteristic of granites, alaskites and leucogranites, as well as granite–gneisses and cataclastic granites. Average anisotropy factors and the intensity of microcrack development in the rocks of both sites show few differences. However, the dynamics of saturation for the same types of rocks in some cases differ greatly. Therefore, this suggests that the geological conditions of their formation, or the nature of subsequent deformation, had some characteristics, which affected the structure of the pore space.

Nearly all types of rocks at the I site (with the exception of the metasomatic rocks) have a higher density than their analogues at the K site. Presumably, these are characterized by lower values of closed porosity. The average values of P-wave velocity for saturated samples of quartz-diorites and tonalites for both sites differ substantially: 5.97 km s^{-1} for K and 6.53 km s^{-1} for I. As follows from equation (5) above, the values of velocities in a water-saturated rock, which has no open cracks, are nearly equal to those in a solid rock. Moreover, at the I site, quartz-diorites and tonalites occur throughout the entire section, while at the K site they are confined to deeper horizons. It is entirely possible that the difference in density and average P-wave velocity in saturated samples can be explained by closed porosity, as the variations in composition are minimal.

For granite–gneisses and cataclastic granites at both sites, the higher values of the a_s anisotropy factor are characteristic. This is controlled

by the orientation of minerals, and it demonstrates the presence of substantial plastic deformation. At the same time it suggests that cataclasis has not resulted in the formation of extensive nets of microcracks that could have been reflected in a meaningful increase in the elastic-wave velocity or in the saturation volume of the rock samples.

With regard to the regularity of changes of the studied properties for every site, the following must be noted. At the K site the coefficient of fluid conductivity increases with depth, and it is determined, mainly, by the increase in the anisotropy of the crack–pore space and the intensity of the microcrack network development (see Fig. 3). The increase in Φ and CPV_{part} values downwards to moderate depths and their further decline (the irregular values at a depth of 674 m are not taken into account) indicates that the volumes of pores of different types correlate with each other. Of importance for the identification of petrophysical zoning is the variation in rock composition. In the upper part of the section, granites prevail, while, at deep horizons, rocks are predominantly quartz-diorites and tonalites characterized by high K_{fl} values.

At the I site, the value of K_{fl} increases with depth. It is determined by the increase in B_{mean}, CPV_{gen} and CPV_{part}, i.e. by the degree of opening of microcracks and their networks. However, the absolute values of these and other parameters included in the formula for the calculation of K_{fl} are noticeably lower for the I site when compared to the K site. This is true both for separate intervals and for the section average value. Accordingly, both the K_{fl} value and the rate of its increase with depth are substantially lower for the I site. Together, the data obtained predict lower values of the rate of fluid flow in rocks and the volume of fluid filtration per time unit, as well as the total volume of the ultimate saturation of rocks.

Thus, it may be concluded that the I site is safer in terms of radionuclide transport in rocks than the K site.

Elastic properties

The elastic properties of the granitoids were studied using core samples, in addition to the work described above on disc samples. Parameters measured on 56 dry core samples included the velocity of ultrasonic P-wave propagation in a vertical direction (V_{Pvert}) and perpendicular (V_{Prad}) to the core axis (at an operating frequency of up to 1.1 MHz).

In addition, the velocities of surface Rayleigh waves (V_R) along the surface of the core sample were also determined. A determination of travel-time values on different distances along the sample surface underlies the method of V_R measurement. Using a set of time values, the velocities of surface Rayleigh waves should be calculated using simple formula $V_R = 1/(dt_n/dl_n)$, where t_n is the travel time on the l_n position of the receiver from the transmitter. The measurements were made at room temperature with samples of 140 to 250 mm length and 35 to 58 mm diameter.

The core size exceeded the length of an ultrasonic wave, and therefore the velocity of P-wave propagation in a vertical direction along the core axis allowed the neutralization of the effect arising from textural peculiarities in some parts of the sample.

The variation of the velocity of P-waves perpendicular to the core axis is a measure of the lack of uniformity in their propagation in any section of a particular core sample and allows the identification of elastic anisotropy oriented relative to the core axis. The comparison of V_{Pvert} and V_{Prad} data enables an assessment to be made of the anisotropy of elastic properties in the vertical and horizontal directions of the borehole section.

The magnitude of the velocity of elastic surface Rayleigh wave propagation gives some information on the expected velocity of S-wave propagation, both for a particular sample and for the entire borehole. To assess the representativeness of the data obtained, measurements were made of shear wave velocities (V_S) in 19 samples.

To enable these measurements to be made, special devices were developed to provide similar conditions of acoustic contact of emitters and receivers with the samples. For the design and manufacture of these devices, as well as for the development of measurement methods, reference materials such as alloyed steel, quartz glass and acrylic plastic, were used. The shape and size of these calibration samples were similar to those of the core samples.

It was shown that the P-wave velocity in rocks increases with saturation with water or other fluids. In this case, the effect of the presence of microcracks is neutralized. To avoid the effect of the saturation of samples, a viscous gel of polysaccharides with low water content was used for the acoustic contacts. Optical microscopy examination indicates that the gel did not penetrate into the pore–crack space. It is important to note that the results of measurements on dry samples enable the assessment of the degree of fracturing on the depth and relaxation behaviour of the cores after they are bought to the surface.

To work out the applied problems it was necessary to consider the anisotropy of the elastic properties of the rocks. The values of anisotropy factors of P-wave velocities were calculated both for the sample in general (A) and for the radial plane (A_{rad}). The calculations were performed using the following formulae:

$$A = [(V_{Pvert} - V_{Prad\ mean})/V_{Prad\ mean}]100\% \quad (8)$$
$$A_{rad} = [(V_{Prad\ max} - V_{Prad\ min})/V_{Prad\ min}]100\% \quad (9)$$

where $V_{Prad\ mean}$ is a simple average of velocity values for 12 radial directions, $V_{Prad\ max}$, and $V_{Prad\ min}$ are the maximal and the minimal values of those measured in this section.

Figure 4 shows various types of polar diagrams of the distributions for P-wave velocities in the radial section for both isotropic and anisotropic cores.

Results

The comparison of P-wave and surface Rayleigh wave velocities in granitoid rocks of borehole sections at the K and I site (Fig. 5a & b) show a clear distinction between these two sites. First, the range of V_P and V_R velocities at the K site is substantially broader than that at the I site. The range of V_{Pvert} velocities in the 1K-700 borehole is from 2.460 km s^{-1} to 5.670 km s^{-1}, while in the 1I-500 borehole it is from 5.000 km s^{-1} to 5.875 km s^{-1}, i.e. the degree of non-uniformity of granitoids of the K site as a function of their elastic properties is substantially higher than that of the I site. P-wave average velocities along the core axis for the K site amount to 4.794 km s^{-1}, while for the I site they are 5.589 km s^{-1}, i.e. higher by 0.800 km s^{-1}. It is significant that along the sections of both boreholes, elastic wave velocities in the radial direction are always higher and change similarly to V_{Pvert}. Lower velocities of elastic waves at the K site can be explained both by the mineral composition of rocks and, presumably, by greater effect of tectonic stress. This assumption is based on greater anisotropy of samples from the K site when compared to those from the I site.

Anisotropy factors A and A_{rad} have a broader data scattering at the K site and, in this case, the A factor is characterized by a broader data scattering than the A_{rad} factor. At the I site these factors have lower values and are characterized by nearly equal data scattering (Fig. 5c & d).

Data on the velocities and anisotropy of elastic waves prove the greater uniformity of physical

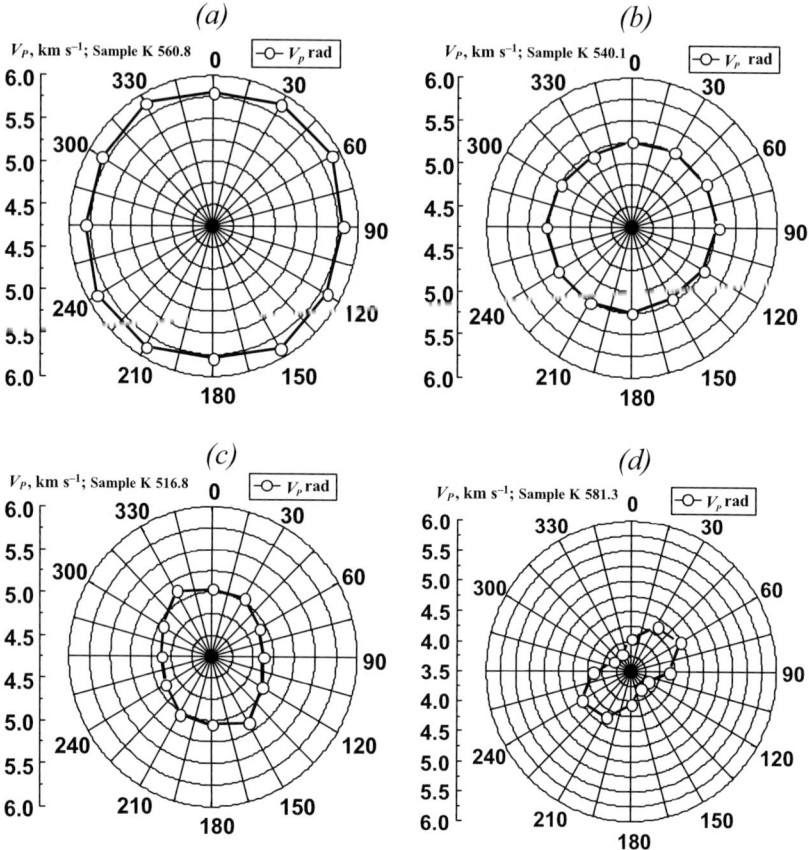

Fig. 4. Typical polar diagrams V_{Prad} for granitoid samples: (**a**) isotropic sample with high velocity; (**b**) isotropic sample with low velocity; (**c**) anisotropic sample with low velocity; (**d**) sample with a high degree of anisotropy.

characteristics of rocks in the geological section of the I site. Rock samples from this site have higher average ratio of the P-wave and the surface Rayleigh wave velocity as compared with the K site. For instance, for the K site the V_{Pvert}/V_R value is within a 1.4–1.9 interval, while for the I site it ranges from 1.7 to 1.95 (Fig. 6a & b). For these velocity values it can be assumed that the granitoids of the I site are relatively less liable to deformational transformation: microfracturing in particular. It is known that rock strength reduction is reflected in the growth of V_P/V_S relation on the one hand, while, on the other hand, there is a direct relationship between V_S and V_R. Therefore it suggests that a similar trend of strength variation will correlate with the V_{Pvert}/V_R relation. Based on this supposition, one can expect a relatively

broader scattering of values of the ultimate strength of rocks at the K site.

Investigation of the anisotropy of the granitoids is based on P-wave velocities in an axial and in two radial directions (with maximum and minimum V_P) on the one hand, and the values of A factor on the other hand (Fig. 6c & d). It has been found that, irrespective of rock composition, the velocity of P-waves in all directions decreases with an increase in the A factor. At the K site the maximum anisotropy factor reaches 35%, while the values of V_{Pvert} decrease to 3.550 km s^{-1}. At the I site, the anisotropy factor reaches only 8%, while the lowest velocity in the vertical direction is 5.250 km s^{-1}.

Thus, the increase in the anisotropy of the granitoid rocks depends directly on the reduction in the P-wave velocities. The effect of micro-

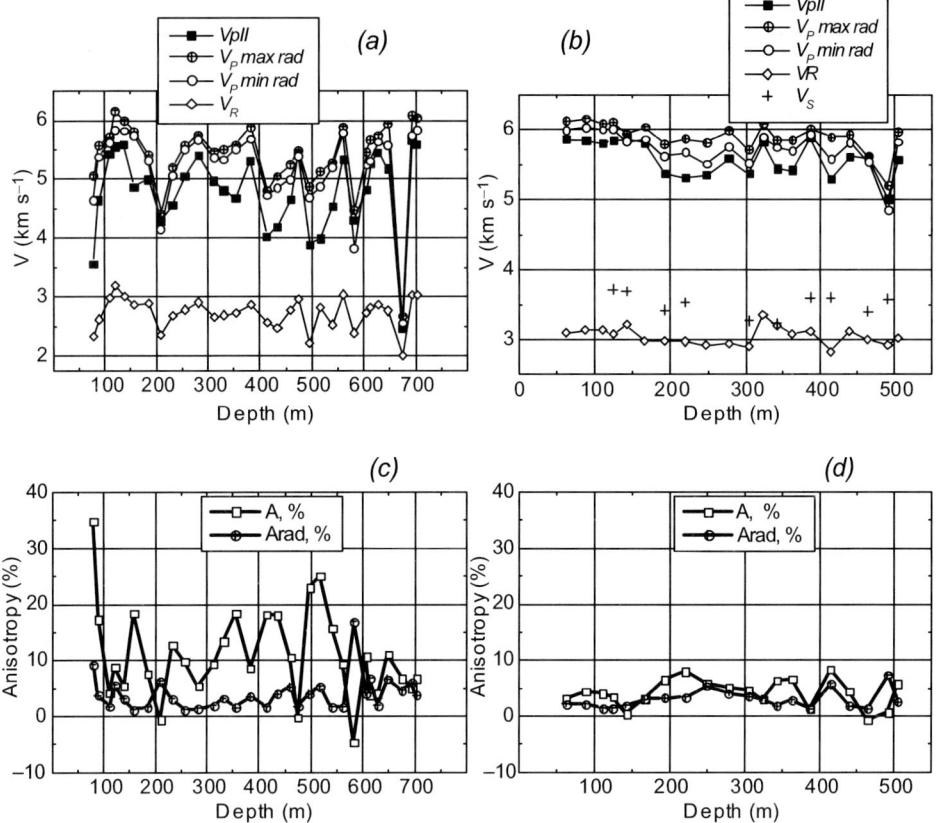

Fig. 5. Variations in the elastic properties along the sections of boreholes for the K (**a** and **c**) and I (**b** and **d**) sites. Upper diagrams are the elastic-wave velocities along the section of the boreholes; the lower ones are the anisotropy factors of the P-wave velocities.

fracturing and weakening of inter-grain contacts may explain this dependence. Therefore, it is suggested that zones of lower velocities and higher anisotropy may be characterized by higher permeability and lower rock strength.

Mechanical properties

Mechanical properties are an important constituent characteristics of the rock massif – both in the selection and in the eventual monitoring of the radioactive-waste disposal system.

For a preliminary assessment of the mechanical properties of the granitoids, uniaxial compression tests were performed on 19 samples. These samples included granites, granite–gneisses and quartz diorites, including cataclastic granites and gneisses. Also tested were samples of granodiorite and vein rocks, including metasomatically altered spessartite.

Tests were carried out using cylinder-shaped samples with a 1:1 height to diameter ratio, with a deviation of at most $\pm 5\%$. The sample diameters were: 42 ± 1 mm, height of 42 ± 3 mm, their end surface was ground down to give a parallelism deviation of at most 0.05 mm. Tests were performed with a press at a 0.5 MPa s^{-1} average rate of load application. To minimize errors, two displacement transducers located opposite each other were used in tests. Data from these transducers were used for the calculation of the average axial strain of a sample. During individual experiments, the force and displacement data were registered every 0.5 s. Figure 7 shows the test results for a number of typical rocks.

Before the strength tests, in the samples of granitoid rocks, the P-wave and S-wave velocities were determined, as well as the density. These data were used for the calculation of

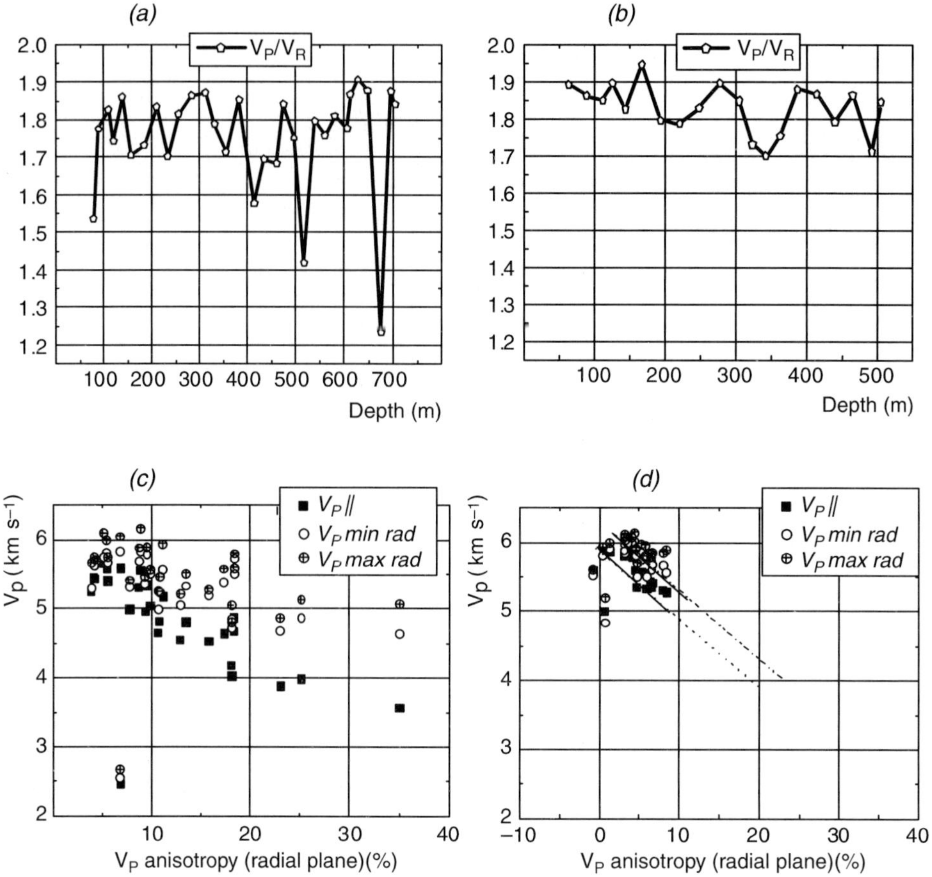

(a)

(b)

Depth (m)

Depth (m)

(c)

(d)

Vₚ anisotropy (radial plane)(%)

Vₚ anisotropy (radial plane)(%)

Fig. 6. Variation of the V_P/V_R relation along the sections of the boreholes for the K (**a**) and I (**b**) sites, as well as diagrams of velocity–anisotropy dependence for the K (**c**) and I (**d**) sites.

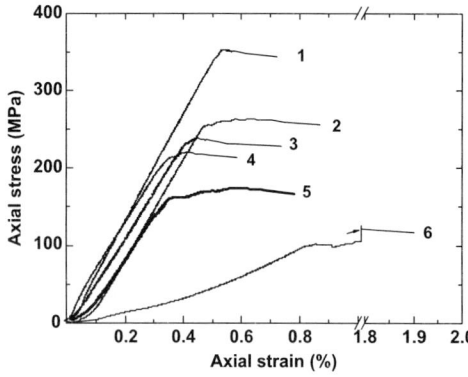

Fig. 7. Typical stress–strain diagrams for rocks of the I and K sites: 1, granite–gneiss; 2, gneiss; 3, granite; 4, quartz diorite; 5, granodiorite; 6, metasomatically altered spessartite.

dynamic bulk and shear moduli, as well as Poisson's ratio. The following formulae, which are well known in the mechanics of continuum solids, were used:

$$E = \rho V_{P^2} - [(1 - \mu)(1 - 2\mu)/(1 - \mu)] \quad (10)$$

$$K = \rho(V_{P^2} - 1.33 V_{S^2}) \quad (11)$$

$$G = \rho V_{S^2} \quad (12)$$

$$\mu = 0.5(R^2 - 2/R^2 - 1) \text{ at } R = V_P/V_S \quad (13)$$

where E is the dynamic modulus of elasticity, 10^4 MPa; K is the dynamic bulk modulus, 10^4 MPa; G is the dynamic shear modulus, 10^4 MPa; μ is Poisson's ratio; ρ is density, g cm^{-3}; V_P is a velocity of ultrasonic P-waves, km s^{-1} and V_S is the velocity of shear ultrasonic waves, km s^{-1}.

Results

Table 4 shows the test results. These data represent the ultimate strength and deformation characteristics for various types of rocks from the K and I sites.

The data show that, for the group of granites, the elastic deformation (ε_1) is on average 0.41%, and, when the ultimate strength (ε_1 amounts to 0.52%) is reached, an explosion-like failure follows. The average ultimate strength in this group (including cataclastic porphyritic granite), at uniaxial compression is 257 MPa, while, without this sample, the average for the granitic samples is 252 MPa. However, the ultimate strength of the cataclastic porphyritic granite is higher at 273 MPa. The ultimate strength limit ranges from 239 to 273 MPa, and the yield strength, σ_{is}, in this case is 92–96% of the ultimate strength, σ_{ic}.

Stress–strain diagrams for granites show that there is a small zone of plastic deformation preceding the failure, and that this zone averages 0.11%. The only exception is a granite sample, I 459.2b, for which the value of plastic deformation is 0.20%. This sample is peculiar for the formation of a large upper fracture cone whose vertex is at two-thirds of the vertical distance from its end surface.

The calculated values of dynamic moduli of elasticity for granites differ from those for all other groups of samples, as their values are lower; the modulus of elasticity is

6.59 ± 0.13 × 10^4 MPa, the bulk modulus is 3.87 ± 0.12 × 10^4 MPa, and the shear modulus is 2.72 ± 0.09 × 10^4 MPa. The average value of Poisson's ratio is 0.21 ± 0.01.

The comparison of these characteristics with similar ones for the cataclastic porphyritic granite K 627.6 shows that all its physical parameters are higher. This may be explained by later crack–pore space sealing processes.

The value of the ultimate strength at uniaxial compression for the group of granite–gneisses ranges from 252 to 352 MPa. The average value of the ultimate strength for a series of five samples is 292 MPa, while the yield strength value derived from the stress–strain diagram, as an average, comes out at 278 MPa (95% of ultimate strength).

The highest strength values are characteristic of adamellitic granite–gneisses (from 308 to 352 MPa). Minor plastic deformation ranging from 0.02 to 0.05% is typical of these samples, and the average plastic deformation preceding the attainment of the ultimate strength is 0.06%.

The samples of metasomatically altered granite–gneiss (I 142.6) and adamellitic gneiss (I 491.7) are characterized by the highest values of plastic deformation – 0.16 and 0.18% respectively. This behaviour of the metasomatically altered gneiss may be explained by a high content of biotite, chlorite, leucoxene-like aggregates, clay minerals and sericite, which make up 25–30% of the rock. Adamellitic

Table 4. *Mechanical properties of granitoid rocks: Nizhnekansky Massif*

Sample no.	Rock	σ_{ic} (MPa)	σ_{is} (MPa)	ε_{is} (%)	ε_{ic} (%)	ε_{pl} (%)
I 459.2a	Granite	239	231	0.44	0.48	0.04
I 459.2b	Granite	266	246	0.47	0.67	0.20
K 613.1	Porphyritic adamellite	250	218	0.37	0.45	0.08
K 627.6	Porphyritic cataclastic granite	273	248	0.38	0.47	0.11
I 330.8a	Adamellitic granite–gneiss	308	286	0.43	0.48	0.05
I 330.8b	Adamellitic granite–gneiss	352	343	0.51	0.53	0.02
I 124.7	Altered granite–gneiss	252	252	0.39	0.39	0.00
I 142.6	Altered granite–gneiss	282	254	0.43	0.59	0.16
I 491.7	Adamellitic gneiss	264	253	0.47	0.65	0.18
K 560.8	Granodiorite	175	149	0.39	0.59	0.20
K 702.9	Quartz diorite	173	149	0.23	0.38	0.15
I 247.6	Quartz diorite	170	139	0.28	0.38	0.10
I 304.2	Quartz diorite	199	149	0.26	0.38	0.12
I 357.2a	Quartz diorite	221	199	0.32	0.41	0.09
I 357.2b	Quartz diorite	193	167	0.29	0.36	0.07
I 189.9a	Quartz diorite–monzodiorite	193	137	0.21	0.38	0.17
I 189.9b	Quartz diorite–monzodiorite	219	164	0.26	0.44	0.18
I 504.8	Quartz diorite	174	158	0.34	0.50	0.16
K 674.0	Altered spessartite	125	99	0.62	1.25	0.63

Note: σ_{ic}, ultimate strength; σ_{is}, yield strength; ε_{is}, deformation at the yield strength; ε_{ic}, deformation at the ultimate strength; ε_{pl}, plastic deformation.

gneiss is characterized by a biotite content of 15–20% and the presence of small zones of blastomylonitic texture. These minerals are weaker, and presumably plastic deformation may progress there at lower levels of axial stress. This is clearly observed on the stress–strain curves for samples in general.

The average of the distribution of dynamic elasticity modules for granite–gneisses is as follows: $E = 7.87 \pm 0.18 \times 10^4$ MPa, $K = 4.54 \pm 0.10 \times 10^4$ MPa, $G = 3.26 \pm 0.08 \times 10^4$ MPa. Poisson's ratio for this group of rocks is 0.21 ± 0.004.

The group of quartz diorites is relatively large. The content of visible quartz in this rock is less than in granites and granite–gneisses: ranging from 5 to 20%. Uniaxial compression tests demonstrate that the quartz diorites are noticeably weaker than the granites. The average ultimate strength of the quartz diorites is 202 MPa, while the average yield strength is 158 MPa. The value of deformation at the yield strength averages 0.27%, while at the attainment of ultimate strength it amounts to 0.40%. Thus, the average value of the area of plastic deformation is 0.13%. The yield strength of quartz diorites averages 78% of the ultimate stress value.

This group of rocks also differs from the other groups by its higher density, which averages 2.75 g cm^{-3}. The average of the distribution of the calculated dynamic moduli of elasticity for the quartz-diorite group is as follows: $E = 7.47 \pm 0.16 \times 10^4$ MPa; $K = 4.90 \pm 0.11 \times 10^4$ MPa, $G = 3.00 \pm 0.08 \times 10^4$ MPa. Poisson's ratio for quartz diorites is 0.25 ± 0.02.

A uniaxial compression test with a granodiorite sample shows that its ultimate strength is lower than that of all the other rocks tested, at 175 MPa. The value of the axial deformation at yield strength reaches 0.39%. After that, deformation continues until failure after an additional 0.20%. At 79% ultimate fracture stress, a minor deviation of linearity occurs, which then returns to the linear dependence, and elastic deformation continues. At 91% ultimate fracture stress, plastic deformation starts to occur. In this connection it seems reasonable to note and to compare the granodiorite yield strength with those of the granites (92% σ_{ic}) and quartz diorites (78% σ_{ic}). For the granodiorite sample, the following values were obtained: $E = 7.64 \times 10^4$ MPa, $K = 4.41 \times 10^4$ MPa, and $G = 3.15 \times 10^4$ MPa. Poisson's ratio is 0.21.

The sample of metasomatically altered spessartite (K 674.0) is notable for the highest deformation values. At all stages of deformation, the sample is deformed two to three times more than the other tested samples. The highest value of plastic deformation is also observed in this sample. The zone of plastic deformation reaches 0.63% after the attainment of yield strength. The stress–strain curve for a uniaxial load is concave. This type of dependence may occur in the presence of microcracks and pores that close gradually with an increase in stress rather than simultaneously. In this case, the effective modulus of elasticity of this type of rock increases with the increase in stress up to the value of the elasticity modulus E. The value of the ultimate strength is 125 MPa, and the yield strength is 99 MPa. Presumably, the yield strength can be at 79% of the ultimate fracture stress. It must be noted that, for this sample, the axial anisotropy factor A for P-wave velocity is 14.3%. Consequently, one of the requirements of the fundamental theory of elasticity in a (quasi)isotropic medium is not met, and so the data for elastic moduli are not given.

This investigation into the ultimate strength and density shows that two main groups of rocks can be identified. These differ radically in their density and mechanical properties. The group of samples with a 2.60–2.64 g cm^{-3} density (granites and granite–gneisses) have ultimate strengths ranging from 238 to 352 MPa. Another group (quartz diorites) are characterized by higher density values (2.74–2.77 g cm^{-3}); however, they differ substantially from the first group by their low values of ultimate strength ranging from 170 to 221 MPa.

Comparing these two groups shows that the quartz diorite group has a greater tendency towards plastic deformation. Thus, on average, the yield strength for granites and granite–gneisses is 94% of the ultimate strength, while for the quartz diorites it is lower at 78% σ_{ic}. The presence of plastic deformation in quartz diorites at earlier stages of load application can probably be explained by the high content of hornblende, biotite and accessory minerals, which have lower ultimate and yield strengths. The composition of plagioclase also changes, and may contribute to the variation in the mechanical properties of these rocks.

The group of rocks with the highest strength: rocks which are liable to more brittle failure, includes granite and granite–gneisses with up to 35% of quartz, the mineral with the highest yield strength and ultimate strength compared to the other rock-forming minerals. In addition, these rocks are characterized by up to 35% K-feldspar, up to 10% biotite, and up to 30% sodic plagioclase.

Quartz diorites have lower quartz contents (up to 15%), K-feldspar (up to 5%) and higher plagioclase (up to 70%) contents and are characterized by lower strength values. The plagioclases in these rocks are slightly more calcic – approaching andesine in composition.

The sample of metasomatically altered spessartite turned out to have the lowest strength.

The static and dynamic elastic moduli were determined for 19 samples. As a rule, the dynamic modulus is higher than the static one, on average by about 18%, with a minimum difference of 5% and a maximum difference of 36%. For these samples, the dynamic modulus of elasticity is higher than the static one, by about 10% for granites, by 18% for the granite–gneisses, and by 19% for the quartz diorites.

No obvious relationship was observed between the elastic-wave velocities and strength, but this may be a result of the small sample size. A more meaningful characteristic can be obtained from the elastic wave velocity V_P/V_S ratio. Given the variation in the composition and texture of the samples, together with a wide observed variation in the V_P/V_S ratio at equal values of strength, a trend is observed towards an increase in the ultimate strength as the V_P/V_S ratio and rock density decrease.

Thermal properties

For long-term storage and/or underground disposal of heat-emitting HLW, it is of paramount importance to be able to forecast the variation in the physical and mechanical properties of rocks resulting from their exposure to (radioactive) heat. Under repository conditions this exposure may result in decompaction of the rocks, and consequently to the changes in their transport characteristics. It has been found experimentally (Zaraisky & Balashov 1994) that the intensity of decompaction naturally increases with temperature, and it increases from ultrabasic through basic to acidic quartz-bearing rocks.

The process of decompaction includes an increase in microcrack length and aperture, the reduction of elastic-wave velocities, and a decrease in the elastic moduli and rock strength. As a result of thermodecompaction, rock permeability may increase by several orders of magnitude. Thus, the investigation of thermophysical properties of the principal petrographic varieties of granitic rocks is of considerable importance.

The thermophysical properties of rocks, minerals and monomineralic aggregates can be characterized by measuring such parameters as heat conductivity (λ), thermal conductivity (α) and heat capacity (C_P). Estimates of heat capacity can be obtained either through direct measurements or, alternatively, by calculation,

provided that the exact mineral composition of the rock and the C_p of rock-forming minerals can be assumed; such physical characteristics are additive. Heat and thermal conductivity cannot be calculated in this way, because factors such as the size of mineral grains, porosity and fluid saturation are too variable.

Thermal conductivity and heat capacity were measured on 27 granitoid samples, using two different methods: method 1, using an optical heat source device (Petrunin & Popov 1994) and method 2 using a resistive heater as a temperature wave generation device in the plane variant of the temperature field (Popov et al. 1981).

The first method enabled measurements to be made up to 200°C. The error of measurement of thermal conductivity and heat capacity coefficients for this method is 3–5%, while the error of heat conductivity measurement is less than 10%. The temperature range for the second method is from 18 to 750°C. The error in the measurement of the thermal conductivity coefficient is 3–5%, while that for the heat capacity is 5–7%. Measurements were made at 18, 100 and 200°C.

Results

The average values of thermal conductivity, thermal capacity and heat conductivity, which were calculated using the well-known Debye formula, are presented in Table 5.

An analysis of these data shows that the highest value of heat capacity is observed for rocks whose composition contains a substantial number of minerals with bound water (hornblende, chlorite, biotite and clay minerals). Hydrogen has the highest heat capacity of all naturally occurring elements. Its contribution to specific heat capacity becomes evident in rocks in which the concentration of water-bearing minerals ranges from 20 to 40% by weight. In contrast, the lowest values of heat capacity are typical of these rocks, which are essentially free of such minerals.

In contrast to heat capacity, which is a function of chemical composition, the values of thermal conductivity and heat conductivity depend on factors such as the mineral composition and the texture of the rock.

A comparison of the average values of heat and thermal conductivity obtained for the different rock groups (see Table 5) shows that the highest values of a and λ are observed in leucogranites and granites containing up to 35% of free quartz. These values are somewhat lower for granite–gneisses, gneisses and adamellites, including altered varieties of these rocks,

Table 5. *Average values of thermophysical parameters for the main lithologies: Nizhnekansky Massif*

Rock and sample no.	Parameter	$T = 18°C$	$T = 100°C$	$T = 200°C$
Granites–leucogranites:	a	15.65 ± 0.4	12.05 ± 0.9	9.5 ± 0.5
I 459.2, K 156.6,	C_p	800 ± 40	900 ± 50	1000 ± 50
K 232.4, K 459.7	λ	3.4 ± 0.15	2.9 ± 0.4	2.5 ± 0.25
Cataclastic granite:	a	14.83 ± 0.7	9.5 ± 0.5	8.9 ± 0.5
K 627.6	C_p	840 ± 40	970 ± 50	1040 ± 50
	λ	3.30 ± 0.33	2.44 ± 0.25	2.20 ± 0.22
Granite–gneisses:	a	14.6 ± 3.0	10.6 ± 1.8	8.6 ± 1.2
I 124.7, I 142.6, I 330.8	C_p	800 ± 40	900 ± 60	990 ± 50
	λ	3.1 ± 0.4	2.5 ± 0.4	2.2 ± 0.2
Gneisses: I 323.1, I 491.7	u	15.0 ± 0.9	11.1 ± 0.5	9.05 ± 0.6
	C_p	765 ± 40	870 ± 50	950 ± 50
	λ	2.97 ± 0.3	2.52 ± 0.4	2.24 ± 0.2
Adamellite: K 120.5,	a	12.4 ± 0.8	11.1 ± 0.7	8.6 ± 1.8
K 184.5, K 613.1	C_p	850 ± 50	960 ± 70	1040 ± 50
	λ	2.75 ± 0.3	2.5 ± 0.3	2.3 ± 0.4
Quartz monzodiorites–tonalites:	a	11.1 ± 1.4	9.6 ± 1.0	8.5 ± 0.7
I 189.9, I 247.6, K 469.3,	C_p	890 ± 70	940 ± 40	1020 ± 30
K 546.7, K 581.3	λ	2.6 ± 0.3	2.4 ± 0.3	2.3 ± 0.3
Quartz diorites:	a	11.1 ± 2.2	8.27 ± 1.6	7.1 ± 1.1
I 88.3, I 193.1, I 304.2,	C_p	860 ± 50	970 ± 70	1040 ± 50
I 357.2, I 504.8, K 647.4, K 702.9	λ	2.54 ± 0.4	2.1 ± 0.2	2.0 ± 0.25
Granodiorites: K 560.8	a	9.98 ± 0.5	7.0 ± 0.35	6.0 ± 0.3
	C_p	920 ± 50	1040 ± 50	1100 ± 50
	λ	2.44 ± 0.25	1.93 ± 0.2	1.75 ± 0.2
Spessartite: K 674.0	a	5.5 ± 0.25	4.5 ± 0.25	4.2 ± 0.25
	C_p	1020 ± 50	1120 ± 50	1150 ± 50
	λ	1.52 ± 0.15	1.46 ± 0.15	1.31 ± 0.15

Note: a, thermal conductivity, $10^{-7}\,\mathrm{m^2\,s^{-1}}$; C_p, heat capacity, $\mathrm{J\,kg^{-1}\,K^{-1}}$; λ, heat conductivity, $\mathrm{W\,m^{-1}\,K^{-1}}$.

in which the quartz content varies from 20 to 30%, and in addition to plagioclase and K-feldspar they may have substantial concentrations of water-bearing minerals with low heat conductivity (biotite, muscovite, chlorite, clay minerals). In the quartz diorites, monzodiorites and tonalites, these minerals, together with hornblende, nearly effectively replace K-feldspar, while the quartz content decreases to 15–20%. For this reason, these rocks are characterized by the lowest values of heat conductivity and thermal conductivity. Finally, for the metasomatically altered spessartite, which is free from quartz, and in which the content of hornblende, chlorite and argillaceous minerals may reach 40%, the heat transfer parameters are the lowest. For the granitic rocks being measured the coefficients of heat conductivity and thermal conductivity decrease, while an increase in temperature thermal capacity increases, with an increase in temperature.

The distribution of the coefficient of heat conductivity, with depth, in sections of the I and K sites are summarized in Figure 8. It is these sites, which are expected to be used eventually for the modelling of the variation of thermal

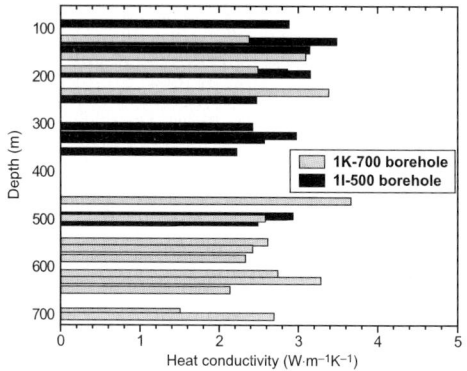

Fig. 8. Variation in the heat conductivity coefficient with depth for rocks in the I and K boreholes.

conditions at different depths of the massif for the underground disposal of HLW.

Conclusions

Two groups of rocks are identified (1 – granites and granite–gneisses, and 2 – quartz diorites and diorites), which differ substantially in density, mechanical and thermal properties. The country rocks are generally metamorphic (amphibolite facies with quartz, feldspars, biotite and amphiboles), locally with hydrothermal–metasomatic alteration (chloritization, sericitization and argillization), and have been deformed in ductile mode (gneiss texture) and brittle mode (cataclastic, brecciated textures and microcracks filled by carbonate, chlorite, sericite and clay minerals).

A range of petrophysical properties (density, filtration, elastic, mechanical and thermal) have been measured on the primary rock types, together with some rocks that have suffered hydrothermal–metasomatic alteration.

With respect to the choice of the I and K sites as possible regions for the disposal of HLW, the following preliminary conclusions can be drawn:

(1) Quartz diorites and diorites of the I site, as well as granites and leucogranites of the K site, occur most widely and are more regular in thickness.
(2) Density values increase with depth at the K site, while at the I site the density reaches maximum values in the central part of the section.
(3) The rocks of the I site are less porous, while at the K site the effective porosity declines with depth. However, at the K site the coefficient of fluid conductivity increases with depth. This is connected with the increase in the anisotropy of the pore space and the intensity of microcrack network development.

In general, the heterogeneity of the elastic, mechanical and thermal properties at the K site is much higher than at the I site. This is connected with both the more diverse mineral composition of the rocks of the K site and their higher intensity of deformation and hydrothermal–metasomatic transformation.

Thus, comparative analysis of the data obtained makes it possible for us to assume that the I site is more favourable for the construction of an underground research laboratory as a precursor to a long-term, monitored, retrievable storage area or HLW disposal repository.

References

ANDERSON, E. B., SHABALEV, S. I., SAVONENKOV, V. G. & LYUBTSEVA, E. F. 1998. Investigations of Nizhnekanskiy granitoid massif (Middle Siberia, Russia) as a promising site for deep geological disposal of HLW. *Procedings. of the International. Conference. on Radioactive. Waste Disposal.*, Hamburg, Germany, 105–110.

DATSENKO, V. M. 1995. [*Nizhnekansky Massif–a Reference Model of the Nizhnekansky Granitoid Complex (Yeniseisky Mountain-Ridge), Novosibirsk*] (in Russian).

KOBRANOVA, V. N. 1986. [*Petrophysics.*] Nedra, Moscow (in Russian).

LAVEROV, N. P., VELICHKIN, V. I., PETROV, V. A., OMELIANENKO, B. I. & PEK, A. A. 2000. *Long-term Solutions to Managing Nuclear Waste in the Russian Federation. The Environmental Challenges of Nuclear Disarmament.* Kluwer Academic Publishers, Dordrecht, Netherlands, 75–83.

PETRUNIN, G. I. & POPOV, V. G. 1994. [Method and instrumentation for high-accuracy measurement of thermophysical characteristics of oceanic sediments.] *Fizika Zemli*, **11**, 78–85 (in Russian).

POPOV, V. G., PETRUNIN, G. I. & NESTEROV, A. G. 1981. [*A System for the Measurement of Temperature Conductivity and Thermal Capacity of Rocks and Minerals within a 300–1000 K Temperature Range (Methods of Calculation of Thermal Characteristics, Experimental Hook-up).*] VINITI, Moscow (in Russian).

STAROSTIN, V. I. 1979. [*Structural–Petrophysical Analysis of Endogenous Ore Fields.*] Nedra, Moscow (in Russian).

VOLOBUEV, M. I. & ZYKOV, S. I. 1961. [On the problem of the absolute age of rocks and minerals of the Yeniseisky Mountain-Ridge.] *In: Materials on Geology and Minerals of the Krasnoyarsk Territory*, KGU Krasnoyarsk, 91–100 (in Russian).

YAKOBASHVILI, O. P. 1992. [*Seismic Methods for the Assessment of the Condition of the Rock Massif in Open Pits.*] IPKON, RAS, Moscow (in Russian).

ZARAISKY, G. P. & BALASHOV, V. N. 1994. Thermal decompaction of rocks. SHNULOVICH, K. I., YARDLEY B. W. D. & GONCHAR, G. G. (eds) *Fluids in the Crust.* Chapman and Hall, London, 253–284.

Effect of compositional and structural variations on log responses of igneous and metamorphic rocks. I: mafic rocks

A. BARTETZKO[1,2], H. DELIUS[3] & R. PECHNIG[2]

[1]*Present address: University of Bremen, Research Center Ocean Margins,*
P.O. Box 330440, 28334 Bremen, Germany
(e-mail: bartetzko@uni-bremen.de)
[2]*Applied Geophysics, RWTH Aachen University of Technology,*
Lochnerstrasse 4-20, 52056 Aachen, Germany
[3]*Department of Geology, University of Leicester, Leicester, LE1 7RH,UK*

Abstract: Well logging has become a standard method in the oil industry for the investigation of subsurface geology. Accordingly, interpretation techniques have been mainly developed for use in sedimentary rocks, and the log responses of sediments are well known. However, this is not the case for igneous and metamorphic rocks. We present a compilation of log responses for mafic rocks from drill-holes in oceanic and continental basement. The holes cover a variety of mafic rocks: mid-ocean ridge basalt (MORB), gabbro, basalt and andesitic basalt from back-arc basins, flood basalt from large igneous provinces (LIPs), and continental metamorphic rocks. The comparison of log responses shows that rocks from the same geological setting have similar *in situ* physical properties. Differences in physical properties between rocks from different geological settings are mainly related to variations in the structure of the rocks, while variations in composition have only a minor effect on the *in situ* physical properties. In volcanic rocks, variations in fracturing and vesicularity related to cooling of the lava strongly influence log responses. Mafic rocks from continental drill-holes were enriched in radioactive elements during regional metamorphism, resulting in higher values in the total gamma-ray compared to the oceanic rocks.

Well-logging data provide continuous information about the physical *in situ* properties of the borehole wall. It is a standard method in oil industry for the investigation of subsurface geology. The development of the technique and interpretation methods has been driven mainly by the oil industry for the use in a specific environment, i.e. sedimentary rocks. As a result, log responses of sedimentary rocks are well known, while those for igneous and metamorphic rocks are relatively poorly constrained. Experience in the use of well logging in these rocks has been increased due to scientific research programmes such as the Deep Sea Drilling Project (DSDP), Ocean Drilling Program (ODP) or International Continental Drilling Program (ICDP).

This study presents a compilation of log responses from mafic rocks encountered in drill-holes in oceanic and continental basement by different scientific research programmes. These holes drilled mafic rocks emplaced in a variety of geological settings. Our primary aim is to compare log responses and to determine typical value ranges of mafic igneous and metamorphic rocks. The influence of various parameters such as porosity, fracturing, alteration, tectonic and metamorphism, but also of composition, on the *in situ* physical properties of mafic rocks shall be revealed by comparing the log responses from the different rock types from different holes. This study is part of a larger compilation of log responses from a wide variety of igneous and metamorphic rocks. Log responses of acid and intermediate rocks are presented by Pechnig *et al.* (2005).

Selection of the boreholes

Data from 19 boreholes have been used in this study (Figs 1 & 2, Table 1). They cover rocks from the following six geological settings:

(1) submarine basalt from upper oceanic crust built at mid-ocean ridges;
(2) gabbro from the lower oceanic crust;
(3) submarine basalt and basaltic andesite from upper oceanic crust built in back-arc basins;

From: HARVEY, P. K., BREWER, T. S., PEZARD, P. A. & PETROV, V. A. (eds) 2005. *Petrophysical Properties of Crystalline Rocks.* Geological Society, London, Special Publications, **240**, 255–278.
0305-8719/05/$15.00   © The Geological Society of London 2005.

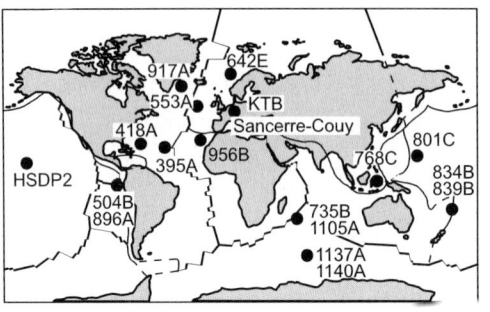

Fig. 1. Location of the boreholes compared in this study.

(4) submarine and subaerial flood basalt from large igneous provinces (LIPs);
(5) resedimented volcaniclastic material within the volcanic apron of an ocean island (OI);
(6) metamorphic rocks from continental basement.

The rocks are mainly of tholeiitic composition, and only a few are more alkaline or more mafic (Fig. 3, Table 2).

In each borehole, one or more depth intervals were identified that reflect the major lithology (Table 1 & Fig. 2). Where possible,

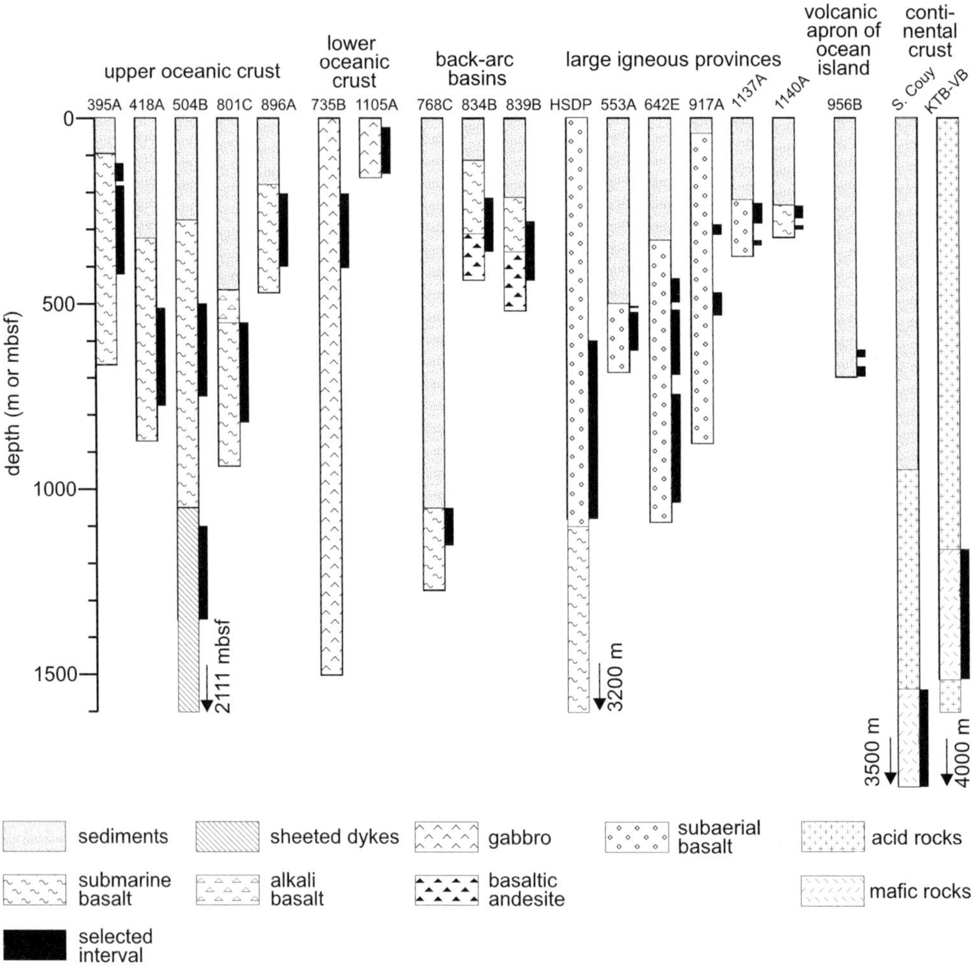

Fig. 2. Schematic lithological overview of the boreholes and selected depth intervals. Depth is given in metres for the continental drill-holes, and in mbsf (metres below sea-floor) for ODP drill holes. S. Couy, Sancerre-Couy; KTB-VB, KTB pilot hole.

Table 1. *Overview of the boreholes included in this study*

Hole	Leg (logging operations)	Total depth	Sediment thickness (m)	Selected depth intervals	Lithology	Reference
				Upper oceanic crust		
395A	174B	664 mbsf	93	120–169, 181–418 mbsf	Pillow basalts, lava flows	Bartetzko et al. (2001)
418A	102	868 mbsf	324	510–775 mbsf	Pillow basalts, lava flows	Bartetzko (1999)
504B	111, 148	2111 mbsf	274.5	500–750, 1100–1350 mbsf	Pillow basalts, lava flows, dykes	Bartetzko et al. (2002)
801C	185	936 mbsf	462	550–820 mbsf	Pillow basalts, lava flows, breccia	Barr et al. (2002)
896A	148	469 mbsf	179	203–350, 381–400 mbsf	Pillow basalts, lava flows, breccia	Bartetzko et al. (2002)
				Lower oceanic crust		
735B	176	1508 mbsf	0	223–403 mbsf	Gabbro, oxide gabbro	Dick et al. (1999)
1105A	179	158 mbsf	0	25–150 mbsf	Gabbro, oxide gabbro	Pettigrew et al. (1999)*
				Back-arc basins		
768C	124	1271 mbsf	1046.6	1050–1140 mbsf	Pillow basalt	Rangin et al. (1990)*
834B	135	435 mbsf	113	216–360 mbsf	Basalt, basaltic andesite	Parson et al. (1992)*
839B	135	517 mbsf	214	279–437 mbsf	Basalt, basaltic andesite	Parson et al. (1992)*
				Large igneous provinces		
HSDP2		3200 m	0	600–1079 m	Subaerial basaltic lava flows	ICDP-database*,†
553A	81	683 mbsf	499	506–510, 522–626 mbsf	Subaerial basaltic lava flows	Bücker et al. (1998)
642E	104	1087 mbsf	328	432–495, 517–691, 743–1036 mbsf	Subaerial basaltic lava flows	Delius et al. (1995)
917A	152	875 mbsf	41.9	288–314, 470–532 mbsf	Subaerial basaltic lava flows	Larsen et al. (1994)*
1137A	183	371 mbsf	220	230–285, 330–343 mbsf	Subaerial basaltic lava flows	Delius et al. (2003)
1140A	183	321 mbsf	235	236–272, 290–300 mbsf	Pillow basalt, lava flows	Delius et al. (2003)
				Volcanic apron of ocean island		
956B	157	697 mbsf	697	625–645, 670–695 mbsf	Volcaniclastic breccia and tuff	Delius et al. (1998)
				Continental crust		
Sancerre-Couy		3500 m	942	1540–1800 m	Amphibolite, dioritic gneiss	Millon & Zwingelberg (1992)
KTB-VB		4000 m	0	1161–1510 m	Amphibolite, metagabbro, meta-ultramafitite	Haverkamp & Wohlenberg (1991), Pechnig et al. (1997)

*Lithology classification by core–log integration; the reference for core description is given.
†*http://www.icdp-online.org/*
mbsf = metres below sea-floor.

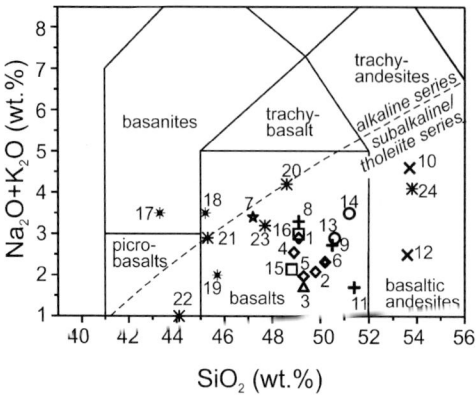

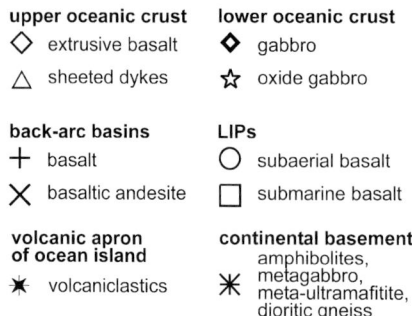

upper oceanic crust
◇ extrusive basalt
△ sheeted dykes

lower oceanic crust
◆ gabbro
☆ oxide gabbro

back-arc basins
+ basalt
✕ basaltic andesite

LIPs
○ subaerial basalt
▢ submarine basalt

volcanic apron
of ocean island
✱ volcaniclastics

continental basement
amphibolites,
✴ metagabbro,
meta-ultramafitite,
dioritic gneiss

Fig. 3. Alkali versus silica diagram for the chemical analysis from boreholes included in this study (modified after Le Bas *et al.* 1986). The dividing line between the alkali and subalkaline/tholeiite series is plotted using co-ordinates from Irvine & Baragar (1971). Most samples fall into the basalt field. Rocks of basaltic andesite composition are from Holes 834B and 839B, and the dioritic gneiss is from Sancerre-Couy. The sample with basanite composition is the volcaniclastic sediment from Hole 956B, and the picro-basalt sample is a meta-ultramafitite from KTB-VB. See Table 2 for numbers and references.

homogeneous sequences with large bed thickness were selected, to avoid integration effects made by down-hole measurements at bed boundaries (e.g. Rider 1996). However, in upper oceanic crust, rapid decimetre-scale changes of rock type (e.g. in lava morphology) are characteristic, and integration effects cannot be excluded. Other critical factors in choosing intervals were the availability and quality of the logging and core data. Depth intervals with above-normal borehole enlargements, high degrees of fracturing or extensively altered zones were generally avoided.

Comparison of log responses from different rock types requires a lithological classification and description of each individual rock type

according to their *in situ* physical properties. Several such studies (see Table 1 for reference) have been undertaken on rocks in these boreholes and, where applicable, they have been used in this study. In a few cases, such studies were not available, and lithological classification was achieved using core–log integration, i.e. by directly comparing core lithostratigraphy with down-hole logs (Holes 768C, 834B, 839B, 917A, 1105A, HSDP). Table 1 details the core description for each borehole.

Basalts from upper oceanic crust

DSDP/ODP-Holes 395A (Mid-Atlantic Ridge), 418A (Bermuda Rise), 504B (Costa Rica Rift), 801C (West Pacific), and 896A (Costa Rica Rift) transect extrusive sections of the oceanic crust consisting of tholeiitic pillow basalts, lava flows and breccias. These three rock types can be distinguished with well-logging data (e.g. Ayadi *et al.* 1998; Brewer *et al.* 1998; Bartetzko *et al.* 2001, 2002; Barr *et al.* 2002; Haggas *et al.* 2002). Massive lava flows and pillow basalts are the most common lithology. Thick intervals of breccias were only identified in Holes 801C and 896A. Transitional features such as thin lava flows or altered lava flows (e.g. Bartetzko *et al.* 2001, 2002) are not studied. Intrusive basalt is represented by the sheeted dyke complex (Anderson *et al.* 1985; Alt *et al.* 1993) in Hole 504B.

Gabbros from lower oceanic crust

ODP Holes 735B and 1105A sampled the gabbroic lower oceanic crust close to a fracture zone south of the Southwest Indian Ridge. Hole 735B is about 1.5 km deep and drilled mostly olivine gabbro and microgabbro. The olivine gabbro is intruded by numerous bodies of iron–titanium-oxide rich or oxide-bearing gabbros (Robinson *et al.* 1989; Dick *et al.* 1999). Hole 1105A is located 1.2 km from Hole 735B, and encountered 158 m of gabbroic rocks similar to those of Hole 735B (Pettigrew *et al.* 1999). The selected depth intervals from Holes 735B and 1105A comprise olivine gabbro and Fe–Ti-oxide rich gabbro (referred to below as oxide gabbro).

Volcanic rocks from back-arc basins

Hole 768C is located in the Sulu Sea, western Pacific. It is 1217 m deep and penetrates 250 m of volcanic rocks (Rangin *et al.* 1990). In contrast to the basalts drilled in upper oceanic crust, basalts from Hole 768C are highly vesicular (30–50%). Vesicles are commonly filled with secondary minerals such as carbonate and clay. The selected depth interval consists of pillow basalt.

Holes 834B and 839B are located in the Lau Basin, western Pacific, and penetrated a succession of primitive basalts and more alkali-rich basaltic andesite (Parson et al. 1992). The rocks in Hole 834B have compositions similar to N-MORB, while the rocks in Hole 839B have more arc-like signatures (Ewart et al. 1994). The rocks are vesicular, although vesicle distribution varies considerably. No distinction between lava morphologies was made (Parson et al. 1992). For this study, basalts and basaltic andesites have been selected.

Basalts from hot-spot volcanoes and Large Igneous Provinces

Hole HSDP-2 was drilled by the ICDP near Hilo, Hawaii. The encountered lithologies are divided into subaerially erupted lavas, which extend to 1079 metres below sea-level (mbsl), and a sequence of submarine erupted lavas down to total depth of 3110 mbsl (DePaolo et al. 2001). Only basaltic lava flows from the subaerial stage are considered in this study (Fig. 2; Table 1).

Several ODP legs investigated LIPs. Holes 553A, 642E, and 917A drilled into the North Atlantic Volcanic Province. Holes 1137A and 1140A are situated in the southern Indian Ocean on the Kerguelen Plateau. The rocks used for this study (Table 1) are tholeiitic basalts (Roberts et al. 1983; Eldholm et al. 1987; Larsen et al. 1994; Coffin et al. 2000). The selected depth intervals from Holes 553A, 642E, 917A, and 1137A contain subaerial lava flows typically characterized by vertical changes in the size and abundance of vesicles resulting in gradual changes in in situ physical properties (e.g. Planke 1994; Delius et al. 1995, 2003; Bücker et al. 1998). Only the end-member varieties are included in this study, i.e. the central massive part which lacks vesicles or contains very few, and the brecciated and highly vesicular top of the lavas. From Hole 1140A, submarine massive and pillow basalts are considered in this study. The lavas form two distinct geochemical groups, with one group being enriched in highly incompatible elements and elevated potassium contents with respect to the other group (Coffin et al. 2000).

Mafic volcaniclastics from the volcanic apron of an ocean island

ODP Site 956 was drilled into the southwestern flank of Gran Canaria, Canary Islands. Hole 956B penetrated 704 metres below the sea-floor

(mbsf), passing through 140 m of basaltic hyaloclastite tuffs, hyaloclastite lapillistones and lithic breccias at the bottom of the hole (Schmincke et al. 1995). These rocks were deposited as submarine, transitional or subaerial deposits during the shield volcano stage and then redeposited by debris flows and interlayered with minor nannofossil oozes and clay. Volcaniclastic breccias and tuff from the lowermost, thickest debrite were selected for this study.

Continental crust

Logging data from two holes in continental basement are included in this study. Metamorphic rocks in these holes are interpreted as ancient oceanic basement rocks that were incorporated into the continental crust during orogenesis (Cabanis & Thiéblemont 1992; Harms et al. 1997).

The German Continental Drilling Project (KTB) is located at the western margin of the Bohemian Massif. Different structural units of the Variscan orogenic belt are exposed in the vicinity of the drill site. Two boreholes, 250 m apart, were drilled as part of the KTB project. The first drill-hole (KTB pilot hole, KTB-VB) was drilled to a total depth of 4001 m, and it was almost completely cored and extensively logged. The selected depth interval (1160 to 1510 m) from the pilot hole contains medium-pressure metamorphic rocks (amphibolites, metagabbros and minor intercalated meta-ultramafitites) that belong to a massive metabasitic unit of igneous origin with enriched-MORB (E-MORB) characteristics (Hirschmann et al. 1997; O'Brien et al. 1997).

Hole Sancerre-Couy was initiated by the Géologie Profonde de la France (GPF) and is located about 20 km south of Sancerre, close to the village of Couy. The 3500-m-deep hole was drilled into a crystalline basement consisting of several variations of metabasites and gneisses (granulite facies) (Autran & Chantraine 1988). The chemical composition of the metabasites indicates an original composition intermediate between normal-MORB (N-MORB) and back-arc basalts (Cabanis & Thiéblemont 1992). The selected depth interval is composed of a dark-green garnet-bearing plagioclase–amphibolite intercalated with plagioclase amphibolite and subordinate dioritic gneisses (Hottin et al. 1988).

Log database

Down-hole measurements were carried out using a variety of tools dependent on the specific scientific targets of the individual drilling projects, the geological environment, the company carrying

Table 2. *References and number of samples for geochemical data used in Figure 3*

Number in Fig. 3	Hole	Rock types	Number of samples	Reference
1	395A	Pillow basalt, lava flows	26	Rhodes *et al.* (1979)
2	504B	Pillow basalt, lava flows	98	Cann *et al.* (1983)
3	504B	Dykes	60	Anderson *et al.* (1985)
4	801C	Pillow basalt, lava flows	33	Plank *et al.* (2000)
5	896A	Pillow basalt, lava flows	34	Alt *et al.* (1993)
6	735B	Oxide gabbro	10	Robinson *et al.* (1989)
7	735B	Olivine gabbro	18	Robinson *et al.* (1989)
8	768C	Pillow basalt	20	Rangin *et al.* (1990)
9	834B	Basalt	9	Parson *et al.* (1992)
10	834B	Basaltic andesite	3	Parson *et al.* (1992)
11	839B	Basalt	5	Parson *et al.* (1992)
12	839B	Basaltic andesite	6	Parson *et al.* (1992)
13	917A	Subaerial basalt	4	Larsen *et al.* (1994)
14	1137A	Subaerial basalt	3	Coffin *et al.* (2000)
15	1140A	Pillow basalt	1	Coffin *et al.* (2000)
16	1140A	Pillow basalt, K-rich	1	Coffin *et al.* (2000)
17	965B	Tuff	1	Schmincke *et al.* (1995)
18	956B	Breccia	1	Schmincke *et al.* (1995)
19	956B	Basalt clast	1	Schmincke *et al.* (1995)
20	KTB-VB	Amphibolite	7	KTB database: *http://icdp.gfz-potsdam.de/html/ktb/*
21	KTB-VB	Metagabbro	7	KTB database: *http://icdp.gfz-potsdam.de/html/ktb/*
22	KTB-VB	Meta-ultramafitite	4	KTB database: *http://icdp.gfz-potsdam.de/html/ktb/*
23	S. -Couy	Amphibolite	9	BRGM
24	S. -Couy	Dioritic gneiss	1	BRGM

S.-Couy, Sancerre-Couy.

out the logging operations, and technical developments. Except for the HSDP borehole, where logging operations were carried out by Geo-Forschungszentrum (GFZ) Potsdam using Antares tools, all logging operations were undertaken by Schlumberger (Table 3). Logging data from the ODP are available at *http://www.ldeo.columbia.edu/BRG/ODP/DATABASE/* and from HSDP and KTB-VB at *http://www.icdp-online.org/*.

Different generations and configurations of logging tools influence the value range of measurements and hence the comparison of lithology-specific log responses. Only a few examples exist that allow a comparison of measurements of the same physical properties carried out with different tools. Bartetzko *et al.* (2001) compared data acquired in ODP Hole 395A during different legs and with tools of different configurations. Although the results show a good correlation, offsets in the values were observed in several measurements. Figure 4 shows another example of repeated measurements from two ODP legs from Hole 735B. Comparing the two data-sets, the electrical

resistivity, density, and photo-electric factor exhibit good reproducibility, while total gamma ray and the neutron log show a good correlation, but offsets in values. Because of these discrepancies, we include a short introduction to the principles of the individual logging tools, with respect to differences between the tools. More detailed information on measuring principles and tools can be found, for example, in Rider (1996) and Borehole Research Group (2000).

Electrical resistivity was measured using two different principles, galvanic measurements (ARI, DLL, SFL; for acronyms see Table 3) and inductive measurements (DIL, DIL3). Inductive measurements are accurate in low-resistivity formations, but become less accurate in formations with resistivities >150 ohm m, whereas galvanic measurements give reliable measurements in formations of high resistivity (Borehole Research Group 2000). Examples from boreholes where different galvanic tools were deployed show a good repeatability and reproducibility of galvanic resistivity measurements, even when tools of different configuration are used (Bartetzko *et al.* 2001). In this study,

Table 3a. *Logging tools* used for the down-hole measurements compared in this study*

Hole	Electrical resistivity	Sonic velocity	Total gamma ray	Density	Neutron porosity
395A	ARI[†]	DSI	NGT	HLDS	APS
418A	DLL[§]	LSS	NGT	LDT	CNT
504B	DLL	SDT	NGT	LDT	CNT
801C	DLL	LSS	NGT	HLDS	APS
896A	DLL	SDT	NGT	—	—
735B	DLL	DSI	NGT	HLDS	APS
1105A	SFL	BHC	NGT	HLDS	APS
768C	DIL	LSS	NGT	LDT	CNT
834B	SFL	LSS	NGT	LDT	CNT
839B	SFL	LSS	NGT	LDT	CNT
HSDP	DIL3	No data	Spectral gamma ray	—	—
553A	SFL	BHC	GR	FDC	CNL
642E	SFL	LSS	NGT	LDT	CNT
917A	SFL	LSS	NGT	LDT	—
1137A	DLL	DSI	HNGS	HLDS	APS
1140A	SFL	LSS	NGT	HLDS	APS
956B	SFL	SDT	NGT	HLDS	CNT
Sancerre-Couy	DLL	LSS	NGT	LDT	CNT
KTB-VB	DLL	LSS	NGT	LDT	CNT

*Except for the HSDP borehole where down-hole logging was carried out using Antares tools, all tools are trademarked by Schlumberger.
[†]Resistivity measurements similar to those of a DLL.
[§]Only shallow laterolog (LLS) available.

Table 3b. *Logging tools used for the down-hole measurements compared in this study*

Acronym	Tool
	Electrical resistivity
ARI	Azimuthal resistivity imager
DLL	Dual latero log
SFL	Spherically focused resistivity log
DIL	Dual induction tool
DIL3	Dual induction tool
	Sonic velocity
DSI	Dipole shear imager
LSS	Long spacing sonic tool
BHC	Borehole compensated sonic tool
SDT	Array sonic tool
	Natural radioactivity
NGT	Natural gamma-ray tool
GR	Gamma ray
HNGS	Hostile environment natural gamma-ray sonde
	Density and photo-electric factor
HLDS	Hostile environment litho density sonde
LDS	Litho density tool
FDC	Formation density compensated tool*
	Neutron porosity
APS	Accelerator porosity sonde
CNL, CNT	Compensated neutron log/tool

*Only density measurement.

galvanic measurements are available for most holes; inductive measurements had to be used in Hole 768C and in HSDP; but resistivity data in these holes are <150 ohm m and thus in a value range where inductive measurements are reliable.

Sonic velocity measurements were carried out with logging tools (LSS, SDT, DSI, BHC) which differed in the number and configuration of transmitters and receivers, and thus varied in their ability to compensate for borehole effects, as well as their vertical resolution, (the latter ranges between 0.6 and 1.2 m; Borehole Research Group 2000).

Measurements of natural radioactivity were performed using four different tools. In most holes, a spectral gamma-ray tool measuring with a sodium iodide crystal was used (NGT, GR, Antares' spectral gamma-ray tool). In some holes, the more advanced HNGS was employed. The HNGS uses a bismuth germanate detector that provides more accurate measurements than sodium iodide crystals (Borehole Research Group 2000). Repeated measurements of total gamma-ray measurements show good correlations with minor offsets in value range (Bartetzko *et al.* 2001; Fig. 4). Such offsets can be caused by differences in the efficiency of the

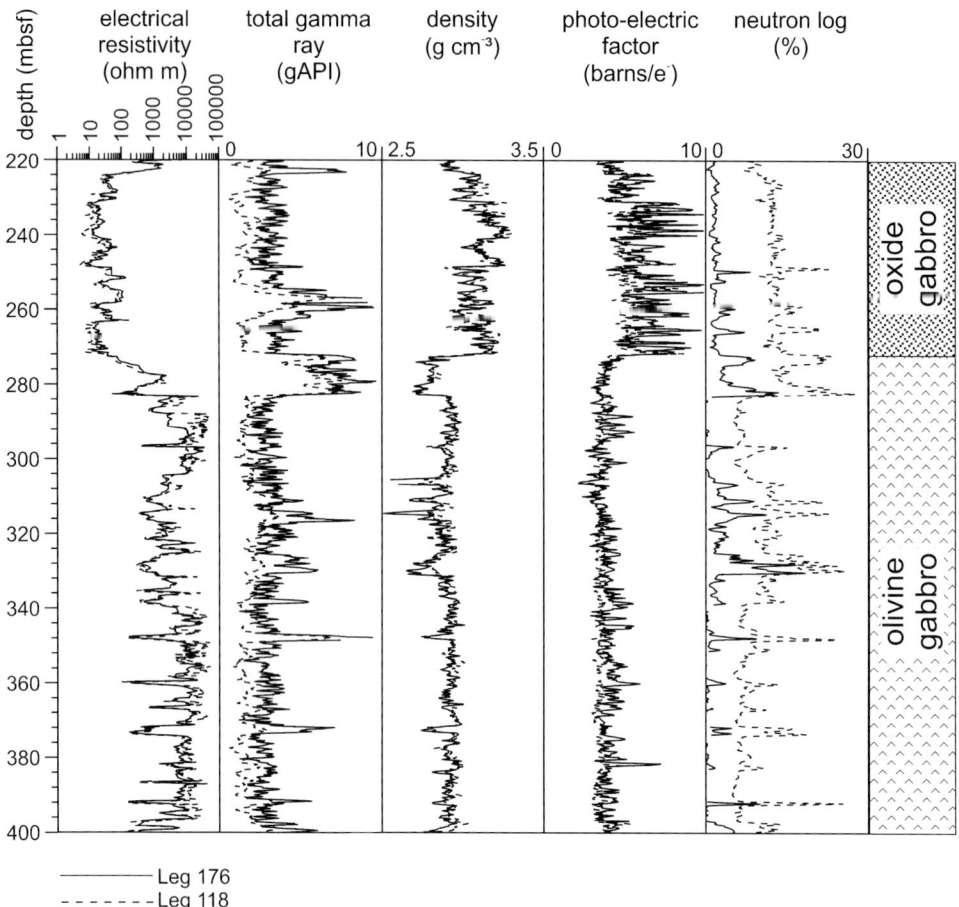

Fig. 4. Comparison of logging data from two different logging operations partly performed with different logging tools in the gabbros of ODP Hole 735B for the depth interval between 220 and 400 mbsf. Logging operations of Leg 118 were carried out in 1987; Leg 176 took place in 1997. Electrical resistivities were measured with a DLL during both legs. Both measurements with the deep laterolog show an excellent agreement. Measurements of total gamma ray were carried out with a Natural Gamma Ray Tool (NGT) during Leg 118, and with the more advanced Hostile Environment Natural Gamma Ray Sonde (HNGS) during Leg 176. Both measurements show a good correlation, however, the readings from the HNGS are higher throughout the displayed depth interval. This may be related to a better resolution of the newer bismuth germanate detector. Density and photo-electric factor data were recorded with a Litho Density Tool (LDT) during Leg 118 and a Hostile Environment Litho Density Sonde (HLDS) during Leg 176, and both show a good agreement. Neutron logging was performed with a Compensated Neutron Tool (CNT) during Leg 118, and with the more advanced Accelerator Porosity Sonde (APS). Although the curve patterns agree very well, measurements with the APS show significantly lower values at low neutron-log readings. This can be attributed to the lower sensitivity of the newer tool towards influences of matrix density effects. This effect is stronger in the more dense oxide gabbros than in the olivine gabbros.

different detectors. Gamma radiations are discrete events, and therefore the detection limit and accuracy of the measurements depend on the time interval of the measurement and the efficiency of the detector. In order to minimize discrepancies between different logging tools, only data from the NGT were used in this study. The only exception is Hole 1137A, where only

HNGS data are available. Measurements of spectral gamma ray provide the total gamma radiation as well as the amounts of the radioactive elements potassium, thorium and uranium.

Density measurements were carried out with three different tools (FDC, LDT and HLDS). In contrast to the oldest tool, the FDC, the newer LDT and HLDS are also able to measure the photo-electric factor of the formation additionally to the density measurements. Differences between the LDT and HLDS mostly concern improvements of the electronic equipment (Borehole Research Group 2000).

Neutron porosity logging was performed with three different tools (CNL, CNT, APS). The basic difference between the CNL/CNT and the newer APS is the use of a neutron accelerator in the APS instead of a radioactive source in the older tools. This enables the measurement of epithermal neutrons and, with a new detector arrangement, minimizes the influence of the formation atom density on the measurement (Schlumberger 1994). Neutron logging is used in sedimentary environments to estimate formation porosity (e.g. Rider 1996). The tool response gives the hydrogen content of the formation, which can be related to porosity. However, hydrogen in hydrous alteration minerals such as clay; the presence of strong neutron absorbers (e.g. boron and gadolinium) in the drilling fluid or the formation; and the presence of minerals of high atomic density, also affect the measurements. Therefore, the neutron log response cannot be used as a porosity log in crystalline rocks, as the tool response (particularly CNL and CNT) gives porosity values that are much higher than the porosity of the rocks (e.g. Lysne 1989; Broglia & Ellis 1990). Among all the logs displayed in Figure 4, the most significant offset in log values occurs in the neutron log, and the offset is larger for the denser oxide gabbros than for the olivine gabbros. The use of the APS instead of the CNT improves the measurements and compensates for the effect of high-density formations. Due to these differences in the measurement principles, neutron porosity is included in the comparison of this study only with restrictions.

Comparison of *in situ* physical properties and their relation to rock type

In Figure 5 we summarize the log responses of the different rock types from the 19 boreholes. Minimum, maximum, mean, and standard deviation are listed in the appendix. Although the chemical and mineralogical composition of the rocks is similar (Fig. 3) they show considerable differences in their *in situ* physical properties, particularly in electrical resistivity, P-wave velocity, total gamma ray, and density. Only minor differences occur in the photo-electric factor log. Neutron-log measurements are difficult to compare due to differences in the measurement principles of the tools (see

above). In order to display the relation between *in situ* physical properties and rock types in more detail, cross plots of the different log properties for the various rock types are analysed in Figures 6 to 9.

Relation between electrical resistivity, density and P-wave velocity

Figure 6a–c shows cross-plots of electrical resistivity, density, and P-wave velocity of well-logging data, while Fig. 6d shows the cross-plot of P-wave velocity versus density for core data. Different lithologies and/or geological settings produce distinct trends in the cross-plots of density versus electrical resistivity (Fig. 6a) and P-wave velocity versus electrical resistivity (Fig. 6b). The cross-plots of P-wave velocity versus density (Fig. 6c & d) show a trend of increasing P-wave velocity with increasing density for all rock types for both, log and core data.

Submarine and subaerial basalts. Basalts from upper oceanic crust, back-arc basins, volcanic islands and LIPs show similar relations in the cross-plots of Figure 6. In these lithologies, electrical resistivity is low to intermediate, with values ranging between 1 and 1000 ohm m, while density and P-wave velocity have broad value ranges of 1.2 to 3 g cm^{-3} and 1.8 to 7.5 km s^{-1}, respectively. Electrical resistivity and P-wave velocity increase with increasing density. This trend can be explained by variations in fracture density, porosity and vesicularity, and is more apparent when the basalts are plotted individually with regard to different lava morphology (Fig. 7a & b). Despite the strong overlap of the data, massive basalts from upper oceanic crust have the highest electrical resistivity, density and P-wave velocity values. Values of subaerial massive basalts from LIPs are only slightly lower. Pillow basalts and breccias from upper oceanic crust have lower electrical resistivities, density and P-wave velocities than the massive basalts. The log responses of electrical resistivity, density and P-wave velocity of the vesicular and brecciated zones of lava flows from LIPs are also lower than those of massive basalts, and marginally lower than those of pillow basalts.

Massive effusive lava flows are characterized by regular fracture patterns caused by thermal contraction during cooling of the lava (e.g. McPhie *et al.* 1993; Gillis & Sapp 1997). However, fracture density in these lavas is low compared to that in pillow basalts and breccias,

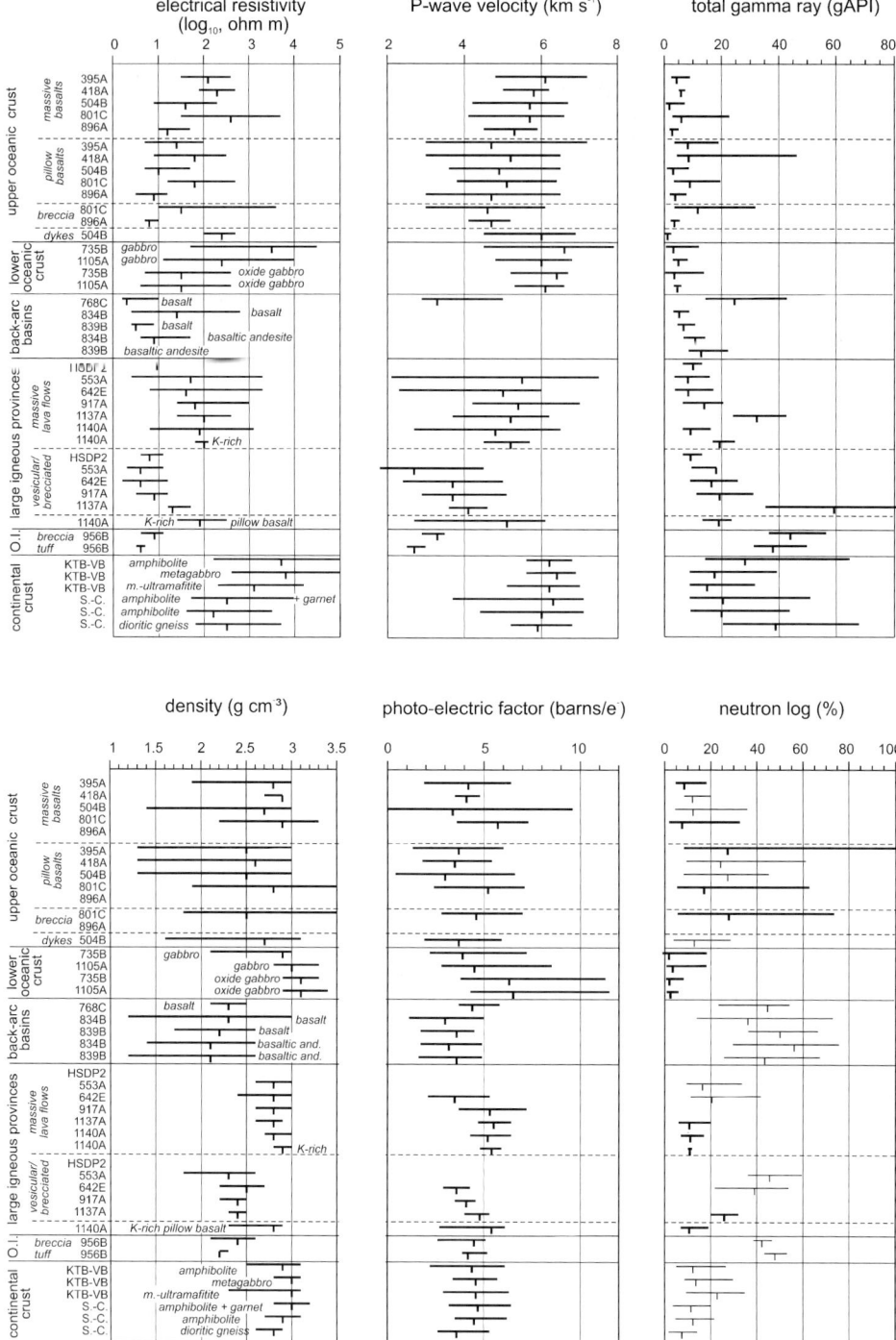

Fig. 5. Log responses (minimum, maximum and average) of the different rock types. In the neutron-log diagram, thin bars represent measurements with the CNL/CNT and thick bars correspond to measurements with the APS. Basaltic and., basaltic andesite; OI, ocean islands; m.-ultramafitite, meta-ultramafitite; S.-C., Sancerre-Couy.

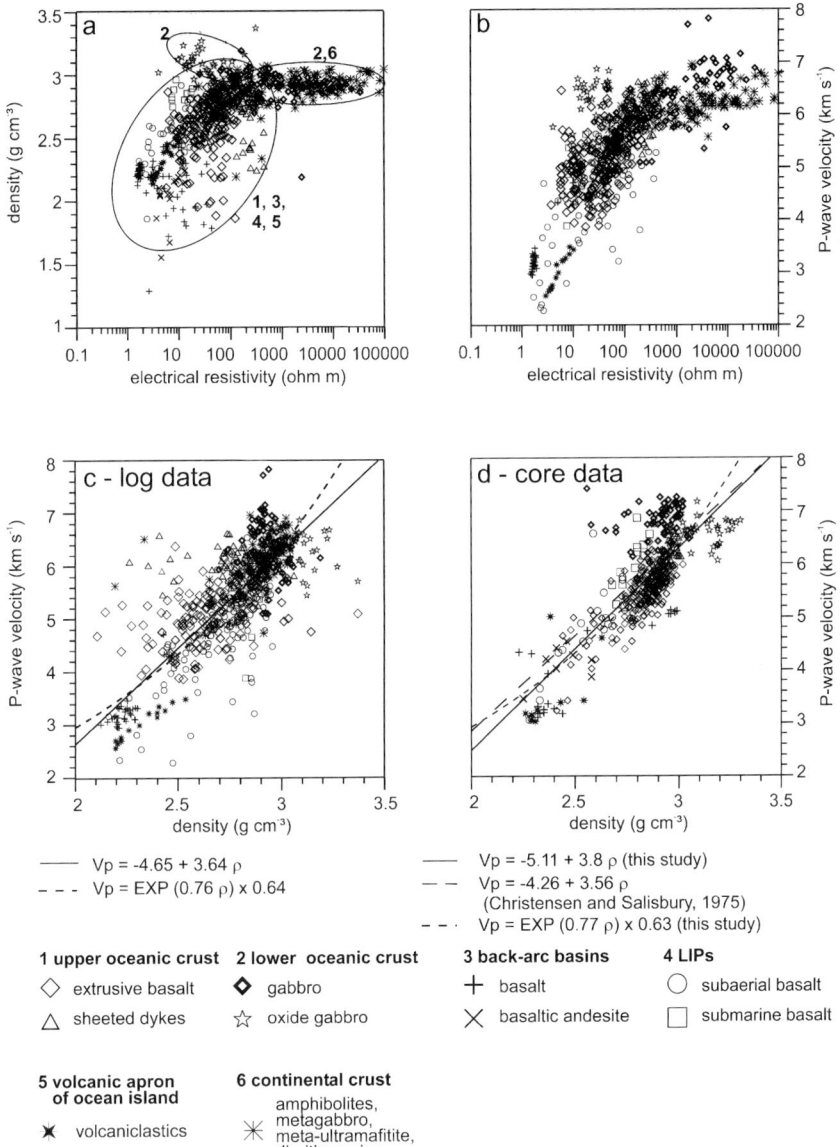

Fig. 6. (a) Cross-plot of electrical resistivity versus density. Most rocks show an increase in density with increasing resistivity, which is mostly controlled by the structure and texture of the rocks. However, the oxide-rich gabbros (ii) show an opposite trend. (b) Cross-plot of electrical resistivity versus P-wave velocity. The relation between electrical resistivity and P-wave velocity is similar to the trend of electrical resistivity versus density. (c) Cross-plot of density versus P-wave velocity. P-wave velocity increases with increasing density in most rock types. (d) Cross-plot of density versus P-wave velocity measured on cores. The relationship between the two properties is similar to that for the log data (Fig. 6c). Core data are compiled from Melson *et al.* (1979), Donnelly *et al.* (1980), Cann *et al.* (1983), Roberts *et al.* (1983), Anderson *et al.* (1985) and the JANUS database (*http://www-odp.tamu.edu/database/janusmodel.htm*). No measurements of P-wave velocity from cores are available for the KTB-VB and Sancerre-Couy boreholes. Parameters for the linear and exponential regression functions are given in Table 4. The linear relation calculated on core data is very similar to that calculated by Christensen & Salisbury (1975) for DSDP basalts. Ellipsoids in (a) mark three major groupings of data points representing different lithologies or rocks from different geological settings. For display in parts a–c, the log data-set is reduced using only every fifth to 50th data point, depending on the amount of data for each borehole.

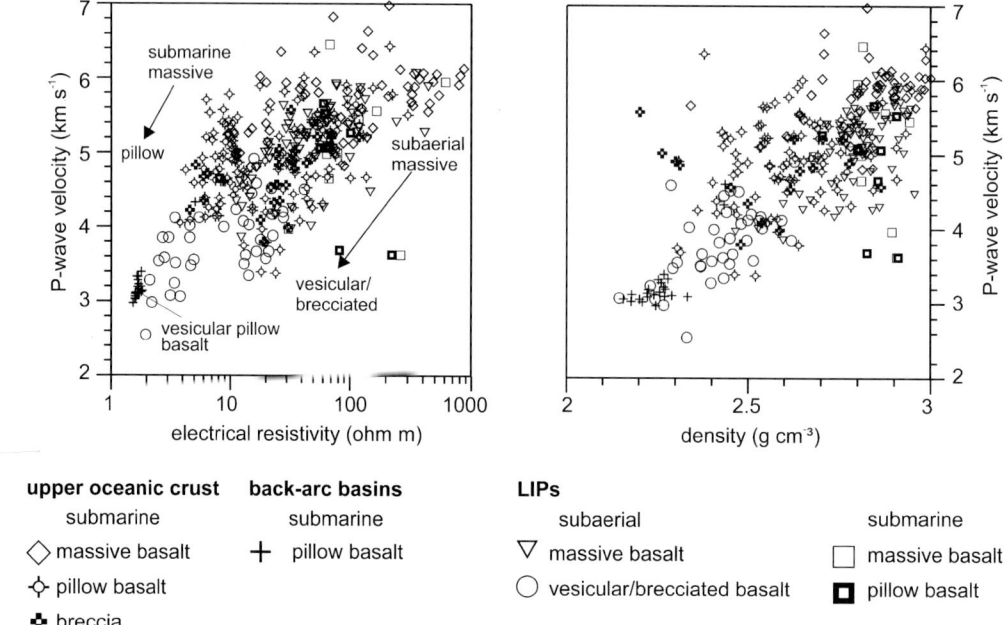

Fig. 7. Cross-plots of (**a**) electrical resistivity versus P-wave velocity, and (**b**) of density versus P-wave velocity for basalts only, show how the gradual increase in fracture density and vesicularity affects *in situ* physical properties. The data-set is reduced for display by using only every fifth to 50th data point.

and this explains the high electrical resistivity, density and P-wave velocity values in these rocks. Slightly lower values for the massive subaerial lava flows are caused by the presence of vesicles that constitute 0.1 to 5% of the rock in Hole 1137A (Coffin *et al.* 2000).

The lower values of electrical resistivity, density and P-wave velocity of the pillow basalts result from intense fracturing (caused by thermal contraction) and a greater void space due to inter-pillow zones with high porosity (Gillis & Sapp 1997). A higher amount of (connected) void space filled with conductive sea-water or conductive alteration minerals (clay minerals) results in a decrease in electrical resistivity (e.g. Pezard 1990), while density decreases due to the presence of water or clay minerals. Within individual boreholes, breccias have typically lower average values of electrical resistivity and P-wave velocity than pillow basalts, but scattering of the data of breccia and pillow basalts is high, and data from both clusters of both lithologies overlap. Submarine basalts from LIPs have intermediate log responses, which fall between those of basalts from upper oceanic crust and subaerial lava sheets. They are characterized by low vesicularity (1–4%). Pillow basalts from Hole 768C, erupted in a back-arc environment, have extremely low electrical resistivities and P-wave velocities, caused by an extremely high vesicularity of 30 to 50% and strong fracturing (Rangin *et al.* 1990).

The brecciated tops of subaerial lava flows are highly vesicular (up to 30%) and they are strongly fractured (Planke 1994; Delius *et al.* 1995, 2003; Coffin *et al.* 2000). The presence of water or clay minerals in the void space results in low density and P-wave velocity readings. Low electrical resistivity values in these rocks indicate that vesicles are connected and filled with conductive material (sea-water or clay minerals).

Resedimented mafic volcaniclastics. The resedimented mafic volcaniclastics have low values of electrical resistivity (<10 ohm m), P-wave velocity (2.5–3.5 km s^{-1}), and density (2.1–2.6 km s^{-1}), with lower values for the tuff than for the volcaniclastic breccia (Figs 5 & 6a–c). This difference in physical properties is related to lithological variation, i.e. crystallinity and grain-size changes within the debris flow. The breccia constitutes the bottom part of the debris flow, and consists of well-cemented basaltic lithic clasts and hyaloclastite lapillistone. This sequence is overlain by a hyaloclastite tuff that is

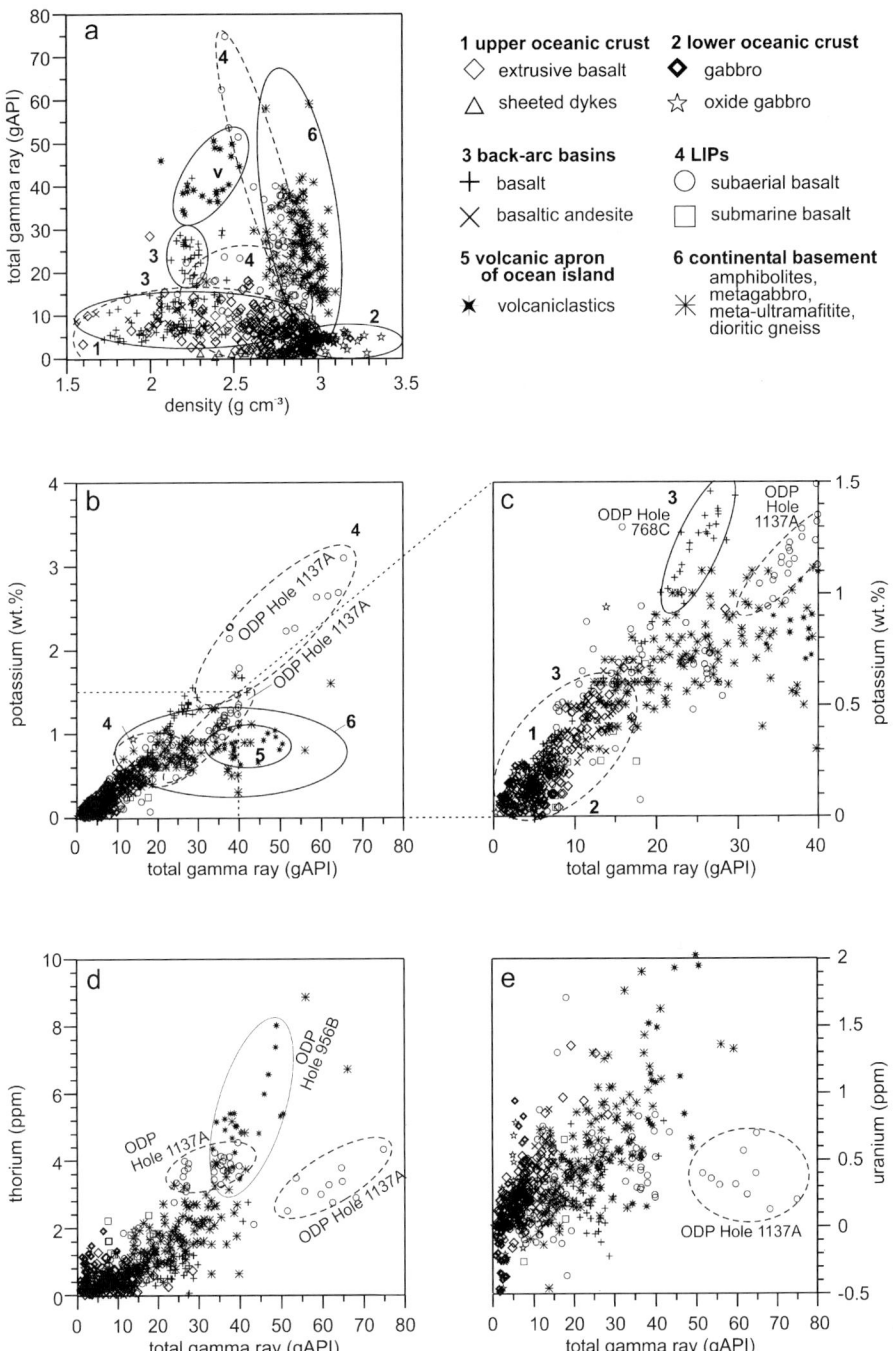

Fig. 8. (**a**) A cross-plot of density versus total gamma-ray distinguishes between mafic rocks from the different environments (i–vi). (**b–e**) Cross-plots of total gamma ray versus potassium, thorium, and uranium content. For most boreholes, a good correlation exists between total gamma ray and potassium. This indicates a large contribution of potassium to the total gamma-ray spectrum, while the contribution of thorium and uranium varies. Ellipsoids mark fields of data points belonging to the same geological setting. In (**b**), dashed lines mark the area zoomed for (**c**). The data-set is reduced for display by using only every fifth to 50th data point.

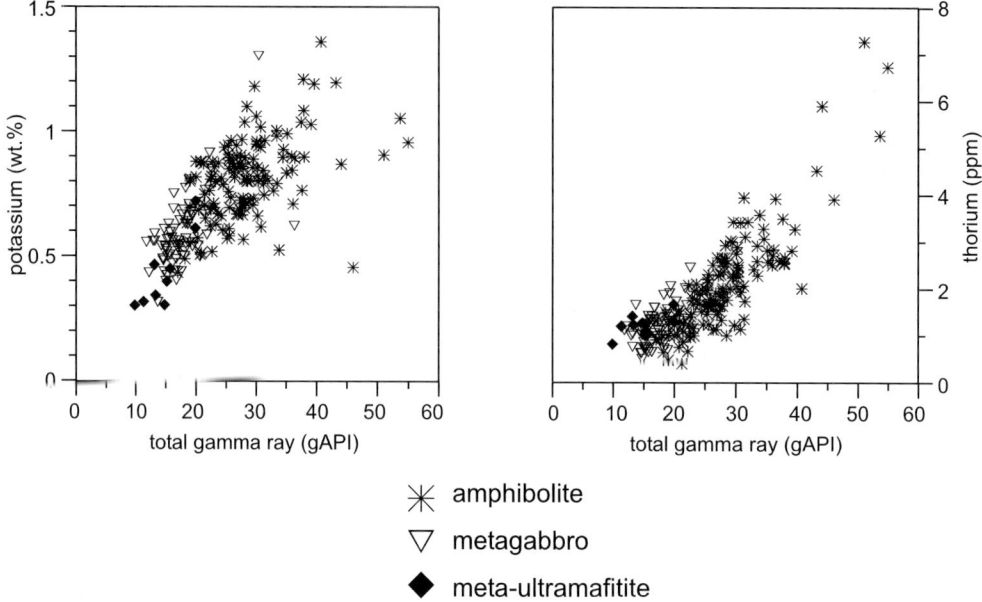

✳ amphibolite

▽ metagabbro

◆ meta-ultramafitite

Fig. 9. Cross-plots of total gamma ray versus potassium and thorium content for the different mafic rock types of the KTB-VB. Amphibolites have generally higher total gamma-ray values and also higher potassium and thorium values than the metagabbros and meta-ultramafitites, due to the overprint of regional metamorphosis. The data-set is reduced for display by using only every 10th data point.

less well consolidated and cemented (Schmincke *et al.* 1995). The better cementation of the breccia results in higher electrical resistivity, density, and P-wave velocity values. Less cementation, the clayey matrix and the small grain size (ash) of the tuffs reduce the P-wave velocity, density, and electrical resistivity.

Oxide-gabbros. Oxide-gabbros are characterized by low to intermediate electrical resistivity values between 3 and 300 ohm m and high densities between 2.9 and 3.4 g cm^{-3}. P-wave velocities are similar to those of the olivine gabbros (5.2–6.7 km s^{-1}). The density increases with decreasing resistivity. We relate this resistivity–density trend to an increasing iron and titanium oxide content in the rocks. High amounts of ilmenite (FeTiO$_3$, density 4.5–5.0 g cm^{-3}) and magnetite (Fe$_3$O$_4$, density 5.2 g cm^{-3}) cause a rise in density and increase the electrical conductivity due to enhanced grain contacts (Pezard *et al.* 1991). Similarly, the high photo-electric factor value in the oxide gabbros (Fig. 5) is caused by the higher photo-electric factor of the heavy elements iron and titanium compared to the lighter rock-forming elements of an olivine gabbro.

Olivine gabbros and metamorphic rocks. Olivine gabbros and metamorphic rocks have the highest electrical resistivity values (up to 100 000 ohm m) and high density (2.8 to 3.1 g cm^{-3}) and P-wave velocity values (6.8–7.9 km s^{-1}) (Figs 5 and 6a–c). Rocks with electrical resistivity <1000 ohm m, show increases in electrical resistivity and P-wave velocity with increasing density, just like the basalts mentioned earlier. At electrical resistivity values >1000 ohm m, electrical resistivity shows a broad value range, while density and P-wave velocity are close to their maximum values. This narrow range in density and P-wave velocity values at high electrical resistivities (>1000 ohm m) indicates that porosity changes are too little to affect density and P-wave velocity readings. The very broad range in resistivity values, however, suggests that the connectivity of (micro-) cracks varies and has an effect solely on electrical resistivity. Petrophysical measurements on cores from Hole 735B give very low porosity values (<1%), and indicate that the porous network is controlled by primary microstructures (grain boundaries) and by plastic foliation (Iledefonse & Pezard 2001).

Low density values (<2.8 g cm^{-3}) were obtained on several gabbro samples from ODP Hole 1105A, and can be related to alteration. In the core data-set, gabbros deviate from the general trend of increasing density with increasing P-wave velocity (Fig. 6d) indicating that the velocity–density relation in the gabbroic rocks is not only controlled by fracturing, but also by compositional and mineralogical changes. Iturrino et al. (1991) illustrated the varying relations between P-wave velocity and density from core with respect to the mineralogical composition of the gabbros. This is particularly important for core measurements, while the logging data integrate over larger volumes and also record fractures.

Slightly higher P-wave velocity values in the gabbros (average value 6.6 km s^{-1}) compared to the metamorphic rocks (average 6.2–6.4 km s^{-1}) and also slightly higher P-wave velocity values in metagabbros compared to the amphibolites (average values 6.4 and 6.2 km s^{-1}, respectively) indicate the influence of anisotropy in the rocks caused by variations in intensity and dip of the metamorphic foliation on P-wave velocity. Large intervals of the gabbros from Holes 735B and 1105A are plastically deformed, and magmatic layering (as indicated by grain size or modal variation) occurs locally (Robinson et al. 1989; Cannat et al. 1991; Pettigrew et al. 1999). However, most of the data from Hole 735B are from depth intervals where foliation is weakly developed (Robinson et al. 1989). In contrast, amphibolites from KTB are locally foliated, while the metagabbros partly preserve their original igneous texture (Hirschmann et al. 1997). Petrophysical analyses on core samples from a deeper depth interval of KTB-VB show that P-wave velocities are influenced by the effect of anisotropy due to metamorphic foliation. Microcracks preferentially spread out in the foliation plane, reducing P-wave velocity perpendicular to it (Berckhemer et al. 1997).

P-wave velocity–density relation. This trend of increasing P-wave velocity with increasing density observed in log and core data (Fig. 6c & d) was approximated using regression analysis. The results for the log data is:

$$V_{P_{log}} = -4.65 + 3.64\rho_{log}$$

using a linear model, or:

$$V_{P_{log}} = \exp(0.76\rho_{log}) \times 0.64$$

using an exponential model, where $V_{P_{log}}$ is the P-wave velocity and ρ_{log} is the density from the down-hole measurements.

The result for the core data is:

$$V_{P_{core}} = -5.11 + 3.8\rho_{core}$$

The exponential model is:

$$V_{P_{core}} = \exp(0.77\rho_{core}) \times 0.63$$

where $V_{P_{core}}$ and ρ_{core} are the P-wave velocity and density from core measurements.

For both core and log data, low values of the residual sums of squares and high values of the coefficient of determination (r^2) indicate that the exponential model gives a better fit to the data (Table 4). The trends observed in the logging and core data are very similar and fit well with those calculated by Christensen &

Table 4. *Parameters for the regression functions in Figure 6c & d*

	Linear model	Non-linear model
	Logging data	
Equation	$V_P = -4.65 + 3.64\rho$	$V_P = \exp(0.76\rho) \times 0.64$
Number of points	7050	7050
Residual sum of squares	2485.5	106.7
Coefficient of determination (r^2)	0.64	0.65
Standard error for constant	0.090	
Standard error for coefficient	0.032	
	Core data	
Equation	$V_P = -5.11 + 3.80\rho$	$V_P = \exp(0.77\rho) \times 0.631$
Number of points	450	450
Residual sum of squares	128.7	4.5
Coefficient of determination (r^2)	0.64	0.68
Standard error for constant	0.382	
Standard error for coefficient	0.134	

Due to limitations in computing capacity, the data-set for logging data was reduced by using only every second data point.

Salisbury (1975) on DSDP basalts (Fig. 6d). The good agreement in the regression functions calculated for log and core data shows that the relationship between density and P-wave velocities in mafic rocks is similar for both *in situ* and laboratory data. Christensen & Salisbury (1975) also observed that a non-linear solution for the density–velocity relation is more appropriate than a linear solution. A non-linear solution suggests that, for the same change in density, the P-wave velocity value range is smaller for low densities than for high densities. This can be interpreted as recording differences in structure and pore space. Vesicular, pillowed, and brecciated basalts are characterized by low density and low velocity values, and show an almost linear increase in density and P-wave velocity values. In contrast, in massive basalts, gabbros and metamorphic rocks where pore space is controlled by fractures, the P-wave velocity values increase more rapidly than the density values (see also Figs 5 & 7b). While density reflects changes in pore space independently from pore space structure, P-wave velocity is affected by the spatial distribution of pore space. Wilkens *et al.* (1991) showed that the aspect ratio of void space strongly influences P-wave velocity; low aspect ratios (fractures) reduce P-wave velocity more strongly than high aspect ratios (vesicles) at the same porosity. Moreover, Christensen & Salisbury (1975) concluded that unaltered basalts have lower velocities because of their finer grain size and hence greater grain boundary porosity than gabbros of the same mineralogical composition.

The P-wave velocity readings of the volcaniclastic rocks are below the regression line. This can be explained by particularly weak grain contacts and deteriorating elastic properties. Data from the oxide gabbros plot off the line because of their much higher density, due to their high content of dense oxide minerals.

Variations in total gamma ray and potassium, thorium, and uranium content

Figure 8a shows a cross-plot of density versus total gamma ray. The data from the different rock types and geological settings plot in distinguishable clusters.

Basalts from upper oceanic crust. Basalts from upper oceanic crust are characterized by low total gamma-ray values, mostly below 20 gAPI (Fig. 8a). Fresh basalts from mid-ocean ridges have particularly low potassium, thorium, and uranium contents, resulting in low total gamma-ray values. Several studies on the geo-

chemistry and mineralogy of oceanic basalts show that potassium is concentrated in secondary minerals such as clays (K-rich smectites, celadonite) during low-temperature oxidizing sea-floor weathering due to circulation of sea-water through the crust (e.g. Gillis & Robinson 1988; Alt 1995). Therefore, total gamma-ray values increase with alteration of these rocks (e.g. Bartetzko *et al.* 2001; Révillion *et al.* 2002). A positive correlation between potassium and total gamma ray confirms that potassium is the largest contributor to the gamma radiation in these rocks (Fig. 8b & c). Thorium and uranium values are very low in oceanic crust, mostly <1 ppm for both logs (Fig. 8d & e). Although these values are very low, Révillion *et al.* (2002) demonstrated that spectral gamma-ray measurements give reliable results in these rocks for uranium. Increased values of up to 1.8 ppm thorium and 1.4 ppm uranium occur locally in Holes 418A and 801C. Farr *et al.* (2001) showed that uranium enrichments in these holes are associated with the precipitation of carbonates and Fe-oxides.

Comparing the boreholes in oceanic crust, variations in natural gamma-ray data are apparent. In particular, the value range of pillow basalts from Hole 418A and of pillow basalts and breccias from Hole 801C are much larger than data from the other holes (Fig. 5). Both holes were drilled into oceanic crust older 100 Ma. Grevemeyer & Bartetzko (2004) showed that scattering of gamma-ray data from ODP holes drilled into oceanic crust increases with the age of the rocks. This is probably related to the increase in the heterogeneity of the crust due to the formation of different generations and types of K-bearing secondary minerals.

Gabbros from lower oceanic crust. Gabbros from lower oceanic crust also have low values of total gamma ray. Similar to the basalts from upper oceanic crust, gabbros from the lower oceanic crust are poor in potassium, thorium, and uranium (Fig. 8a–e). The gabbros of Holes 735B and 1105A represent a lower-crustal hydrothermal alteration regime (Alt 1995) where no accumulation of potassium, thorium and uranium occurred.

Basalts from back-arc basins. Primitive basalts from the Lau Basin (ODP Holes 834B and 839B) have similar low values of total gamma ray (3–11 gAPI) and low contents of radioactive elements, just like the basalts from upper oceanic crust (Fig. 8a–e). Basaltic andesites have slightly higher values (6–22 gAPI), which are probably caused by a higher primary potassium content of the parent magma and/or extensive alteration.

Basalts from Hole 768C are characterized by higher total gamma-ray values (20–30 gAPI; Fig. 8a) and elevated potassium values between 1 and 1.5 wt% (Fig. 8c), while contents of thorium and uranium remain low. These rocks show a strong alteration, associated with the replacement of plagioclase by K-feldspar (Rangin *et al.* 1990) that causes an increase in the gamma-ray values.

Volcanic islands and LIPs. These rocks show very variable values of total gamma ray. Rocks from the HSDP2 borehole and ODP holes from the North Atlantic (Holes 553A, 642E and 917A) have low total gamma-ray values between 2 and 30 gAPI, while basalts from the Kerguelen Plateau (ODP Holes 1137A and 1140A) have the highest gamma-ray values, of up to 80 gAPI (Fig. 8a & b).

In Hole 1140A, elevated total gamma-ray values in the K-rich lava flows and pillow basalts (Fig. 5) are related to a higher potassium content of the magma (Coffin *et al.* 2000). Basalts from ODP Hole 1137A divide into two groups with high total gamma-ray values (Fig. 8b). One group shows very high total gamma ray (40–75 gAPI) and high potassium values (1.5–3.5 wt%; Fig. 8c) and corresponds to the high gamma-ray–low thorium (2–4.5 ppm) field in the total gamma-ray versus thorium cross-plot (Fig. 8d). The other group shows intermediate total gamma-ray values (30 and 45 gAPI) and potassium contents of 1 to 1.5 wt%. These basalts are characterized by higher thorium values (3–6.6 ppm). In Hole 1137A, a glauconite-bearing sandy packstone overlies the basalts and marks a transgression. In addition, volcaniclastic sediments and a fluvial conglomerate are sandwiched between lava flows (Coffin *et al.* 2000). These potassium- and thorium-bearing sediments partly infill the rough surface of the underlying vesicular and brecciated top part of the lava flows, and cause the higher thorium values seen in the rocks of Hole 1137A. Post-magmatic alteration probably mobilized potassium in these sediments, resulting in the observed high potassium level in these lavas. In the brecciated top parts of the subaerial basalts, alteration has led to a more substantial secondary enrichment of potassium and the highest values in total gamma-ray and potassium logs.

Resedimented volcaniclastics. Mafic volcaniclastic rocks from the volcanic apron of Gran Canaria are characterized by total gamma-ray values between 31 and 57 gAPI (Fig. 8a). Compared to these high total gamma-ray values, potassium values are relatively low (0.4–1.2 wt%). Thorium values are high, and range between 3.5 and 8 ppm (Fig. 8d). Uranium values are highly variable and range between 0.2 and >2 ppm (Fig. 8e). The high thorium values may be explained by synsedimentary mixing of the basalt clasts with thorium-bearing minerals in volcaniclastic sediments derived from more evolved volcanism of the neighbouring island of La Gomera (Schmincke & Segschneider 1998).

Metamorphic rocks from continental basement. Rocks from continental basement are characterized by high total gamma-ray values (31 to 57 gAPI). Figure 8 shows a good correlation between the three radioactive elements potassium, thorium, uranium and total gamma-ray. Amphibolites have higher total gamma-ray, potassium, and thorium values than the metagabbros and ultramafitites (Fig. 9). However, although the amphibolites and metagabbros were derived from similar source rocks of enriched E-MORB composition (Harms *et al.* 1997) they exhibit differences in fabrics that point to a strongly inhomogeneous deformation during metamorphism. While the original igneous texture is widely preserved in the metagabbros, amphibolites exhibit a wide range of metamorphic fabrics, ranging from homogeneous, weakly foliated amphibolites, through fine-grained amphibolites with pronounced compositional layering, to coarse-grained amphibolites, usually enriched in felsic minerals. Quartz–feldspar mobilization suggests that partial melting has taken place as a consequence of an influx of water (Schalkwijk 1991; Schalkwijk & Stöckhert 1992). By comparing down-hole data with rock fabric and mineralogical composition, Pechnig *et al.* (1997) identified a positive correlation between the grade of metamorphic overprint and the amount of radioactive elements: regional metamorphism modified the structure and composition more in amphibolites than in the metagabbros and ultramafitites. As a consequence, metagabbros and ultramafitites remained closer to the original composition and thus have lower gamma-ray values.

The dioritic gneisses from hole Sancerre-Couy are characterized by higher total gamma-ray values (39 gAPI on average) and slightly lower density values (average value 2.8 g cm^{-3}) than the amphibolites from the same hole (Fig. 5). These physical properties can be related to their more intermediate chemical composition (Fig. 3) and an enrichment in felsic minerals of lower density (e.g. feldspar). Log responses of intermediate igneous and metamorphic rocks are discussed in more detail by Pechnig *et al.* (2005).

Summary and conclusions

The *in-situ* physical properties of the various mafic rocks presented here constitute the first comprehensive compilation of such data. The rocks are mostly of basaltic composition and occur as different lava morphologies (submarine and subaerially erupted lava flows or pillows), as intrusives (dykes and gabbros), metamorphic rocks (amphibolite, metagabbro, meta-ultramafitite) or resedimented volcaniclastic sediments (breccia and tuffs in a debris flow). Although their composition is similar, they show large differences in the value ranges of the different physical properties. Differences in electrical resistivity, P-wave velocity and density can be explained by structural and textural variations in the rocks, as these log properties are particularly sensitive towards fracturing and porosity. These variations are mostly caused by the differing cooling histories of the magmas in the different geological settings (e.g. slow cooling in lower oceanic crust, rapid quenching in submarine environments). Differences in photo-electric factor values and total gamma ray are primarily related to variations in composition and to the alteration of the rocks.

These relations between log responses and variations in structure and composition are summarized in Table 5. The most important observations are as follows.

- Fracturing is an important factor explaining the differences in electrical resistivity, density and P-wave velocity. In igneous rocks, fracturing is strongly related to the cooling history of the magmas and intrusive rocks that cooled slowly and are characterized by a low fracture density (e.g. gabbros) and have high values of these physical properties. Extrusive rocks that quenched rapidly (e.g. pillow basalts, brecciated tops of lavas) are strongly fractured and are therefore characterized by low values of electrical resistivity, density and P-wave velocity.

- The vesicularity of a rock is controlled by the primary composition of the magma (volatile content), and the cooling history of the lava flow and vesicularity is the key factor for changes in the physical properties of basalts from LIPs and back-arc basins. The presence of vesicles increases the porosity of the rock and, in dependence of the filling material (water, void space, alteration minerals), causes the density and P-wave velocity to decrease. In most boreholes drilled into back-arc basins and LIPs, low electrical resistivity values in vesicular rocks indicate that vesicles are interconnected and filled with conductive material (sea-water, clay minerals).

- Metamorphic foliation causes an anisotropy in the rocks that particularly influences P-wave velocity. Therefore, oceanic gabbros from intervals with low foliation have higher P-wave velocities than the metamorphic rocks from KTB.

- Variations in grain size and cementation of resedimented volcaniclastic material result in changes in electrical resistivity, P-wave velocity and density. The coarser the grain size and the better cemented the rock, the higher are the values in these physical properties.

- Primary variations in magma composition may be observed (1) in higher values of total gamma ray, for example in basaltic andesites from back-arc basins and (2) in increased values of the photo-electric factor

Table 5. *Importance of the different log properties for distinguishing the various effects of variations in structure and composition of mafic rocks*

	Electrical resistivity	P-wave velocity	Density	Photo-electric factor	Total gamma ray
Variations in structure					
Fracturing	•	•	•	—	—
Vesicularity	•	•	•	—	—
Metamorphic foliation	—	•	—	—	—
Resedimentation	•	•	•	—	—
Variations in composition					
Primary composition	(•)	—	•	•	•
Alteration	•	—	•	—	•
Metamorphic overprint	—	—	—	—	•
Resedimentation	—	—	—	—	•

• Important.
— Not relevant.

and density in the iron- and titanium-rich gabbros.

- Weathering, subsurface alteration, metamorphism and resedimentation often result in an uptake of highly mobile potassium. The total gamma-ray log may therefore be used as an indicator for such processes.

We have shown that cross-plots of one physical property reflecting variations in composition (e.g. total gamma ray), versus one physical property related to structural variations (e.g. density), can be used to distinguish rocks from different geological settings. The overall similarity of log properties from the same lithology and geological setting confirms the reliability of the logging data. This is important, as the logging tools were originally not designed for igneous or metamorphic environments. Therefore, our presented data provide a good basis for quality control of logging data from future boreholes drilled in similar geological settings. Moreover, this study may present basic data for geoscientific modelling where *in situ* data are required.

This study used logging data and core data from various boreholes, collected and composed from different sources. The Ocean Drilling Program (ODP) provided part of the data. The ODP is sponsored by the US National Science Foundation (NSF) and participating countries under the auspices of Joint Oceanographic Institutions (JOI), Inc. The Hawaii Scientific Drilling Project was drilled by the International Continental Drilling Project (ICDP). Data were made available by GeoForschungszentrum (GFZ) Potsdam, Germany. The German Continental Drilling Project (KTB) was funded by the German Ministry of Research and Technology (BMFT). Logging data and core data from Sancerre-Couy were kindly made available by the Bureau de Recherches Géologiques et Minières (BRGM), France. We are grateful to the German Science Foundation (DFG), which supported most of the individual studies, and with it, the establishment of a database. Many thanks to R. England for his constructive comments on the manuscript, and to R. Brown for proofreading. Constructive comments by two anonymous reviewers helped to significantly improve the manuscript. We especially thank J. Wohlenberg, who initiated and supported these studies.

References

ALT, J. C. 1995. Subseafloor processes in Mid-Ocean Ridge hydrothermal systems. *In*: HUMPHRIES, S. E., ZIERENBERG, R. A., MULLINEAUX, L. S. & THOMSON, R. E. (eds) *Seafloor Hydrothermal Systems–Physical, Chemical, Biological, and Geological Interactions.* AGU, Geophysical Monograhs, **91**, 85–114.

ALT, J. C., KINOSHITA, H., STOKKING, L. B. *et al.* 1993. *Proceedings of the Ocean Drilling Program, Initial Reports*, **148**, College Station, TX (Ocean Drilling Program).

ANDERSON, R. N., HONNOREZ. J. & BECKER, K. *et al.* 1985. *Initial Reports of the Deep Sea Drilling Project*, **83**, Washington (US Government Printing Office).

AUTRAN, A. & CHANTRAINE, J. 1988. L'anomalie du bassin de Paris et le sondage de Sancerre-Couy. Discussion de quelques hypothèses possibles pour la nature du corps magnétique. *In*: CHANTRAINE, J. (ed.) *Géologie Profonde de la France–Forage scientifique de Sancerre-Couy (Cher). Socle: données preliminaires.* Documents du BRGM, **137**, 253–272.

AYADI, M., PEZARD, P. A., LAVERNE, C. & BRONNER, G. 1998. Multi-scalar structure at DSDP/ODP Site 504, Costa Rica Rift, I: stratigraphy of eruptive products and accretion processes. *In*: HARVEY, P. K. & LOVELL, M. A. (eds) *Core–Log Integration.* Geological Society, London, Special Publications, **136**, 297–310.

BARR, S. R., RÉVILLION, S., BREWER, T. S., HARVEY, P. K. & TARNEY, J. 2002. Determining the inputs to the Mariana Subduction Factory: using core–log integration to reconstruct basement lithology at ODP Hole 801C. *Geochemisty, Geophysics, Geosystems*, **3** (11), 8901, doi: 10.1029/2001GC000255.

BARTETZKO, A. 1999. *Aufbau und Entstehung ozeanischer Kruste an mittelozeanischen Rücken – Eine Interpretation von Bohrlochmessungen.* PhD thesis, Technical University, Aachen, Germany.

BARTETZKO, A., PECHNIG, R. & WOHLENBERG, J. 2002. Interpretation of well-logging data to study lateral variations in young oceanic crust: DSDP/ODP Holes 504B and 896A, Costa Rica Rift. *In*: LOVELL, M. & PARKINSON, N. (eds) *Geological Application of Well Logs.* AAPG, Methods in Exploration Series, **13**, 213–228.

BARTETZKO, A., PEZARD, P., GOLDBERG, D., SUN, Y.-F. & BECKER, K. 2001. Volcanic stratigraphy of DSDP/ODP Hole 395A: an interpretation using well-logging data. *Marine Geophysical Researches*, **22**, 111–127.

BERCKHEMER, H., RAUEN, A. & WINTER, H. *et al.* 1997. Petrophysical properties of the 9-km-deep crustal section at KTB. *Journal of Geophysical Research*, **102**, 18 337–18 361.

BOREHOLE RESEARCH GROUP 2000. *ODP Logging Manual: an electronic guide to ODP logging services*, CD-ROM v2.0, Lamont–Doherty Earth Observatory of Columbia University, Palisades, NY, USA.

BREWER, T. S., HARVEY, P. K., LOVELL, M. A., HAGGAS, S., WILLIAMSON, G. & PEZARD, P. 1998. Ocean floor volcanism: constraints from the integration of core and downhole logging measurements. *In*: HARVEY, P. K. & LOVELL, M. A. (eds) *Core–Log Integration.* Geological Society, London, Special Publications, **136**, 341–362.

BROGLIA, C. & ELLIS, D. 1990. Effect of alteration, formation absorption, and stand-off on the response of the thermal neutron porosity log in gabbros and basalts: examples from Deep Sea Drilling Project–Ocean Drilling Program Sites. *Journal of Geophysical Research*, **95**, 9171–9188.

BÜCKER, C. J., DELIUS, H., WOHLENBERG, J. & LEG 163 SHIPBOARD SCIENTIFIC PARTY, 1998. Physical signature of basaltic volcanics drilled on the north-east Atlantic volcanic rifted margin. *In*: HARVEY, P. K. & LOVELL, M. A. (eds) *Core–Log Integration*. Geological Society, London, Special Publications, **136**, 363–374.

CABANIS, B. & THIÉBLEMONT, D. 1992. Forage scientifique de Sancerre-Couy. Géochimie des protolithes (formations orthodérivées). Mise en évidence de deux événements paléomagmatiques au cours de l'évolution antévarisque. *In*: CHANTRAINE, J., LORENZ, C., MÉGNIEN, C., MILLION, R. & LIENHART, M.-J. (eds) *Géologie de la France – Forage scientifique de Sancerra-Couy (Cher) Synthèse des études 1986–1922*. Mémoire GPF, **3**, 123–128.

CANN, J. R., LANGSETH, M. G., HONNOREZ, J., VON HERZEN, R. P., WHITE, S. M. *et al.* 1983. *Initial Reports of the Deep Sea Drilling Project*, **69**, Washington (US Government Printing Office).

CANNAT, M., MÉVEL, C. & STAKES, D. (1991). Normal ductile shear zones at an oceanic spreading ridge: tectonic evolution of Site 735 gabbros (southwest Indian Ocean). *In*: VON HERZEN, R. P., ROBINSON, P. T. *et al.* (eds) *Proceedings of the Ocean Drilling Program, Scientific Results*, **118**, College Station, TX (Ocean Drilling Program), 415–429.

CHRISTENSEN, N. I. & SALISBURY, M. H. 1975. Structure and constitution of the lower oceanic crust. *Reviews of Geophysics and Space Physics*, **13**, 57–86.

COFFIN, M. F., FREY, F. A., WALLACE, P. J. *et al.* 2000. *Proceedings of the Ocean Drilling Program, Initial Reports*, **183**, (CD-ROM). Available from: Ocean Drilling Program, Texas A & M University, College Station, TX, USA.

DELIUS, H., BREWER, T. S. & HARVEY, P. K. 2003. Evidence for textural and alteration changes in basaltic lava flows using variations in rock magnetic properties (ODP Leg 183). *Tectonophysics*, **371**, 111–140.

DELIUS, H., BÜCKER, C. & WOHLENBERG, J. 1995. Significant log responses of basaltic lava flows and volcaniclastic sediments in ODP Hole 642E. *Scientific Drilling*, **5**, 217–226.

DELIUS, H., BÜCKER, C. & WOHLENBERG, J. 1998. Determination and characterization of volcaniclastic sediments by wireline logs: Sites 953, 955, and 956, Canary Islands. *In*: WEAVER, P. P. E., SCHMINCKE, H.-U., FIRTH, J. V. & DUFFIELD, W. (eds) *Proceedings of the Ocean Drilling, Scientific Results*, **157**, College Station, TX (Ocean Drilling Program), 29–37.

DEPAOLO, D. J., STOLPER, E. & THOMAS, D. M. 2001. Deep drilling into a Hawaiian volcano. *EOS, Transactions, American Geophysical Union*, **82**, 150–155.

DICK, H. J. B., NATLAND, J., MILLER, D. J., *et al.* 1999. *Proceedings of the Ocean Drilling Program, Initial Reports*, **176**, (CD-ROM). Available from: Ocean Drilling Program, Texas A&M University, College Station, TX, USA.

DONNELLY, T., FRANCHETEAU, J., BRYAN, W., ROBINSON, P., FLOWER, M., SALISBURY, M. *et al.* 1980. *Initial Reports of the Deep Sea Drilling Project*, **51, 52, 53**, Washington (US Government Printing Office).

ELDHOLM, O., THIEDE, J., TAYLOR, E., *et al.* 1987. *Proceedings of the Ocean Drilling Program, Initial Reports*, **104**, College Station, TX (Ocean Drilling Program).

EWART, A., BRYAN, W. B., CHAPPELL, B. W. & RUDNICK, R. L. 1994. Regional geochemistry of the Lau–Tonga arc and backarc systems. *In*: HAWKINS, J., PARSON, L., ALLAN, J. *et al. Proceedings of the Ocean Drilling Program, Scientific Results*, **135**, College Station, TX (Ocean Drilling Program), 385–425.

FARR, L. C., PLANK, T., KELLEY, K. & ALT, J. C. 2001. U mineral host and enrichment processes in altered oceanic crust. *EOS, Transactions American Geophysical Union*, **82**, Fall Meeting Supplement, Abstract T22C-0926.

GILLIS, K. M. & ROBINSON, P. T. 1988. Distribution of alteration zones in the upper oceanic crust. *Geology*, **16**, 262–266.

GILLIS, K. M. & SAPP, K. 1997. Distribution of porosity in a section of upper oceanic crust exposed in the Troodos Ophiolite. *Journal of Geophysical Research*, **102**, 10 133–10 149.

GREVEMEYER, I. & BARTETZKO, A. (2004) Hydrothermal ageing of oceanic crust: inferences from seismic refraction and borehole studies. *In*: DAVIS, E. E. & ELDERFIELD, H. (eds) *Hydrogeology of Oceanic Lithosphere*. Cambridge University Press, 128–150.

HAGGAS, S. L., BREWER, T. S, & HARVEY, P. K. 2002. Architecture of the volcanic layer from the Costa, Rica Rift, constraints from core–log integration. *Journal of Geophysical Research*, **107**, 10.1029/2001JB000147.

HARMS, U., CAMERON, K. L., SIMON, K., & BRÄTZ, H. 1997. Geochemistry and petrogenesis of metabasites from the KTB ultradeep borehole, Germany. *Geologische Rundschau*, supplement to volume **86**, 155–166.

HAVERKAMP, S. & WOHLENBERG, J. 1991. EFA-Log-Rekonstruktion kristalliner Lithologie anhand von bohrlochgeophysikalischen Mesungen für die Bohrungen URACH 3 und KTB-Oberpfalz VB. *KTB Report*, **91–4**.

HIRSCHMANN, G., DUYSTER, J., HARMS, U., KONTNY, A., LAPP, M., DE WALL, H., & ZULAUF, G. 1997. The KTB superdeep borehole: petrography and structure of a 9-km-deep crustal section. *Geologische Rundschau*, supplement to volume **86**, 3–14.

HOTTIN, A. M., LAFORET, C., VEZAT, R., WYNS, R., BALE, P. & BOUTIN, R. 1988. Description pétrographique des roches du socle dans le forage de Sancerre-Couy. *In*: CHANTRAINE, J. (ed.) *Géologie Profonde de la France–Forage scientifique de Sancerre-Couy (Cher). Socle: données preliminaries*. Documents du BRGM, **137**, 99–114.

ILEDEFONSE, B. & PEZARD, P. 2001. Electrical properties of slow-spreading ridge gabbros from ODP Site 735, Southwest Indian Ridge. *Tectonophysics*, **330**, 69–92.

IRVINE, T. N. & BARAGAR, W. R. A. 1971. A guide to the chemical classification of the common volcanic rocks. *Canadian Journal of Earth Sciences*, **8**, 523–548.

ITURRINO, G. J., CHRISTENSEN, N. I., KIRBY, S. & SALISBURY, M. H. 1991. Seismic velocities and elastic properties of oceanic gabbroic rocks from Hole 735B. *In*: VON HERZEN, R. P. & ROBINSON, P. T. *et al.* (eds) *Proceedings of the Ocean Drilling Program, Scientific Results*, **118**, College Station, TX (Ocean Drilling Program), 227–244.

LARSEN, H. C., SAUNDERS, A. D., CLIFT, P. D. *et al.* 1994. *Proceedings of the Ocean Drilling Program, Initial Reports*, **152**, College Station, TX (Ocean Drilling Program).

LE BAS, M. J., LE MAITRE, R. W., STRECKEISEN, A. & ZANETTIN, B. 1986. A chemical classification of volcanic rocks based on the total alkali–silica diagram. *Journal of Petrology*, **27**, 745–750.

LYSNE, P. 1989. Investigation of neutron-porosity log uncertainties: Ocean Drilling Program Hole 642E. *In*: ELDHOLM, O., THIEDE, J., TAYLOR, E. *et al.* (eds) *Proceedings of the Ocean Drilling Program, Scientific Results*, **104**, College Station, TX (Ocean Drilling Program), 973–977.

MCPHIE, J., DOYLE, M. & ALLEN, R. 1993. *Volcanic Textures – A Guide to the Interpretation of Textures in Volcanic Rocks*. Centre for Ore Deposit and Exploration Studies, University of Tasmania.

MELSON, W. G., RABINOWITZ, P. D. *et al.* 1979. *Initial Reports of the Deep Sea Drilling Project*, **45**, Washington (US Government Printing Office).

MILLON, R. & ZWINGELBERG, F. 1992. Forage scientifique de Sancerre-Couy: les diagraphies dans le socle. *In*: CHANTRAINE, J., LORENZ, C., MÉGNIEN, C., MILLION, R. & LIENHART, M.-J. (eds) *Géologie de la France – Forage scientifique de Sancerra-Couy (Cher) Synthèse des études*, 1986–1922. Mémoire GPF, **3**, 83–92.

O'BRIEN, P. J., DUYSTER, J., GRAUERT, B., SCHREYER, W., STÖCKHERT, B. & WEBER, K. 1997. Crustal evolution of the KTB drill site: from oldest relics to the late Hercynian granites. *Journal of Geophysical Research*, **102**, 18 203–18 220.

PARSON, L., HAWKINS, J., ALLAN, J. *et al.* 1992. *Proceedings of the Ocean Drilling Program, Initial Reports*, **135**, College Station, TX (Ocean Drilling Program).

PECHNIG, R., DELIUS, H., & BARTETZKO A. (2005). Effect of compositional and structural variations on log responses of igneous and metamorphic rocks II: Acid and intermediate rocks. *In*: HARVEY, P. K., BREWER, T. S., PEZARD, P. A. & PETROV, V. Y. (eds) *Petrophysical properties of crystalline rocks*. Geological Society London, Special Publications, **240**, 279–300.

PECHNIG, R., HAVERKAMP, S., WOHLENBERG, J., ZIMMERMANN, G. & BURKHARDT, H. 1997. Integrated log interpretation in the German Continental Deep Drilling Program: Lithology, porosity, and fracture zones. *Journal of Geophysical Research*, **102**, 18 363–18 390.

PETTIGREW, T. L., CASEY, J. F., MILLER, D. J. *et al.* 1999. *Proceedings of the Ocean Drilling Program, Initial Reports*, **179**, (CD-ROM). Available from: Ocean Drilling Program, Texas A&M University, College Station, TX, USA.

PEZARD, P. A. 1990. Electrical properties of Mid-Ocean Ridge Basalt and implications for the structure of the upper oceanic crust in Hole 504B. *Journal of Geophysical Research*, **95**, 9237–9264.

PEZARD, P. A., HOWARD, J. J. & GOLDBERG, D. 1991. Electrical conduction in oceanic gabbro, Hole 735B, Southwest Indian Ridge. *In*: Von HERZEN, R. P. & ROBINSON, P. T. *et al.* (eds) *Proceedings of the Ocean Drilling Program, Initial Reports*, **118**, College Station, TX (Ocean Drilling Program), 323–331.

PLANK, T., LUDDEN, J. N. & ESCUTIA, C. *et al.* 2000. Proceedings of the Ocean Drilling Program, Initial Reports, **185**, (CD-ROM). Available from: Ocean Drilling Program, Texas A&M University, College Station, TX, USA.

PLANKE, S. 1994. Geophysical response of flood basalts from analysis of wire line logs: Ocean Drilling Program Site 642, Voring volcanic margin. *Journal of Geophysical Research*, **99**, 9279–9296.

RANGIN, C., SILVER, E. A., VON BREYMANN, M. T. *et al.* 1990. *Proceedings of the Ocean Drilling Program, Initial Reports*, **124**, College Station, TX (Ocean Drilling Program).

RÉVILLION, S., BARR, S. R., BREWER, T. S., HARVEY, P. K. & TARNEY, J. 2002. An alternative approach using integrated gamma-ray and geochemical data to estimate the inputs to subduction zones from ODP Leg 185, Site 801. *Geochemistry, Geophysics, Geosystems*, **3** (12), 8902, doi: 10.1029/2002GC000344.

RHODES, J. M., BLANCHARD, D. P., DUNGAN, M. A., RODGERS, K. V. & BRANNON, J. C. 1979. Chemistry of leg 45 basalts. *In*: MELSON, W. G. & RABINOWITZ, P. D. *et al.* (eds) *Initial Reports of the Deep Sea Drilling Project*, **45**, Washington (US Government Printing Office), 447–459.

RIDER, M. 1996. *The geological Interpretation of Well Logs*. 2nd edition, Whittles Publishing, Caithness.

ROBERTS, D. G., SCHNITKER, D. *et al.* 1983. *Initial Reports of the Deep Sea Drilling Project*, **81**, Washington (US Government Printing Office).

ROBINSON, R. T., VON HERZEN, R. *et al.* 1989. *Proceedings of the Ocean Drilling Program, Initial Reports*, **118**, College Station, TX (Ocean Drilling Program).

SCHALKWIJK, G. 1991. *Metabasites in the pilot borehole of the German Continental Drilling Project, Windischeschenbach, eastern Bavaria. Structures and fabrics as documents of crustal evolution.* PhD thesis, University of Bochum, Germany.

SCHALKWIJK, G. & STÖCKHERT, B. 1992. Metabasites in the pilot borehole of the KTB – structures and fabrics. *In*: VOLLBRECHT, A. & WEBER, K. (eds) *KTB-Report 92–4*, 7–14.

SCHLUMBERGER 1994. Neutron porosity logging revisited. *Oilfield Review*, **October**, 4–8.

SCHMINCKE, H.-U. & SEGSCHNEIDER, B. 1998. Shallow submarine to emergent basaltic shield volcanism of Gran Canaria: evidence from drilling into the volcanic apron. *In*: WEAVER, P. P. E., SCHMINCKE, H.-U., FIRTH, J. V. & DUFFIELD, W. (eds) *Proceedings of the Ocean Drilling, Scientific Results*, **157**, College Station, TX (Ocean Drilling Program), 141–182.

SCHMINCKE, H.-U., WEAVER, P. P. E. & FIRTH J. V. *et al.* 1995. *Proceedings of the Ocean Drilling*

A. BARTETZKO *ET AL.*

Program, Initial Reports, **157**, College Station, TX (Ocean Drilling Program).

WILKENS, R. H., FRYER, G. J. & KARSTEN, J. 1991. Evolution of porosity and seismic structure of upper oceanic crust: importance of aspect ratios. *Journal of Geophysical Research.* **96**, 17 981–17 995.

Appendix 1. *Minimum, maximum, mean, standard deviation (SD), and number of depth points (n) for electrical resistivity (logarithmic scale) and P-wave velocity of the different rock types from the single boreholes*

	Hole	Electrical resistivity (log, ohm m)					P-wave-velocity (km s^{-1})				
		Min	Max	Mean	SD	*n*	Min	Max	Mean	SD	*n*
		Upper oceanic crust									
Massive	395A	1.5	2.6	2.1	0.26	173	4.8	7.2	6.1	0.62	173
basalt	418A	1.9	2.7	2.3	0.24	64	5.0	6.2	3.8	0.24	64
	504D	0.9	2.3	1.6	0.29	442	4.2	6.7	5.7	0.42	442
	801C	1.5	3.7	2.6	0.5	433	4.1	6.6	5.7	0.36	438
	896A	1.0	1.7	1.2	0.19	85	4.5	5.9	5.3	0.25	85
Pillows	395A	0.7	2	1.4	0.21	1463	3.0	7.2	4.7	0.63	1461
	418A	0.9	2.5	1.8	0.22	1580	3.0	6.5	5.2	0.44	1580
	504B	0.7	1.7	1.0	0.16	984	3.6	6.5	4.9	0.45	984
	801C	1.2	2.7	1.8	0.26	667	3.8	6.4	5.1	0.38	667
	896A	0.5	1.2	0.9	0.13	703	3.0	6.5	4.7	0.39	703
Breccia	801C	1.0	3.6	1.5	0.3	465	3.0	6.1	4.6	0.49	465
	896A	0.7	1.0	0.8	0.04	51	4.1	5.2	4.7	0.25	51
Dykes	504B	2.0	2.7	2.4	0.15	1208	4.5	6.9	6.0	0.31	1208
		Lower oceanic crust									
Gabbro	735B	1.7	4.5	3.5	0.56	860	4.5	7.9	6.6	0.46	859
	1105A	1.1	4.0	2.4	0.52	596	4.8	6.8	6.0	0.38	596
Oxide gabbro	735B	0.7	2.6	1.5	0.30	325	5.2	6.7	6.4	0.29	325
	1105A	0.6	2.6	1.5	0.42	125	5.3	6.6	6.1	0.28	125
		Back-arc basins									
Basalt	768C	0.2	1.0	0.3	0.20	656	2.9	5.0	3.3	0.41	656
	834B	0.4	2.8	1.4	0.53	633			*No data*		
	839B	0.4	0.9	0.5	0.01	383			*No data*		
Basaltic	834B	0.6	1.7	0.9	0.24	313			*No data*		
andesite	839B			*No data*					*No data*		
		Hot-spot volcanoes and large igneous provinces									
Massive	HSDP2	0.7	1.3	0.96	0.12	2808			*No data*		
basalt	553A	0.4	3.3	1.7	0.44	276	2.1	7.5	5.5	0.89	258
	642E	0.8	3.3	1.6	0.38	727	2.3	6.0	5.0	0.59	719
	917A	1.4	3.0	1.8	0.33	345	4.2	7.0	5.4	0.42	342
	1137A	1.4	2.6	2.0	0.33	154	3.7	6.2	5.2	0.47	152
	1140A	0.8	3.1	1.9	0.36	111	2.7	6.5	4.8	0.91	102
K-rich	1140A	1.8	2.1	2.0	0.14	13	4.5	5.7	5.2	0.45	9
Vesicular/	HSDP2	0.6	1.1	0.8	0.01	1953			*No data*		
brecciated											
	553A	0.3	1.1	0.6	0.23	65	1.8	4.5	2.7	0.68	65
	642E	0.2	1.2	0.6	0.22	102	2.4	5.0	3.7	0.54	102
	917A	0.5	1.2	0.9	0.25	81	2.9	5.1	3.7	0.50	81
	1137A	1.2	1.7	1.3	0.12	63	3.6	4.6	4.1	0.30	63
Pillows	1140A	1.4	2.5	1.9	0.21	66	2.7	6.1	5.1	0.67	63
		Mafic volcaniclastics from the volcanic apron of an ocean island									
Breccia	956B	0.6	1.1	0.9	0.12	164	2.9	3.5	3.3	0.15	65
Tuff	956B	0.5	0.7	0.6	0.01	127	2.5	3.0	2.7	0.14	127
		Continental crust									
Amphibolite	KTB	2.2	5.0	3.7	0.57	1281	5.6	6.8	6.2	0.20	1316
Metagabbro	KTB	2.6	5.0	3.8	0.53	649	5.6	6.9	6.4	0.25	654
M.-ultra-m.	KTB	2.3	4.2	3.1	0.38	86	5.1	7.0	6.2	0.23	86
Amph. + gar.	S. Couy	1.7	4.0	2.5	0.36	288	3.7	7.1	6.3	0.34	689
Amphibolite	S. Couy	1.6	3.5	2.2	0.38	689	4.4	7.1	6.0	0.42	288
Dioritic gn.	S. Couy	1.8	3.7	2.5	0.51	84	5.2	6.8	5.9	0.32	84

m.-ultra-m., meta-ultramafitite; amph. + gar., garnet-bearing amphibolite; dioritic gn., dioritic gneiss; S. Couy, Sancerre-Couy.

Appendix 2. *Minimum, maximum, mean, standard deviation (SD), and number of depth points (n) for total gamma ray and density of the different rock types from the single boreholes*

	Hole	Total gamma ray (gAPI)					Density (g cm^{-3})				
		Min	Max	Mean	SD	n	Min	Max	Mean	SD	n
Massive	395A	2.4	9	4.3	1.2	173	1.9	3.0	2.8	0.17	170
basalt	418A	5	7.3	5.8	0.6	64	2.7	2.9	2.9	0.04	64
	504B	0.5	7.2	1.8	1.0	442	1.4	3.0	2.7	0.26	442
	801C	2.8	22.7	5.9	2.3	388	2.2	3.3	2.9	0.11	438
	896A	1.8	5	2.7	0.5	85			*No data*		
Pillows	395A	3.5	19	8.2	2.1	1463	1.3	3.0	2.5	0.28	1449
	418A	4.3	46.3	8.5	3.6	1580	1.3	3.0	2.6	0.26	1580
	504B	0.8	8.6	3.0	0.1	984	1.3	3.0	2.5	0.24	984
	801C	3.3	19.6	8.9	2.8	667	1.9	3.5	2.8	0.21	667
	896A	1.8	7.7	3.7	0.9	703			*No data*		
Breccia	801C	3.4	31.8	11.6	4.7	448	1.8	3.5	2.5	0.27	465
	896A	2.1	5.4	3.4	0.8	51			*No data*		
Dykes	504B	0	2.4	1.1	0.4	1208	1.6	3.1	2.7	0.25	1208
			Lower oceanic crust								
Gabbro	735B	0.5	12.1	3.06	2.1	860	2.1	3	2.9	0.09	860
	1105A	2.9	8.1	4.9	1.0	596	2.8	3.3	3.0	0.06	596
Oxide gabbro	735B	0.3	13.9	3.5	1.9	325	2.9	3.3	3.1	0.08	325
	1105A	3.2	5.9	4.5	0.6	125	2.9	3.4	3.1	0.09	125
			Back-arc basins								
Basalt	768C	14.3	42.8	24.5	4.6	656	2.1	2.5	2.3	0.08	656
	834B	3.1	8.7	5.1	1.1	633	1.2	3	2.3	0.46	633
	839B	4.5	10.8	6.7	0.94	532	1.7	2.6	2.2	0.12	532
Basaltic	834B	6.6	14.2	10.8	1.7	317	1.4	2.6	2.1	0.19	317
andesite	839B	8.4	22.3	12.8	1.6	505	1.2	2.6	2.1	0.23	505
			Hot-spot volcanoes and large igneous provinces								
Massive	HSDP2	6.3	13.2	10.0	1.5	2808			*No data*		
basalt	553A	3.4	15.9	8.2	2.5	276	2.6	3	2.8	0.11	276
	642E	3.6	17.2	8.3	2.0	739	2.4	3	2.8	0.10	739
	917A	6.3	20.6	13.9	3.5	345	2.6	3	2.8	0.09	345
	1137A	23.9	42.6	32.3	5.3	154	2.6	2.9	2.8	0.05	154
	1140A	6.3	16.2	9.1	1.8	111	2.7	3	2.8	0.06	111
K-rich	1140A	16.8	24.7	19.3	2.3	13	2.8	3	2.9	0.03	13
Vesicular/brecciated	HSDP2	6.3	13.2	9.1	1.5	1983			*No data*		
	553A	9.4	18.2	18.0	4.6	65	1.8	2.6	2.3	0.17	65
	642E	8.9	25.6	16.4	4.4	102	2.2	2.7	2.5	1.12	102
	917A	11.1	31	19.2	5.2	81	2.2	2.5	2.4	0.10	81
	1137A	35.2	82.4	59.3	11.6	63	2.3	2.5	2.4	0.07	63
Pillows	1140A	13.2	23.5	18.9	2.4	66	2.3	2.9	2.8	0.14	66
			Mafic volcaniclastics from the volcanic apron of an ocean island								
Breccia	956B	36.4	56.5	43.9	4.9	164	2.1	2.6	2.4	0.08	143
Tuff	956B	31.1	49.7	37.8	3.5	127	2.2	2.3	2.2	0.03	127
			Continental crust								
Amphibolite	KTB	14.1	64.7	28.1	6.3	1316	2.5	3.1	2.9	0.07	1316
Metagabbro	KTB	8.7	39.2	17.4	3.8	654	2.8	3.1	3.0	0.05	654
M.-ultra-m.	KTB	8.7	31.6	14.8	4.1	86	2.3	3.1	3.0	0.12	86
Amph. + gar.	S. Couy	8.7	50.9	20.4	7.7	689	2.8	3.2	3.0	0.06	689
Amphibolite	S. Couy	9.0	43.7	19.9	6.1	288	2.7	3.1	2.9	0.06	288
Dioritic gn.	S. Couy	20.4	67.9	38.7	15.0	84	2.6	2.9	2.8	0.08	84

m.-ultra-m., meta-ultramafitite; amph. + gar., garnet-bearing amphibolite; dioritic gn., dioritic gneiss; S. Couy, Sancerre-Couy.

Appendix 3. *Minimum, maximum, mean, standard deviation (SD), and number of depth points (n) of photo-electric factor and neutron log responses of the different rock types from the single boreholes*

	Hole	Photo-electric factor (barns/e⁻)					Neutron log (%)				
		Min	Max	Mean	SD	n	Min	Max	Mean	SD	n
Upper oceanic crust											
Massive	395A	1.9	6.4	4.2	0.6	173	4.7	18.2	8.5	3.1	173
basalt	418A	3.5	4.8	4.1	0.3	64	8.5	20.1	12	3.0	64
	504B	0	9.6	3.4	0.8	442	4.5	35.9	12.3	5.3	442
	801C	3.6	7.3	5.7	0.6	438	1.9	32.7	7.5	3.5	438
	896A	*No data*					*No data*				
Pillows	395A	1.3	6	3.7	0.8	1463	8.3	100	27.2	14.5	1463
	418A	1.8	5.4	3.5	0.5	1580	9.3	61.2	24.2	7.7	1580
	504B	0.4	6.6	3.0	0.5	984	8.1	45.2	27.3	6.1	984
	801C	2.4	7.1	5.2	0.7	667	5.2	62.7	17.0	8.2	667
	896A	*No data*					*No data*				
Breccia	801C	2.8	7.0	4.6	0.8	465	5.5	73.5	27.8	12.5	465
	896A	*No data*					*No data*				
Dykes	504B	1.9	5.9	3.7	0.64	1208	3.8	28.6	12.8	4.9	1208
Lower oceanic crust											
Gabbro	735B	2.2	7.2	3.9	0.6	860	−1	18.2	1.8	2.9	860
	1105A	2.8	8.5	4.5	2.8	596	0.7	18.1	3.4	0.7	596
Oxide gabbro	735B	3.8	11.3	6.3	1.5	325	0.6	8.2	2.0	1.1	325
	1105A	4.3	11.5	6.5	4.3	125	0.8	6.0	2.5	0.8	125
Back-arc basins											
Basalt	768C	3.7	5.8	4.4	0.4	656	23.2	54.1	44.6	5.6	656
	834B	1.1	5.0	3.0	0.9	633	13.8	72.9	36.1	13.9	633
	839B	1.7	4.5	3.6	0.6	526	36.2	66.4	49.9	4.9	532
Basaltic	834B	1.7	4.9	3.2	0.6	317	29.5	75.6	56.1	7.7	317
andesite	839B	1.6	4.9	3.6	0.7	505	25.7	67.1	43.4	7.5	505
Hot-spot volcanoes and large igneous provinces											
Massive	HSDP2	*No data*					*No data*				
basalt	553A	*No data*					9.4	33.5	16.3	5.4	276
	642E	2.1	5.3	3.5	0.4	739	11.2	41.7	20.4	6.0	739
	917A	3.7	7.2	5.3	0.6	345	*No data*				
	1137A	4.7	6.4	5.5	0.4	154	6	20	11.0	3.4	154
	1140A	4.3	6.4	5.2	0.5	111	7	17	11.1	2.8	111
K-rich	1140A	4.8	5.9	5.4	0.3	13	10	12	10.9	6.9	13
Vesicular/brecciated	HSDP2	*No data*					*No data*				
	553A	*No data*					36.1	59.5	45.5	5.1	65
	642E	2.9	4.3	3.6	0.3	102	21.7	53.8	39	6.7	102
	917A	3.5	4.6	4.1	0.24	81	*No data*				
	1137A	4.0	5.3	4.9	0.3	63	20	32	25.9	3.0	63
Pillows	1140A	2.7	6.1	5.4	0.7	66	7	19	10.7	2.6	66
Mafic volcaniclastics from the volcanic apron of an ocean island											
Breccia	956B	2.6	5.1	4.5	0.3	143	38.5	46.6	42.2	1.8	104
Tuff	956B	3.9	5.2	4.2	0.3	127	43.3	53.1	47.8	2.0	127
Continental basement											
Amphibolite	KTB	2.2	6.1	4.4	0.4	1316	5.1	26.6	12.3	2.4	1316
Metagabbro	KTB	3.4	5.7	4.6	0.4	654	8.6	29.7	13.6	3.0	654
M.-ultra-m.	KTB	2.9	6.3	4.6	0.5	86	9.3	34.8	22.8	6.8	86
Amph. + gar.	S. Couy	3.2	6.4	4.7	0.5	688	3.6	20.2	11.4	2.9	689
Amphibolite	S. Couy	3.5	6.2	4.5	0.4	288	4.7	21.6	12.3	3.1	288
Dioritic gn.	S. Couy	2.6	5.3	3.6	0.6	84	1.7	14.1	7.5	2.8	84

m.-ultra-m., meta-ultramafitite; amph. + gar., garnet-bearing amphibolite; dioritic gn., dioritic gneiss; S. Couy, Sancerre-Couy.

Effect of compositional variations on log responses of igneous and metamorphic rocks. II: acid and intermediate rocks

R. PECHNIG[1], H. DELIUS[2] & A. BARTETZKO[1]

[1]Angewandte Geophysik, RWTH Aachen University of Technology,
Lochnerstrasse 4-20, 52056 Aachen, Germany
(e-mail: r.pechnig@geophysik.rwth-aachen.de)
[2]Department of Geology, University of Leicester, Leicester, LE1 7RH, UK

Abstract: An extensive data-set of petrophysical down-hole measurements exists for boreholes drilled into continental crystalline crust. We selected boreholes covering a range of different types of plutonic rocks and gneisses in amphibolite or high-grade metamorphic rocks. According to Serra's concept of electrofacies, a specific set of log responses should characterize one rock type. Here, we concentrate on the detection of compositional variations between rock types. Bulk composition of the protoliths influences the mineralogical composition of the metamorphic rock, and we demonstrate how this impacts on the down-hole measurements. Integration of logging data with geochemical core data and mineralogical descriptions allows the calibration of the log responses to rock types. The relationship of the log responses with core data shows a remarkably good correlation, and diagnostic trends are detected. From the logs, potassium and neutron porosity are particularly helpful in distinguishing different types of gneisses and igneous rocks with respect to their protoliths. The proportions of amphibole/pyroxene, mica + K-feldspar and feldspar + quartz in the rocks seem to control the direction of correlation in a cross-plot, i.e. positive or negative, depending on increasing or decreasing mineral proportions. This is true for all boreholes, and a generalized classification scheme could be developed for these crystalline rocks.

Introduction

During the last few decades, several boreholes were drilled into igneous and metamorphic basement rocks. Exploration drilling has been performed for the purposes of mineral mining, tapping geothermal energy or disposal of wastes as well as in the framework of research drilling in continental or oceanic crust. In contrast to well logging in the traditional hydrocarbon environment, information about log responses in igneous and metamorphic rocks is scarce. This is especially true for the acid and intermediate rocks of the continental basement. Besides some early compilations of log responses from the principal igneous and metamorphic rocks (Keys 1979; Desbrandes 1982), most studies performed hitherto have focused on single boreholes drilled into continental basement (Daniels *et al.* 1983; Sattel 1986; Paillet 1991; Pratson *et al.* 1992; Traineau *et al.* 1992; Nelson & Johnston 1994; Pechnig *et al.* 1997). These studies showed that the complex geological conditions of crystalline rocks, particularly metamorphic rocks, are not easily determined

from the logs, because of the superposition of log responses produced by the varying compositional and structural variations of crystalline rocks. Hitherto, no systematic interpretation or classification charts are available for crystalline rocks. Therefore, this study is focused on a comparison of igneous and metamorphic rock types from drill-holes in continental basement. It comprises, besides foregoing electrofacies analyses individually performed for the different holes, a core log integration to develop classification charts, which include all available information on rock chemistry and mineralogy.

Log and lithological data compilation

This study is based on a comparison of wireline data with mineralogical and geochemical data from core and cuttings samples. Log, core, and cuttings data from the following boreholes were compiled and analysed: KTB pilot hole, Leuggern, Schafisheim, Böttstein, Soultz-sous-Forêts GPK1, Moodus and Cajon Pass (Fig. 1). These boreholes cover a wide range of different

From: HARVEY, P. K., BREWER, T. S., PEZARD, P. A. & PETROV, V. A. (eds) 2005. *Petrophysical Properties of Crystalline Rocks*. Geological Society, London, Special Publications, **240**, 279–300.
0305-8719/05/$15.00   © The Geological Society of London 2005.

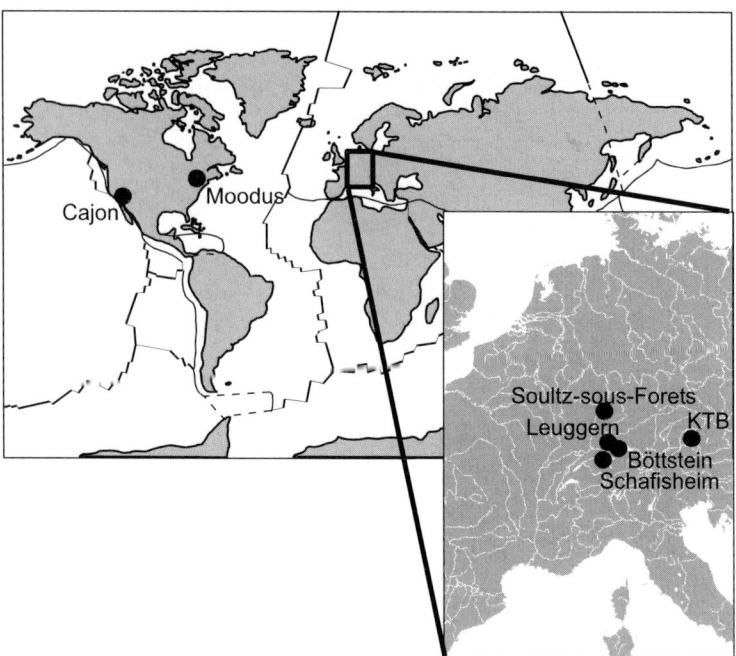

Fig. 1. Location of the boreholes compared in this study.

types of acid to intermediate plutonics, orthogneisses and paragneisses. From each borehole, one or several intervals were chosen to be representative for the lithology. Intervals were selected using the following criteria:

(1) general availability and quality of the logging data;
(2) availability of core/cuttings information;
(3) intervals with good borehole conditions to avoid logging tool failure sources from borehole enlargements, and
(4) homogeneous sequences and large bed thickness, to minimize log integration effects on bed boundaries. Since the focus of this study is concerned with the effect of the rock composition on the tool responses, fractured, brecciated and strongly altered intervals were generally excluded.

Geological setting and lithology of the selected boreholes

Metamorphic rocks were drilled in the boreholes KTB (Emmermann & Lauterjung 1997; Hirschmann *et al.* 1997), Moodus (Naumhoff 1988), Leuggern (Peters *et al.* 1989*a*) and Cajon Pass (Silver *et al.* 1988). The rocks

drilled in these boreholes mainly comprise paragneisses, orthogneisses and metabasites overprinted by amphibolite-facies up to granulite-facies metamorphism. Plutonic rocks were drilled in the boreholes Soultz-sous-Fôrets (Traineau *et al.* 1992), Böttstein (Peters *et al.* 1986) and Schafisheim (Matter *et al.* 1988), and in the deeper parts of the Leuggern borehole (Peters *et al.* 1989*a*). The drilled rocks include granites to granodiorites, syenites and monzonites. An overview of the boreholes, their rock content and petrogenetic evolution is given in Table 1 and Figure 2, and is summarized in the following section.

The German Deep Continental Drilling Project KTB drilled two deep holes at the western margin of the Bohemian Massif. The drill site itself lies within the Zone of Erbendorf–Vohenstrauss (ZEV), a small crustal segment of the Variscan orogenic belt. Two boreholes, separated by only a short distance of 200 m from each other, were drilled. The pilot hole (KTB-VB) reached a final depth of 4001 m, and was almost completely cored. The drilled crustal segment consists of an alternating sequence of three main lithological units: paragneisses, metabasites and 'variegated' units of paragneiss–metabasite alternations (Emmermann & Lauterjung 1997). All these units have suffered pervasive metamorphism to

Table 1. *Overview of general borehole data, lithology and petrogenetic evolution*

Borehole	Total depth (m)	Core recovery in basement (%)	Selected depth intervals (m)	Rock content of the selected intervals	Igneous and metamorphic evolution
Schafisheim	2006	90	1495–1560, 1600–1680 1710–1730, 1760–1960	Biotite-granite, monzonite, syenite, aplite	316 Ma: crystallization age of rocks 300–230 Ma: 1. Pervasive hydrothermal stage 280–260 Ma: 2. tectono-hydrothermal overprint
Böttstein	1501	75	595–800	Biotite-granite	330–320: granite intrusion 300–230 Ma: 1. pervasive hydrothermal stage 280–260 Ma: 2. tectono-hydrothermal overprint
Leuggern	1690	98	300–500, 720–750, 770–850, 960–1200, 1450–1600	Biotite-granite, sillimanite-bearing biotite-paragneiss, hornblende gneiss	500 Ma: 1. amphibolite-facies overprint during Caledonian orogeny 380 Ma: 2. amphibolite-facies overprint 330–320 Ma: granite intrusions followed by 1. hydrothermal overprint and cataclastic deformation of gneisses 280–260 Ma: 2. hydrothermal overprint
Soultz GPK1	2000	7	1450–1990	Biotite-granite	316 ± 7 Ma: granite intrusion and following 1. hydrothermal stage during cooling of the batholith 280–260 Ma: 2. tectono-hydrothermal overprint
Cajon Pass	3550	<3	2650–3100	Acid to mafic orthogneisses	81–75 Ma: amphibolite-facies overprint (T 600–750°C; P 3–6 kbar)
Moodus	1460	<2	200–400, 580–625, 825–955, 1100–1200	Biotite-paragneisses, acid orthogneisses, hornblende gneiss	390 Ma: amphibolite-facies overprint
KTB-VB	4000	93	570–900, 2500–2650, 3410–3440	Biotite-paragneisses, hornblende gneisses	375–405 Ma: higher amphibolite-facies overprint

References for petrogenetic data: Leuggern, Böttstein and Schafisheim: Thury *et al.* (1994), Soultz GPK1: Chevremont & Genter (1988), Cajon Pass: (3) Silver *et al.* (1988); Moodus: Wintsch & Aleinikoff (1987); and KTB-VB: O'Brien *et al.* (1997).

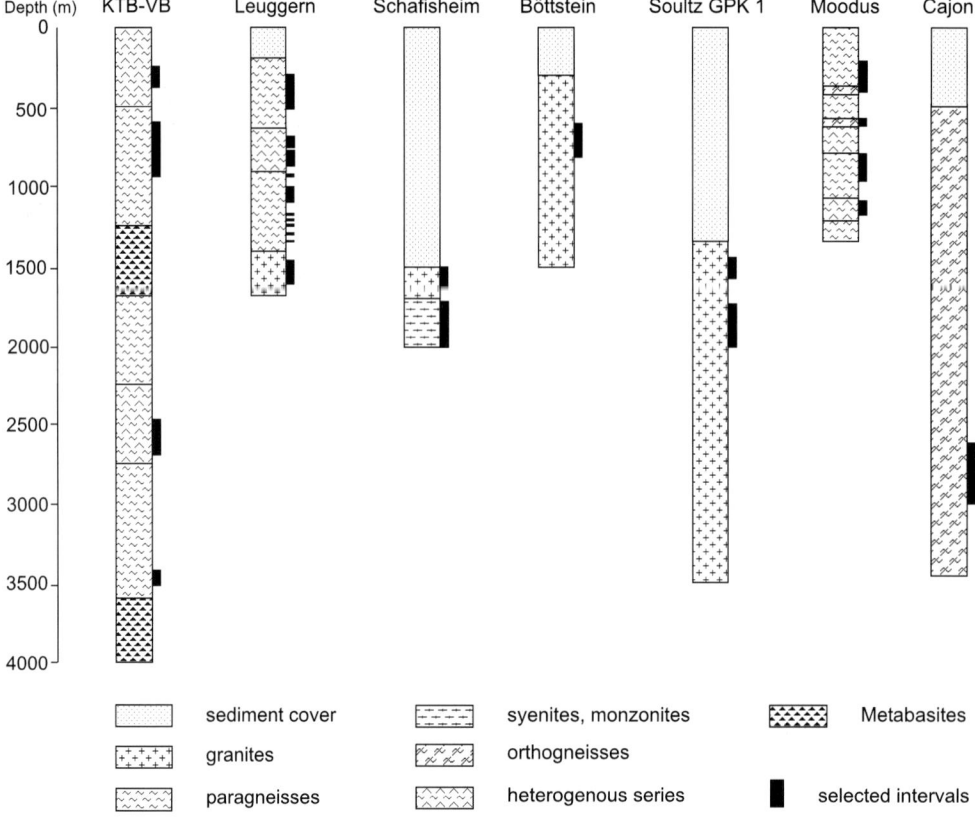

Fig. 2. Schematic lithological overview of the boreholes and the intervals selected for this study.

upper amphibolite facies, dated at between 405 and 375 Ma ago (O'Brien *et al.* 1997). The whole section is characterized by steeply dipping foliation (50–70°) and abundant fault zones resulting from multi-stage brittle deformation (Hirschmann *et al.* 1997).

As part of an extensive Swiss research programme focused on the evaluation of possible radioactive waste repositories, seven deep boreholes were drilled in northern Switzerland. Three of these boreholes were selected for this study: Leuggern, Schafisheim and Böttstein. The boreholes reached final depths between 1300 and 2500 m. In total, about 3000 m were drilled through crystalline rocks, most of it completely cored and thus extremely well documented. The boreholes revealed the crystalline basement as a complex mixture of metamorphic and igneous rocks. The metamorphic units are mainly composed of metasediments with minor intercalations of mafic rocks (Peters *et al.* 1986, 1989*a*, 1989*b*). These rocks underwent high-grade amphibolite-facies to granulite-facies

conditions about 500 Ma ago, followed by different stages of ductile deformation (Thury *et al.* 1994). During the final stage of the Variscan orogeny (330–315 Ma) large volumes of granitic melt were intruded into the gneiss series. The intrusions were accompanied by strong cataclastic deformation and hydrothermal overprinting of the metamorphic series. A later tectonohydrothermal event during the Lower Permian strongly altered the granites (Thury *et al.* 1994).

The borehole GPK1 is located on the western side of the Rhine Graben near Soultz-sous-Forêts (France), in a region with an unusually high geothermal gradient. The well was drilled as part of the Franco-German 'Hot Dry Rock' project, in order to test the feasibility of establishing a deep geothermal heat exchanger. GPK1 was drilled in 1988, reaching a depth of 2000 m, through 1376 m of sedimentary cover into the underlying granite. Only a few cores, with a total length of 43 m, were taken. The granites crystallized about 316 ± 7 Ma ago (Chevremont & Genter 1988). The massif is mainly composed of porphyroid granite with intercalated fine-grained leucogranitic intrusions, both affected by pervasive alteration. Hydrothermally altered zones of several metres in thickness are frequently observed along major fracture zones, where the granite is significantly altered (Genter 1989; Genter et al. 1989; Traineau et al. 1992).

The Moodus borehole extends to a depth of 1460 m near the town of Moodus in central Connecticut, USA. In 1987 the well was drilled into the crystalline rocks of the eastern Appalachians, with the objective of evaluating seismic hazard risks for the installation of power plants. Nine cores, roughly 2.5 m in length, were recovered at 150 m intervals, and cuttings were collected at 6 m intervals (Naumhoff 1988). The drilled lithology consists of paragneisses and orthogneisses of acid to intermediate composition, with minor intercalations of mafic metamorphic rocks. Two different terranes of the Appalachian mountain belt were encountered. In the upper part, the hole penetrates rocks of the Early Palaeozoic Merrimack terrane, and passes at about 820 m into the underlying Avalon terrane (Naumhoff 1988). Based on radiometric age determinations, these rocks are Late Precambrian in age (620 Ma; Wintsch & Aleinikoff 1987). The Merrimack and the Avalon terranes are separated by a major ductile thrust fault system called the Honey Hill fault zone. Mineral assemblages associated with the fault zone indicate that the dominant ductile fabric developed under amphibolite-grade metamorphic conditions about 390 Ma ago (Naumhoff 1988). In contrast to the other boreholes described, the rocks drilled in the Moodus borehole only locally show a greenschist-facies overprint, and brittle deformation is rare.

The Cajon Pass scientific borehole, 4 km north of the San Andreas Fault in California, was drilled to measure the heat flow and state-of-stress at seismogenic depth to explain the 'heat-flow-stress paradox' observed in the San Andreas fault zone (Lachenbruch and Sass 1988). The borehole was drilled to 3.55 km into a series of plutonic rocks and orthogneisses. Information on lithology is mostly based on cuttings, since only 3% of the borehole was cored. The rocks are orthogneisses of gabbroic to granitic composition, and were overprinted under amphibolite-facies conditions 75–81 Ma ago almost sychronously with dioritic intrusions (Silver et al. 1988). A Late Miocene tectonic phase resulted in the formation of 'low angle fractures', which, during the Pliocene, acted as pathways for hydrothermal fluid circulation and corresponding zeolite-facies alteration (Vincent and Ehlig 1988).

Mineralogical and geochemical composition of the rocks

Geochemical data-sets and information on rock modal composition were collected from the literature (Matter et al. 1988; Peters et al. 1989a, b; Peters et al. 1986; Genter et al. 1989; Silver et al. 1988; Ambers 1989; KTB-Data CD). The rock modal data comprise results from point counting and X-ray diffraction analyses. The average geochemical and mineralogical compositions for 25 rock types are displayed in Tables 2 & 3. The bulk composition of the rocks can be summarized as follows.

The biotite granites and aplites of the boreholes: Schafisheim, Leuggern, Böttstein and Soultz-sous Forêts GPK1 show comparable granitic chemical signatures, with tendencies towards alkaline composition for the Schafisheim granite and aplite (Fig. 3). The main chemical differences concern the K_2O content, which corresponds to variations in the average potassium-feldspar volume between 17% and 41%. The variations in K-feldspar contents are caused by the substitution of K-feldspar by plagioclase, while quartz and biotite proportions remain almost constant in the granites. Compared to the granites, the aplites have lower biotite contents.

The monzonites and syenites of the Schafisheim borehole were classified after Streckeisen with regard to their modal composition (Matter et al. 1988). Low quartz contents and high amphibole contents separate them from the other plutonic rocks compiled for this study. Although the monzonites and syenites show considerable differences in their modal composition, their average chemical composition is quite similar, both plotting into the syeno-diorite field (Fig. 3). Larger differences between syenites and monzonites concern the potassium and sodium concentrations controlling the varying potassium-feldspar/plagioclase ratio of these rocks (Table 3).

The rocks of the orthogneiss group span a suite from almost gabbroic composition to leucocratic gneisses of granitic composition. Their chemical signature is widely controlled by the stage of magmatic differentiation of the plutonic precursor rocks as visible for the Cajon Pass orthogneisses (Fig. 3). The average element contents and modal composition of the Cajon granitic gneiss are in the range of the biotite granites from the other boreholes. Higher quartz contents are documented for the Moodus granitic gneiss. This observation is in agreement with the higher SiO_2 values measured in this rock type.

Table 2. *Rock mineralogical composition compiled from petrographic and XRD analyses*

Lithology	Hole	n	Quartz		K-feldspar		Plagioclase		Biotite		Muscovite		Chlorite (sillimanite)		Garnet (cordierite)		Amphibole (pyroxene)		Ref.
			Range	Mean	Range	Mean	Range	Mean	Range	Mean	Range	Mean	Range	Mean	Range	Mean	Range	Mean	
Plutonic rocks																			
Biotite granite	Schaf.	22	15–32	23	27–53	41	10–30	22	2–14	9	—	—	—	—	—	—	—	—	(1)
	Leug.	10	15–30	23	25–45	35	20–40	31	5–15	11	—	—	—	—	<4	(1)	—	—	(2)
	Bött.	12	15–33	27	30–45	38	20–40	28	5–15	8	—	—	—	—	<4	(1)	—	—	(3)
	GPK1	3	20–38	28	15–27	19	35–45	40	<10	8	—	—	—	—	—	—	—	—	(4)
Monzonite	Schaf.	8	1–15	3	16–37	28	16–34	25	11–27	21	—	—	—	—	—	—	20–32	23	(1)
Syenite	Schaf.	21	0–5	2	20–50	39	3–13	7	13–33	26	—	—	—	—	—	—	13–33	23	(1)
Aplite	Schaf.	5	25–30	28	20–35	27	26–43	34	2–10	5	—	—	—	—	—	—	—	—	(1)
	Leug.	9	34–40	38	8–35	17	25–50	35	<5	2	0–10	5	—	—	—	—	—	—	(2)
Orthogneisses																			
Gabbroic	Cajon	*	NA	—	NA	—	NA	42	NA	4	NA	—	NA	—	NA	—	NA	50	(5)
Dioritic	Cajon	*	NA	20	NA	—	NA	55	NA	20	NA	—	NA	—	NA	—	NA	5	(5)
Granodioritic	Cajon	*	NA	21	NA	16	NA	50	NA	7	NA	—	NA	—	NA	—	NA	5	(5)
Granitic	Cajon	*	NA	29	NA	41	NA	23	NA	7	NA	—	NA	—	NA	—	NA	—	(5)
K-feldspar	Mood.	3	30–35	34	30–32	31	31–33	32	1–5	3	<1	0	—	—	—	—	—	—	(6)
	Mood.	1	30	30	35	35	35	35	—	—	—	—	—	—	—	—	—	—	(6)
Quartz–feldspar	Mood.	3	30–33	32	34–35	34	30–35	34	—	—	—	—	—	—	—	—	—	—	(6)
Paragneisses																			
(Sillimanite)–biotite	Leug. a	†	30–50	40	<8	4	25–45	35	15–45	28	<3	1	—	—	<5	2	—	—	(2)
	Leug. b	†	15–30	23	—	—	25–45	35	25–35	30	5–30	17	(10–20)	(15)	<10	5	—	—	(2)
	Leug. c	†	15–40	28	0–35	18	10–25	18	0–30	15	—	—	(0–10)	(5)	(15–50)	(32)	—	—	(2)
	KTB	393‡	25–55	40	—	—	15–35	26	0–25	6	3–28	12	5–30	12	<5	—	—	—	(7)
	Mood.	9	30–40	34	—	—	32–43	36	20–25	23	<10	2	—	—	—	—	(<10)	(3)	(6)
Hornblende	Leug.	†	15–40	28	<30	15	25–40	33	10–20	15	<10	1	0–17	7	—	—	5–30	17	(1)
	KTB	72‡	15–35	24	0–15	5	24–48	33	0–15	3	<10	1	0–17	7	0–7	1	7–34	18	(7)
	Mood.	11	30–40	36	—	—	38–52	43	5–20	11	—	—	0–18	10	—	—	0–20	9	(6)
K-feldspar	KTB	16	9–21	12	12–37	22	20–50	44	0–18	5	0–14	5	0–18	10	<2	0	0–12	5	(7)
Quartz–plag.	KTB	12	30–40	35	0–7	1	42–60	52	0–10	3	—	—	0–11	5	—	—	0–12	—	(7)

References (1) Matter *et al.* (1988); (2) Peters *et al.* (1989); (3) Peters *et al.* (1986); (4) Genter *et al.* (1989); (5) Silver *et al.* (1988); (6) Ambers (1989); and (7) KTB-Data CD.

Leug. a: plagioclase–biotite gneiss (metagreywacke); Leug. b: sillimanite–biotite gneiss (metapelite); Leug. c: cordierite–sillimanite gneiss (migmatic metapelite).

*Typical modes of the main rock types of the Cajon Pass borehole are extracted from a total of 120 rock samples.

†Typical value ranges of the Leuggern gneisses summarized from modal analyses; number of analyses is not given.

‡Values combine petrographic and RDA analyses.

NA: Information is not available.

Table 3. *Mean major-element values of the selected rocks*

Lithology	Hole	n	SiO$_2$	Al$_2$O$_3$	K$_2$O	Na$_2$O	CaO	Fe$_2$O$_3$	FeO	MgO	TiO$_2$	Ref.
			Plutonic rocks									
Biotite granite	Schaf.	8	69.0	13.9	5.7	3.4	1.6	9.2	0.8	1.2	0.5	(1)
	Leug.	10	69.1	14.7	5.2	2.9	1.5	0.4	2.1	1.3	0.5	(2)
	Bött.	12	71.1	14.3	4.9	2.7	1.4	0.5	1.6	0.9	0.4	(3)
	GPK1	3	66.8	14.2	4.1	3.9	2.2	4.3		1.5	0.7	(4)
Monzonite	Schaf.	8	50.1	14.3	5.1	2.7	5.7	1.8	5.9	7.7	1.5	(1)
Syenite	Schaf.	19	49.3	12.5	6.2	1.5	6.1	1.8	5.9	10.1	1.5	(1)
Aplite	Schaf.	7	69.7	14.3	4.6	4.5	1.5	0.7	1.1	0.7	0.3	(1)
	Leug.	13	74.1	13.6	4.3	3.4	0.9	0.5	0.6	0.4	0.1	(2)
			Orthogneisses									
Gabbroic	Cajon	1	53.8	14.8	1.7	2.9	7.7		9.5	5.1	1.0	(5)
Dioritic	Cajon	5	60.3	16.9	2.6	4.1	5.3		5.3	2.6	0.8	(5)
Granodioritic	Cajon	6	61.5	15.7	3.4	3.5	5.1		5.2	3.0	0.7	(5)
Granitic	Cajon	1	71.1	14.0	4.3	3.4	2.0		1.4	0.5	0.2	(5)
	Mood.	*298	77.6	13.8	5.4	NA	2.3	0.5		0.1	0.1	
K-feldspar	Mood.	*577	73.5	12.9	4.5	NA	3.7	3.0		1.6	0.5	
Quartz–feldspar	Mood.	*586	74.4	12.4	4.8	NA	2.0	1.6		4.6	0.3	
			Paragneisses									
(Sillimanite)–biotite	Leug. a	10	74.2	12.0	2.4	2.6	1.4	0.8	2.4	1.2	0.5	(2)
	Leug. b	18	60.3	19.5	4.2	0.9	2.5	1.6	4.7	1.9	0.8	(2)
	KTB	32	64.4	15.8	2.6	2.7	1.3	6.0		2.4	0.9	(6,7)
	Mood.	*2541	70.8	12.3	3.4	NA	4.8	4.7		2.7	0.8	
Hornblende	Leug.	3	60.3	14.7	2.4	3.5	3.5	3.2	4.8	3.5	1.4	(1)
	KTB	7	56.0	15.0	2.0	2.9	5.2	8.3		3.8	1.3	(6,8,9)
	Mood.	*2707	68.3	14.0	2.7	NA	5.2	3.5		5.6	0.6	
K-feldpar	KTB	8	52.4	15.5	3.3	3.7	5.2	8.6		3.1	1.1	(6,8)
Quartz–plag.	KTB	2	66.8	16.0	1.7	4.9	2.1	3.9		2.7	0.5	(6,10)

References (1) Matter *et al.* (1988); (2) Peters *et al.* (1989*b*); (3) Peters *et al.* (1986); (4) Traineau *et al.* (1992); (5) Silver *et al.* (1988); (6) KTB-Data CD; (7) Heinschild *et al.* (1988); (8) Stroh & Tapfer (1988); (9) Tapfer *et al.* (1989); Wittenbecher *et al.* (1989).
Leug. a: plagioclase-biotite gneiss (metagreywacke).
Leug. b: cordierite-bearing sillimanite–biotite gneiss (metapelite).
*Mean values calculated from geochemical logging data.
NA: Information is not available.

Compared to the selected plutonic rocks and orthogneisses, the paragneisses show a more complex mineralogical composition. Besides the main components (quartz, feldspar, biotite and amphibole in varying amounts), white mica, sillimanite, cordierite and garnet also occur. The present mineralogical composition is mainly caused by the original sediment bulk composition, but is also affected by the degree of metamorphism. The biotite gneisses show strong variations in quartz content (15–55%) and the amount of biotite, white mica and sillimanite (Hirschmann *et al.* 1997; Naumhoff 1988; Peters *et al.* 1989*a*). Besides the migmatitic version of the Leuggern paragneisses, all other types contain almost no K-feldspar. Pyroxene is observed only in the Moodus biotite gneisses. The chemical composition of the biotite gneisses of the different boreholes varies significantly, corresponding with changes in the modal composition.

Hornblende paragneisses were drilled in the boreholes Leuggern, KTB and Moodus. The precursor rocks are siliciclastic sediments mixed with mafic rock components of volcanic origin, such as ashes or volcaniclastics (Harms *et al.* 1997; Hirschmann *et al.* 1997; Naumhoff 1988; Peters *et al.* 1989*a*). The relative proportions of siliciclastics and volcanic components determine the overall chemistry and mineralogy of the hornblende gneisses. Thus, they show considerable differences in their mean mineralogical composition, particularly regarding the quartz, amphibole and biotite contents.

Petrophysical characteristics of the rocks

A wide spectrum of logging operations were carried out in the selected boreholes, including the recording of acoustic, electric and nuclear data. The logs that were evaluated for this study (Table 4) were performed by the following Schlumberger tools: the dual laterolog (DLL); litho density tool (LDT); natural gamma-ray spectroscopy tool (NGT); compensated neutron

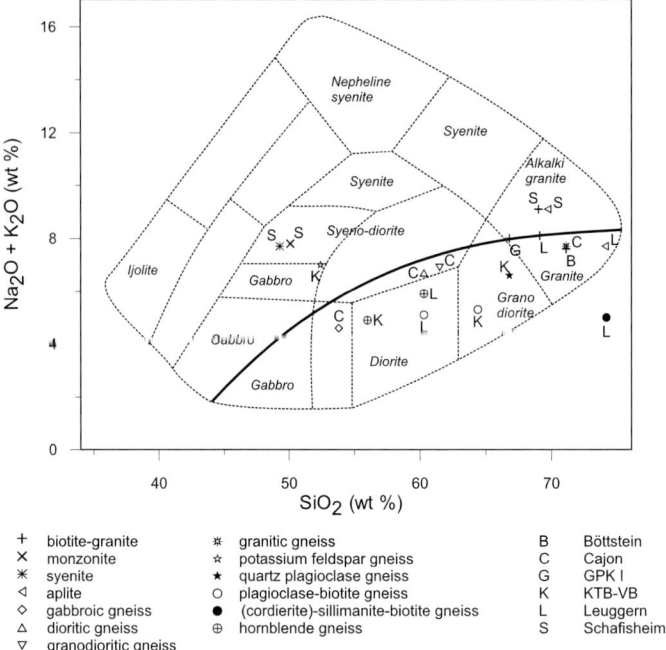

+	biotite-granite	✳	granitic gneiss	B	Böttstein
×	monzonite	☆	potassium feldspar gneiss	C	Cajon
✻	syenite	★	quartz plagioclase gneiss	G	GPK I
◁	aplite	○	plagioclase-biotite gneiss	K	KTB-VB
◇	gabbroic gneiss	●	(cordierite)-sillimanite-biotite gneiss	L	Leuggern
△	dioritic gneiss	⊕	hornblende gneiss	S	Schafisheim
▽	granodioritic gneiss				

Fig. 3. Total alkali versus silica diagram for the chemical analyses of rock types selected for this study (modified after Wilson 1989).

tool (CNT); and different sonic tools, such as the digital sonic array tool (DST) and the compensated sonic tool (CST). General information on logging tools and their measurement principles can be found in Ellis (1987), Rider (1996) and Hearst & Nelson (2000).

Log analysis was carried out by applying the electrofacies method, originally introduced by Serra (1984) for log interpretation in sedimentary rocks. Like a lithofacies description, which comprises several petrographic and chemical

characteristics, the electrofacies method combines the petrophysical value ranges recorded by different logging tools. An electrofacies characterizes a rock type by a specific set of log responses, and thus allows it to be distinguished from others. The concept is not restricted to the use of electrical measurements. Any log property suitable for the characterization of a rock type may be used. This study incorporates the log curves for total gamma-ray, potassium, bulk density, neutron porosity, electrical resis-

Table 4. *Overview of tools and logs used for this study*

Tool	Log		Principle
DLL (dual laterolog)	LLD (ohm m)	Laterolog deep	Electrical resistivity
SDT (Sonic digital tool)	DTCO (μs m^{-1})	Compressional-wave travel time	Travel time of sound
DSI (Dipole shear sonic imager)			
CNT (compensated neutron tool)	NPHI (%)	Neutron porosity	Absorption of neutrons
LDT (litho density tool)	RHOB (g cm^{-3})	Bulk density	Absorption/scattering of gamma-rays
NGT (natural gamma spectrometry tool)	SGR (API*)	Spectral (total) gamma-ray	Natural gamma-ray emissions
	POTA (%)	Potassium content	
	THOR (ppm)	Thorium content	
	URAN (ppm)	Uranium content	

*API: American Petroleum Institute.

tivity and P-wave velocity as being standard logging parameters.

Log analysis was carried out separately for each borehole. All available information on core and cuttings descriptions was collected and, using core–log-correlation, the *in situ* logging data were assigned to the different magmatic and metamorphic rock types. A total of 25 electrofacies were defined. Tables 5–7 display the log value statistics for the rock types separated within the different boreholes. The minimum, maximum, mean values and standard deviation are given for each electrofacie. The data are calculated from the borehole sections marked in Figure 2, taking into account the selection criteria defined on p. 280. Figure 4 displays these values graphically. The similarities and differences in the borehole geophysical characteristics can be summarized as follows.

The gamma-ray values cover a large value range (40–550 API) within the plutonic rocks

and gneisses. The average gamma-ray values of the plutonic rocks are about 250 API for the granites/aplites and about 350 API for the syenites/monzonites, and are thus significantly higher than the gamma-ray mean values of all the gneisses. Highest values occur in the Schafisheim monzonites, with about 550 API. Besides the granitic gneiss and the K-feldspar gneiss of the Moodus borehole, all of the other gneisses have lower gamma-ray values (<210 API) than the igneous rocks. The potassium log responses in general correlate with the gamma-ray ones, showing on average higher potassium contents for the plutonic rocks than for the gneisses.

The investigated rocks have bulk density values from 2.30 g cm^{-3} to 2.95 g cm^{-3} with strong variation within the three groups of plutonics, ortho- and paragneisses. The average bulk density of granites and aplites ranges between $2.59 \pm 0.02 \text{ g cm}^{-3}$ and $2.65 \pm 0.02 \text{ g cm}^{-3}$, which is in excellent agreement

Table 5. *Electrofacies statistics displayed for the different rock types. Minimum, maximum mean, and standard deviation of gamma-ray and potassium log responses*

Lithology	Hole	Total gamma ray (API)					Potassium (%)				
		Min	Max	Mean	SD	*n*	Min	Max	Mean	SD	*n*
		Plutonic rocks									
Biotite	Schafisheim	187	356	259	32	462	3.55	5.82	4.79	0.40	462
Granite	Leuggern	163	312	257	17	984	4.31	6.92	5.85	0.38	984
	Böttstein	181	304	230	19	662	4.37	6.59	5.35	0.40	662
	GPK1	196	308	260	18	1005	3.73	7.93	4.96	0.50	1005
Leucogranite	GPK1	130	280	225	35	320	3.87	6.36	5.30	0.50	320
Monzonite	Schafisheim	127	548	333	64	456	4.31	8.91	6.30	0.90	456
Syenite	Schafisheim	215	510	332	58	532	4.47	8.44	6.84	0.74	532
Aplite	Schafisheim	168	329	244	33	242	4.06	6.62	5.29	0.39	242
	Leuggern	157	333	252	32	406	3.58	6.79	5.57	0.59	406
		Orthogneisses									
Gabbroic	Cajon	44	100	73	11	61	1.73	2.51	1.95	0.18	61
Granodioritic	Cajon	66	210	97	20	720	2.36	4.07	3.21	0.28	720
Dioritic	Cajon	59	198	91	16	1028	2.03	3.57	2.68	0.21	1028
Granitic	Cajon	80	165	106	15	343	3.35	4.52	4.03	0.28	343
	Moodus	156	433	202	31	298	3.43	7.99	4.57	0.65	298
K-feldspar	Moodus	111	363	176	45	448	2.57	5.75	3.62	0.55	448
Quartz–feldsp.	Moodus	64	134	109	12	264	1.77	5.31	4.05	0.78	264
		Paragneisses									
(Sillimanite)	Leuggern*	82	192	141	16	903	2.33	5.76	3.90	0.54	903
Biotite gneiss	Leuggern[†]	90	211	151	19	1546	1.99	6.14	4.32	0.61	1546
	KTB-VB	72	136	99	11	1856	1.44	3.34	2.39	0.38	1856
	Moodus	98	167	121	10	755	2.06	4.22	2.91	0.33	755
Hornblende gneiss	Leuggern	75	169	133	22	130	1.50	4.04	2.53	0.63	130
	KTB-VB	53	99	77	8	664	1.33	2.57	1.88	0.23	664
	Moodus	35	135	81	16	1116	1.17	3.91	2.34	0.46	1116
K-feldspar	KTB-VB	79	185	138	26	101	1.83	5.62	3.85	1.02	101
Quartz–plag.	KTB-VB	66	100	76	7	109	1.35	2.46	1.8	0.05	109

*Plagioclase–biotite gneiss (metagreywacke).
[†]Cordierite-bearing sillimanite–biotite gneiss (metapelite).
SD: standard deviation.

Table 6. *Electrofacies statistics displayed for the different rock types. Minimum, maximum mean, and standard deviation of density and neutron log responses.*

	Hole	Density (g cm^{-3})					Neutron porosity (%)				
		Min	Max	Mean	SD	*n*	Min	Max	Mean	SD	*n*
		Plutonic rocks									
Biotite	Schafisheim	2.55	2.64	2.61	0.02	462	0.6	4.0	1.7	0.6	462
Granite	Leuggern	2.52	2.66	2.60	0.02	984	0.8	7.5	3.4	1.3	984
	Böttstein	2.45	2.64	2.59	0.02	662	1.4	13.0	3.7	1.6	662
	GPK1	2.58	2.73	2.65	0.02	1005	0.7	5.9	1.8	0.8	1005
Leucogranite	GPK1	2.54	2.70	2.60	0.02	320	0.3	2.9	0.9	0.3	320
Monzonite	Schafisheim	2.58	2.90	2.79	0.05	456	6.5	16.1	9.8	1.8	456
Syenite	Schafisheim	2.61	2.94	2.80	0.07	532	3.2	20.5	10.4	2.9	532
Aplite	Schafisheim	2.55	2.66	2.61	0.02	242	1.2	2.8	1.8	0.4	242
	Leuggern	2.51	2.71	2.61	0.03	406	0.8	8.9	2.9	1.8	406
		Orthogneisses									
Gabbroic	Cajon	2.58	2.92	2.73	0.09	61	−0.1	9.6	2.7	2.2	61
Granodioritic	Cajon	2.31	2.79	2.58	0.09	720	−0.1	4.9	0.2	0.9	720
Dioritic	Cajon	2.30	2.91	2.68	0.10	1028	−0.2	8.3	1.6	1.4	1028
Granitic	Cajon	2.38	2.71	2.55	0.07	343	−1.3	2.4	0.5	0.5	343
	Moodus	2.59	2.70	2.62	0.01	298	−0.6	0.9	0.1	0.2	298
K-feldspar	Moodus	2.57	2.81	2.71	0.05	448	−0.6	4.3	2.1	1.0	448
Quartz–feldsp.	Moodus	2.53	2.85	2.68	0.07	264	−0.9	2.0	0.1	0.6	264
		Paragneisses									
(Sillimanite)	Leuggern*	2.49	2.81	2.68	0.04	903	3.6	19.6	8.0	2.5	903
Biotite gneiss	Leuggern[†]	2.50	2.81	2.7	0.04	1546	4.6	24.4	12.5	3.9	1546
	KTB-VB	2.40	2.92	2.74	0.04	1856	2.8	14.6	8.7	2.3	1856
	Moodus	2.68	2.82	2.76	0.02	755	1.4	5.8	3.2	0.8	755
Hornblende gneiss	Leuggern	2.47	2.85	2.72	0.06	130	0.9	19.7	6.0	4.5	130
	KTB-VB	2.68	3.01	2.86	0.05	664	6.1	15.9	11.1	1.5	664
	Moodus	2.54	2.95	2.73	0.06	1116	−0.6	5.6	1.1	0.9	1116
K-feldspar	KTB-VB	2.58	2.87	2.69	0.06	101	1.1	15.6	5.9	3.1	101
Quartz–plag.	KTB-VB	2.63	2.74	2.68	0.02	109	3.8	9.7	6.3	1.2	109

*Plagioclase–biotite gneiss (metagreywacke).
[†]Cordierite-bearing sillimanite–biotite gneiss (metapelite).

with granite density data from laboratory measurements (2.60 ± 0.07 g cm^{-3}; Landolt-Börnstein 1982). Monzonites and syenites are distinctly separated from the acid plutonics by higher-density values of 2.79 ± 0.05 g cm^{-3} and 2.80 ± 0.07 g cm^{-3}. Most of the gneissic rock types show broad value ranges, frequently skewed by very low-density values. This is particularly the case for the Cajon orthogneisses with density records lower than 2.45 g cm^{-3}. Although we restricted the data selection to intervals of good borehole conditions (caliper deviations <10% of bit size) and a general low tectonic overprint, low-density values were registered. They are explained as being related to small joints and fissures or borehole wall irregularities not detected prior to data analysis.

The P-wave velocity values of the selected rocks are between 3.2 km s^{-1} and 6.9 km s^{-1}. Mean values of the different rocks do not scatter significantly, and are between 5.1 ±

0.03 km s^{-1} and 6.1 ± 0.02 km s^{-1}. Surprisingly, P-wave velocity values are lower in the monzonites/syenites than in the granites/aplites, although the monzonites and syenites have higher density values. Values of the orthogneisses are high, with mean values of more than 5.8 ± 0.02 km s^{-1} and minimum values not below 5.0 km s^{-1}. Besides the hornblende gneisses, mean values of the paragneiss group are below 5.7 km s^{-1} and in a similar range to the plutonic rocks. Very low values (<4 km s^{-1}) are only observed for the Leuggern biotite gneisses and the KTB potassium-feldspar gneiss.

Beside the gamma-ray and the potassium log the neutron porosity exhibits the strongest differences between the rocks selected for this study. Most paragneisses are significantly separated from the orthogneisses, granites and aplites by their high neutron porosity values of up to almost 25%. Only the monzonites/syenites reach comparable value ranges. Granites,

Table 7. *Electrofacies statistics displayed for the different rock types. Minimum, maximum mean, and standard deviation of P-wave velocity and electrical resistivity data*

Lithology	Hole	Electrical resistivity (log ohm m)					P-wave velocity (km s^{-1})				
		Min	Max	Mean	SD	n	Min	Max	Mean	SD	n
Plutonic rocks
Biotite granite	Schafisheim	1.6	2.6	2.0	0.1	462	4.7	5.7	5.3	0.2	462
	Leuggern	2.4	4.8	3.9	0.7	984	4.9	5.9	5.6	0.2	984
	Böttstein	2.6	4.4	3.5	0.4	662	5.3	5.9	5.6	0.2	662
	GPK1	1.8	4.7	3.3	0.6	1005	5.3	6.3	5.8	0.2	1005
Leucogranite	GPK1	2.3	4.6	3.3	0.5	320	5.3	6.1	5.8	0.2	320
Monzonite	Schafisheim	1.3	3	2.1	0.4	456	4.9	5.9	5.5	0.2	456
Syenite	Schafisheim	1.8	1.4	2.9	0.2	532	4.4	5.9	5.4	0.2	532
Aplite	Schafisheim	2.6	3.9	3.3	0.2	242	5.2	6.0	5.7	0.1	242
	Leuggern	2.6	4.7	3.5	0.5	406	5.0	5.9	5.5	0.2	406
Orthogneisses
Gabbroic	Cajon	2.4	3.8	3.2	0.3	61	5.5	6.5	6.1	0.2	61
Granodioritic	Cajon	2.2	4.6	3.9	0.6	720	5.0	6.5	5.8	0.2	720
Dioritic	Cajon	2.0	4.6	3.5	0.6	1028	5.0	6.9	5.8	0.2	1028
Granitic	Cajon	2.8	4.6	4.5	0.3	343	5.2	6.4	5.9	0.2	343
	Moodus	3.0	4.6	4.2	0.2	298	5.6	6.4	5.9	0.2	298
K-feldspar	Moodus	3.3	4.0	3.8	0.1	448	5.2	6.0	5.6	0.1	448
Quartz–plag.	Moodus	2.9	4.8	3.8	0.4	264	5.7	6.4	6.1	0.2	264
Paragneisses
(Sillimanite)	Leuggern*	1.2	4.6	3.1	0.5	903	3.8	6.3	5.3	0.2	903
Biotite gneiss	Leuggern†	1.3	4.7	3.0	0.5	1546	3.2	6.0	5.1	0.3	1546
	KTB-VB	1.5	4.6	3.4	0.4	1856	4.8	6.2	5.6	0.2	1856
	Moodus	3.5	4.5	4.0	0.2	755	5.0	6.0	5.5	0.1	755
Hornblende	Leuggern	2.5	4.7	3.8	0.6	130	5.0	5.8	5.6	0.2	130
Gneiss	KTB-VB	1.7	4.9	3.6	0.5	664	5.6	6.6	6.1	0.2	664
	Moodus	2.8	5.0	4.0	0.4	1116	5.3	6.6	6.1	0.2	1116
K-feldspar	KTB-VB	2.3	3.4	2.9	0.4	101	3.9	5.8	5.4	0.3	101
Quartz–plag.	KTB-VB	2.3	3.8	3.4	0.3	109	5.5	5.8	5.7	0.1	109

*Plagioclase–biotite gneiss (metagreywacke).
†Cordierite-bearing sillimanite–biotite gneiss (metapelite).

aplites and orthogneisses generally do not exceed 10%. It is noticeable that the neutron porosity values of the Moodus borehole also show low values of less than 6% for the paragneisses.

Electrical resistivity values vary by several orders of magnitude, from 10^1 to 10^5 ohm m. They do not show any significant differences between the different rock types. In most cases mean resistivity values are above 10^3 ohm m, which is characteristic for massive igneous and crystalline rocks known from laboratory measurements (Landolt-Börnstein 1982). Significantly lower value ranges are observed for the Schafisheim granite, monzonite and syenites that point to borehole-specific influences, such as an overall higher microcrack density or stronger alteration.

Integrated analysis of log and rock data

The geochemical, petrographic and borehole geophysical data were compared in order to extract significant relationships between rock composition and log responses. This was performed using correlation analysis. Pearson correlation coefficients were calculated using the averages of logging, geochemical and mineralogical values of the different rock types. Besides the single mineral phases, correlation coefficients were also calculated for the mineral groups: (1) quartz, plagioclase and K-feldspar; (2) the total amount of biotite, muscovite, chlorite and amphibole; and (3) biotite, muscovite and K-feldspar. The results are given in Table 8. Results of correlation analysis between the log data and mineralogical and geochemical rock composition will be discussed and interpreted in the following sections.

Effects of rock mineralogy on log responses

The correlation between mineralogical rock data and wireline data reveals the following systematic observations.

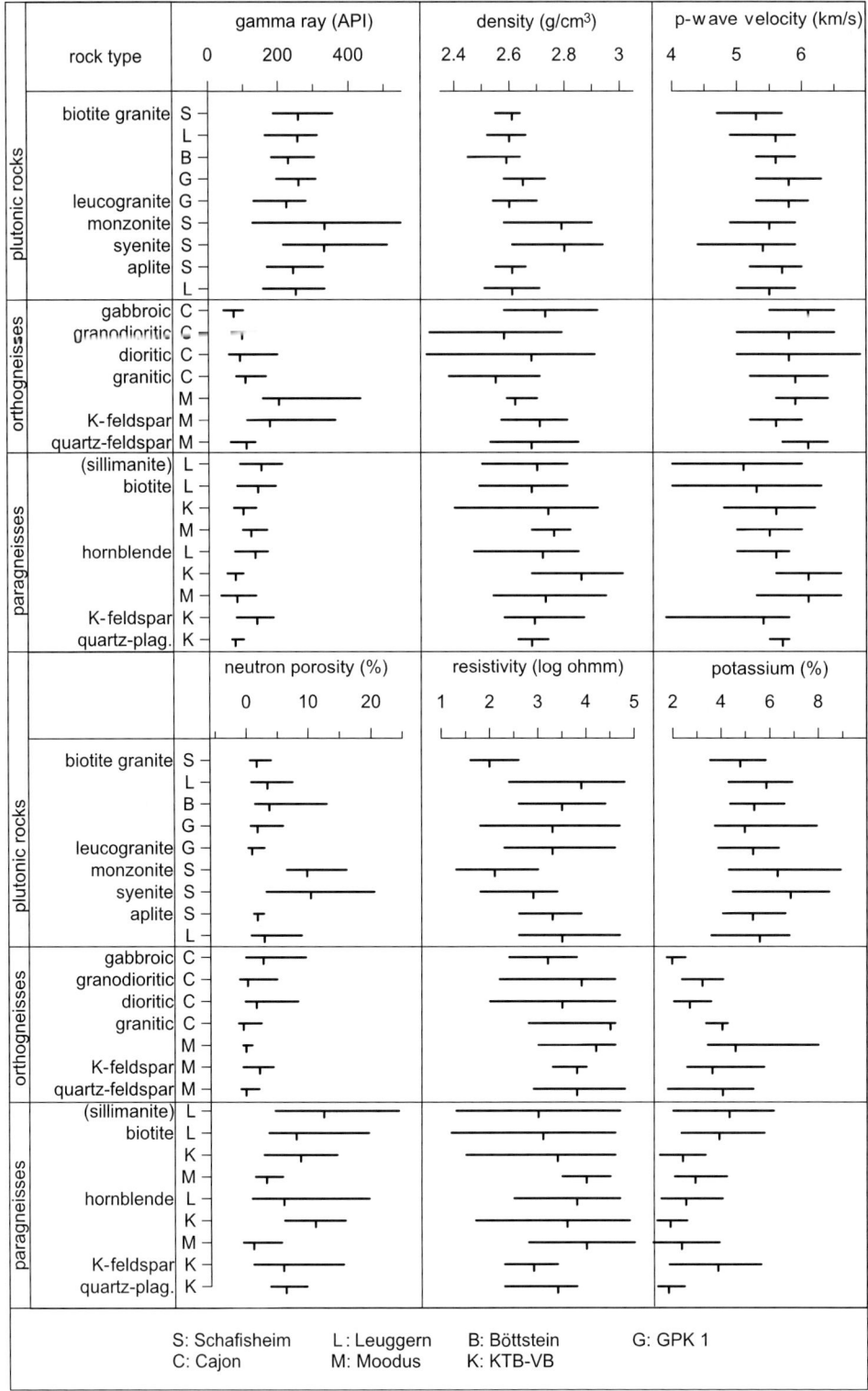

Fig. 4. Electrofacies of the different rock types. Displayed are minimum, mean and maximum values for the standard logs.

Table 8. *Pearson correlation coefficients calculated between mean values of geochemical, mineralogical and borehole geophysical data-sets*

Oxide/mineral	Gamma ray	Density	P-wave velocity	Neutron porosity	Potassium	Resistivity
SiO_2	−0.057	*−0.579	0.094	*−0.582	0.032	0.488
Al_2O_3	−0.282	−0.072	−0.139	0.307	−0.259	−0.105
K_2O	†0.807	−0.287	−0.315	−0.146	†0.874	−0.213
CaO	−0.201	*0.597	0.268	0.155	−0.278	−0.116
MgO	0.125	†0.709	0.102	0.410	0.060	−0.332
TiO_2	0.057	†0.762	−0.144	†0.665	−0.099	−0.417
Quartz	−0.304	−0.300	0.059	−0.265	−0.267	−0.496
K-feldspar	†0.634	−0.369	−0.092	−0.296	†0.697	−0.092
Plagioclase	*−0.591	−0.171	0.311	−0.315	*−0.568	0.239
Biotite	0.201	0.216	*−0.588	*0.500	0.239	−0.275
Muscovite	−0.172	0.117	−0.418	*0.510	−0.144	−0.119
Chlorite	−0.342	0.330	−0.012	0.406	−0.391	−0.108
Sillimanite	−0.044	0.049	−0.437	0.445	0.053	−0.159
Garnet	−0.117	0.109	−0.472	*0.610	−0.026	−0.152
Amphibole	−0.056	*0.546	0.286	0.254	−0.174	−0.245
Mica + Chl. + Amp.	−0.042	*0.619	−0.242	†0.704	−0.108	−0.401
Mica + K-feldspar	†0.746	−0.227	*−0.500	0.065	†0.825	−0.272
Quartz + K-feldspar + Plag	0.019	†−0.674	0.153	†−0.671	0.120	0.410

*Correlation is different from zero at a significance level of 0.05.
†Correlation is different from zero at a significance level of 0.01.

The most sensitive log for rock matrix variations for all gneisses and plutonic rocks is the gamma-ray log and, in particular, the potassium log, showing the highest correlation coefficients ($r > 0.8$) between rock and log data. Both parameters show comparable correlation coefficients with rock chemistry and mineralogy. Significant positive correlations are calculated between the K-feldspar content and the content of the mineral group K-feldspar plus biotite and muscovite. The relation to rock mineralogy is obvious, since K-feldspar, biotite and muscovite contain a considerable amount of potassium in their crystal lattice. Figure 5b shows the correlation between the potassium log and the total amount of K-feldspar plus biotite and muscovite. With regard to rock type and mineralogical composition (Table 2) the amount of K-bearing minerals controls the potassium and gamma-ray data to different degrees. In the acid orthogneisses, the rocks' potassium content is mainly related to the K-feldspars, while potassium content and gamma activity in the intermediate to basic gneisses is dependent on the amount of biotite.

The only significant negative correlation is calculated for the potassium, respectively the gamma ray versus the plagioclase content (Table 8). This can be explained by the substitution of K-feldspar by plagioclase in igneous rocks during magmatic differentiation, bearing in mind the rock classification after Streckeisen (1967).

Table 8 also reveals significant correlations between neutron porosity and rock composition. The highest positive correlation ($r = 0.7$) was calculated for the neutron log versus the amphibole plus biotite and muscovite content. Neutron porosity increases with the increasing amounts of these minerals (Fig. 5d). Rocks with high mica and amphibole contents, such as hornblende gneisses, biotite paragneisses, syenites and monzonites (see Table 2), plot in the upper right corner of Figure 5d, while granites, aplites and granitic rocks plot in the lower left-hand corner. In contrast, the neutron log is negatively correlated with the quartz plus K-feldspar and plagioclase content (Fig. 5c). Neutron porosity decreases with increasing amounts of quartz plus feldspar. In consequence, most granites, aplites and gneisses of granitic composition plot into the lower right edge of the cross-plot, while syenites/monzonites and most paragneisses have higher neutron and lower quartz/feldspar values. The high neutron-porosity values of some rock types and the observed correlation to rock mineralogy can be explained by the physical principles of the neutron-log measurement. Here, the hydrogen content is the most important factor controlling the neutron-log responses of a formation, since fast-emitted neutrons are slowed down predominantly by hydrogen nuclei occurring in the formation. Hydrogen can be kept in a formation in three different ways (Rider 1996):

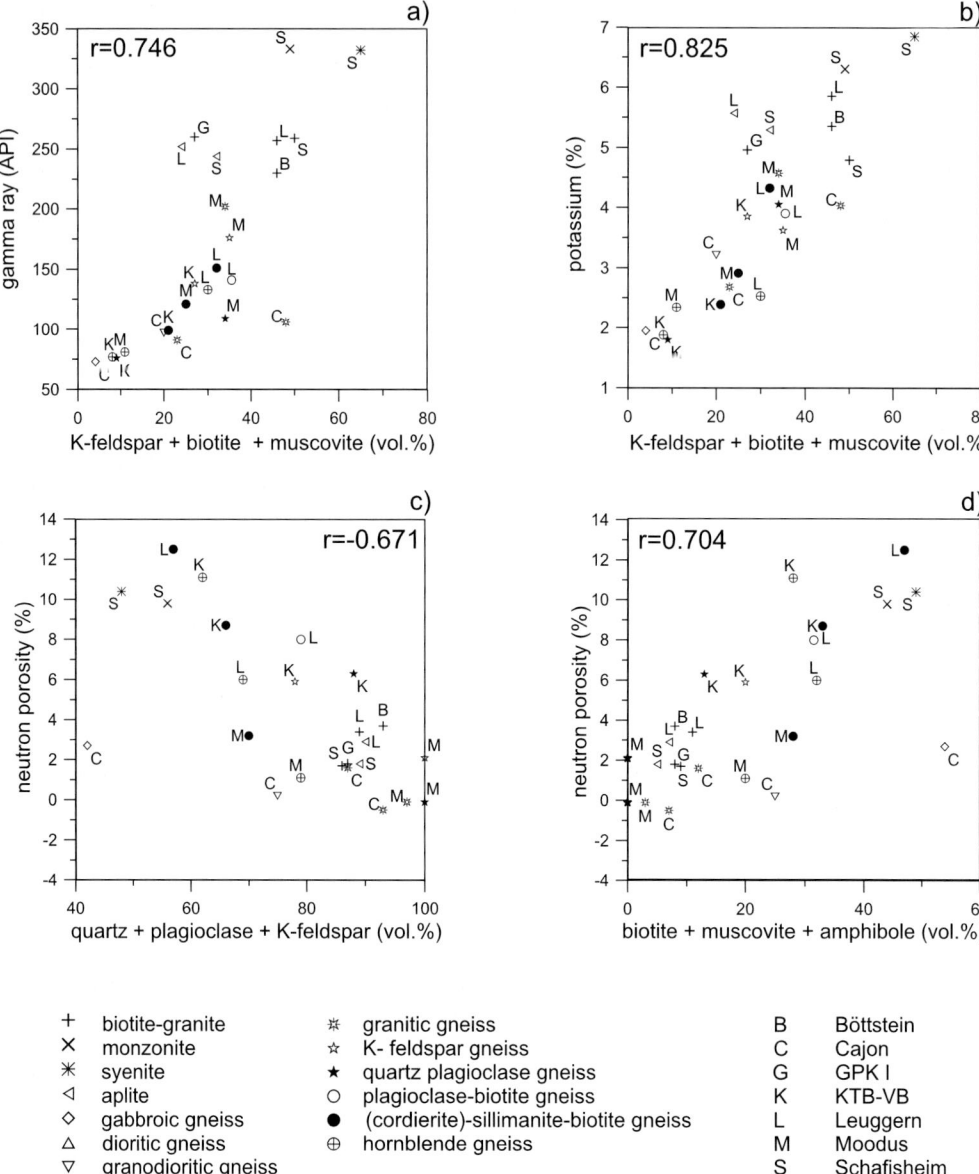

Fig. 5. Cross-plots of log responses of gamma ray, potassium, and neutron porosity, versus mineralogical composition of the rocks. Displayed are the average data from Tables 2 and 4. (**a**) & (**b**) Show positive trends between the gamma-ray/potassium logs with the total amount of micas and potassium feldspar within the rocks. (**c**) Shows the strong negative correlation between the neutron-porosity log and the quartz plus feldspar content. (**d**) Reveals the positive correlation between the neutron log and the total amount of OH-group bearing minerals. The corresponding Pearson correlation coefficients are given in Table 8.

(1) as free water in pores, fissures and fractures;

(2) as bound water incorporated within minerals; and

(3) as part of the crystal lattice in the form of OH-groups.

Since data selection was restricted to fresh, tectonically almost undisturbed crystalline rocks with effective porosities usually <3%, the effect on the neutron log of free water in pores and fissures becomes negligible. This makes it evident that the recorded high neutron porosities in the crystalline rocks are not real in the sense of indicating rock porosity, but they are the result of rock matrix effects. Minerals with bound water are clay minerals such as montmorillonite and minerals of the zeolite group; however, neither occur in the investigated rocks. Therefore, only OH-bearing minerals can contribute significantly to the total hydrogen content in crystalline rocks. This fits with the results of Table 8, since amphibole and micas have OH-groups incorporated in their lattice and their concentration is positively correlated with the neutron log. On the contrary, quartz and feldspars are free of hydrogen and the occurrence of this mineral group is negatively correlated with neutron porosity.

There might be an argument that, besides the hydrogen in the formation, the atomic weight of the chemical elements also influences the (thermal) neutron log response (Schlumberger 1994), as there is a trend that elements with high atomic weights capture neutrons more effectively. Furthermore, the presence of certain rare-earth and trace elements with particularly large capture cross-sections (e.g. boron, lithium and cadmium) can have a significant effect on the neutron-log readings (Harvey et al. 1996). Rare-earth elements occur in acid/intermediate rocks, sometimes making quite a significant contribution. It can be assumed that the high neutron-log readings in crystalline rocks are produced by superposition of the different effects of hydrogen content, atomic weight and rare-earth element concentrations. Further investigations would be needed to separate the neutron-porosity readings for the different effects.

Correlation analysis has also revealed correlations between the density and rock mineralogy. A positive correlation ($r = 0.546$) is calculated for the relation between density and the amphibole content. A more significant negative correlation was calculated between density and the total amount of quartz and feldspar. The calculated results are explained by the different mineral densities, increasing from potassium feldspar (2.56 g cm^{-3}) to quartz (2.65 g cm^{-3}), to plagioclases ($2.61-2.76 \text{ g cm}^{-3}$) and micas ($2.7-3.1 \text{ g cm}^{-3}$), up to the mafic minerals of the amphibole group ($3.15-1.25 \text{ g cm}^{-3}$) and the pyroxene group ($2.9-3.5 \text{ g cm}^{-3}$) groups (Wohlenberg 1982; Schön 1996). Therefore, quartz/feldspar-rich rocks show low bulk density values, and mafic rocks with considerable amounts of amphibole/pyroxene have high rock densities.

Correlation analysis shows no significant correlations between P-wave velocity and mineral composition. A weak correlation was calculated only for the relationship between P-wave vel-

ocity and biotite and mica contents, respectively. These correlations are negative, indicating that average P-wave velocity decreases with mica content. At the moment there is no plausible explanation for this. In this study, correlation analysis reveals that the P-wave velocity log is not indicative of rock composition. However, it is known from laboratory measurements that rock-building minerals have characteristic P-wave velocities (Schön 1996) and that a relationship with rock modal composition does exist. Missing this relationship in our study could be explained by structural effects such as micro-cracks and foliation. Metamorphic rocks are characterized by pronounced foliation, which is known to influence the P-wave anisotropy considerably. In the KTB borehole, up to 15% P-wave anisotropy was measured within strongly foliated biotite gneisses, even under simulated in situ conditions (Berckhemer et al. 1997). The highest P-wave velocity values are measured parallel to foliation, while the lowest values are measured perpendicular to the foliation. Therefore, rock anisotropy could be an explanation for the strong scatter of in situ V_P data for the different rocks.

The resistivity log does not show any significant correlation with any of the main mineralogical components. This was expected, as the pore space of the selected crystalline rocks is very low and rock-forming minerals usually act as electrical isolators.

Effects of rock chemistry on log responses

The highest correlation coefficient ($r = 0.87$) between log data and rock chemical data is calculated between the potassium log and the K_2O content of the rocks (Table 8 & Fig. 6a). Small differences between the two types of data-set are caused by the principal differences in integration volume and sampling method between the continuous log data and the discrete core or cutting analyses. Taking into account the strong difference in the amount of data between core and log data, the high correlation between the two data sources is remarkable. Nevertheless, high potassium values in the log data may be biased, due to the calculation of potassium content from the total energy spectra of counted gamma-rays or by accumulation of potassium in the borehole mud. The latter was observed in mica-rich gneisses in the KTB-VB hole (Pechnig et al. 1997). Mica crystals became detached from the borehole wall and were kept in suspension in the mud. This caused gamma-ray log responses to increase by up to 10%.

A slightly smaller correlation coefficient ($r = 0.81$) was calculated for the gamma-ray and the K_2O content of the rocks. While the relation potassium log versus K_2O from cores represents variations in only one chemical element, the relation between gamma ray and versus K_2O from cores is also influenced by

thorium and uranium. These two elements contribute to the total gamma-ray responses and thus change the gamma-ray/potassium ratio.

Significant correlations exist between the density log and the TiO_2 and MgO content. Titanium and magnesium have higher atomic weights than the other rock-forming elements

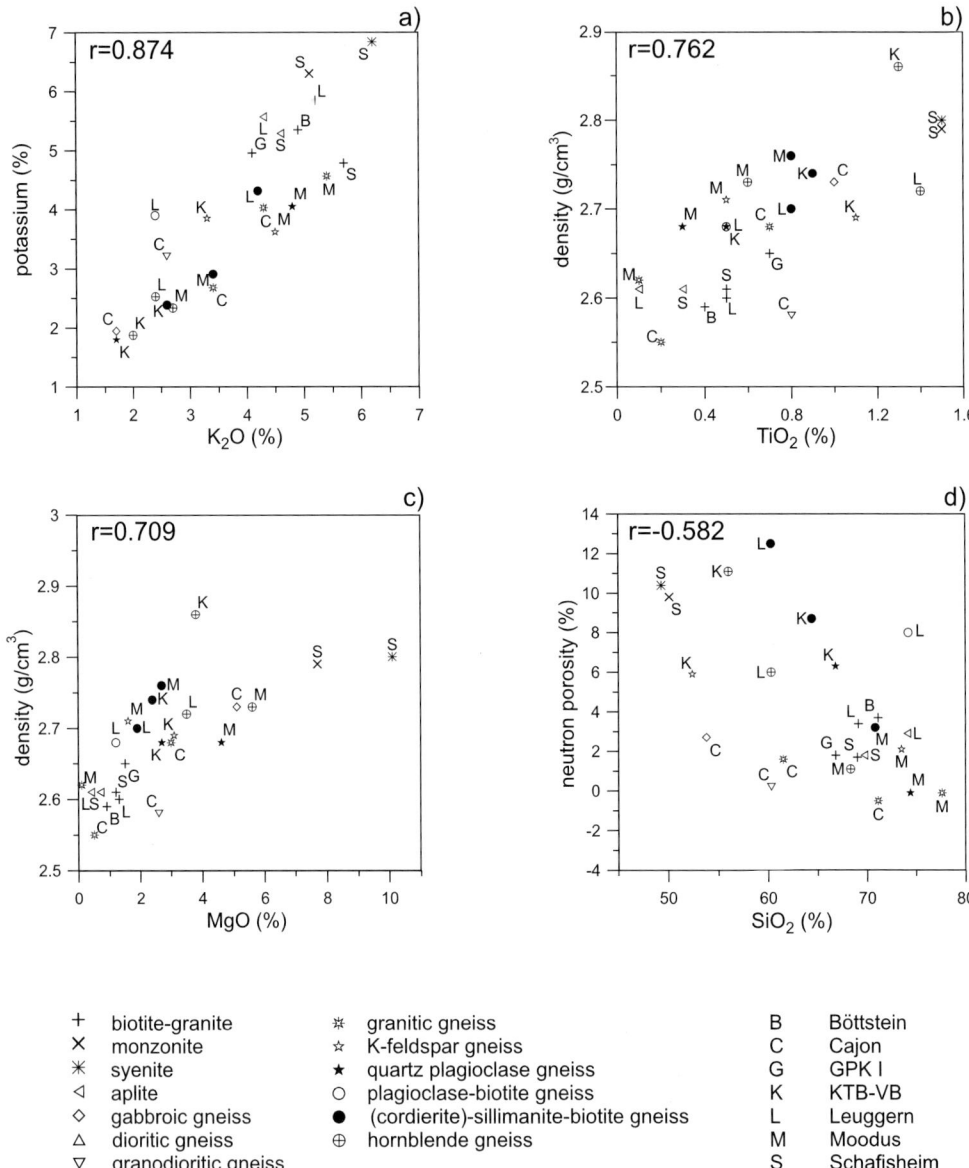

					B	Böttstein
+	biotite-granite	✷	granitic gneiss		C	Cajon
×	monzonite	☆	K-feldspar gneiss		G	GPK I
✳	syenite	★	quartz plagioclase gneiss		K	KTB-VB
◁	aplite	○	plagioclase-biotite gneiss		L	Leuggern
◇	gabbroic gneiss	●	(cordierite)-sillimanite-biotite gneiss		M	Moodus
△	dioritic gneiss	⊕	hornblende gneiss		S	Schafisheim
▽	granodioritic gneiss					

Fig. 6. Cross-plots of log responses of potassium, density, and neutron porosity versus the chemical composition of the rocks. Displayed are the average data from Tables 3 & 4. The corresponding Pearson correlation coefficients are given in Table 8.

silicon, aluminium, potassium and calcium, and they are preferentially incorporated into mafic minerals. High TiO$_2$ contents and high density values are therefore measured for the most mafic rocks: the Schafisheim monzonites/syenites, followed by the hornblende gneisses of Leuggern and KTB (Fig. 6b). The correlation between density and TiO$_2$ content is stronger than the one between density and MgO content. The latter is mainly caused by the very high MgO values of the monzonites and syenites, thus plotting separately from the general trend (Fig. 6c). A weak negative correlation ($r = -0.58$) occurs between density and SiO$_2$ content. A possible reason is a coupling of a decrease in mafic constituents with an increase in both quartz and feldspar.

The neutron porosity shows correlations comparable to those calculated between the density and the chemical elements. The highest coefficient is calculated for the relation between neutron porosity and the TiO$_2$ content. This might be due to different effects of:

(1) OH-groups incorporated in some mafic TiO$_2$-rich constituents;
(2) higher atomic weights of elements in mafic rocks; and
(3) the occurrence of trace elements.

The negative trend between the neutron porosity and the silicon content is also visible, although a strong scatter occurs between the data points.

The P-wave velocity and electrical resistivity logs show no significant correlation with any of the chemical elements. This is a further indication that these logging parameters are not significantly influenced by the chemical composition of the selected rocks.

Log correlation trends as result of mineralogical variability

Although significant correlations between log readings and rock composition exist, correlations between the log parameters are not detectable on

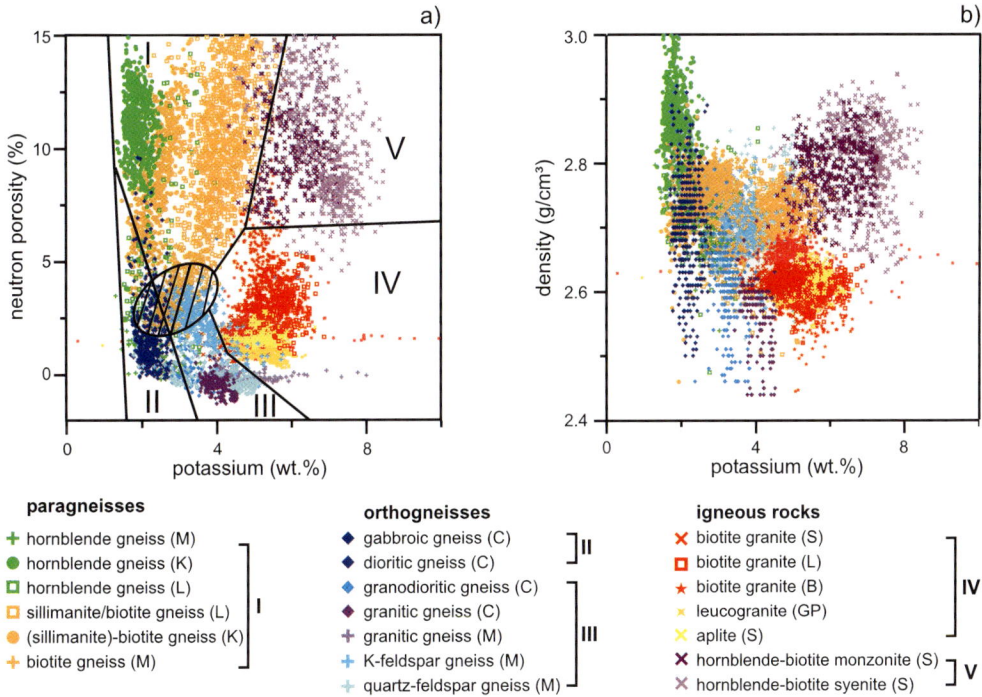

Fig. 7. Cross-plots of (**a**) potassium versus neutron porosity and (**b**) potassium log versus density, both showing the selected rock types. The potassium versus neutron porosity cross-plot makes it possible to span the classification fields to distinguish between the main rock groups: granites/aplites, monzonites/syenites, paragneisses, and acid to intermediate orthogneisses. Similar relationships between are also visible for the potassium–density cross-plot, but the separation of the groups is not as distinct as for the neutron–potassium plot. L, Leuggern; B, Böttstein; S, Schafisheim; GP, GPK1; C, Cajon Pass; M, Moodus; K, KTB.

the first viewing. When displayed as a cross-plot, the logging data show a large scatter (Fig. 7a & b). However, the cross-plot neutron porosity versus the potassium log allows the separation of the main rock types, since comparable rocks from different boreholes plot into the same plot areas (Fig. 7a). This is especially the case for the granites and aplites. More than 90% of these data plot between 0 and 7% neutron porosity and 4 and 7 wt% potassium. Also, the plot area of the orthogneisses is quite limited. The largest data scatter is produced by the paragneisses, especially for the neutron-porosity values, which range between 2% and > 15%. The monzonites and syenites also exhibit a large data scatter. These rocks are clearly separated from the granite field and the paragneiss field. Significant overlaps between the main rock types exist between the paragneiss and the orthogneiss fields in the neutron porosity versus potassium plot. Rocks plotting in this area cannot be distinguished. Data clustering is also visible in the cross-plot density versus potassium, and the separation of the main rock types is not as sharp as in Figure 7a. This is especially the case for the orthogneisses, biotite paragneisses and hornblende gneisses, which show strong overlaps in the density field 2.6–2.8 g cm^{-3}. Granites and aplites are better separated by their high potassium values. This is also the case for the monzonites/syenites.

By separating the rocks into paragneisses and plutonic rocks, significant trends between log parameters become visible. The most prominent trends exist for the paragneisses, where two correlation trends occur. Within the large group of biotite gneisses, the gamma ray is positively correlated with the neutron porosity (Fig. 7) and the density log (Fig. 8). This correlation is clearly observed for all the biotite gneisses of the different boreholes, although offsets exist between the Leuggern and the Moodus/KTB data. The trends can be related to the rock mineralogical variability and will be explained by examples from the KTB biotite paragneiss. The mineralogical and chemical composition of this gneiss type varies considerably, with quartz concentrations ranging between 25% and 55%; biotite contents between 5% and 40%; and muscovite contents between 3% and 38% (Table 2). The varying quartz–mica ratios are reflected in the chemical composition: i.e. SiO$_2$ values between 45% and 75% and K$_2$O contents between 2.2% and 3.6% (Table 3). Petrographic studies revealed two end-members of the KTB biotite gneisses: a quartz-rich type with high SiO$_2$ values, and a mica-rich rich variety with increased potassium contents (Müller *et al.* 1989). By comparing

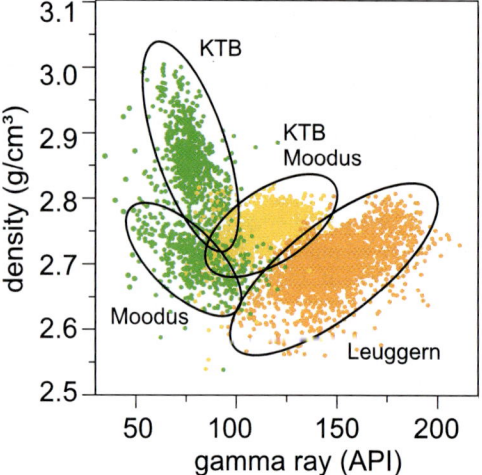

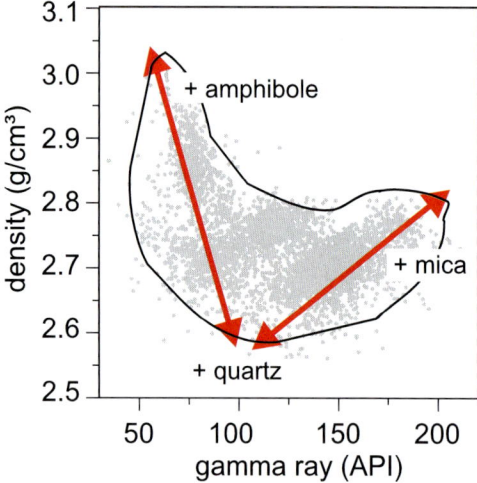

Fig. 8. Diagram of the mineralogical end-members controlling the relationship between the density and gamma array in paragneisses. For signatures, see Figure 7.

mineralogical and chemical data with logging data it was demonstrated (Haverkamp & Wohlenberg 1991; Pechnig *et al.* 1997) that the mica-rich paragneisses have much higher gamma-ray, potassium, density and neutron-porosity values than the quartz-rich types. This can be explained by mineral physics and chemistry. Micas are the most important potassium-bearing minerals in the biotite paragneisses, and thus they contribute significantly to the measured total gamma activity. Micas have higher densities than quartz (muscovite: 2.83 g cm^{-3}, biotite: 3.01 g cm^{-3}, quartz: 2.65 g cm^{-3}), and they have incorporated OH-groups in their lattice. After Serra (1986) a rock consisting of 100% of mica would produce 20–25% apparent porosity in the neutron-log readings, while pure quartz would produce a neutron-log response of -2%.

The relationship between the quartz/mica ratio and the logging data exists also for the Leuggern and Moodus gneisses. In all boreholes, the end-members of the observed trends are, as marked by the arrows in Figure 8, quartz-rich gneisses with low density and low gamma-ray values on the one hand, and mica-rich gneisses with high density and high gamma-ray values on the other hand. In particular, the Leuggern biotite gneisses show a pronounced modal variability. Quartz contents vary between 15% and 50%, and total mica plus sillimanite content can be more than 60% (Table 2). This produces total gamma-ray values varying from 80 to more than 200 API (Table 5a), and log density ranging between 2.49 and 2.81 g cm^{-3}.

The hornblende gneisses of the Moodus and the KTB borehole show divergent trends. In contrast to the biotite gneisses, density and gamma ray are negatively correlated (Fig. 8). The main components of hornblende gneisses are quartz, feldspar, biotite and amphibole, in variable amounts (Table 2). Gneisses with low amphibole contents have high quartz contents of, and vice versa. Enrichment in quartz corresponds with the change to a more acid geochemical composition of the rocks, which is also accompanied by a potassium increase. Since amphiboles have a significant higher density than quartz, the compositional changes produce a trend with the following end-members: amphibole-rich hornblende gneisses with low gamma-ray values and high density values on the one hand, and quartz- and plagioclase-rich hornblende gneisses with low densities on the other (Fig. 8).

Plotting data from the orthogneisses and plutonic rocks in the cross-plot density versus potassium does not reveal significant trends (Fig. 9) as observed for the paragneisses. A negative correlation between density and potassium is visible for the orthogneisses, which can be explained by mineralogical variations similar to those of the hornblende gneisses. Mineralogical end-members of this trend are biotite–amphibole-rich orthogneisses of gabbroic/dioritic composition with the highest densities, and quartz–plagioclase-rich orthogneisses of granodioritic/granitic composition with the lowest densities. Gamma activity and potassium values increase significantly in the field of granites and aplites. Here, only a weak trend is visible. This trend is caused by the enrichment of K-feldspar in these rocks. Due to the substitution of K-feldspar and plagioclase in granites,

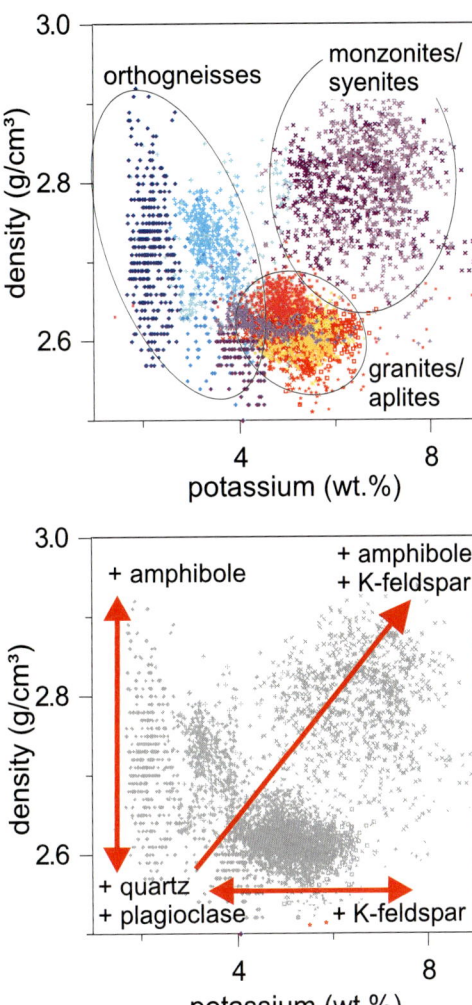

Fig. 9. Diagram of the mineralogical end-members controlling the relation between the density and the potassium log in orthogneisses and plutonic rocks. For signatures, see Figure 7.

potassium values vary over a considerable range. Since the density of K-feldspar, plagioclase and quartz (the main components of the granites and aplites, Table 2), is more or less the same, rock density is not affected by the mineralogical variability.

The monzonites and syenites are separated from the granites and orthogneisses, and show a data scatter without any trends visible. These rocks exhibit high variability in the potassium and density values, which is related to the changing amounts of their components. High density and high potassium values in syenites/

monzonites are related to the enrichment of both minerals, hornblende and potassium feldspar.

In summary, for the orthogneisses and plutonic rocks, the relation between the density and potassium log is controlled by the relative amount of the mineral assemblages: amphibole plus biotite, quartz plus plagioclase, and potassium feldspar.

Conclusions

Mineralogical, geochemical and *in situ* petrophysical data-sets were compiled for this study in order to explain log responses in relation to rock composition for acid to intermediate gneisses and plutonic rock. Correlation analysis between the different data types reveals significant correlations between some of the standard logs and rock mineralogy. The most prominent are the gamma-ray and potassium logs as indicators for the potassium feldspar and mica content. In addition, the density and neutron-porosity log are indicative of the rock composition. They are strongly controlled by the relative amount of amphibole plus mica versus quartz plus feldspar. P-wave velocity and electrical resistivity are essentially independent of rock composition.

These results show that in the investigated rock types the standard logging parameters are of a different importance for lithology prediction. Density-log and neutron-porosity devices, which are used in the sedimentary environment as porosity predictors, are indicative of mineralogical changes in the crystalline environment. The most powerful discrimination between different rock types is given by a combination of the gamma-ray and the potassium logs on the one hand versus the neutron and density logs on the other. The combination of neutron versus potassium logs is suitable for discriminating between the main rock types.

Strong correlations between logging parameters are also observed within the major rock groups. Integration of petrophysical log data with information on rock composition reveals three major mineralogical background parameters which are controlling the log responses in gneisses and acid to intermediate plutonic rocks. In paragneisses, log correlations follow the ratios of the amphibole, quartz and mica. In the orthogneisses and granites/aplites this is biotite plus amphibole versus K-feldspar plus quartz, following the magmatic differentiation from mafic to acid rocks. Syenites and monzonites strongly separate from this trend, since potassium feldspar occurs together with amphibole and biotite.

This study used logging data and core data from various boreholes, collected and composed from different sources. The German Continental Drilling Project (KTB) was funded by the German Ministry of Research and Technology (BMFT). KTB data were made available by the GeoForschungszentrum (GFZ) Potsdam, Germany. Logging data from the Leuggern, Böttstein and Schafisheim boreholes were kindly made available by NAGRA (National Co-operative for the Disposal of Radioactive Waste). Data from the borehole GPK1 Soultz-sous-Forêts were made available by the Bureau de recherches géologiques et minières, and the digital logging data-sets of the Moodus and Cajon Pass borehole were kindly provided by the Borehole Research Group of the Lamont Doherty Earth Observatory, NY, USA. Parts of this study were funded by the German Science Foundation (DFG, Wo-159/11; Wo-159/14). We would also like to thank Prof. J. Wohlenberg for initiating these studies and for making them possible.

References

ADAMS, J. A. S., OSMOND, J. K. & ROGERS, J. J. W. 1959. The geochemistry of thorium and uranium. *In*: AHRENS, L. H., PRESS, F., RANKAMA, K. & RUNCORN, S. K. (eds) *Physics and Chemistry of the Earth*, **3**, 298–348.

AMBERS, C. P. 1989. *Core and Cuttings Descriptions of the 1987 Moodus 4764 Foot Deep Well*, Connecticut. Connecticut Geological and Natural History Survey, Open File Report **89-1**, 207 pp.

BERCKHEMER, H., RAUEN, A. *et al.* 1997. Petrophysical properties of the 9-km-deep crustal section at KTB. *Journal of Geophysical Research*, **102**, 18 337–18 361.

CHEVREMONT, P. & GENTER, A. 1989. *Etude Pétrologique des Échantillons de Granite de Soultz-sous-Forêts, Forage GPK1*. Note interne 89/IRG, 20.

DANIELS, J. J., SCOTT, J. H., & OLHOEFT, G. R. 1993. Interpretation of core and well log physical property data from drill hole UPH-3, Stephenson County, Illinois. *Journal of Geophysical Research*, **88**, 7346–7354.

DESBRANDES, R. 1982. *Encyclopedia of Well Logging*, Edition Technips, Paris, 548 pp.

ELLIS, D. V. 1987. *Well Logging for Earth Scientists*. Elsevier, Amsterdam, 532 pp.

EMMERMANN, R. & LAUTERJUNG, J. 1997. The German Continental Deep Drilling Program KTB: overview and major results. *Journal of Geophysical Research*, **102**, 18 179–18 201.

GENTER, A. 1989. *Géothermie Roches Chaudes Sèches: le granite de Soultz-sous-Forêts (Bas-Rhine, France). Fracturation naturelle, altération hydrothermale et interaction eau–roche*. Thèse de doctorat de l'Université d'Orléans, France, 201 pp.

GENTER, A., BEAUFORT, A., MEUNIER, A. & CHEVREMONT, P. 1989. Etude des altérations hydrothermales du granite de Soultz-sous-Forêts—Forage GPK1. *Rapport BRGM 89*, SGN 295 EEE/IRG.

HARVEY, P. K., LOVELL, M. A., BREWER, T. S., LOCKE, J. & MANSLEY, E. 1996. Measurements of thermal neutron absorption cross section in selected geothermal reference material. *Geostandard Newsletter*, **20**, 79–85.

HAVERKAMP, S. & WOHLENBERG, J. 1991. EFA-Log-Rekonstruktion kristalliner Lithologie anhand von bohrlochgeophysikalischen Messungen für die Bohrungen URACH 3 und KTB-Oberpfalz VB. *KTB Report*, **91-4**, 213 pp.

HEARST, J. R., NELSON, P. H. & PAILLET, F. L. 2000. *Well Logging for Physical Properties. A Handbook for Geophysicists, Geologists and Engineers*, 2nd edn, Wiley, Chichester, UK.

HEINSCHILD. H. J., HOMANN, K., STROH, A. & TAPFER, M. 1988. Tiefbohrung KTB Oberpfalz VB, Ergebnisse der geowissenschaftlichen Bohrungsbearbeitung im KTB Feldlabor, Teufenbereich 480 bis 992 m: C. Geochemie. *In*: EMMERMANN, R., DIETRICH, H. G., HEINISCH, M. & WÖHRL, T. (eds) *KTB-Report*, **88-2**, C1–C107, Hannover, Germany.

HIRSCHMANN, G, DUYSTER, J., HARMS, U., KOTNY, A., LAPP, M., DE WALL, H., ZULAUF, G. 1997. The KTB superdeep borehole: petrography and structure of a 9-km-deep crustal section. *Geologische Rundschau*, **86**, 3–14.

KEYS, W. S. 1979. Borehole geophysics in igneous and metamorphic rocks. *Transactions, SPWLA, 20th Annual Logging Symposium*, Tulsa, Oklahoma, paper OO.

LACHENBRUCH, A. H. & SASS, J. H. 1988. The stress heat-flow paradox and thermal results from Cajon Pass. *Geophysical Research Letters*, **15**, 981–984.

LANDOLT-BÖRNSTEIN 1982. Zahlenwerte und Funktionen aus Naturwissenschaften und Technik. *Neue Serie Vol. 1: Physical Properties of Rocks*, Subvol. b. – 604 S., Springer, Berlin, Heidelberg and New York.

MATTER, A., PETERS, T., BLÄSI, H., SCHENKER, F. & WEISS, H. P. 1988. Sondierbohrung Schafisheim – Geologie; Text- und Beilagenband. *Nagra Technische Berichte NTB 86-03*, Nagra, Wettingen, Switzerland.

NAUMHOFF, P. 1988. Lithology and structure identified in a 1.5 km borehole near Moodus, Connecticut. *EOS*, **69**, p. 491.

NELSON, P. H. & JOHNSTON, D. 1994. Geophysical and geochemical logs from a copper oxide deposit, Santa Cruz Project, Casa Grande, Arizona. *Geophysics*, **59** (12), 1827–1838.

O'BRIEN, P. J., DUYSTER, J., GRAUERT, B., SCHREYER, W., STÖCKHERT, B. & WEBER, K. 1997. Crustal evolution of the KTB drill site: from oldest relics to the late Hercynian granites. *Journal Geophysical Research*, **102**, 18 203–18 220.

PECHNIG, R., HAVERKAMP, S., WOHLENBERG, J., ZIMMERMANN, G. & BURCKHARDT, H. 1997. Integrated log interpretation in the German Continental Deep Drilling Program: lithology, porosity, and fracture zones. *Journal Geophysical Research*, **102**, 18 363–18 390.

PETERS, T., MATTER, A., BLÄSLI, H.-R. & GAUTSCHI, A. 1986. Sondierbohrung Böttstein – Geologie; Text- und Beilagenband, *Nagra Techn. Ber. NTB 85-02*, Nagra, Wettingen, Switzerland.

PETERS, T. J., MATTER, A., BLÄSLI, H.-R., ISENSCHMID, CH., KLEBOTH, P., MEYER, CH. & MEYER, J. 1989a. Sondierbohrung Leuggern – Geologie; Text- und Beilagenband, *Nagra Techn. Ber. NTB 86-05*, Nagra, Wettingen, Switzerland.

PETERS, T. J., MATTER, A., MEYER, J., ISENSCHMID, CH. & ZIEGLER, H.-J. 1989b. Sondierbohrung Kaisten – Geologie; Text- und Beilagenband, *Nagra Techn. Ber. NTB 86-04*, Nagra, Wettingen, Switzerland.

PRATSON, E. L., ANDERSON, R. N., DOVE, R. E., LYLE, M., SILVER, L. T., JAMES, E. J. & CHAPPELL, B. W. 1992. Geochemical logging in the Cajon Pass drill hole and a new, oxide, igneous rock classification scheme. *Journal Geophysical Research*, **97**, 5167–5180.

RIDER, M. 1996. *The Geological Interpretation of Well Logs*. Whittles Publishing, Caithness, Scotland, UK, 280 pp.

ROGERS, J. J. W. & ADAMS, J. A. S. 1969a. Thorium. *In*: WEDEPOHL, K. H. (eds) *Handbook of Geochemistry*, **2** (4), 90B–90O, Springer, Berlin.

ROGERS, J. J. W. & ADAMS, J. A. S. 1969b. Uranium. *In*: WEDEPOHL, K. H. (eds). *Handbook of Geochemistry*, **2** (4), 92B–92O, Springer, Berlin.

SATTEL, G. 1986. FACIOLOG – Anwendung eines Programmsystems zur Charakterisierung der Kristallin-Lithologie. *Nagra Informiert*, **1**, 18–24.

SCHLUMBERGER 1994. Neutron porosity logging revisited. *Oilfield Review*, October, 4–8.

SCHÖN, J. H. 1996. Physical properties of rocks – fundamentals and principles of petrophysics. HELBIG, K. & TREITEL, S. (eds) *Handbook of Geophysical Exploration, Section I, Seismic Exploration*, Vol. 18. Pergamon, Oxford, 583 pp.

SERRA, O. 1984. *Fundamentals of Well-Log Interpretation*. Vol. 1, Elsevier, Amsterdam.

SERRA, O. 1986. *Fundamentals of Well-Log Interpretation*. Vol. 2, Elsevier, Amsterdam.

SILVER, L. T., JAMES, E. W. & CHAPPELL, B. W. 1988. Petrological and geochemical investigations at the Cajon Pass Deep Drillhole. *Geophysical Research Letters*, **15**, 961–964.

STRECKEISEN, A. 1967. Classification and nomenclature of igneous rocks. *Neues Jahrbuch Mineralogische Abhandluagen*, **107**, 144–214.

STROH, A. & TAPFER, M. 1988. Tiefbohrung KTB Oberpfalz VB, Ergebnisse der geowissenschaftlichen Bohrungsbearbeitung im KTB Feldlabor, Teufenbereich 0–480 m: C. Geochemie. *In*: EMMERMANN, R., DIETRICH, H. G., HEINISCH, M. & WÖHRL, T. (eds) *KTB-Report*, **88-1**, C1–C73, Hannover.

TAPFER, M., HEINSCHILD. H. J., STROH, A., WITTENBECHER, M. & ZIMMER, M. 1989. Tiefbohrung KTB Oberpfalz VB, Ergebnisse der geowissenschaftlichen Bohrungsbearbeitung im KTB Feldlabor, Teufenbereich 2500–3009,7 m: C. Geochemie. *In*: EMMERMANN, R., DIETRICH, H. G., HEINISCH, M. & WÖHRL, T. (eds) *KTB-Report*, **89-4**, C1–C50, Hannover.

THURY, M., GAUTSCHI, A. *et al.* 1994. Geology and hydrology of the crystalline basement of northern Switzerland. *Geologische Berichte Nr.*

18, Bundesamt für Umwelt, Wald und Landschaft, Bern, Switzerland.

TRAINEAU, H. A., GENTER, J. P., CAUTRU, F., FABRIOL, H. & CHEVREMONT, P. 1992. Petrography of the granite massif from drill cutting analysis and well log interpretation in the geothermal HDR borehole GPK1 (Soultz, Alsace, France). *In*: BRESEE, J.C. (eds). *Geothermal Energy in Europe*, 1–29, Gordon and Breach Science Publishers, Switzerland.

VINCENT, M. W. & EHLIG, P. L. 1988. Laumontite mineralization in rocks exposed of San Andreas Fault at Cajon Pass, Southern California. *Geophysical Research Letters*, **15**, 977–980.

WILSON, M. 1989. *Igneous Petrogenesis.* Unwin Hyman, London

WINTSCH, R. P. & ALEINIKOFF, J. N. 1987. U–Pb Isotopic and geological evidence for Late Paleozoic anatexis, deformation, and accretion of the Late Proterozoic Avalon Terrane. *American Journal of Science*, **287**, 107–126.

WITTENBECHER, M., HEINSCHILD, H. J., STROH, A. & TAPFER, M. 1989. Tiefbohrung KTB Oberpfalz VB, Ergebnisse der geowissenschaftlichen Bohrungsbearbeitung im KTB Feldlabor, Teufenbereich 3009, 7–3500 m: C. Geochemie. *In*: EMMERMANN, R., DIETRICH, H. G., HEINISCH, M. & WÖHRL, T. (eds) *KTB-Report*, **89-5**, C1–C62, Hannover, Germany.

WOHLENBERG, J. 1982. Dichte der Minerale und Gesteine. *In*: ANGENHEISTER, G. (eds) *Landolt Börnstein – Physikalische Eigenschaften der Gesteine.* Subvolume A, 66–119, Springer, Berlin, Heidelberg and New York.

Laboratory investigations for the evaluation of *in situ* geophysical measurements in a salt mine

J. KULENKAMPFF[1], A. JUST[2], L. ASCHMANN[2] & F. JACOBS[2]

[1]*GeoForschungsZentrum Potsdam, Telegrafenberg, D-14473 Potsdam, Germany*
(e-mail: hannes@gfz-potsdam.de)
[2]*Institute for Geophysics and Geology, Talstr. 35, D-04103 Leipzig, Germany*

Abstract: Problem zones in a sylvinite mine were explored by underground measurements using a set of geophysical methods (geosonar, seismics, georadar, EM methods and geo-electrics), which yield parameter sets for disturbed and undisturbed rock. Underground geo-electrical methods are most suitable for examining humidity content and structures for the transport of brines, although they have to be constrained by wave-propagation methods. Additional laboratory experiments on rock salt samples were carried out in order to provide quantitative evaluation methods. A titration method was used to determine the amount of water. This was applied simultaneously with a four-electrode resistivity measuring system for samples with very high resistivity. These measurements were made on samples at different scales, yielding relationships between water content and resistivity, as well as information about the parameter distribution in larger core samples. Together, the field and laboratory results show that geo-electrics is a suitable method for the detection and evaluation of problem zones of the geological barrier of hazardous waste repositories, and a criterion for the risk assessment based upon resistivity measurements is defined.

High plasticity, high thermal conductivity, and ease of mining are the almost ideal properties of salt bodies acting as host rocks and cap rocks for underground repositories of chemotoxic and radioactive waste. There are several locations in layered salt deposits and salt domes in northern Germany (Warren 1999) that are potentially suited for storing hazardous wastes, but, since they occur in populated areas, the requirements for the safety analysis are very high. Reliable measurement and evaluation methods for non-destructive geophysical measurements are needed urgently in order to investigate the integrity of the geological barrier and to detect problem zones. Conventional geological exploration using drill-holes is not possible without adversely affecting the geological barrier.

The occurrences of faults and brines in the cap rocks constitute the most severe types of problem zones. Other risks are related to geological inhomogeneities and occurrences of clay, anhydrite and salt minerals containing considerable amounts of water of crystallization. Such problem zones were explored using a set of geophysical methods (geosonar, seismics, georadar, EM methods and geo-electrics) in a sylvinite mine in Germany (Kulenkampff *et al.* 2002; Kurz *et al.* 2002), yielding parameter sets for disturbed and undisturbed rock. In this mine in the Südharz region, potash was produced from a 30-m-thick bedded sylvinite layer (Stassfurt K2) at about 500 m depth.

Both geo-electrical measurements and the computation of their results are particularly intricate, but result in a unique method yielding a spatial parameter distribution, including lateral effects, which shows very strong variations and can be evaluated quantitatively. Based upon existing experience (Yaramanci & Flach 1992; Yaramanci 1994, 2000) geo-electrical measuring methods were optimized in order to obtain an investigation depth comparable to seismics and georadar.

The sparsity and poor accuracy of laboratory data on the petrophysical properties of salt rocks pose a problem for the interpretation of resistivity surveys. Due to the fragile nature of the rocks, the effects of humidity, and possible mineralogical alterations, the samples have to be prepared and handled very carefully. In an earlier study, Kulenkampff & Yaramanci (1993) determined the water content by weighing. But, taking into account the low water content, which is generally less than 0.1% by weight, this method is not accurate enough. The resistivity measurement method is even more demanding, considering the extremely

From: HARVEY, P. K., BREWER, T. S., PEZARD, P. A. & PETROV, V. A. (eds) 2005. *Petrophysical Properties of Crystalline Rocks*. Geological Society, London, Special Publications, **240**, 301–306.
0305-8719/05/$15.00 © The Geological Society of London 2005.

high resistivity and the very small contact area between the electrodes and conducting brine. Thus, a four-electrode set-up must be used in the frequency range relevant for the interpretation of geo-electrical measurements (0.1 Hz to 10 Hz). The usual sizes of laboratory samples are only slightly larger than the salt crystals, which are not representative for the rock properties. Due to the requirement for intact samples, this causes resistivity values which are systematically too high. Samples from more porous or fractured zones which are more conductive usually do not remain intact and cannot be measured. Small-scale *in situ* surveys and measurements on larger drill cores (diameters of 10 cm) can bridge the gap between laboratory-scale measurements and large-scale underground surveys.

Field methods

Problems with underground geo-electrical measurements and inversion have a variety of causes such as:

- the prevailing resistivity is very high compared with more common measuring objects;
- parameter contrasts are very strong (several orders of magnitude);
- the electrode configurations are constrained by limited accessibility;
- three-dimensional electrode positions and a full space resistivity distribution have to be considered.

Instruments that can measure exceptionally resistive rocks were used (RESECS (DMT) and SIP Fuchs (Radic Research, Berlin)), as well as single-channel transient recorders (Texan (REFTEK)) with a portable high-power current source (Brunner 2001). Even with such specialized instruments, some resistivity values still deviate by a factor of two between interchanged current and potential electrode arrangements. That would mean an error of 100% on a linear scale. However, this error is acceptable, because the resistivity of the salt rocks ranges from 10 ohm m to 100 Mohm m and corresponds rather to a logarithmic distribution. Nevertheless, the methods have been optimized for a large depth of investigation and high resolution at lower depths, using dipole–dipole, Wenner, and pole–dipole arrangements (Friedel 2000). Three-dimensional geo-electrical configuration factors were computed with the help of an exact 3D finite-element model (ANSYS 5.6) of the drift, to take into account the irregularity and three-dimensional nature of the electrode

positions, the sketchy geometry of the drift and the influences of known cavities in the vicinity. Then a 2D-inversion program (Weller 1986; Weller *et al.* 1996) was used together with these 3D-configuration factors to invert more than 5000 data points on a profile with a length of 250 m. The inversion result is a resistivity section with a span of five orders of resistivity and reasonable sensitivities (= investigation depth) up to 40 m from the hanging wall (Fig. 1).

However, underground geo-electrical measurements respond to the complete three-dimensional environment. The 2D-inversion scheme projects anomalies that are situated below and in lateral distance from the profile into the measuring plane. The application of a 3D-inversion scheme (LaBreque *et al.* 1999) was not promising, because 2D-electrode arrangements were required, but only the 1D-profiles along the drift were accessible. Nevertheless, although lateral effects are possible, most geo-electrical anomalies are in accordance with results from wave propagation methods. These methods, seismics and georadar, are focused into the same vertical section. Because the main features of wave-propagation methods and geo-electrics are matching, these features are not supposed to be artifacts, e.g. lateral inhomogeneities reflected into the measuring plane. The maximum resistivities between 1 Mohm m and 10 Mohm m are related to undisturbed salt rocks, in accordance with the laboratory results. A low-resistivity zone with a resistivity of 100 ohm m in the central part of the hanging wall corresponds to a fault zone detected with seismics (Fig. 1). From geological and geomechanical data, it was interpreted as a fracture zone containing considerable quantities of brine, but, due to safety issues, it could not be explored by drilling.

Laboratory methods

Salt rocks usually contain only traces of free pore water, which only contributes to bulk conductivity. However, the constituent minerals often contain high amounts of water of crystallization, which may be mobilized at room temperature, and they may be extremely hygroscopic. Therefore, different water binding types have to be distinguished. The Karl–Fischer method, the most sensitive titration method for trace water, with a resolution in the range of a few micrograms (Scholz 1983), was used to determine the water content and to distinguish between the binding types, based upon the analysis of evaporation curves.

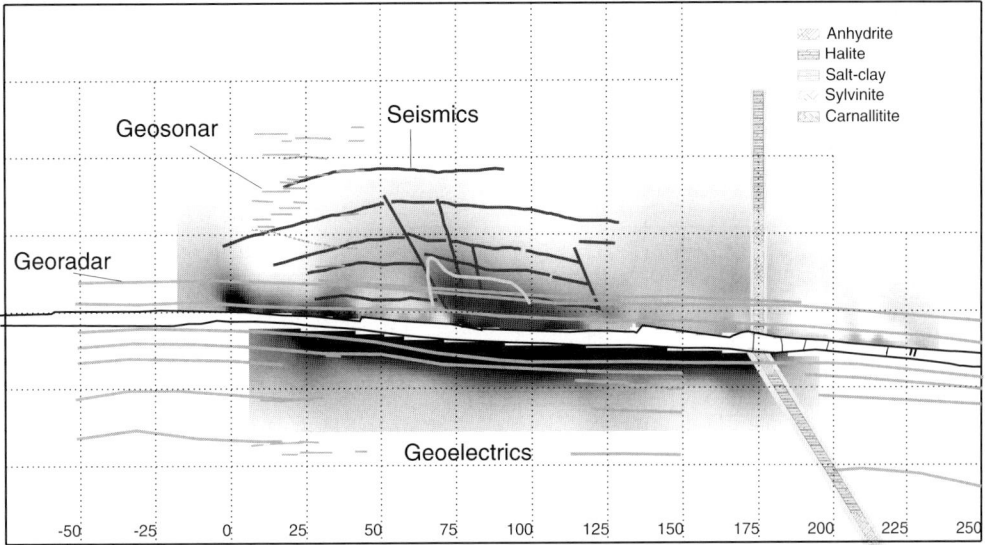

Fig. 1. Underground geophysical results: interpreted P-wave seismics, georadar reflectors, ultrasonic reflectors (geosonar), and geo-electrical inversion results (light: >1 Mohm m, dark: <1 k ohm m). Geo-electrical data processing included 2D-inversion with 3D-geometry modelling. Both seismics and geo-electrics detected a previously unknown brine-containing fault zone in the centre of the profile. Strong reflections from 10 m above the hanging wall (salt clay T3) hampered the georadar wave penetration.

Mini-plugs (<20 g) were heat-sealed in the mine in order to keep the initial water content. The water content was evaporated in a Karl–Fischer oven at 60°C, while recording the amount of water emitted. Three phases of evaporation can be distinguished (Fig. 2):

• an initial peak in the evaporation curve due to free water on the sample's surface;

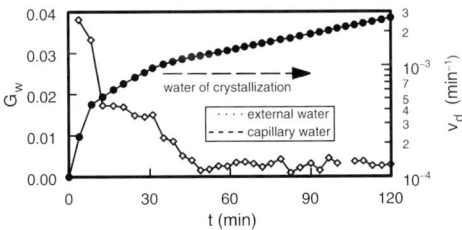

Fig. 2. Drying curve of a sylvinite mini-plug in the Karl–Fischer oven at 60°C. Dots: evaporated water mass referred to the sample mass (G_w). Diamonds: evaporation rate referred to the sample mass (v_d). Water from the external sample surface is evaporated within 10 minutes; the plateau up until 30 minutes is caused by free pore water. After about 20 minutes the water of crystallization is activated and emitted at a rate of 0.1 mg/(min g^{-1}).

• a constant evaporation rate (0.5 mg/(min g), referred to sample mass) for 15 to 60 minutes due to the free pore water;
• a constant rate (0.1 mg/(min g)) for a long period due to water of crystallization or no further evaporation.

At the same time, four-electrode resistivity measurements were made. The current, supplied by a frequency generator (0.001 to 10 Hz), was injected across the end faces of the sample. Two ring electrodes were used for potential measurements with a high-impedance pre-amplifier. All signals were recorded with a transient recorder. The resistivity data were computed by correlation between voltage and current. Due to the high impedance, a sophisticated algorithm was necessary, because the signal-to-noise ratio was poor although the sample container was shielded. The range of resistivities in the initial state (10 k ohm m to 100 Mohm m) are about one order higher than the field results. The two-electrode resistivity, including the coupling resistance of the current electrode, was a factor of 10 to 100 higher and thus should be considered only as a control parameter, not as a rock parameter.

The free pore water was extracted within, at the most, one hour, provided that the sample

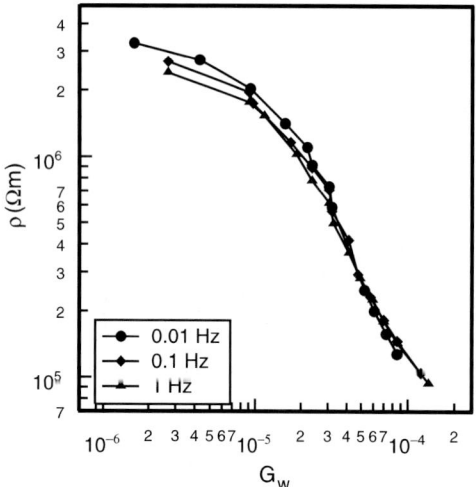

Fig. 3. Four-electrode resistivity versus water content, by weight of sample containing the water of crystallization. Measuring frequencies: 0.01 Hz, 0.1 Hz and 1 Hz. The resistivity value remains below 5 Mohm m. (cf. Fig. 5).

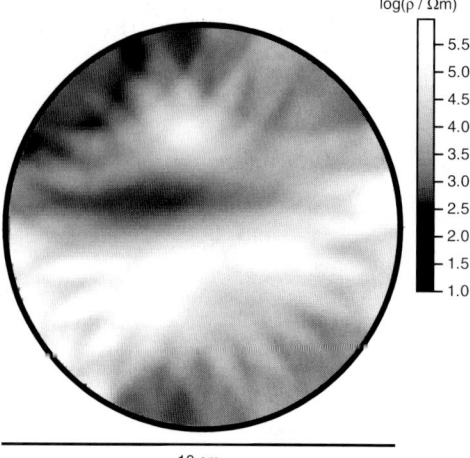

Fig. 4. Electrical resistivity tomogram of a sylvinite core (diameter 10 cm). Resistivity range from 10 ohm m (dark) to 1 Mohm m (light).

did not contain water of crystallization. At the same time, the resistivity increased rapidly. Otherwise, when the sample contained water of crystallization, a constant rate of evaporation was attained after 30 min, while the resistivity did not vary considerably during the extraction of water (Fig. 3). This means that the extracted pore water is replaced from the reservoir of chemically bound water.

These laboratory resistivity values are still 10 to 100 times higher than the field results (cf. Figs 5 & 6). This discrepancy could be caused by scaling effects, which were examined with electrical resistivity tomography on 10 cm cores. The measuring method was developed at Leipzig University (Just *et al.* 1998; Just 2001) for the investigation of high-resistivity crystalline rocks.

The measuring array consists of two dipoles, A–B and M–N on the mantle of the core. Current is injected at the source dipole A–B into the core, and is measured with a lock-in amplifier. The difference in voltage is measured at the potential dipole M–N with small-band pre-amplifiers and a second lock-in amplifier. In order to apply a tomographic reconstruction algorithm, apparent resistivities are measured at several configurations of the dipoles in one plane. For this purpose, a special apparatus was constructed which allows exact positioning of the four electrodes at each position on the

core's mantle. The tomographic analysis is based on the simultaneous iterative reconstruction technique (SIRT) and includes forward modelling based on Weller (1986).

This electrical core tomography yielded results (Fig. 4) which were compatible with the mini plug measurements because equally high-resistivity zones within the tomograms are present, but, at the same time, very strong variations occur over ranges of 1 cm, comparable to the global variations in the metre range of the geo-electrical inversion results.

Discussion

The field results demonstrate the usefulness of geo-electrical methods for the detection and evaluation of conducting inhomogeneities in salt rocks, which are caused by the presence of brines. Most single geophysical methods produce ambiguous results in such potential risk zones for underground waste deposits, because variations in the generally very low-porosity only cause slight alterations of parameters concerning wave propagation (like seismic or electromagnetic velocities), and because underground DC geo-electrical measurement results are ambiguous by themselves. Nevertheless, the combination of different methods yields a powerful means to evaluate such barriers.

One advantage of geo-electrics is the extraction of a spatial parameter distribution. However, the geological interpretation of the results is still tentative, because of the lack of laboratory data relating physical parameters to geological properties. Earlier laboratory results (Kulenkampff & Yaramanci 1993) appear to be partially doubtful in comparison with our results, because no suitable laboratory method for measurements in the frequency range of the field measurements was available. Such prerequisites are very sensitive methods for the determination of the water content and a resistivity measurement set up with extremely high impedance and noise suppression. An overview showing the new results and discrepancies with former results is given in Figure 5. The methods used in the former investigations resulted in overlarge resistivity values, due to the two-electrode set-up and water contents that were too high because of the weighing method.

The laboratory resistivity results for mini-plugs are systematically much higher than the field results. The mini-plug size is not representative in two respects:

- the size is too small (comparable to the grain size);
- the selection is not representative (only intact zones).

A feasible alternative to four-electrode measurements on mini-plugs is the electrical tomography of medium-sized drill cores which are both representative in size and in selection, because they are mechanically more stable. The

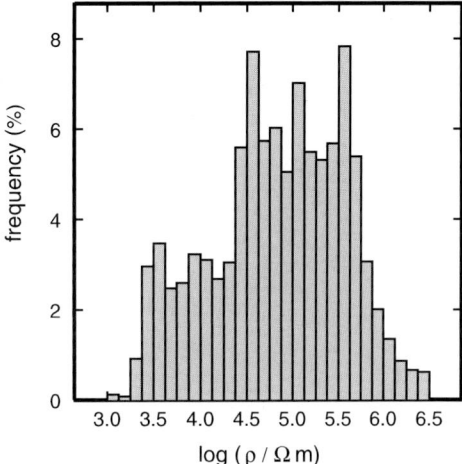

Fig. 6. Resistivity distribution of the field measurements (Fig. 1).

water content and mineral composition are then investigated after non-destructive tomography. The tomography shows strong resistivity variations within small ranges which are due to visible small fracture zones and lithological variations in the core. These variations are in the same range as the variations observed in the field and are thus more trustworthy (Figs 6 & 7) than the measurements on mini-plugs.

The laboratory results show that the electrical resistivity may be used as a criterion for the risk

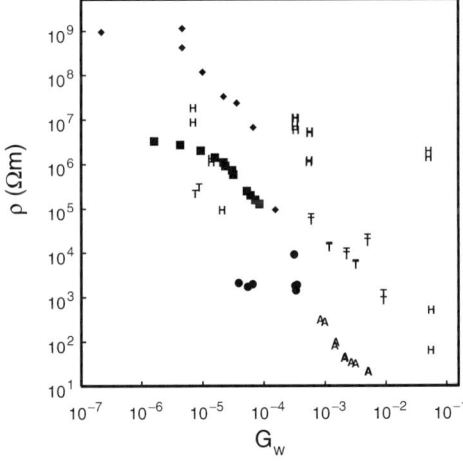

Fig. 5. Resistivity of evaporites versus water content by weight. Dots: four-electrode resistivity of sylvinite (partially carnallite) mini-plugs (diameter 1.5 cm, measuring frequency 0.1 Hz). Letters: H: halite; A: anhydrite; T: salt–clay, measured with a two-electrode method at 10 Hz.

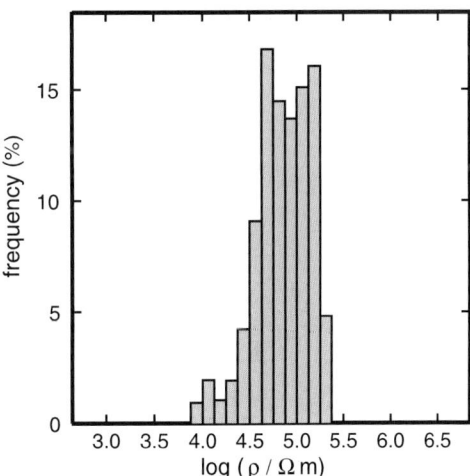

Fig. 7. Resistivity distribution of the laboratory resistivity tomogram (Fig. 4).

assessment of underground deposits. Zones with resistivities less than 1 Mohm m always contain brines and thus generally represent risk zones. The physical cause for the low resistivity (fault zones, diffuse brine occurrences, lithology, water of crystallization) may be assessed with the aid of other geophysical and geological methods.

The authors would like to thank the German Federal Ministry for Education and Research for financial support (02C 0558) and the GVV mbH Sondershausen for logistical aid.

References

BRUNNER, I. 2001. *Beiträge zur geoelektrischen Potentialtomographie für die Ermittlung von Leitfähigkeitsunterschieden im Untergrund.* PhD thesis, Leipzig University, Shaker, Aachen, Germany.

FRIEDEL, S. 2000. *Über die Abbildungseigenschaften der geoelektrischen Impedanztomographie unter Berücksichtigung von endlicher Anzahl und endlicher Genauigkeit der Meßdaten.* PhD thesis, Leipzig University, Shaker, Aachen, Germany.

JUST, A. 2001. *Bestimmung der räumlichen Verteilung des spezifischen elektrischen Widerstandes an zylinderförmigen Körpern mit Hilfe der elektrischen Widerstandstomographie.* PhD thesis, Leipzig University, Germany.

JUST, A., KÜPPER, T., KÜRSCHNER, D. & JACOBS, F. 1998. Core imaging by electrical resistivity tomography. *Proceedings of the 4th SEGJ International Symposium*, Tokyo, 231–237.

KULENKAMPFF, J. M. & YARAMANCI, U. 1993. Frequency-dependent complex resistivity of rock-salt samples and related petrophysical parameters. *Geophysical Prospecting*, 41, 995–1008.

KULENKAMPFF, J. M., ASCHMANN, L. *et al.* 2002. *Komplexes Mess- und Auswerteinstrumentarium für die untertägige Erkundung von Problemzonen der geologischen Barriere von Endlagern und Untertagedeponien (UTD) im Salinar.* Institut für Geophysik und Geologie, Leipzig, available at the TIB Hannover. World Wide Web Address: *http://edok01.tib.uni-hannover.de/edoks/e01fb02/359733271.pdf*

KURZ, G., Moïse, E. *et al.* 2002. Auswertekonzept zur geophysikalischen Erkundung von Problemzonen im Salinar. *Zeitschrift für Angewandte Geologie*, **48** (2), 56–62.

LABRECQUE, D. J., MORELLI, G., DAILY, W. & RAMIREZ, P. L. 1999. Occam's Inversion of 3-D Electrical resistivity tomography. *In*: ORISTAGLIO, M. & SPIES, B. (eds) *Three-Dimensional Electromagnetics*. SEG Tulsa, Geophysical Developments, **7**, 575–590.

SCHOLZ, E. 1983. *Karl Fischer Titration*, Springer, Berlin.

WARREN, J. 1999. *Evaporites – their Evolution and Economics*. Blackwell Science, Oxford.

WELLER, A. 1986. Berechnung geoelektrischer Potentialfelder mit dem Differenzenverfahren. *Freiberger Forschungshefte C*, **405**, 86–122.

WELLER, A., SEICHTER, M. & KAMPKE, A. 1996. Induced-polarization modelling using complex electrical conductivities. *Geophysical Journal International*, **127**, 387–398.

YARAMANCI, U. 1994. Relation of *in-situ* resistivity to water content in salt-rocks. *Geophysical Prospecting*, **41**, 229–239.

YARAMANCI, U. 2000. Geoelectric exploration and monitoring in rock salt for the safety assessment of underground waste disposal sites. *Journal of Applied Geophysics*, **44**, 181–196.

YARAMANCI, U. & FLACH, D. 1992. Resistivity of rock-salt in Asse (Germany) and petrophysical aspects. *Geophysical Prospecting*, **40**, 85–100.

Deformation of metavolcanics in the Karachay Lake area, Southern Urals: petrophysical and mineral–chemical aspects

V. A. PETROV[1], V. V. POLUEKTOV[1], A. V. ZHARIKOV[1], V. I. VELICHKIN[1], R. M. NASIMOV[2], N. I. DIAUR[2], V. A. TERENTIEV[2], V. M. SHMONOV[3] & V. M. VITOVTOVA[3]

[1]*Institute of Geology of Ore Deposits, Petrography, Mineralogy and Geochemistry (IGEM), Russian Academy of Sciences, 35 Staromonetny per., Moscow 109017, Russia (e-mail: vlad@igem.ru)*
[2]*United Institute of Physics of the Earth (OIFZ), Russian Academy of Sciences, 10 B, Gruzinskaya Street, Moscow 123810, Russia*
[3]*Institute of Experimental Mineralogy (IEM), Russian Academy of Sciences, Chernogolovka, Moscow Region 142432, Russia*

Abstract: This paper is the result of laboratory tests aimed at comparative assessment of the filtration, elastic and mechanical properties of metavolcanics, as well as a preliminary investigation of the nature of characteristic variations in $P–T$ parameters, simulating the conditions in the vicinity of an underground storage facility for heat-generating radioactive waste. The properties have been studied in cores collected from the deepest (1200 m) hole in the metavolcanics, focusing on intervals subjected to different strain intensities. Interpretation of the data obtained suggests that the identification of the most probable pathways for fluid flow and radionuclide transport depends on estimates of the retardation characteristics of zones of disjunctive dislocation. However, the widespread occurrence of linear zones of schistosity, mylonitization and brecciation may not be a decisive negative factor in the long-term performance assessment of the underground storage facility, because these minerals, which concentrate radionuclides such as epidote, chlorite, sericite, and Fe oxyhydroxides are widely distributed in these zones. Moreover, open microfractures and pores affected by heat will be filled with neoformed minerals (epidote, chlorite, carbonate) characterized by high sorption capacity with respect to radionuclides, due to release of intergranular water, as well as carbonatization reactions. Simultaneously, the system of open macro- and microfractures that resulted from stress affecting relatively undisturbed blocks may be one of the factors dramatically reducing country rock retention capability. Therefore, the results of petrophysical and mineral–chemical investigations must be taken into account in feasibility studies of designs for the construction of underground systems for final disposal of high-level wastes and/or long-term storage of spent nuclear fuel in the PA Mayak area.

Engineering context

In the area for restricting contaminant release (SPZ) of the Production Association (PA) Mayak, in the Southern Urals, work is in progress on the identification of geological blocks and rock types that are most favourable for the underground intermediate-life, high-level wastes (HLW), containing caesium and strontium radionuclides (Glagolenko *et al.* 1996) at depths of 0.5–1.0 km below the surface (Laverov *et al.* 2000). Geological, geophysical and hydrogeological studies have contributed to the identification of a series of blocks that seem most promising for the construction of the HLW repository (Velichkin *et al.* 1997). At present, with regard to the assessment of the long-term safety of the repository, great importance is attached to the detailed characterization of the rocks that, by their texture, mineral–chemical and petrophysical characteristics, are favourable for the establishment of an underground laboratory and eventually, a monitored retrievable storage or underground disposal facility.

The area under investigation (Fig. 1) is composed of gneiss–amphibolite ($PR_3–PZ_1$), metavolcanic ($S–D_1$) and carbonate–terrigenous (C_{1-2}) complexes. The gneiss–amphibolite complex contains granite–granodiorite and

From: HARVEY, P. K., BREWER, T. S., PEZARD, P. A. & PETROV, V. A. (eds) 2005. *Petrophysical Properties of Crystalline Rocks*. Geological Society, London, Special Publications, **240**, 307–322.
0305-8719/05/$15.00   © The Geological Society of London 2005.

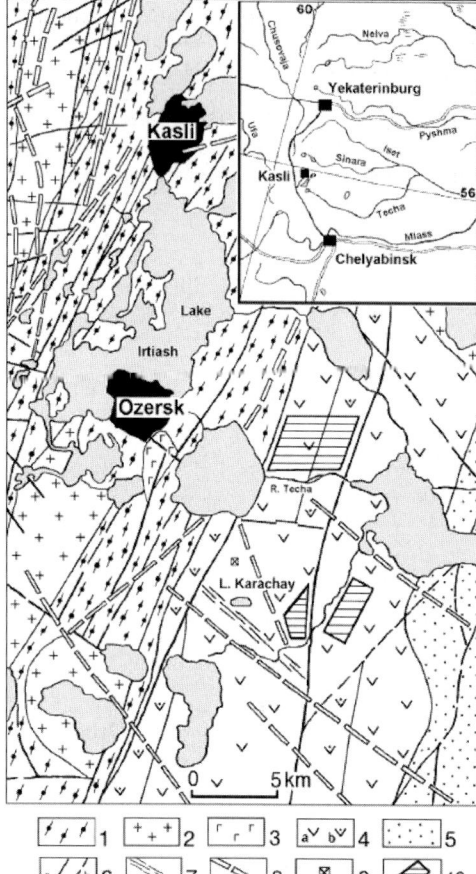

Fig. 1. Generalized map showing the location and geology of the PA Mayak area. (1) gneiss–amphibolite complex (PR$_3$–PZ$_1$), including (2) granite–granodiorite and (3) gabbro bodies; (4) upper (a) and lower (b) series of the metavolcanic complex (S-D$_1$); (5) carbonate–terrigenous complex (C$_{1-2}$); (6) principal faults investigated (a) and proposed (b); (7) linear zones of schistosity; (8) neoformed faults (lineaments); (9) location of the investigated borehole; (10) blocks most promising for the construction of the HLW repository.

gabbro bodies of PZ$_1$ age. The most detailed research was undertaken on the metavolcanic complex within the Lake Karachay area. The complex forms the main part of the SPZ, to which the search for the underground site laboratory is limited, for social and political reasons.

The decision to construct the underground laboratory in a particular block within the metavolcanics may be taken only after completion of a series of standard procedures (Chapman & McKinley 1987; Savage 1995), including

detailed comparative studies of rock blocks, demonstration and performance assessment. However, it is now feasible to determine trends in the behaviour of an underground disposal system in the context of specific features of the geological structure, mineral–chemical composition and petrophysical properties of the metavolcanic strata. These data will be useful for the identification of a more reasonable approach for the more detailed investigations within promising geological blocks.

Hence, the field studies and laboratory tests were aimed at assessing comparatively the deformation related changes within the rocks (which are coupled with modification of their petrophysical and mineral–chemical properties), as well as investigating the nature of the characteristic variation in P–T parameters simulating the conditions in the vicinity of an underground storage facility.

Geological and mineral–chemical context

The maximum thickness of the metavolcanic complex, estimated on the basis of geological–structural and geophysical data, amounts to 3 km (Velichkin *et al.* 1995). The complex is divided into two series: a lower one, made up of volcanogenic–sedimentary rocks, and a upper one of predominantly volcanogenic rocks. Discontinuities are widely distributed in both series, and can be grouped according to their morphogenetic features, time of formation and scale of occurrence as follows: (1) deep faults accompanied by the silicification of rocks and by systems of convergent quartz veins; (2) zones of schistosity; (3) mylonite sutures and brecciation zones with chlorite or chlorite–epidote–quartz cement; (4) numerous systems of fractures with quartz–carbonate and carbonate sealing. Deep faults are inter-block structures, while the linear zones of schistosity, mylonite sutures and brecciation zones are intra-block structures. Geomorphological analysis of the area identified (Sysoev *et al.* 1999) a series of neoformed faults (lineaments), which either inherited the ancient inter-block discontinuities, or are represented by flexures in loose sediments.

Regional greenschist-facies metamorphism of the original volcanogenic–sedimentary rocks has homogenized their mineral–chemical composition (Table 1). The present-day mineral composition of the rocks is determined by the association of both relict primary minerals (plagioclase, pyroxene and hornblende) and

Table 1. *Average chemical composition of the metavolcanic complex (upper and lower series)*

Rock-forming oxides (wt%)	Upper series ($n = 83$)	Lower series ($n = 10$)
SiO_2	50.17	54.86
TiO_2	0.85	0.54
Al_2O_3	17.11	14.9
Fe_2O_3	5.65	2.71
FeO	4.44	5.10
MnO	0.17	0.22
MgO	6.25	7.18
CaO	7.53	4.40
Na_2O	2.60	4.06
K_2O	0.35	1.90
H_2O^-	0.45	0.41
H_2O^+	4.26	3.32
CO_2	0.29	0.05

secondary minerals (actinolite, albite, epidote, chlorite, calcite and sericite) (Petrov & Poluektov 1999).

From their content of rock-forming oxides, the rocks tholeiitic basalts and andesitic basalts. In terms of alkalinity, the upper series (US) belongs to the sodic group ($Na_2O/K_2O = 3.12–7.43$) and the lower series (LS) ($Na_2O/K_2O = 2.14$), while according to the aluminiferrous ($Al_2O_3/\Sigma Fe + MgO$) coefficient (0.65–1.01 (US); 0.99 (LS)) they are classed as low to moderately aluminiferrous basalts. From their femicity ($\Sigma Fe + MgO + TiO_2$) (17.96–22.12 (US); 15.74 (LS)) they belong to mesocratic and transition forms, as well as to melanocratic and leucocratic varieties of basalts and andesitic basalts. The agpaicity ($Na_2O + K_2O/Al_2O_3$) is <1, while the ferruginosity ($\Sigma Fe/MgO + \Sigma Fe$) (52.4–59.6 (US) and 52.1 (LS)) correlates with the relative amount of melanocratic and leucocratic minerals. For the rocks of the upper series, the $K_2O/TiO_2 = 0.41–1.13$ relationship is typical of tholeiitic basalts, while for the volcanics of the lower series, higher values (3.32) are more common.

The greenschist metamorphism has also homogenized the rocks' petrophysical characteristics; this is most clearly shown outside the zones of dislocation. The undisturbed, unweathered intact andesite–basalt porphyrites, tuffs, porphyritic lava-breccias and tuff lavas show high density, elasticity and strength as well as minimal values of effective porosity and permeability. However, the presence of linear zones of schistosity, mylonitization and brecciation means that the data obtained cannot be applied to the entire rock massif. The discontinuities are accompanied by low-temperature hydrother-mal–metasomatic alteration (sericite, carbonate and chlorite). Here, the concentrations of macro-components are significantly different, with depletion in SiO_2, MgO and Na_2O, and enrichment of CO_2, K_2O, CaO, TiO_2, and redistribution of ΣFe and Al_2O_3. In addition, the rocks are characterized by higher effective porosity and permeability, and lower values of elasticity and strength parameters. In zones of schistosity, where rocks have been transformed into epidote–actinolite, epidote–chlorite and chlorite–carbonate schists, the highest values of effective porosity, permeability and Poisson's ratio were observed (Petrov *et al.* 2000).

Analysis of the spatial relationship between discontinuities and the results of hydrogeological investigations (Solodov & Zotov 1995) shows that linear zones of schistosity, mylonite sutures and brecciation zones, as well as various systems of fractures, can be considered as the most probable pathways for migration of radionuclide contamination.

As the search for a prospective site for the underground facility is limited to a small area, it is difficult to completely avoid linear schistosity zones within the boundaries of blocks with dimensions suitable for the infrastructure of the underground storage facility. Hence, laboratory tests were aimed at the detailed assessment of filtration and elastic properties of variously deformed varieties of volcanogenic rocks under various $P–T$ conditions. The samples were collected from the borehole section of the upper series, where a relatively thick zone of schistosity occurs. Sampling was done at relatively regular intervals (about 50 m) down to a depth of 1200 m. The investigations were carried out on intact samples without fractures and veinlets.

Porosity and permeability

Porosity (Φ, %) and permeability (K, m^2) of 27 rock samples were studied. Cylindrical specimens 1.5 cm in length and 2.5 cm in diameter were cut parallel to the core axes. Hence, vertical flow was simulated, except for one sample with pronounced schistosity. This sample was tested with the flow parallel and normal to the foliation plane. The measurements were carried out at room temperature and atmospheric pressure. The pulse decay method was used with argon as a flowing medium. Measurement errors were 10–25% at $K > n \cdot E$–$20 \, m^2$ and 20–35% at $K < n \cdot E$–$20 \, m^2$. The experimental data obtained are presented in Table 2 and Figure 2.

Porosity values as a rule are very low: ranging from 0.07 to 0.69%. The mean value is 0.26%. It is clear that porosity values increase for sample

Table 2. *Effective porosity and permeability of the samples from the borehole section of the upper series at room temperature and atmospheric pressure*

No. of sample	Rock	Effective porosity (%)	Permeability (m^2)
37	Lava-breccia	0.35	$1.40 \cdot E-19$
50	Tuff lava	0.43	$3.30 \cdot E-19$
102	Schist	0.34	$4.38 \cdot E-21$
150	Lava-breccia	0.09	$2.39 \cdot E-20$
175	Tuff lava	0.11	$3.06 \cdot E-20$
200	Andesite–basaltic porphyrite	0.17	$1.01 \cdot E-20$
250	Lava-breccia	0.30	$1.13 \cdot E-19$
305	Tuff lava	0.24	$1.10 \cdot E-19$
400	Tuff	0.14	$6.10 \cdot E-19$
454	Tuff, lava-breccia	0.18	$3.00 \cdot E-19$
515	Lava-breccia	0.18	$4.77 \cdot E-21$
554	Lava breccia	0.21	
600	Tuff	0.16	$6.65 \cdot E-21$
650	Lava-breccia	0.20	$4.26 \cdot E-21$
701	Tuff	0.13	$4.09 \cdot E-20$
752	Andesite–basaltic porphyrite	0.16	$1.03 \cdot E-19$
798	Tuff	0.50	$2.40 \cdot E-20$
834	Schist, mylonite: parallel to schistosity	0.66	$1.89 \cdot E-19$
	normal to schistosity	0.65	$1.10 \cdot E-19$
850	Tuff	0.69	$4.80 \cdot E-19$
900	Andesite–basaltic porphyrite	0.25	$1.90 \cdot E-18$
950	Tuff	0.32	$7.77 \cdot E-20$
1000	Tuff	0.25	$1.05 \cdot E-20$
1050	Lava-breccia	0.07	$9.21 \cdot E-20$
1100	Tuff lava	0.20	$1.71 \cdot E-21$
1150	Lava-breccia	0.18	$2.52 \cdot E-19$
1170	Andesite–basaltic porphyrite	0.21	$4.18 \cdot E-20$
1195	Lava-breccia	0.19	$8.30 \cdot E-20$

The sample number correlates with the depth of its occurrence.

No. 834, collected from the zone of schistosity, as well as for adjoining samples No. 798 and No. 850.

Very low permeability values are typical of the majority of the studied samples. Seventeen samples have permeabilities lower than $1.00 \cdot E-19$ m^2. The permeability of the samples collected from and near the zone of schistosity increases to $4.80 \cdot E-19$ m^2 in accordance with porosity values. However, the maximum value of permeability ($1.90 \cdot E-18$) is observed for sample No. 900.

Schistose sample No. 834 was tested with flow parallel and normal to the foliation plane. The porosity and permeability values obtained are practically the same. This supports the assumption that in schistose varieties the interconnected microcrack systems occur parallel and normal to the foliation plane. The system of microchannels normal to the foliation plane was found during investigation of the pore space structure by ultrasonic methods (see below).

The values obtained are typical of the bulk permeability of solid low-porosity crystalline rocks. With a permeability value of $n \cdot E-19$ m^2

and a hydrodynamic gradient of 15 Pa m^{-1} the filtration rate will be only $c.5 \times 10^{-5}$ m/a (Petrov *et al.* 1998). However, flow rates obtained in the area under investigation – during well logging – are considerably higher (Drozhko *et al.* 1997). Underground flow and contaminant transport must therefore take place through structures at a scale larger than rock samples, namely fractures, faults and dislocation zones. If so, then it is necessary to characterize and quantify the fracture systems at the macroscopic scale using borehole data. Such an analysis has been developed, but the results are not the subject of this paper. However, for the identification of the possible mechanism of fracture sealing under conditions in the vicinity of the proposed repository the results of an investigation into the dynamics of the permeability of an artificial crack in porphyrite during water filtration are of particular interest (Zaraisky 1994). These tests demonstrated that at *T* of up to 250°C and *P* of up to 290 bar the crack permeability decreased from $n \cdot E-13$ m^2 to $2.5 \cdot E-15$ m^2 over 38 days. Furthermore, the crack was filled with fine-grained aggregates of carbon-

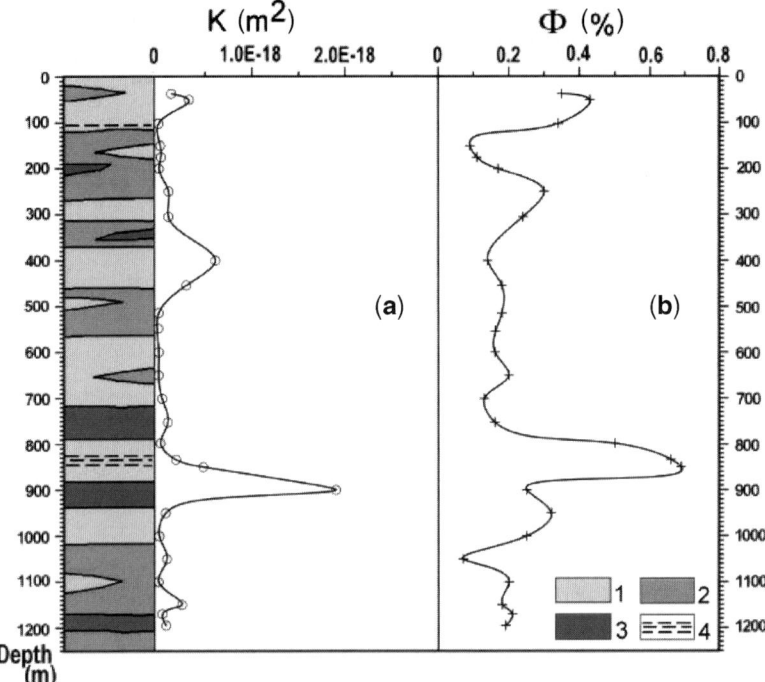

Fig. 2. Simplified geological structure of the borehole and distribution of permeability (**a**) and effective porosity (**b**) as a function of depth. Petrographic units of the upper series: (1) tuff, tuff lava, (2) lava-breccia, (3) porphyritic andesite–basalt, (4) schistose rock.

ate, chlorite and epidote. Assuming that the zones of schistosity, mylonitization and brecciation, as well as the fractures investigated in the area under consideration, are sealed with sericite, chlorite, epidote, carbonate (which have high adsorption capacities for Cs and Sr), these data must be taken into account for the assessment of the long-term safety of an underground disposal system.

The diffusion of material in the near-fracture space is one of the most important mechanisms of radionuclide transport (Neretnieks 1980), while the porosity and permeability of the rock matrix in combination with the characteristics of pore space structure, are the parameters required for the assessment of the retention capability of metavolcanics.

Pore space characterization

The pore space structure was studied by ultrasonic structural analysis of intact rocks (Starostin 1979). Ten disc-like rock samples taken in metavolcanics at depths ranging from 50 to 1100 m, were studied. Comparative analysis of the anisotropy factors of P-waves in dry (a_d) and water-saturated (a_s) samples (saturated for

over seven days), showed a sharp decrease in a_s for the samples from 400, 554, 701 and 798 m depths. The analysis of stereograms of V_P indicatrix distribution showed the occurrence of two types of microvoids: plane-parallel microcracks and linearly oriented systems of micropores and pore capillaries. In schistose varieties, a microcrack system perpendicular to the foliation plane was found. Petrographic investigations identified these structures as Riedel microshears (Fig. 3). A more detailed description of the method is presented in Petrov *et al.* (2005).

The effect of heat may results in the decompaction of rocks and the rearrangement of the structure of their pore space (Fredrich & Wong 1986). The process of decompaction includes the expansion of the aperture and length of the microcracks; the decrease in elastic wave velocities; and lower values of elasticity and strength moduli. As a result of thermodecompaction, the permeability of rocks may increase by several orders of magnitude. This has been proved by permeability tests in conditions of thermal expansion and decompaction. Within our research project, these additional tests were carried out by G. P. Zaraisky at temperatures of

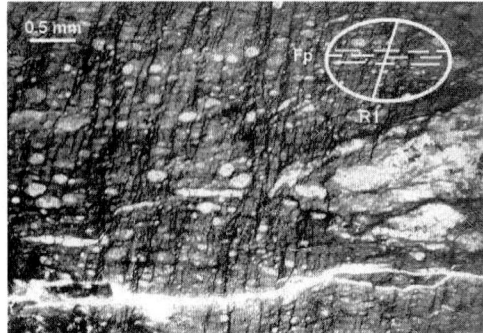

Fig. 3. Development of Riedel microshears (R_1) in schists perpendicular to the foliation plane (F_p). Microphotograph. Nicols are crossed.

up to 400°C and atmospheric pressure, as well as in an autoclave at water pressure of *c.*1 kbar in Institute of Experimental Mineralogy, Russian Academy of Science (IEM RAS). The analysis was performed on samples of massive andesite–basaltic porphyrite (No. 752), and weakly (No. 798) and intensively (No. 834) schistose tuffs. In sample No. 752, the effect of thermal expansion or thermodecompaction was registered with a temperature rise,
resulting in an increase in permeability to $2.2 \cdot E-18$ m² at T of 350°C, while its original permeability values were $1.03 \cdot E-19$ m². However, at about 200°C, permeability reached its minimum, and this can probably be explained by the complete disruption of the original system of pore space resulting from non-uniform thermal expansion of mineral grains. Thermodecompaction of sample No. 834 with an original permeability of $1.89 \cdot E-19$ m² (parallel to schistosity) and $1.10 \cdot E-19$ m² (normal to schistosity) after heating to 400°C in atmospheric conditions resulted in an increase in permeability to $9.8 \cdot E-17$ m² and $2.2 \cdot E-15$ m², respectively. In the same sample, after heating to 300°C, but at P_{H_2O} ~1 kbar, permeability increased by an order of magnitude. The same was typical of sample No. 798. Higher values of rock permeability in the direction perpendicular to the foliation plane can be explained by the widespread presence of Riedel microshears. Such behaviour of intact rock proves exactly during heating.

Thus, under increased temperatures in the vicinity of the proposed repository, a network interconnected non-mineralized microcracks can be formed, and this will result in a greater permeability of metavolcanics and a decrease in their retention capability with regard to radionuclides. The probability of the development of stress-related macrocracks in these microheterogeneities depends in many respects on the elastic and strength characteristics of the metavolcanics.

Elastic properties

The velocities of P-waves (V_P) and plane-polarized shear waves (V_S), as well as their waveforms in three orthogonal directions, were experimentally studied within cubic rock samples, using the ultrasonic pulse method. The tested samples were collected from the zone of schistosity, of which the central, most strained part, was at 834 m depth. The samples were selected in order to obtain elastic characteristics at different strain intensities, including strongly (No 834) to weakly (No. 827 and No. 838) schistose tuffs and virtually massive (No. 798 and No. 850) tuffs. Every core sample was cut into two or three cubic specimens. One of the cube axes was oriented in parallel with the core axis.

The P-wave velocities in the cubes were measured in three orthogonal directions. The velocity of the S-waves was measured in every specimen, with the change of their polarization plane, in turn, parallel with the other two faces of a cube. For sample No. 834 the velocities of the S-waves were measured every 22.5°, around a full rotation of the polarization plane. Digital recordings of oscillations passing through the acoustic system, including the sample, were analysed, and the S-mode arrival was identified. The whole system was calibrated using S-wave velocity measurements in quartz glass etalons.

In contrast to studying P-wave velocities, the measurements for S-waves require the description of the orientation of the polarization plane relative to the co-ordinate frame fixed to the specimen (Gorbatsevich 1999). Figure 4 illustrates the pattern of S-wave velocity measurements, within a cubic sample. At least two measurements were taken for one direction. The polarization plane of the transducers was oriented parallel with the remaining two cubic faces. The $1-1'$ index shows the directions where the transmitter was placed on the face 1, while the receiver is placed on the opposite face $1'$. Studies with the rotation of the S-wave polarization plane relative to the schistosity of the samples from group 834, were performed under uniaxial stress at 5 MPa. Diameters dividing the circles on the sample planes into sectors reflect the rotation of the orientation of the polarization plane. The initial orientation of the polarization plane was assumed to be 'zero', which, in the $1-1'$ direction is parallel to the 3 and $3'$ faces; in the $2-2'$ direction it is

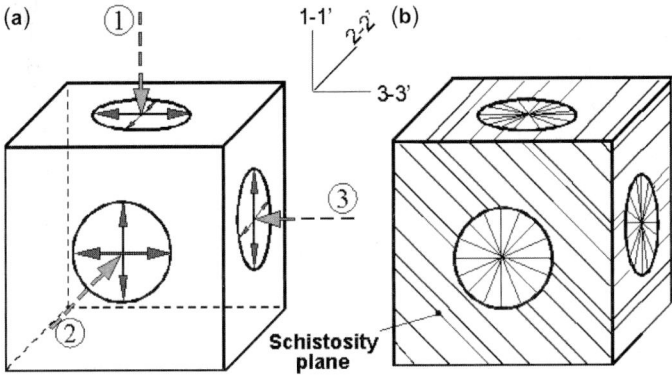

Fig. 4. Schematic representation of V_S measurement in cubic samples: (**a**) general case where arrows 1, 2 and 3 show directions of wave propagation, and orthogonal arrows in circles at each face show the two directions of orientation of the polarization plane utilized in all measurements; (**b**) location of the polarization plane in the case of measurements on schistose samples from group 834.

parallel to the 3 and 3' faces, and in the 3–3' direction it is parallel to the 1 and 1' faces.

The results of the velocity measurements are presented in Table 3. The V_S values for the crossways-oriented polarization plane of the same wave direction are also presented.

In the virtually massive tuff group 798 (a, b, c) the samples are characterized by high S-wave velocities. The mean V_S values for directions 1, 2 and 3 are 4.371, 4.383 and 4.397 km s^{-1} respectively. It should be noted that the maximum P-wave velocities and densities were also obtained for this series of samples (Fig. 5). The samples of this group are elastically isotropic, as shown by the V_P and V_S values.

The samples from group 827 (a, b) are weakly schistose tuff, and they are characterized by relatively low V_S values. In sample 827b, for all directions except the 1–1' direction, a high attenuation of S-waves is observed. The highest attenuation is noted for direction 3, with the polarization plane of S-waves parallel to face 2. Substantial attenuation did not allow the identification of the S-mode for the 1–1' and 3–3' directions in this sample. Such attenuation can be explained by the textural peculiarities of this rock and its relatively low density $\sim$2.80 g cm^{-3}. According to the available velocity values for three directions in sample 827a, V_S anisotropy is observed. For instance,

Table 3. *Result of* V$_S$ *and* V$_P$ *analyses in cubic samples for 1, 2 and 3 directions*

No. of sample	Density (g cm^{-3})	V_{S1}	V_{S1} (+)	V_{P1}	V_{S2}	V_{S2} (+)	V_{P2}	V_{S3}	V_{S3} (+)	V_{P3}
798a	2.977	4.396	4.305	6.665	4.363	4.410	6.795	4.363	4.363	6.795
798b	2.971	4.363	4.318	6.795	4.402	4.355	6.783	4.394	4.302	6.770
798c	2.983	4.355	4.310	6.896	4.402	4.402	6.673	4.460	4.412	6.799
827a	2.815	3.383	3.197	5.370	3.320	—	6.231	3.627	3.174	5.722
827b	2.809	—	3.474	5.279	3.689	3.326	5.483	—	3.537	6.071
834a	2.774	3.625	3.210	6.094	3.678	3.846	6.551	3.373	3.457	5.809
834b	2.774	3.430	3.402	6.083	3.782	3.887	6.285	3.469	3.557	6.539
834c	2.762	3.481	3.540	5.968	3.851	3.715	5.830	3.638	3.578	6.698
838a	2.862	3.894	3.823	6.276	3.873	3.946	6.337	3.865	3.830	6.105
838b	2.864	3.898	3.793	6.379	3.846	3.918	6.076	3.791	3.860	6.011
838c	2.849	3.759	3.862	6.191	3.827	3.793	6.014	3.851	3.851	6.360
850a	2.947	4.271	4.271	6.904	4.366	4.276	6.690	4.289	4.379	6.820
850b	2.985	4.289	4.245	6.710	4.318	4.230	6.880	4.339	4.165	6.718
850c	2.962	4.121	4.247	6.824	4.230	4.145	7.025	4.160	4.119	6.710

Values are expressed in km s^{-1}. Symbol (+) means a measurement for the same direction, with the orientation of S-wave polarization plane perpendicular to the preceding 'zero' measurement. Numbers, which accompany symbols in the table heading, indicate measurements for directions 1–1', 2–2' and 3–3'. Symbol (−) indicates no data.

(a)　　　　　　　　**(b)**

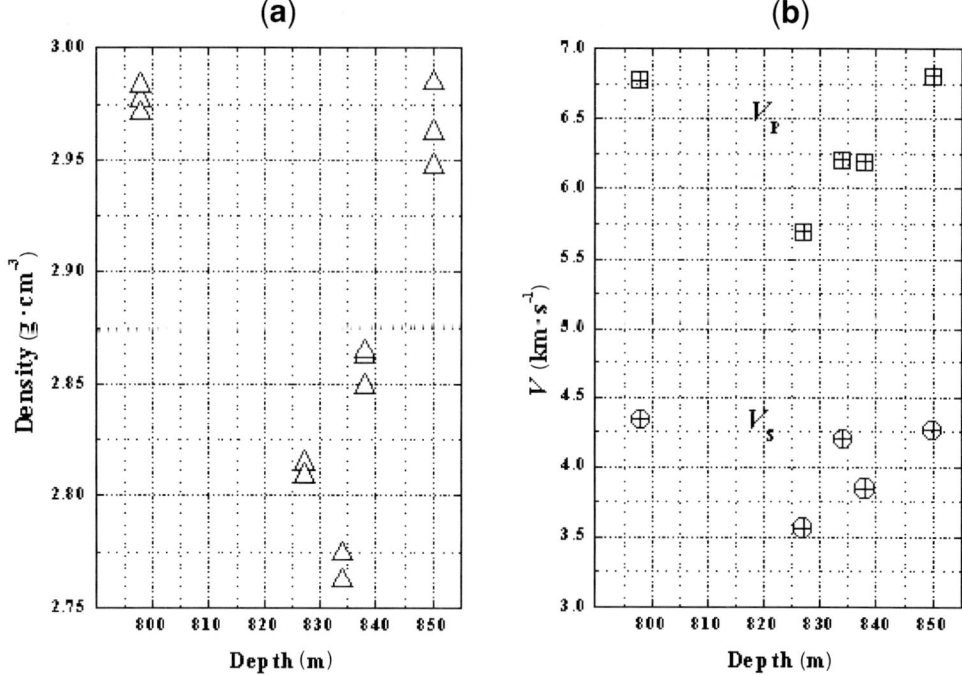

Fig. 5. Distribution of the density (**a**) and average values of elastic-wave velocities (**b**) in the samples of metavolcanics taken in the zone of schistosity at a depth of 798 to 850 m (groups 798, 827, 834, 838 and 850).

for directions 1–1′, 2–2′ and 3–3′, the velocity values were 3.380, 3.320 and 3.627 km s^{-1} respectively. In this group of rocks, V_P anisotropy is also observed.

The samples of the intensively schistose tuff group 834 (a, b and c) differ from other series in their clearly indicated schistosity. The cubic samples were cut in such a way that the direction 2–2′ is parallel to the schistosity plane, while the directions 1–1′ and 3–3′ intersect it at an angle of 45°. Low V_S values in the 1–1′ and 3–3′ directions characterize the samples; however, a higher V_S value in the 2–2′ direction along the schistosity is observed. The mean V_S values for the 1–1′ and 3–3′ directions are 3.438 and 3.459 km s^{-1} accordingly, while, for the 2–2′ direction, it is 3.765 km s^{-1}. This group of samples has the highest anisotropy factors of P-wave velocities.

As it was not sufficient to identify the maximal and minimal axes of V_S anisotropy (as well as their relation with the orientation of schistosity, based only on measurements for two perpendicular directions on each face of the cube), additional detailed studies were carried out. On each face of samples of group 834, the V_S velocities were measured with the rotation of the S-wave polarization plane from 0 to 360°, discre-

tely, every 22.5° (samples 834a and 834b) and every 45° (sample 834c). The dependence of the V_S data on the turning angle is shown in the polar diagrams (Fig. 6). The schistosity direction vectors shown on the diagrams indicate the projection of 3D vectors on to the 2D surface of S-velocity measurement.

The diagrams thus obtained illustrate the dependence of the V_S value on the orientation of the S-wave polarization plane. The anisotropic properties of schistose tuffs are most clearly shown on the diagrams of the 1–1′ and 3–3′ directions, and less clearly indicated on the diagram of the 2–2′ direction. As already mentioned, for the 2–2′ direction (along the schistosity) the highest mean values were registered, while, for the 1–1′ and 3–3′ directions the V_S values were low. The rotation of the polarization plane when S-waves are propagated along the foliation results in lower or even insignificant variations in V_S than in the case of propagation at a 45° angle to the schistosity plane. The comparison of the diagram curves illustrates their similarity for both directions 1–1′ and 3–3′.

The analysis of first arrival amplitudes of the S-mode was made simultaneously with the examination of the V_S value. The amplitude

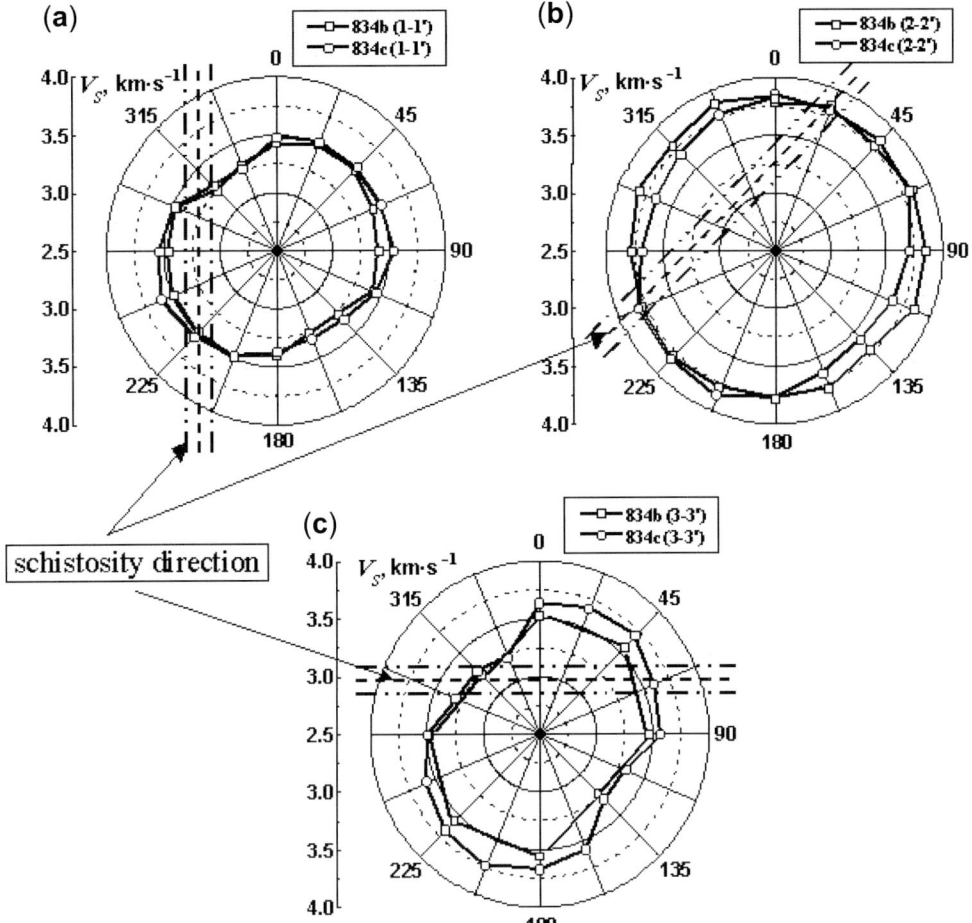

Fig. 6. Polar diagrams of V_S in two samples from group 834 with the application of plane-polarized waves: **(a)** and **(c)** – directions 1–1′ and 3–3′ of wave propagation at a 45° angle to schistosity; **(b)** – direction 2–2′ of wave propagation along schistosity.

values are characterized by periodicity and depend on the orientation of the polarization plane. In some cases, their values may change by an order of magnitude. Bipolar variation in amplitude values supports the assumption of the presence of two orthogonal systems of microcracks in these rocks. Moreover, one of the systems of microcracks should be parallel to the schistosity, while the other (Riedel microshears), which is more developed, intersects it at an angle of nearly 90°. This assumption is supported by the orientation of the maximal axes of the V_S and amplitude diagrams (Fig. 7). The numerical values of V_S for samples 834b and 834c, dependent on the rotation (azimuth) angle of S-wave polarization plane and the direction of wave propagation, are given in Table 4.

The samples from the weakly schistose tuff group 838 (a, b and c) actually show isotropic V_S velocities. Mean V_S values for each of the three directions are in close agreement (3.850, 3.848 and 3.835 km s^{-1}) and the deviations correlate with admissible measurement error. The values of V_S are in the middle of the range of the whole series of investigated rocks. The values of P-wave velocity belong to the same intermediate position in relation to the whole series.

In the virtually massive tuff group 850 (a, b and c), minimal scattering of V_S mean values for directions 1–1′, 2–2′ and 3–3′ (4.227, 4.304 and 4.242 km s^{-1}) may indicate slight V_S anisotropy (up to 1.8%). Similar results were obtained for P-wave velocities. The rock is

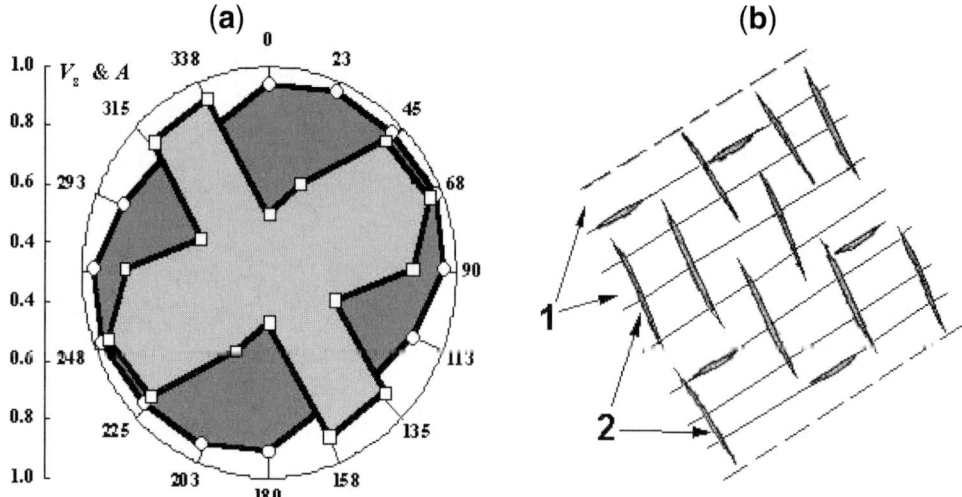

Fig. 7. Typical polar diagram of a schistose sample: (**a**) plane-polarized S-wave normalized velocities (grey pattern and open circles) and amplitudes (light-grey pattern and open squares); (**b**) corresponding texture with schistose layers (1) and Riedel microshears (2) perpendicular to schistosity.

characterized by high density-values and the highest values of P-wave velocities. The values of S-wave velocity are close to the maximal values for all investigated samples.

For one sample from each group, the effect of compression on P-wave velocities was studied. Experiments were performed at stress levels up to 5 MPa. Figure 8 shows the effect of stresses on sample 827b, which has relatively high V_P

anisotropy. It was found that under the effect of uniaxial stress the anisotropy remains as it was, but the sample underwent elastic deformation, with the recovery of its properties following resetting to the initial state.

Thus, massive and strain-free varieties of metavolcanics are characterized by isotropic elastic properties, while the maximum aniso-tropy is typical of intensively strained schistose

Table 4. *Azimuthal distribution of S-wave velocities in samples 834b and 834c*

Angle	834b			834c		
	V_{S^1}	V_{S^2}	V_{S^3}	V_{S^1}	V_{S^2}	V_{S^3}
0	3.430	3.782	3.557	3.481	3.851	3.638
22	3.487	3.851	—	3.511	3.816	3.670
45	3.487	3.782	3.469	3.511	3.851	3.702
67	3.430	3.846	—	3.511	3.811	3.607
90	3.402	3.851	3.229	3.540	3.715	3.576
112	3.430	3.851	—	3.452	3.650	3.323
135	3.270	3.715	3.557	3.342	3.588	3.297
158	3.270	3.782	—	3.315	3.650	3.576
180	3.402	3.782	3.557	3.369	3.782	3.670
203	3.487	3.782	—	3.481	3.851	3.735
225	3.487	3.782	3.469	3.540	3.816	3.670
248	3.487	3.782	—	3.601	3.816	3.576
270	3.459	3.782	3.229	3.540	3.682	3.488
293	3.459	3.816	—	3.481	3.650	3.297
315	3.224	3.782	3.527	3.289	3.682	3.271
337	3.295	3.887	—	3.264	3.782	3.221
360	3.430	3.715	3.557	3.481	3.816	3.638

Values are expressed in km s^{-1}. Numbers indicate measurements for directions 1–1′, 2–2′ and 3–3′ accordingly. Symbol (—) indicates no data.

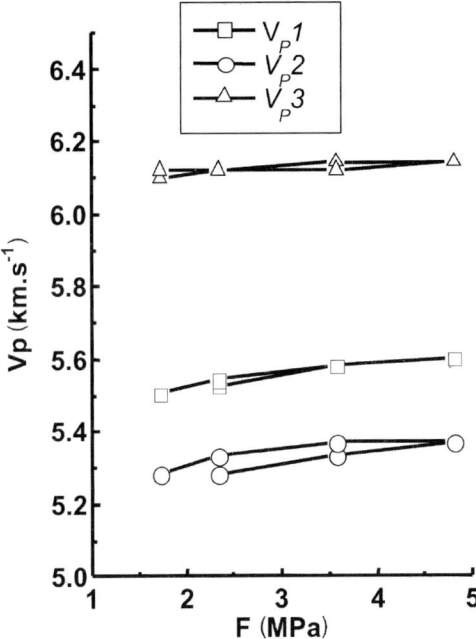

Fig. 8. Variation of the V_P velocity for three orthogonal directions (1, 2 and 3) in sample 827b, under the effect of uniaxial stress (loading and unloading).

varieties. In the latter rocks, anisotropy of properties is determined by systems of cracks, one of which is parallel to the plane of schistosity, while the second, which is more developed, intersects this plane at a 90° angle. As has been stated above, this system is formed by contiguous Riedel microshears, allowing higher values of permeability in the direction perpendicular to the plane of schistosity.

Acoustic emission data

Acoustic emission (AE) is a phenomenon of ultrasonic or acoustic wave emission occurring during fracture formation, without finite destruction of the material (Jones *et al.* 1997; Zietlow & Labuz 1998). Elastic waves from acoustic emission are characterized by the frequency of emitted waves (F) with activity A (the number of events per time unit), intensity I (energy characteristic of a separate event) and some other particular features. The oscillation frequency of a wave depends on the degree of disintegration; elastic wave velocity in a given material; type of crack (shear or tensile); rock texture; and size of mineral grains, etc.

   The samples from groups 827, 834 and 838 were tested in order to obtain AE data at different strain intensities, including strongly (No. 834) to weakly (No. 827 and No. 838) schistose tuffs. The first two are cubic, and the last one is a cylinder. To remove moisture from open pores, the samples were initially rinsed with ethyl alcohol and dried at 50°C for three hours. A schematic representation of the device used for AE investigation is shown in Figure 9. The acoustic emission detector (1 MHz resonance frequency) is placed on the upper rod (acoustic duct), which is in contact with the surface of the tested sample. Satisfactory acoustic contact is provided by the compression of ground-in surfaces of the sample and the rod. A thermocouple was placed on the sample surface. The heating rate during the test was $0.06°C \, s^{-1}$, and with this

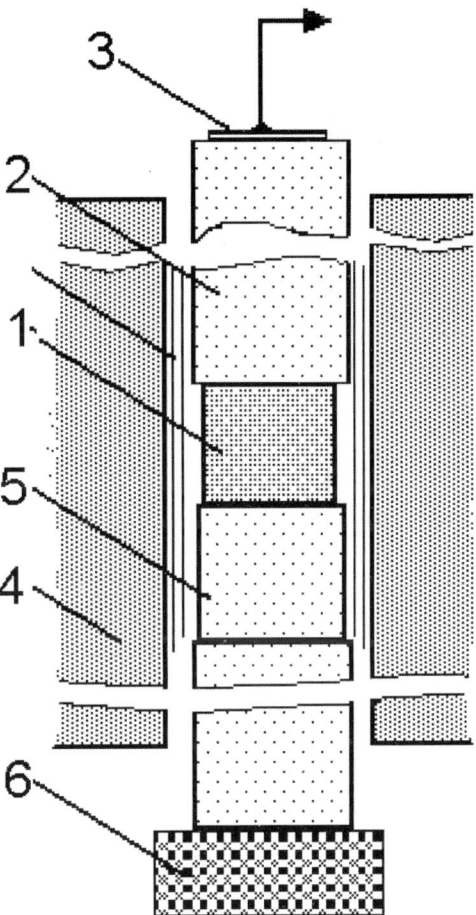

Fig. 9. Schematic diagram of the device for the measurement of acoustic emission in rock samples at high temperatures: 1, sample; 2, quartz acoustic duct; 3, acoustic emission detector; 4, heater; 5, quartz temperature damper; 6, acoustic buffer.

sample size and the heat conductivity values of this type of rock being studied it managed to uniformly heat the whole volume. Moreover, due to the application of quartz glass rods at the top and bottom of the sample, as well as a special heat buffer, the temperature gradient was minimized ($0.2°C\ mm^{-1}$).

A set of equipment was used for the detection of AE. The set included a calibrated broadband sensor and a signal recorder, which is connected to a PC through a special interface card. With original software, the activity of AE was recorded and stored in the PC every 10 seconds. Using high- and low-frequency filters, the effect of industrial noise was eliminated. An acoustic buffer was also used for the same

purpose. Elastic pulses with amplitudes below the established threshold (0.1 μm) were assigned as insignificant, and they were not recorded. Acoustic events were short impulse signals with an approximate duration of about 2 ms and from 0.4 to 2 MHz frequencies. The temp-erature rise to 300°C was provided over 4600 seconds, and after that the temperature was decreased.

During the experiments, no strong acoustic signals were detected: suggesting that intensive fracturing did not occur with temperature rise and fall, The AE activity may be characterized as weak for all investigated samples, because the temperature intervals, in which numerous events could take place within a short time

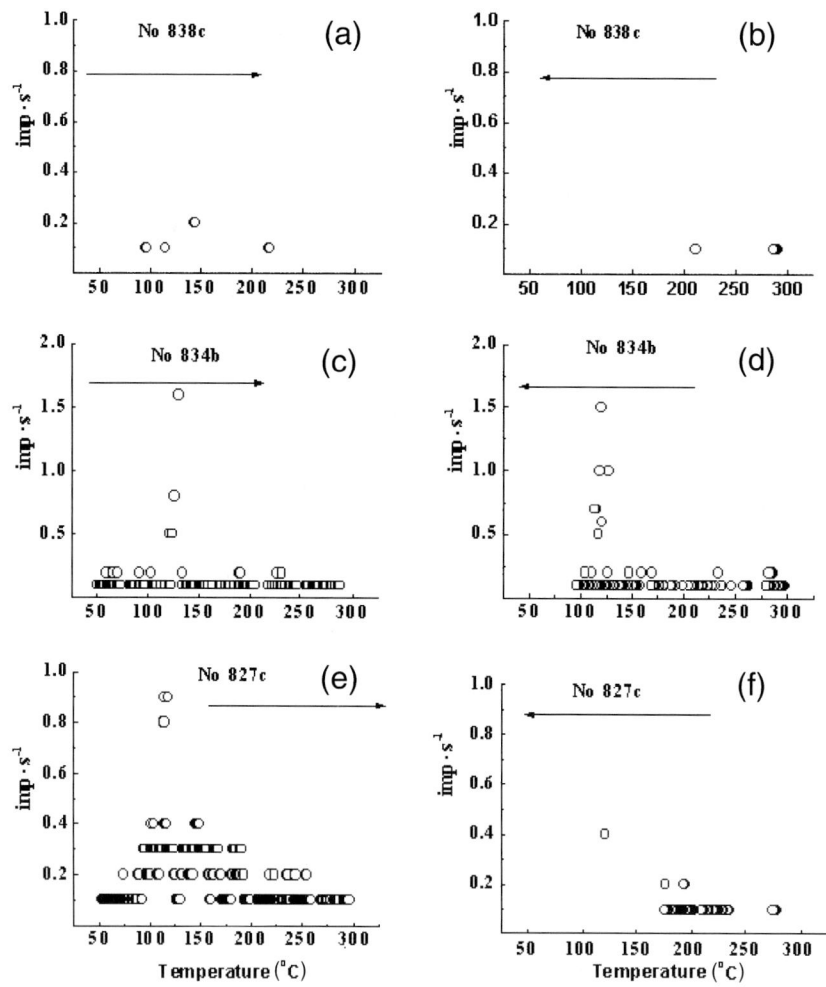

Fig. 10. Acoustic emission ($imp\ s^{-1}$) in samples 838c, 834c and 827c with the application of heat. The arrows on the diagrams indicate increases (a, c, e) and decreases (b, d, f) in temperature.

period, were not observed. The activity value did not exceed two events per second. It supports the absence of local concentrations of stresses, which may result in intensive fracturing.

However, a temporary increase of activity was observed at temperatures within the 120–150°C range, and a reduction in activity with the temperature reaching nearly 180°C (Fig. 10). With further heating up to 300°C, the activity lowers and stabilizes. This phenomenon could be explained by the presence of fine-grained mineral aggregates, in which temperature stresses in crystals of various minerals relax evenly, while, due to the presence of 'soft' interlayers, interganular contacts contribute to the formation of more uniform stress fields over the entire volume. On the other hand, the relative increase in activity can be explained by the release of intergranular water, as well as the inter-layer water of aluminosilicate minerals such as sericite, chlorite and epidote. With the temperature fall, the diagram of acoustic emission was actually the same as it was with the temperature rise, i.e. no meaningful fracturing was observed.

Thus, these tests show that the frequency, activity and intensity of acoustic events depend on the mineral composition, texture and density of metavolcanics. With the application of heat, the highest level of acoustic emission is observed, and therefore, the extensive of transformation of the pore space structure is expected for the intensively strained varieties of metavolcanics in the zones of schistosity.

Mechanical properties

Investigation of the mechanical properties of massive and schistose tuffs of andesite–basaltic porphyrite were performed in standard conditions using cylindrical samples from groups 827, 834 and 838, with a 1 : 1 height to diameter ratio (with a deviation of at most $\pm 5\%$). The samples had a diameter of 32 ± 1 mm; a height of 32 ± 2 mm; and their end surfaces were ground down to be parallel to each other, to within 0.05 mm. The perpendicular distance from the end-wall to the cylinder–generator lines was at most 0.05 mm.

The main aim was to identify the deformation behaviour of rocks subjected to oriented stress effects. It was found that with an equal rate of compression (0.15 MPa/h) the mechanism of deformational transformations at consecutive stages of stress (55–60, 75 and 100% of the $\sigma_{1_{max}}$ ultimate strength) varied substantially and depended upon their texture and the mineral composition of detritus and cementing materials, as well as upon the orientation of compressive

stresses in relation to the texture. In general, the stages of elastic deformation, deformational hardening, plastic deformation and rock failure were observed from the stress–strain curves. The greatest plastic deformation was found in the schistose varieties of group 834, with maximum compressive stresses oriented orthogonal to schistosity rather than parallel to schistosity.

Analysis of the results shows that under natural conditions the lowest mechanical strength is typical of rocks in the zones of schistosity. Where these zones are affected by tectonism, irreversible brittle deformation will appear, and this will result in a dramatic increase in permeability. The situation when stress is oriented parallel to the schistosity is most critical. Therefore it is important to undertake detailed research into the modern stress–strained state of the geological environment of the PA Mayak area, and to identify the mechanical properties of rocks and displacements in the zones of dislocation.

Conclusions

Relationships were found between the intensity of deformational transformations, the mineral–chemical and the petrophysical characteristics of the rocks in the PA Mayak area. Investigations into the nature of the variation of these characteristics under P–T conditions simulating those in the vicinity of a proposed underground HLW facility were carried out. Transformations of the original metavolcanics during regional greenschist metamorphism have resulted in the homogenization of rock textures, petrochemical and petrophysical characteristics.

Outside the zones of intense deformation, the intact rock shows high density, elasticity and strength, as well as minimal values of effective porosity and permeability. At mean values of permeability (n · E−19 m^2) and effective porosity (0.26%) and using a mean value for the hydrodynamic gradient of underground water flow of $\sim$15 Pa m^{-1}, the rate of filtration, according to the Darcy's law, will be low – amounting to only $\sim$5 × 10^{-5} m/a. Therefore, the transport of radionuclides by underground waters at the depth of an underground facility (0.5–1.0 km) will be focused along the discontinuities, occurring as linear zones of schistosity, mylonitization and brecciation with a widespread development of albite, chlorite, sericite, carbonate and Fe and Ti oxides and hydroxides. Under the effects of directed stress, the rocks behave as elastic–plastic bodies, with the consecutive formation of systems of tensile and shear fractures, but the schistose rocks exhibit the most complex deformation behaviour.

In conditions approximating the area around an underground disposal system for heat-generating HLW, the rocks undergo thermode-compaction, which is accompanied by the restructuring of the original pore space; the formation of neoformed microheterogeneities; and an increase in solid matrix permeability. However, as a result of alterations, the neo-formed crack–pore space is filled with minerals (epidote, chlorite, carbonate), which demonstrate a high sorptive capacity with respect to Cs and Sr radionuclides.

As radionuclides are characterized by a selective tendency for sorption on mineral aggregates of a certain composition (for example, Cs to biotite, chlorite, Sr to carbonates, epidote), the widespread occurrence of these mineral-concentrators of radionuclides in zones of blast-esis, cataclasis and brecciation must be taken into account when predicting the long-term safety of an underground repository. At the same time, due account must be given to the fact that systems of open macro- and microcracks in schistosity zones may form channels of underground water filtration, and can become a factor contributing to the reduction of the retaining capacity of rocks in relation to radionuclides.

For the assessment of the consequences of hydro-thermomechanical processes, it is necessary not only to proceed with the laboratory research under $P-T$ conditions imitating the state of rocks in the vicinity of the prospective underground repository, but also to work out a programme of *in situ* thermophysical and hydro-thermomechanical tests. For the identification of the transport characteristics of the dislocation zones, it is necessary to undertake *in situ* investigations such as electrometry, resistivity, gamma-logs, radar and differential tomography in combination with tracer tests. These data will be used as the basis for the design of the underground repository.

We are grateful to Dr E. G. Drozhko and Dr S. I. Rovny from PA Mayak for help in undertaking the fieldwork. The authors also thank N. N. Grigorieva for translating the paper, as well as two anonymous reviewers who made useful comments that have helped us clarify and improve the text.

References

CHAPMAN, N. A. & MCKINLEY, I. G. 1987. *The Geological Disposal of Nuclear Waste*. John Wiley & Sons, Chichester.

DROZHKO, E. G., GLAGOLENKO, Y. U., MOKROV, Y. G. *et al.* 1997. Joint Russian–American hydrogeological–geochemical studies of the Karachay–Mishelyak system, South Urals, Russia. *Environmental Geology*, **24**, 216–227.

FREDRICH, J. T. & WONG, T.-F. 1986. Micromechanics of thermally induced cracking in three crustal rocks. *Journal of Geophysical Research*, **91**, 12 743–12 764.

GLAGOLENKO, YU. V., DZEKUN, E. G., DROZHKO, E. G. ET AL. 1996. [The strategy for radioactive waste management at the PO Mayak.] *Voprosy radiatsionnoi bezopasnosti*, **1**, 3–10 (in Russian).

GORBATSEVICH, F. F. 1999. Acoustic polarization method for determining elastic symmetry and constants of anisotropy in solid media, *Ultrasonics*, **37**, 309–319.

JONES, C., KEANEY, G., MEREDITH, P. G. & MURRELL, S. A. 1997. Acoustic emission and fluid permeability measurements on thermally cracked rocks. *Physics and Chemistry of the Earth*, **22** (1–2), 13–17.

LAVEROV, N. P., VELICHKIN, V. I., OMELIANENKO, B. I. *et al.* 2000. [New approaches for underground disposal of high-level wastes in Russia.] *Geoekologiya*, 1, 3–12 (in Russian).

NERETNIEKS, I. 1980. Diffusion of the rock matrix: an important factor in radionuclide retardation? *Journal of Geophysical Research*, **85**, 4379–4397.

PETROV, V. A. & POLUEKTOV, V. V. 1999. On the issue of retentivity of metavolcanites at the PA 'Mayak' territory (Russia): Mineral–chemical and structural–petrophysical aspects. *Proceedings of the 7th International Conference on Radioactive Waste Management and Environmental Remediation, ICEM' 99, Nagoya, Japan* (CD-ROM).

PETROV, V. A., POLUEKTOV, V. V. & ZHARIKOV, A. V. 1998. Peculiarities of fluid transport in the metavolcanic strata at the PA Mayak territory, Russia. *Proceedings of the 8th International Conference on HLRW Management*, Las Vegas, Nevada, 68–70.

PETROV, V. A., VELICHKIN, V. I., POLUEKTOV, V. V. & TARASOV, N. N. 2000. Integration of geological, geomechanical and petrophysical data for the preliminary prediction of the HLW disposal system behaviour at the Southern Urals, Russia. *Proceedings of the International Conference on Radioactive Waste Disposal*, Berlin, Germany, 218–223.

PETROV, V. A., POLUEKTOV, V. V., ZHARIKOV, A. V. *et al.* 2005. Microstructure, filtration, elastic and thermal properties of granite rock samples: implications for HLW disposal. *In*: HARVEY, P. K., BREWER, T. S., PEZARD, P. A. & PETROV, V. A. (eds). *Petrophysical Properties of Crystalline Rocks*. Geological Society, London, Special Publications, **240**, 237–253.

SAVAGE, D. 1995. *The Scientific and Regulatory Basis for the Geological Disposal of Radioactive Waste*. John Wiley & Sons, Chichester.

SOLODOV, I. N. & ZOTOV, A. V. 1995. Geochemistry and flow structure of contaminated subsurface waters in fissured bedrock of the Lake Karachay area, Southern Urals, Russia. *Proceedings of the 5th International Conference on Radioactive Waste*

Management and Environmental Remediation, Berlin, Germany, 1321–1328.

STAROSTIN, V. I. 1979. [*Structural–Petrophysical Analysis of Endogenous Ore Fields*.] Nedra Publishers, Moscow (in Russian).

SYSOEV, A. N., PETROV, V. A., IVANOV, I. A. *et al.* 1999. Geomorphological and tectonophysical approach to the identification of sites for radioactive waste disposal at the PA Mayak territory, Russia. *Proceedings of the 7th International Conference on Radioactive Waste Management and Environmental Remediation, ICEM'99, Nagoya, Japan* (CD-ROM).

VELICHKIN, V. I., OMELIANENKO, B. I., TARASOV, N. N. *et al.* 1997. Geological aspects of solidified high-level radioactive waste (HLW) disposal problem at PA 'Mayak', Chelyabinsk district, Russia. *Proceedings of the 6th International Con-*

ference on Radioactive Waste Management and Environmental Remediation, Singapore, 399–402.

VELICHKIN, V. I., PETROV, V. A., TARASOV, N. N. *et al.* 1995. Appraisal of the physical and dynamic state of the Mayak operations geological environment with a view to underground radwaste disposal. *Proceedings of the 5th International Conference on Radioactive Waste Management*, Berlin, Germany, 823–826.

ZARAISKY, G. P. 1994. [*Experimental Modelling of Single Crack Sealing With Water Filtration on Porphyrite*.] Articles on Physical–Chemical Petrology, Moscow, Nauka Publishers, 139–165 (in Russian).

ZIETLOW, W. K. & LABUZ, J. F. 1998. Measurement of the intrinsic process zone in rock using acoustic emission. *International Journal of Rock Mechanics and Mining Science*, **35** (3), 291–299.

Non-linearity in multidirectional P-wave velocity: confining pressure behaviour based on real 3D laboratory measurements, and its mathematical approximation

R. PŘIKRYL[1], KAREL KLÍMA[2], T. LOKAJÍČEK[2] & Z. PROS[2]

[1]Institute of Geochemistry, Mineralogy and Mineral Resources, Faculty of Science, Charles University in Prague, Albertov 6, 128 43 Prague, Czech Republic
(e-mail: prikryl@natur.cuni.cz)

[2]Institute of Rock Structure and Mechanics, Academy of Sciences of the Czech Republic, V Holešovičkách 41, 180 00 Prague, Czech Republic

Abstract: Experimental laboratory measurements of P-wave velocity confirm the superposition of linearity over non-linearity by a progressive increase in confining pressure. The increase in confining pressure diminishes the influence of microcracks that are partly or totally closed. At a certain stress level, the trend of P-wave velocity with applied confining pressure approaches that of a solid without cracks, and a linear increase in elastic wave velocity occurs under high confinement.

Several studies have focused on the problem of mathematical approximation of this phenomenon (Carlson & Gangi 1985; Wepfer & Christensen 1991; Greenfield & Graham 1996; Meglis *et al.* 1996). Although successful within a certain degree of error, they provide neither a multidirectional solution nor the comparison of results with rock fabric. In this study, an analytical relation was applied to describe the P-wave velocity–confining pressure behaviour of quasi-isotropic rocks (granites) and their anisotropic equivalents (orthogneisses). Two parameters of this relationship reflect the elastic properties of the rock matrix, and two others are related to the presence of microcracks, their density and genesis. The results of a mathematical approximation of the P-wave velocity–confining pressure behaviour show a favourable correlation to the measured data-sets. Comparison of individual fitted parameters with the rock fabric provides an improved understanding of the material's mechanical behaviour.

Any isotropic homogeneous solid subjected to external loading should exhibit a linear response over the applied pressure in any direction, as expressed by Hooke's law. The presence of any fabric defect (Lama & Vutukuri 1978) significantly modifies the linear elasticity. Natural rocks do not present an ideal model, being more likely to be anisotropic due to their intrinsic properties. Firstly, most rocks contain microcracks (planar defects) formed during the formation of the rock – e.g. during submagmatic stages (Bouchez *et al.* 1992) or as a result of brittle deformation (Simmons & Richter 1976; Kranz 1983) and/or weathering (Winkler 1994). Secondly, the isotropic elastic behaviour of natural rocks is rarely observed, because the mineral skeleton fabric is anisotropic in most cases. Both phenomena shift the mechanical response of rocks to non-linear behaviour.

From the systematic work of Birch (1960, 1961) on the effect of hydrostatic pressure on elastic wave velocities, the experimental observation noted a characteristic P-wave velocity–confining pressure curve that is explained, based on the progressive closure of microcracks superposed by elastic response of rock matrix. The so-called 'closure pressure' (here expressed as microcrack closing pressure) occurs at the point where the initial non-linear curve dramatically changes its shape and slope (e.g. Christensen 1974; Greenfield & Graham 1996). When microcrack closure pressure is reached, all cracks are assumed to be closed, and the P-wave velocity–confining pressure curve shows a linear or quasi-linear trend with moderate slope.

A number of studies have attempted to describe mathematically the increase of P-wave velocity with applied confining pressure in rocks containing defects, i.e. cracks and pores (Warren & Tiernan 1981; Stesky 1985). Four-parameter functions form the basis of various models: e.g. physical models of crack closure

From: HARVEY, P. K., BREWER, T. S., PEZARD, P. A. & PETROV, V. A. (eds) 2005. *Petrophysical Properties of Crystalline Rocks*. Geological Society, London, Special Publications, **240**, 323–334.
0305-8719/05/$15.00 © The Geological Society of London 2005.

presented by Carlson & Gangi (1985), Greenfield & Graham (1996) or Meglis *et al.* (1996), the purely empirical relation used by Wepfer & Christensen (1991), and the analytical formulae evaluated by Pros *et al.* (1998*a*). In those studies, the mathematical description of P-wave velocity–confining pressure behaviour was limited to one-dimensional dependence. Evidently, the previously proposed models have assumed the homogeneity and isotropy of a material, properties that were not confirmed by multidirectional measurements of P-wave velocity distribution in rocks (e.g. Birch 1961; Thill *et al.* 1969; Babuška & Pros 1984; Siegesmund & Dahms 1994; Zang *et al.* 1996).

This study attempts to present a mathematical approximation of the P-wave velocity–confining pressure behaviour and to derive fitting parameters reflecting material behaviour under confining compression in many directions. The mathematical function is based on the evaluation of the intrinsic properties of the solid matrix, as well as on the presence of low-aspect-ratio fabric defects (microcracks) that are closed due to the applied confining pressure. The mathematical approximation of P-wave velocity is run in the three-dimensional regular net conform with directions of measured P-wave velocity. The resulting spatial distribution of individual fitting parameters is compared with the experimental P-wave velocity data determined on spherical rock samples, and with the measured rock fabric parameters. This approach helps in interpreting processes leading to the non-linearity of P-wave velocity–confining pressure behaviour, and to evaluate the contribution of individual rock fabric parameters. The comparison of analytical and experimental data provides high accuracy of the employed mathematical approximation. This approximation seems to be valid for igneous and metamorphic rocks containing planar defects. Materials with equidimensional pores were not studied. The method presented is not limited to isotropic or quasi-isotropic materials, as shown by an example involving highly anisotropic rocks.

Experimentally derived P-wave velocities

Technique

Laboratory P-wave velocity measurement was carried out using a pulse transmission technique employing two electro-acoustic transducers (Hughes *et al.* 1949) later modified for rock spherical samples 50 mm in diameter (Pros *et al.* 1969) that are investigated by unique

apparatus at different levels of confining pressure (Pros *et al.* 1998*a* and references therein).

The equipment consists of a pressure vessel connected with a two-step pressure generator; a sample positioning unit equipped with a pair of ultrasonic piezoceramic transducers (transmitter and receiver); a device for generating ultrasonic pulses, and a travel-time measurement and data acquisition unit.

The technique can measure P-wave velocity in any selected direction (except the area near the vertical axis of rotation) with the same accuracy. In practice, the measurement is conducted on the net dividing the sphere in steps of $15°$ defining 132 independent measuring directions (Fig. 1). The measurement of P-wave velocity starts at atmospheric pressure conditions, and continues through several levels of hydrostatic pressure (commonly 10 or 20, 50, 100, 200 MPa) up to 400 MPa, which is the limit pressure for the apparatus. The travel-times of ultrasonic signals are measured for all 132 directions at each confining pressure level.

Experimental results

The data obtained are analysed using special software packages, allowing the selection of parameters on time of signal arrival and the value of the first amplitude (Pros *et al.* 1998*a*). From these data, P-wave velocities were computed.

The P-wave velocity data from all independent 132 directions of given pressure level are projected from the surface of the sphere on to the lower hemisphere in Lambert's (equal-area) polar projection (Fig. 1). The region of a certain P-wave velocity interval is then enclosed by an isoline on the plot. The resulting plot provides knowledge of the spatial distribution of P-wave velocities in the sample. The procedure for computation of P-wave velocity isolines is described by Pšenčík (1975).

Samples

Rock samples for the laboratory measurements of P-wave velocities were extracted in several granitoid plutons and orthogneiss bodies in the Bohemian Massif (Czech Republic). The mineralogical composition (based on quantitative analysis using image processing of thin sections – see Přikryl 2001) of those rocks corresponds to that of true granites. Rock-forming minerals in studied granites do not exhibit any significant shape or crystallographic preferred orientation. The orthogneiss studied, on the other hand, is a fine-grained rock exhibiting strong shape- and crystallographic-preferred orientation

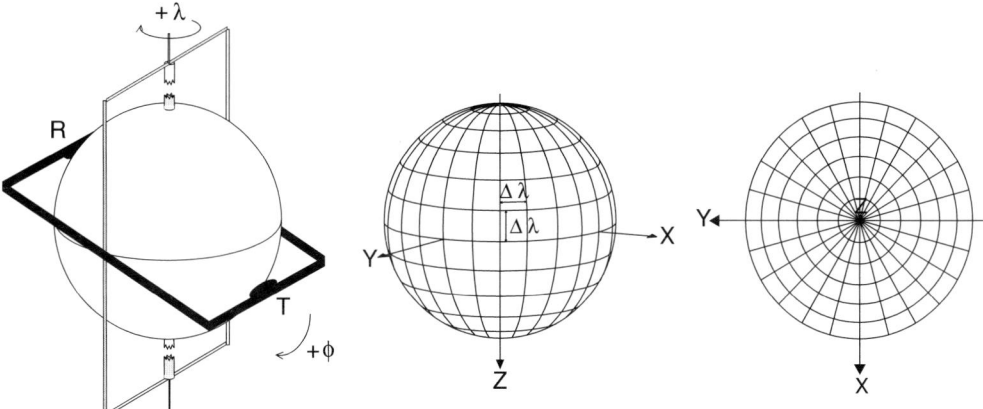

Fig. 1. Kinematic diagram (on the left), showing the rotation of a spherical sample along its vertical axis (angle λ), and the rotation of the frame holding the transducer (T) and the receiver (R). These can rotate along indicated angle ϕ. The diagram in the centre shows the measuring grid on the surface of the sphere, and the diagram on the right depicts a grid of measured points projected on to the lower hemisphere (in Lambert's projection).

(e.g. Přikryl et al. 1996). Details of the rock composition are given in Table 1.

Mathematical approximation

Function

The model of P-wave velocity–confining pressure behaviour employed in this study presumes the superposition of two basic phenomena of material behaviour. The intrinsic properties of a rock matrix depend on the elasticity of the constituent minerals. The general assumption on the linearity of rocks' elastic moduli over applied hydrostatic pressure range is generally assumed (Birch 1938) and applied to estimate elastic wave velocities with confining pressure in rocks. The presence of microcracks is the second important factor influencing P-wave velocity–confining pressure behaviour. The empirical consideration assumes the decreasing effect of low-aspect-ratio microcracks on elastic wave velocity with increasing pressure. This behaviour has been described by an inverse exponential function and applied to mathematical models (e.g. Eberhart-Phillips et al. 1989; Greenfield & Graham 1996).

Our mathematical approximation of P-wave velocity dependence on applied confining pressure is based on considerations mentioned above. The P-wave velocity–confining pressure relationship is expressed by an analytical four-parameter P-wave velocity–confining pressure equation (adopted and modified after Pros et al. 1998a):

$$V_P = V_0 + k_v \times P - V_{\text{dif}} \times 10^{(-P/P_0)} \quad (1)$$

where V_P is velocity of the P-wave at confining pressure P, V_0 is the P-wave velocity in the mineral skeleton of an ideal sample without microcracks, extrapolated to the atmospheric pressure level from the high confining pressure interval ($c.200$–400 MPa), k_v is the increasing P-wave velocity coefficient of such a sample (the numerical expression of the slope of a linear part of the P-wave velocity–confining pressure behaviour); V_{dif} is the velocity difference between V_0 and the measured P-wave velocity at the atmospheric pressure level; and P_0 is the confining pressure at which the difference between the P-wave velocity of an ideal sample and a real sample decreases to 10% v_{dif} (Fig. 2).

Data processing

The equation proposed in the previous section can be applied for any desired direction if the multidirectional measurement of P-wave velocity is employed. In this study, the P-wave velocities were approximated in all 132 directions that coincide with directions in which the P-wave velocity was measured. P-wave velocity mathematical approximation is the sum of two functions for one direction. Both functions show an increase of elastic wave velocity with confining pressure. In such a case, the presence

Table 1. *Microfabric characteristics of studied samples (after Přikryl 2001)*

Sample	Microfabric	Minerals	Microstructures	Microdefects and alteration
RP1 Granite	Magmatic, heterogranular Average grain size: 0.13 mm	Quartz (36)	Equigranular grains with straightforward boundaries	Very low intragranular microcracking, low undulatory extinction
		Plagioclase (26)	Sub- to anhedral shapes	Low density of intragranular microcracks, 20% of grains' area sericitized
		K-feldspar (33)	Common anhedral shapes	Very low microcracking (almost crack-free)
		Micas (5)	Platy grains	Low cleavage cracks
G6 Granite	Magmatic, heterogranular Average grain size: 0.33 mm	Quartz (36)	Anhedral grains, lobate grain boundaries	High intragranular and multigrain microcracking, common undulatory extinction
		Plagioclase (32)	Subhedral porphyry grains, small grains show anhedral to subhedral shapes	Low intragranular cracking, 10% of grains' area sericitized
		K-feldspar (20)	Sub- to anhedral grains	Very low microcracking
		Micas (12)	Platy to irregular grains	Cleavage cracks
RP3 Orthogneiss	Linear, metamorphic, heterogranular Average grain size: 0.12 mm	Quartz (29)	Anhedral grains with lobate boundaries	Low microcracking, intense undulatory extinction
		Plagioclase (22)	Anhedral grains	Low microcracking, 5% of grains' area sericitized
		K-feldspar (34)	Anhedral grains	Intense fracturing of porphyry grains – healed cracks
		Micas (15)	Platy grains	Intense cleavage cracks

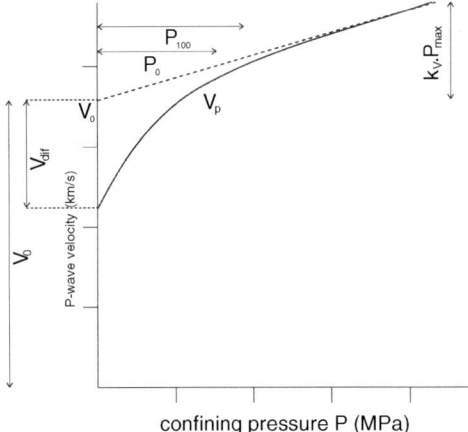

Fig. 2. Graphic presentation of P-wave velocity–confining pressure parameters based on equations (1) and (9). V_0 is the P-wave velocity in the mineral skeleton of an ideal sample without microcracks; k_v is the increasing P-wave velocity coefficient of such a sample; P is the confining pressure in MPa; v_{dif} is the velocity difference between v_0 and the measured P-wave velocity at the atmospheric pressure level; P_0 is the confining pressure at which v_{dif} decreases to 10%, and P_{100} is the confining pressure fixed to the 100 m s^{-1} P-wave velocity difference. The thick curve presents the idealized P-wave velocity–confining pressure behaviour.

of a random error in a primary (measured) data-set can significantly influence individual parameters of the P-wave velocity–confining pressure equation (compare equation 1). This is more pronounced if the P-wave velocity data are determined at a few confining pressure levels.

The suggested method for the laboratory measurement of P-wave velocity employs six confining pressure levels for computation of four parameters in each measured direction. The scatter of results in such a case is obvious. In order to diminish the influence of this factor and random errors, the approximated data were derived from the calculated P-wave velocities that were computed from elastic constants. These represent smoothed experimentally measured P-wave velocities.

Computation of elastic constants

The computation procedure (Klíma 1973) of elastic constants adopts a data-set of measured P-wave velocities.

A common anisotropic body is characterized by 21 independent elastic parameters. Propagation of elastic waves in anisotropic material

is characterized by the equation:

$$\Gamma_{ij} U_j = \rho v^2 U_i \qquad (i = 1, 2, 3) \qquad (2)$$

where ρ is density, v is phase velocity of elastic wave and Γ_{ij} is Cristofell's matrix tensor derived from the elastic constant:

$$\Gamma_{ij} = c_{ikjl} a_k a_l \qquad (3)$$

where c_{ikjl} is the elastic constant tensor and $a_k a_l$ are the directions of the normal of the wave front. The solution of those equations exists, if:

$$|\Gamma_{ij} - \delta_{ij} \rho v^2| = 0 \qquad (4)$$

This cubic equation determines the relationship between elastic constants and elastic wave velocities in the direction of the normal of the propagating wave front. The computation procedure of elastic constants from measured P-wave velocities is based on this equation.

The set of cubic equations and 21 unknown elastic constants in each direction can be solved numerically. The set is linearized by putting Γ_{ii} on the left, and the rest of the equation on the right:

$$\Gamma_{ii} = \rho v^2 + Z_i \qquad (5)$$

where Z_i is a correction based on elastic constants and on ρv^2. This equation is valid for all three body waves. The solution of this equation can be found by an iteration process if $Z_i \ll \rho v^2$. Low values of Z_i are ensured if the orientation of the co-ordinate system approaches the direction of displacement of body waves for a given direction. Choosing the directions of the co-ordinate system axes to be strictly parallel with the direction of displacement, and especially the x_1-axis parallel to the displacement of P-wave, then the following solution exists in a x_1 direction:

$$\Gamma_{11} = \rho V_P^2$$
$$\Gamma_{22} = \rho V_{S1}^2 \qquad (6)$$
$$\Gamma_{33} = \rho V_{S2}^2$$

Orientation of displacement vectors in common anisotropic media is unknown, and previous simplification is not valid. For P-waves in a weak anisotropic medium, the displacement is almost parallel to the direction of its propagation. The elastic constants can be computed using an iteration process. Transforming the co-ordinate system axis x_1 to be equal to the analysed direction, i.e. $\mathbf{a} = \begin{pmatrix} 1 \\ 0 \\ 0 \end{pmatrix}$ (eq. 3), the equation (4)

is given by:

$$*c_{11} = \rho V_P^2 + [(*c_{66} - \rho V_P^2)*c_{15}^2$$
$$+ (*c_{55} - \rho V_P^2)*c_{16}^2 - 2*c_{16}*c_{15}*c_{56}]/$$
$$[(*c_{66} - \rho V_P^2)(*c_{55} - \rho V_P^2) - *c_{56}^2] \qquad (7)$$

Going back to the initial co-ordinate system, $*c_{11}$ represents the linear combination of all elastic constants. Considering the elastic constants as unknown, then only 15 independent equations exist for more directions, because the elastic constant possesses the same coefficients, dependent on the direction of wave propagation, in six cases:

$$*c_{11} = a_1^4 c_{11} + 2a_1^2 a_2^2 (c_{12} + 2c_{66})$$
$$+ 2a_1^2 a_3^2 (c_{13} + 2c_{55})$$
$$+ 4a_1^2 a_2 a_3 (c_{14} + 2c_{56}) + 4a_1^3 a_3 c_{15}$$
$$+ 4a_1^3 a_2 c_{16} + 4a_2^4 c_{22}$$
$$+ 2a_2^2 a_3^2 (c_{23} + 2c_{44}) + 4a_2^3 a_3 c_{24}$$
$$+ 4a_2^2 a_1 a_3 (c_{25} + 2c_{46})$$
$$+ 4a_2^3 a_1 c_{26} + a_3^4 c_{33} + 4a_3^3 a_2 c_{34}$$
$$+ 4a_3^3 a_1 c_{35} + 4a_3^2 a_1 a_2 (c_{36} + 2c_{45}) \qquad (8)$$

This system of linear equations of P-waves is not suitable for determination of all 21 elastic constants. A further equation can be obtained through the knowledge of S-waves. In the case when S-waves are unknown, appropriate estimates must be adopted, because all 21 constants must be used for computation of the correction factor Z_i in each iteration step. The estimates are given by the ratio of given pairs of elastic constants like $c_{66} = k c_{12}$, where k is a unique constant for each pair. Conventionally, the ratio $k = 1.5$ is used. The correct values of elastic constants will be given for nine constants, for 12 constants the correct sums will be given and the values are estimates. Nevertheless, the P-wave velocities calculated from these constants are correct.

The P-wave velocities calculated from these elastic constants are smoothed in comparison with measured ones and with respect to the minimal anisotropy symmetry of material. Thus, most of the local anomalies and measured errors are neglected (Fig. 3). It is evident that calculated P-wave velocities are more convenient

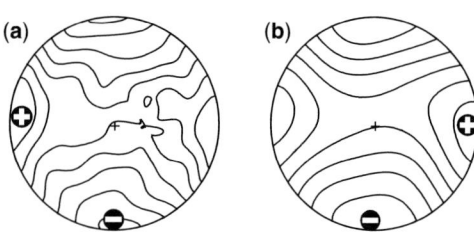

Fig. 3. Comparison of experimentally measured (**a**) and computed (**b**) P-wave velocities expressed in the form of P-wave velocity isolines. The data from 132 measured directions are projected on the lower-hemisphere equal-area projection, and the P-wave velocity isolines are then computed. Computed P-wave velocities (**b**) are processed according to the set of equations (2–8). Example of granite (sample G6), P-wave velocity measured at atmospheric pressure.

for the analysis of P-wave velocity–confining pressure behaviour. The P-wave velocity spatial distribution calculated with the aid of elastic constants has a similar configuration to that one determined experimentally. The only significant difference lies in the smoothed appearance of the calculated isolines.

Results and discussion

Spatial distribution of the fitting parameters

Detailed spatial analysis of the above-mentioned parameters was applied on granitic rocks exhibiting diverse fabric symmetry. The first group – granites – represent material where the influence of microcracks on the anisotropy of P-wave velocities prevails. This fact is deduced from the decrease in their degree of anisotropy (e.g. Pros *et al.* 1998*b*). The symmetry of their P-wave velocity spatial distribution (and rock fabric as well) ranges from transversal isotropic or orthorhombic, under atmospheric pressure conditions, to quasi-isotropic fabric symmetry under hydrostatic pressure conditions. The second group of rocks, orthogneisses, show an example of a material with a well-developed anisotropic fabric due to the presence of both shape- and crystallographic-preferred-orientation of minerals. The symmetry of their P-wave velocity spatial distribution is markedly orthorhombic at all confining pressure levels.

The V_0 parameter in granites exhibits a quasi-isotropic pattern (Fig. 4). The pattern of the slope of straight line k_V shows more pronounced orthorhombic pattern (e.g. sample G6, Fig. 4A) but can be also almost isotropic (sample RP1, Fig. 4B). The lowest velocity difference V_{dif} of

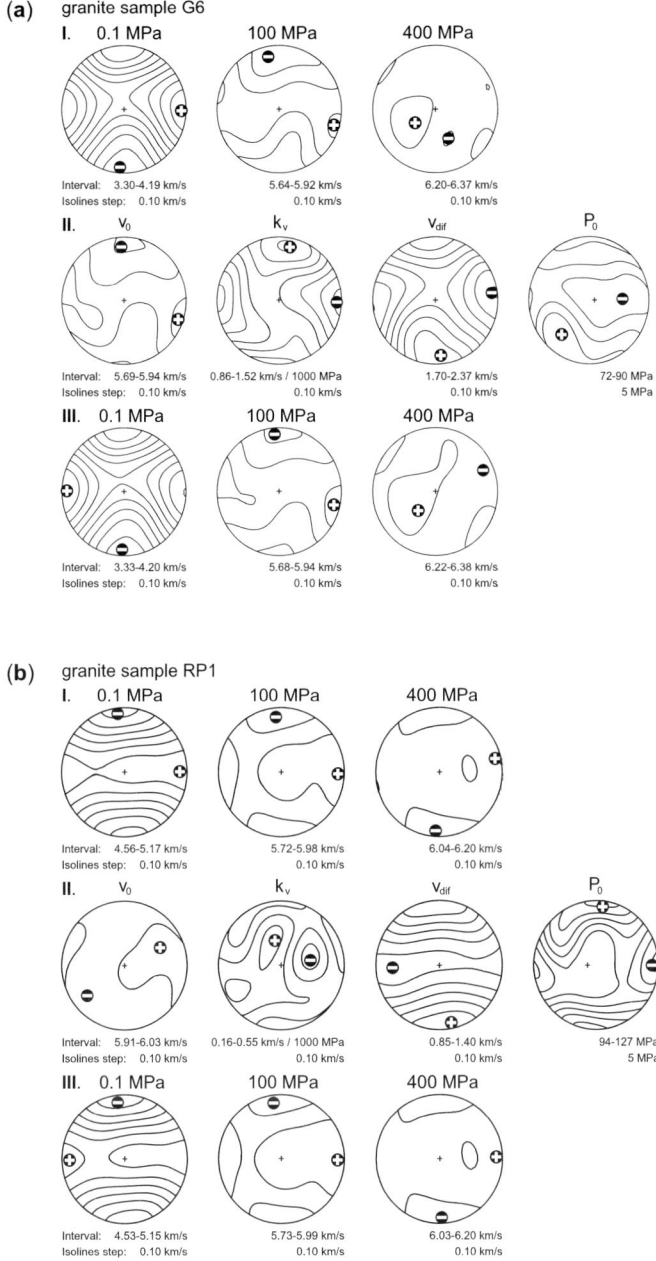

Fig. 4. Representative spatial distributions of P-wave velocities calculated with the help of elastic constants from experimentally measured data-sets (**I**), parameters derived by mathematical approximation of P-wave velocities (**II**) and P-wave velocity spatial distribution (**III**) calculated using the proposed equation (equation 1). Samples G6 (**a**) and RP1 (**b**) are granites, and sample RP3 (**c**) is orthogneiss. All data are projected on to the lower hemisphere in equal-area projection.

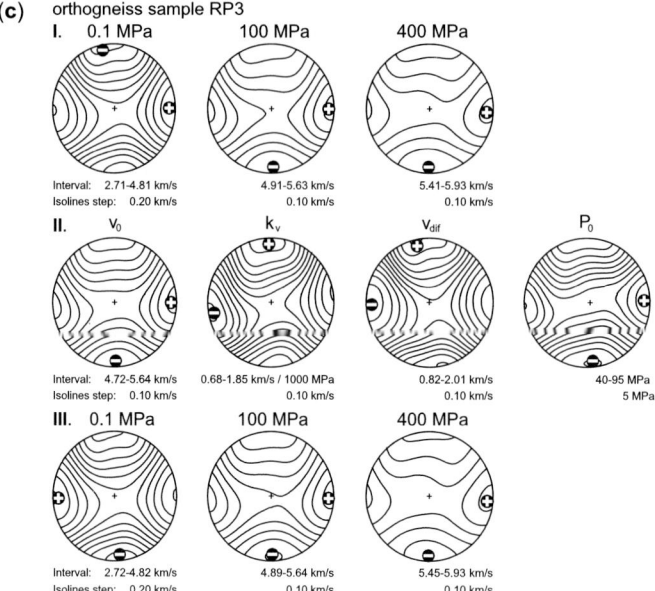

Fig. 4. Continued

all samples under study is in the direction of maximum measured P-wave velocity (V_{Pmax}), while the position of the maximum V_{dif} corresponds to the minimum measured P-wave velocity (V_{Pmin}). The minimum microcrack closing pressure P_0 coincides roughly with the V_{Pmax} direction and maximum microcrack closing pressure with V_{Pmin} (Fig. 4). This is probably due to the fact that longer cracks show more undulated faces, and thus they are closed at higher confining pressure than short cracks occurring in a V_{Pmax} direction.

The orthorhombic pattern of P-wave velocity–confining pressure fitted parameters faithfully reflects the anisotropic rock fabric of orthogneisses (e.g. sample RP3). The model sample with no microcracks still exhibits high anisotropy of P-wave velocity, V_0, and linear compressibility k_v at atmospheric pressure (Fig. 4c). In this rock, the maximum microcrack closing pressure P_0 coincides with V_{Pmax} direction and minimum P_0 with V_{Pmin} direction. In comparison with granites, the extremes of the P_0 fitting parameter exhibit reverse positions. This is due to the fact that most of the microcracks are of intra-granular nature in micas (cleavage cracks). These have very smooth faces and are easily closed at low confining pressures.

The validity of the calculated parameters was evaluated by calculation of P-wave velocity

spatial distribution at different levels of confining pressure, and their comparison to experimentally derived P-wave velocities. The graphical comparison of calculated and measured P-wave velocity values produced almost identical pattern that reflects the validity of calculated data.

Comparison of fitting parameters and rock fabric

The spatial distribution of individual approximation parameters is primarily affected by the arrangement of rock fabric elements and by the whole-rock fabric symmetry.

In granites, both parameters V_0 and k_v show a non-symmetrical spatial distribution that coincides with generally assumed ideas on isotropic or quasi-isotropic symmetry of their mineral skeleton (e.g. Ramamurthy 1993). The V_{dif} parameter shows transversally isotropic (e.g. sample RP1) or orthorhombic symmetry (e.g. sample G6) with a maximum oriented in the direction of measured v_{Pmin}. Transversally isotropic distribution (Fig. 5) results from the presence of parallel systems of intragranular and grain-boundary microcracks that have probably formed due to the exfoliation in a parent granitoid pluton. The orthorhombic symmetry of the v_{dif} parameter probably results

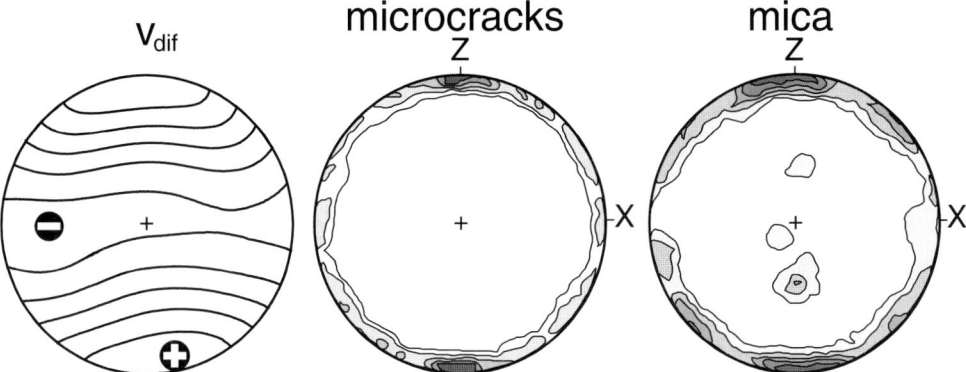

Fig. 5. Comparison of the spatial distribution of mathematically derived P-wave velocity difference v_{dif} fitting parameter of granite (sample RP1) to the microcrack and mica fabric. Those fabrics were determined by measurement with an optical microscope equipped with a U-stage. The preferred orientation of the poles of microcracks is contoured at 2.9, 5.8, 8.7, 11.6, 14.5 and 17.4%. The preferred orientations of the poles to the basal planes of the micas are contoured at 2.0, 4.0, 6.1, 8.1, 10.1 and 12.1%. All data are projected on to the lower hemisphere in an equal-area projection.

from the presence of several sets of microcracks formed due to the denudation of the granitoid body and by the acting regional stresses. The symmetry of P_0 is not so pronounced, but tends to show an orthorhombic pattern. Its maximum coincides with the direction of maximum increase of P-wave velocity. The symmetrical distribution of the last two parameters in quasi-isotropic rocks is thus strongly connected with a microcrack fabric that is rarely isotropic in crystalline rocks (e.g. Walsh 1993). The non-linearity of P-wave velocity–confining pressure behaviour of quasi-isotropic rocks like granites can therefore be explained in terms of progressive closure of high-aspect-ratio microcracks (Batzle *et al.* 1980; Wepfer & Christensen 1991).

In orthogneisses, the spatial distribution of all four fitting parameters is more likely to be affected by the presence of micas (about 20–25 vol. %) that show strong crystallographic preferred orientation. Micas exhibit the most pronounced elastic anisotropy (Babuška 1981) of all rock-forming minerals present in granitoid rocks. The well-ordered orthorhombic symmetry of the V_0 fitting parameter exhibits the position of the minimum in the direction of maximum distribution of poles to the (001) planes of micas (Fig. 6). The orientation of V_{Pmin} in the direction normal to the preferred arrangement of mica (001) planes was confirmed by numerous experimental studies on low- to high-grade tectonites (Kern & Wenk 1990; Burlini & Fountain 1993; Hrouda

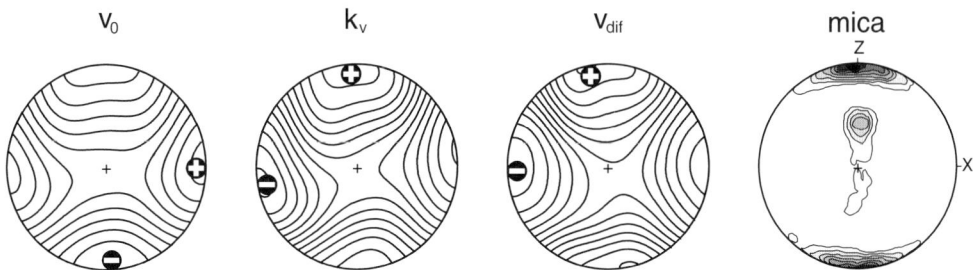

Fig. 6. Comparison between the anisotropic fabric of orthogneiss (here expressed by the preferred orientation of poles to mica (001) planes) and the orthorhombic symmetry of the spatial distribution of mathematically derived P-wave velocity fitting parameters. Mica fabric was determined by measurement using an optical microscope equipped with a U-stage. The preferred orientations of the poles to the basal planes of the micas are contoured at 2.7, 5.5, 8.2, 11.0, 13.7, 16.4 and 19.2%. All data are projected on to the lower hemisphere in equal-area projection.

et al. 1993; Ji *et al.* 1993; Johnston & Christensen 1995). The spatial distribution of the k_v parameter shows the same symmetry. Its maximum lies in the direction of maximum concentration of poles to the (001) planes of micas, i.e. in the direction of maximum compressibility of the sample. The same behaviour was found for the behaviour of the V_{dif} parameter. Its maximum coincides with minimum V_0 and maximum k_v. This means that the maximum increase of P-wave velocity occurs in the direction oriented perpendicularly to the planes of foliation that are defined by the alignment of platy micas, intragranular microcracks in (001) planes of micas, grain-boundary microcracks and grain-boundary bonds between minerals showing a strong shape-preferred orientation. This direction is evidently more susceptible to being compressed by the acting confining pressure than the direction of stretching lineation that is the direction of minimum V_{dif} (Fig. 6). The studied rock exhibits a lower crack density of cracks oriented perpendicular to the lineation than in the foliation plane. The coinciding lineation and V_{Pmax} direction represent the P-wave velocity extreme in anisotropic rocks (Johnson & Wenk 1974; Kern 1974, 1993; Kern & Fakhimi, 1975; Kern & Wenk 1990; Burlini & Fountain 1993; Hrouda *et al.* 1993; Ji *et al.* 1993). The symmetry of P_0 copies symmetry of previous fitting parameters. The minimum microcrack closing pressure coincides with the measured V_{Pmin} and V_0 minimum and the V_{dif} maximum, i.e. with the easiest compressible direction.

Problem of microcrack closing pressure

Although the V_{dif} and P_0 fitted parameters are mainly influenced by the microcrack content, the rocks studied show some significant differences in the behaviour of both parameters. In granites, the V_{dif} maximum coincides with the P_0 maximum. In orthogneisses, the V_{dif} maximum lies in the direction of the P_0 minimum. This can be caused by the diverse nature of microcracks present in both of the rock groups studied. In orthogneiss, the prevalent microcracks are of intragranular nature in micas that show more smoothed faces and therefore are easy compressible under lower confining pressures (Babuška & Pros 1984). In granites, on the other hand, the intragranular and grain-boundary microcracks show more uneven faces (e.g. Carlson & Gangi 1985) with fracture bridges, and are less susceptible to being compressed. As a result they stay open at higher confining pressure levels (Wepfer & Christensen 1991). A similar observation on the effect of uneven fracture faces on P-wave velocity–confining pressure behaviour was done,

for example, on granodiorites (Babuška & Pros 1984). The observed phenomena can be also influenced by the diverse nature and degree of sample anisotropy and/or the relationship of microcracks to the bulk rock fabric.

Another important problem of microcrack closing pressure P_0 determination arises from its significant dependence on v_{dif} parameter and, consequently, on microcrack density. Evidently, a higher velocity difference can seriously decrease the value of the microcrack closing pressure. This leads to the 'false' determination of microcrack closing pressure that can occur not directly in the passage from non-linear to linear P-wave velocity–confining pressure relationship, but deeply in the non-linear region. This was the reason for the application of a second microcrack closing pressure P_{100} that is fixed to the 100 m s^{-1} velocity difference and is calculated according to the following equation:

$$P_{100} = P_0 (\log V_{dif} - \log 100) \qquad (9)$$

This fitted parameter shows a greater scatter of values in different directions in one rock – as well as between rocks with varying densities of microcracks – than the P_0 microcrack closing pressure. This suggests that P_{100} reflects the real density of the microcracks. The P_0 fitting parameter, on the other hand, probably expresses the character of the microcracks, e.g. the quality of their contacts and the type of microcrack.

The microcrack closing pressures P_0 of the rocks being studied varied between 100 and 150 MPa. The wide range of the V_{dif} parameter significantly influenced the calculated value of microcrack closing pressure P_0. The lower values of P-wave velocity at atmospheric pressure, and the higher level of P-wave velocity in samples with closed microcracks, seriously decrease the value of the microcrack closing pressure, P_0.

The P_{100} microcrack closing pressure offers another view on the process of microcrack closure in rocks exhibiting a great variability in the influence of cracks at the atmospheric pressure level. The P_{100} microcrack closing pressure shows a greater scatter in comparison to the P_0 values. The microcrack closing pressures P_{100} drop down to about 50–80 MPa for rocks with low crack density. The increase in microcrack closing pressures P_{100} to the range 150–200 MPa was observed in densely cracked granites.

Conclusions

A proposed mathematical four-parameter function provides an approximation of the P-wave velocity dependence on the applied confining pressure. The individual parameters of the equation (1) reflect the physical behaviour of various rock fabric elements. The V_0 parameter is the P-wave velocity in a mineral skeleton without microcracks. The k_v parameter is the slope of the straight-line dependence of P-wave velocity on confining pressure. The steeper the slope, the greater the linear deformability of the material or some of its constituents. This parameter therefore reveals the most compliant direction of the mineral skeleton. Both of these parameters and their symmetry are affected only by the degree of elastic anisotropy of the minerals present and their preferred arrangement. The third parameter, V_{dif}, characterizes the crack influence at atmospheric pressure, and indicates the most deformable directions due to the presence of microcracks. The microcrack closing pressures P_0 and P_{100} are influenced by the nature and density of microcracks.

The non-linearity in P-wave velocity–confining pressure behaviour and spatial distribution of its fitting parameters are then interpreted in terms of the microcracks present in the case of quasi-isotropic rocks – granites, and the preferred orientation of highly anisotropic rock-forming minerals like micas in the case of anisotropic rocks – orthogneiss.

When plotted spatially, these fitting parameters show quasi-isotropic to anisotropic distribution. Their symmetries can be correlated with measured P-wave velocities on real rock samples. The symmetry and extremes of individual fitting parameters can be also related to the preferred orientation of individual rock fabric elements. The v_0 and k_v parameters depend on the fabric of a mineral skeleton without microcracks. They show isotropic to quasi-isotropic distribution in rocks that possess the quasi-isotropic fabric of the mineral skeleton (e.g. granites), but both parameters can develop anisotropic (e.g. orthorhombic) symmetry in rocks where minerals tend to show crystallographic and shape-preferred orientation (e.g. micas in orthogneiss).

The spatial distribution of v_{dif} and P_0, P_{100} fitting parameters depends on the presence and orientation of microcracks, but can be also affected by the preferred orientation of highly anisotropic minerals in some rocks (e.g. micas in orthogneiss). The minimum P-wave velocity difference occurs in the direction of maximum measured P-wave velocity that correlates with the direction of linear fabric in rocks if present. The maximum velocity difference corresponds to the direction normal to the structural bedding (foliation, dominant set of microcracks) and to the direction of minimum measured P-wave velocity. The spatial pattern of this fit is similar to the laboratory measured

P-wave velocity spatial distribution recorded at atmospheric pressure.

The closure of microcracks due to increasing confinement was simulated with the help of two fitting parameters P_0 and P_{100}. Firstly, the P_{100} microcrack closing pressure fit depends mainly on the density of microcracks. Secondly, P_0 fit presents the qualitatively proportional microcrack closing pressure for rocks with significantly different crack densities. This parameter expresses the more likely character of the microcracks.

This work is being supported by grants from the Grant Agency of the Czech Republic (Grant Nos 205/04/0088, 205/03/0071, 205/01/1430 and 103/01/1058). The support for this study was also provided by the project MSM 113100005 'Material and energy flows in the upper parts of the Earth'. The experimental work would not be possible without the help of L. Padjen, J. Vyhnal and J. Štís. The authors appreciated the comments of two anonymous reviewers, which greatly improved this paper.

References

BABUŠKA, V. 1981. Anisotropy of V_P and V_S in rock-forming minerals. *Journal of Geophysics*, **50**, 1–6.

BABUŠKA, V. & PROS, Z. 1984. Velocity anisotropy in granodiorite and quartzite due to the distribution of microcracks. *Geophysical Journal of the Royal Astronomical Society*, **76**, 121–127.

BATZLE, M. L., SIMMONS, G. & SIEGFRIED, R. W. 1980. Microcrack closure in rocks under stress: part I, direct observation. *Journal of Geophysical Research*, **85**, 7072–7090.

BIRCH, F. 1938. The effect of pressure upon the elastic parameters of isotropic solids, according to Murnaghan's theory of finite strain. *Journal of Applied Physics*, **9**, 279–288.

BIRCH, F. 1960. The velocity of compressional waves in rocks to 10 kilobars, part 1. *Journal of Geophysical Research*, **65**, 1083–1102.

BIRCH, F. 1961. The velocity of compressional waves in rocks to 10 kilobars, part 2. *Journal of Geophysical Research*, **66**, 2199–2224.

BOUCHEZ, J.-L., DELAS, C., GLEIZES, G., NÉDÉLEC, A. & CUNEY, M. 1992. Submagmatic microfractures in granites. *Geology*, **20**, 35–38.

BURLINI, L. & FOUNTAIN, D. M. 1993. Seismic anisotropy of metapelites from the Ivrea–Verbano zone and Serie dei Laghi (northern Italy). *Physics of the Earth and Planetary Interiors*, **78**, 301–317.

CARLSON, R. L. & GANGI, A. F. 1985. Effect of cracks on pressure dependence of P-wave velocities. *Journal of Geophysical Research*, **90**, 8675–8684.

CHRISTENSEN, N. I. 1974. Compressional wave velocities in possible mantle rocks to pressure of 30 kilobars. *Journal of Geophysical Research*, **79**, 407–412.

EBERHART-PHILLIPS, D., HAN, D. H. & ZOBACK, M. D. 1989. Empirical relationship among seismic velocity, effective pressure, porosity, and clay content in sandstone. *Geophysics*, **54**, 82–89.

GREENFIELD, R. J. & GRAHAM, E. K. 1996. Application of simple relation for describing wave velocity as a function of pressure in rocks containing microcracks. *Journal of Geophysical Research*, **101**, 5643–5652.

HROUDA, F., PROS, Z. & WOHLGEMUTH, J. 1993. Development of magnetic and elastic anisotropies in slates during progressive deformation. *Physics of the Earth and Planetary Interiors*, **77**, 251–265.

HUGHES, D. S., PONDROM, W. L. & MIMS, R. L. 1949. Transmission of electric pulses in metal rods. *Physical Reviews*, **75**, 1552–1556.

JI, S., SALISBURY, M. H. & HANMER, S. 1993. Petrofabric, P-wave anisotropy and seismic reflectivity of high grade tectonites. *Tectonophysics*, **222**, 195–226.

JOHNSON, L. R. & WENK, H.-R. 1974. Anisotropy of physical properties in metamorphic rocks. *Tectonophysics*, **23**, 79–98.

JOHNSTON, J. E. & CHRISTENSEN, N. I. 1995. Seismic anisotropy of shales. *Journal of Geophysical Research*, **100**, 5991–6003.

KERN, H. 1974. Gefügeregelung und elastische Anisotropie eines Marmors. *Contributions to Mineralogy and Petrology*, **73**, 47–54.

KERN, H. 1993. P- and S-wave anisotropy and shear-wave splitting at pressure and temperature in possible mantle rocks and their relation to the rock fabric. *Physics of Earth and Planetary Interiors*, **78**, 245–256.

KERN, H. & FAKHIMI, M. 1975. Effect of fabric anisotropy on compressional-wave propagation in various metamorphic rocks for the range 20–700 °C at 2 kbars. *Tectonophysics*, **28**, 227–244.

KERN, H. & WENK, H.-R. 1990. Fabric related velocity anisotropy and shear wave splitting in rocks from the Santa Rosa mylonite Zone, California. *Journal of Geophysical Research*, **95**, 11 213–11 223.

KLÍMA, K. 1973. The computation of the elastic constants of an anisotropic medium from the velocities of body waves. *Studia Geophysica et Geodetica*, **17**, 115–122.

KRANZ, R. L. 1983. Microcracks in rocks: a review. *Tectonophysics*, **100**, 449–480.

LAMA, R. D. & VUTUKURI, V. S. 1978. *Handbook on Mechanical Properties of Rocks. Vol. 4 – Testing Techniques and Results*. Trans Tech Publications, Clausthal, Germany.

MEGLIS, I. L., GREENFIELD, R. J., ENGELDER, T. & GRAHAM, E. K. 1996. Pressure dependence of velocity and attenuation and its relationship to crack closure in crystalline rocks. *Journal of Geophysical Research*, **101**, 17 523–17 533.

PROS, Z. & BABUŠKA, V. 1967. A method for investigating the elastic anisotropy on spherical rock samples. *Zeitschrift für Geophysik*, **33**, 289–291.

PROS, Z., VANĚK, J., KLÍMA, K. & BABUŠKA, V. 1969. Experimentelle Untersuchung des Wellenbildes bei der Ultraschall-Durchstrahlung einer Kugel. *Zeitschrift für Geophysik*, **35** (3), 287–296.

PROS, Z., LOKAJÍČEK, T. & KLÍMA, K. 1998a. Laboratory study of elastic anisotropy on rock samples. *Pure and Applied Geophysics*, **151**, 619–629.

PROS, Z., LOKAJÍČEK, T., PŘIKRYL, R., ŠPIČÁK, A., VAJDOVÁ, V. & KLÍMA, K. 1998b. Elastic parameters of West Bohemian granites under hydrostatic pressure. *Pure and Applied Geophysics*, **151**, 631–646.

PŘIKRYL, R. 2001. Some microstructural aspects of strength variation in rocks. *International Journal of Rock Mechanics and Mining Sciences and Geomechanical Abstracts*, **38**, 671–682.

PŘIKRYL, R., SCHULMANN, K. & MELKA, R. 1996. Perpendicular fabrics in the Orlické hory orthogneisses (western part of the Orlice–Sněžník dome, Bohemian massif) due to high temperature E–W deformational event and late lower temperature N–S overprint. *Journal of the Czech Geological Society*, **41**, 156–166.

PŠENČÍK, I. 1975. Continuous computer contouring. *Studia Geophysica et Geodetica*, **19**, 184–187.

RAMAMURTHY, T. 1993. Strength and modulus responses of anisotropic rocks. *In*: BROWN, E. T. (ed.) *Comprehensive Rock Engineering. Principles, Practice and Projects. Vol. 1. Fundamentals*. Pergamon Press, Oxford, 313–329.

SIEGESMUND, S. & DAHMS, M. 1994. Fabric-controlled anisotropy of elastic, magnetic and thermal properties of rocks. *In*: BUNGE, H. J., SIEGESMUND, S., SKROTZKI, W. & WEBER, K. (eds) *Textures of Geological Materials. Deutsche Gesellschaft für Materialkunde – Informationsgesellschaft –Verlag, Oberursel*, 353–379.

SIMMONS, G. & RICHTER, D. 1976. Microcracks in rocks. *In*: STRENS, R. G. J. (ed.) *The Physics and Chemistry of Minerals and Rocks*. John Wiley & Sons, London, 105–137.

STESKY, R.M. 1985. Compressional and shear velocities of dry and saturated jointed rock: a laboratory study. *Geophysical Journal of the Royal Astronomical Society*, **79**, 4383–4385.

THILL, R.E., WILLARD, R. J. & BUR, T. R. 1969. Correlation of longitudinal velocity variation with rock fabric. *Journal of Geophysical Research*, **74**, 4897–4909.

WALSH, J.B. 1993. The influence of microstructure on rock deformation. *In*: BROWN, E. T. (ed.) *Comprehensive Rock Engineering. Principles, Practice and Projects. Vol. 1. Fundamentals*. Pergamon Press, Oxford, 243–254.

WARREN, N. & TIERNAN, M. 1981. Systematics of crack controlled mechanical properties for a suite of Conway granites from the White Mountains, New Hampshire. *Tectonophysics*, **73**, 295–322.

WEPFER, W. W. & CHRISTENSEN, N. I. 1991. A seismic velocity-confining pressure relation, with applications. *International Journal of Rock Mechanics and Mining Science & Geomechanical Abstracts*, **28**, 451–456.

WINKLER, E. M. 1994. *Stone in Architecture. Properties, Durability*. Springer-Verlag, Berlin, 3rd edition, 313 pp.

ZANG, A., BERCKHEMER, H. & LIENERT, M. 1996. Crack closure pressures inferred from ultrasonic drill-core measurements to 8 km depth in the KTB wells. *Geophysical Journal International*, **124**, 657–674.

The ^{14}C-polymethylmethacrylate (PMMA) impregnation method and image analysis as a tool for porosity characterization of rock-forming minerals

E. OILA[1], P. SARDINI[2], M. SIITARI-KAUPPI[1] & K.-H. HELLMUTH[3]

[1]Laboratory of Radiochemistry, Department of Chemistry, PO Box 55, 00014,
University of Helsinki, Finland

[2]UMR 6532 CNRS HYDRASA, University of Poitiers, 40 Avenue du
Recteur Pineau, 86022 Poitiers, France
(e-mail: paul.sardini@hydrasa.univ-poitiers.fr)

[3]Radiation and Nuclear Safety Authority, PO Box 14, Helsinki, Finland

Abstract: Accurate knowledge of porosity is essential for understanding the links between basic petrophysical parameters, such as diffusion coefficients, permeability and conductivity. Standard methods used to determine porosity quantify the bulk porosity and the distribution of pore sizes. Crystalline rocks are rarely monomineralic, and the porosity of polyphasic rocks is considered heterogeneous at the mineral grain scale. Calculation of bulk petrophysical parameters must take into account porosity and mineral-phase microstructures, as well as connectivity. The polymethylmethacrylate (PMMA) method uses radioactively (^{14}C)-labelled methylmethacrylate (^{14}C-MMA) liquid to impregnate the rock sample, which is then polymerized by irradiation, cut and autoradiographed. Porosity is quantified by digitizing the autoradiograph and subsequent densitometry. Staining of the same rock surface uses chemical agents that rapidly reveal the primary minerals of unaltered and altered crystalline rocks: mainly quartz, K-feldspar, plagioclase and dark minerals. The images of PMMA autoradiographs and stained rock surfaces are combined to quantify mineral-specific porosities.

The methodology has been applied here to a granite core from Palmottu (central Finland) representing coarse-grained granite adjacent to a potential water-conducting fracture. Imaging of the porosity relative to mineralogy is presented and complemented by mineral specific porosities.

Many geological sites in different countries have been evaluated extensively as potential locations for nuclear waste disposal. Their geological settings differ in that the impermeable formations are either crystalline rocks, clay formations or salt domains (Gascoyne et al. 1995). Over geologically long time periods, radioactive substances may be released from a nuclear waste repository and percolate along permeable fault zones. Radio-element migration within a rock matrix under natural long-term conditions is a complex process controlled by various parameters. Radio-elements are retarded by diffusion into the surrounding rock (Neretnieks 1980). Diffusion is influenced by physical properties of the rock, such as pore size distribution, connectivity, tortuosity, constrictivity, and the petrological and chemical nature and ionic charge of the rock pore surfaces (Katsube & Kamineni 1983; Katz & Thompson 1987; Clennell 1997). The degree of structural heterogeneity in the natural rock matrix and its various mineral phase compositions will determine its influence on radionuclide retardation. Therefore, understanding the retardation behaviour of rocks requires a robust research methodology.

The aim of this work was to test how the porosity of a granite rock could be quantitatively linked to mineralogical and structural variations. Firstly, autoradiography was used to visualize the granite as a porous medium, followed by quantitative evaluation of the porosity distribution using image analysis tools. Secondly, a mineral staining procedure was used to obtain a mineral map of the rock surface at the same scale as the autoradiograph. The superimposition of images from these two techniques enabled assessment of the mean porosity and porosity distribution for all the major rock-forming minerals in the studied rock.

From: HARVEY, P. K., BREWER, T. S., PEZARD, P. A. & PETROV, V. A. (eds) 2005. *Petrophysical Properties of Crystalline Rocks.* Geological Society, London, Special Publications, **240**, 335–342.
0305-8719/05/$15.00   © The Geological Society of London 2005.

Materials and method

A 42 × 150 mm granite rock core from borehole R384 in the Palmottu Natural Analogue Site, (central Finland) adjacent to a water-conducting fracture at a depth of 74 m, was used in this study. The existence of naturally occurring uranium has been intensively studied at Palmottu (Blomqvist *et al.* 1998, 2000). The mineralogical characterization of water-conducting fractures in borehole R384 has been reported by Ruskeeniemi *et al.* (1998). The main petrographic features of our sample are summarized in Table 1. The average grain size of the major K-feldspar phenocrysts forming the rock is a few centimetres. From optical microscopy the intragranular, cleavage and grain-boundary micro cracks were found mainly in quartz, K-feldspar and plagioclase. Porous zones with large internal surface areas were observed in hematized plagioclase and chloritized biotite (Siitari-Kauppi *et al.* 1999).

^{14}C-PMMA impregnation method

The ^{14}C-PMMA method developed by Hellmuth *et al.* (1993, 1994), uses a vacuum to impregnate decimetre-scale rock cores with ^{14}C-MMA. The subsequent procedures are irradiation polymerization, autoradiography and optical densitometry by digital image processing.

The ^{14}C-MMA, which wets the silicate surfaces well and can be fixed by polymerization, provides unique information on the accessible pore space in crystalline rock. MMA is a monomer with very low viscosity, 0.00584 Pa s (20°C), when compared to water, 0.01005 Pa s (20°C). Because the contact angle of MMA on silicate surfaces is lower than for water, impregnation of bulk rock specimens by capillary forces is rapid. The MMA molecule is small (molecular weight 100.1), it is non-electrolytic and it has low polarity with respect to water. The low β-energy of carbon-14 (maximum 155 keV) is optimal for autoradiographic measurements.

The initial ^{14}C-MMA concentration used in this study was 925 kBq ml^{-1}. The rock sample was initially dried at 110°C for two weeks in a vacuum chamber, and then cooled to 18°C. The ^{14}C-MMA was put into a 50-ml reservoir and transported under vacuum to the chamber for an impregnation time of two weeks.

The impregnated sample was irradiated with γ-rays from a ^{60}Co source to polymerize the monomer in the rock matrix; the total dose was 50 kGy. During irradiation, the temperature increased slightly in the system. However, this was minimized by cooling the system with MMA-saturated water surrounding the sample.

After the irradiation, the sample was heated at 120°C for three hours to eliminate irradiation-induced luminescence of the feldspar minerals. Luminescence emissions would have exposed the autoradiographic film and so masked the detection of β-radiation originating from ^{14}C-PMMA.

The core was cut in half along its long axis, the planar sawn surfaces were polished with aluminium oxide powder (400 meshes) and were then cleaned ultrasonically for two minutes. In a dark room the polished surfaces were placed on an autoradiographic film (Kodak® Biomax), shielded from luminescence emissions by an aluminium-coated Mylar foil. Exposure times depend on tracer activity and on the mean sample porosity; the exposure time for this study was seven days. The ^{14}C-PMMA standards with 370–462 kBq ml^{-1} activity concentrations were used to establish the calibration function for quantitative analysis.

The autoradiographs were digitized with a table scanner (Ricoh FS2) using an 8-bit grey-level mode to produce images for quantitative analysis. In this work, 400-dpi resolution was used (a pixel is 63.5 × 63.5 μm²). The photo-image of the stained rock surface (see below) and the corresponding autoradiograph are illustrated in Figure 1. Different shades of grey on the autoradiograph represent different porosities: the darker the shade, the higher the porosity.

Table 1. *Mineral content and texture of R384 74 m rock sample*

Primary mineral	Mineral content (vol. %)	Secondary phases	Deformation
Quartz	30	—	Grain-boundary pores and intragranular microcracks
K-feldspar	35	Not observed in the studied sample	Open cleavage and intragranular microcracks
Plagioclase	30	Hematite	Intragranular microcracks
Biotite	5	Chlorite	Open cleavages

The main petrographic features of the cored sample are typical for Palmottu granite. However, the compositions of primary minerals and the presence and nature of secondary phases such as uranium and calcite can vary greatly from one Palmottu sample to another.

(a) (b)

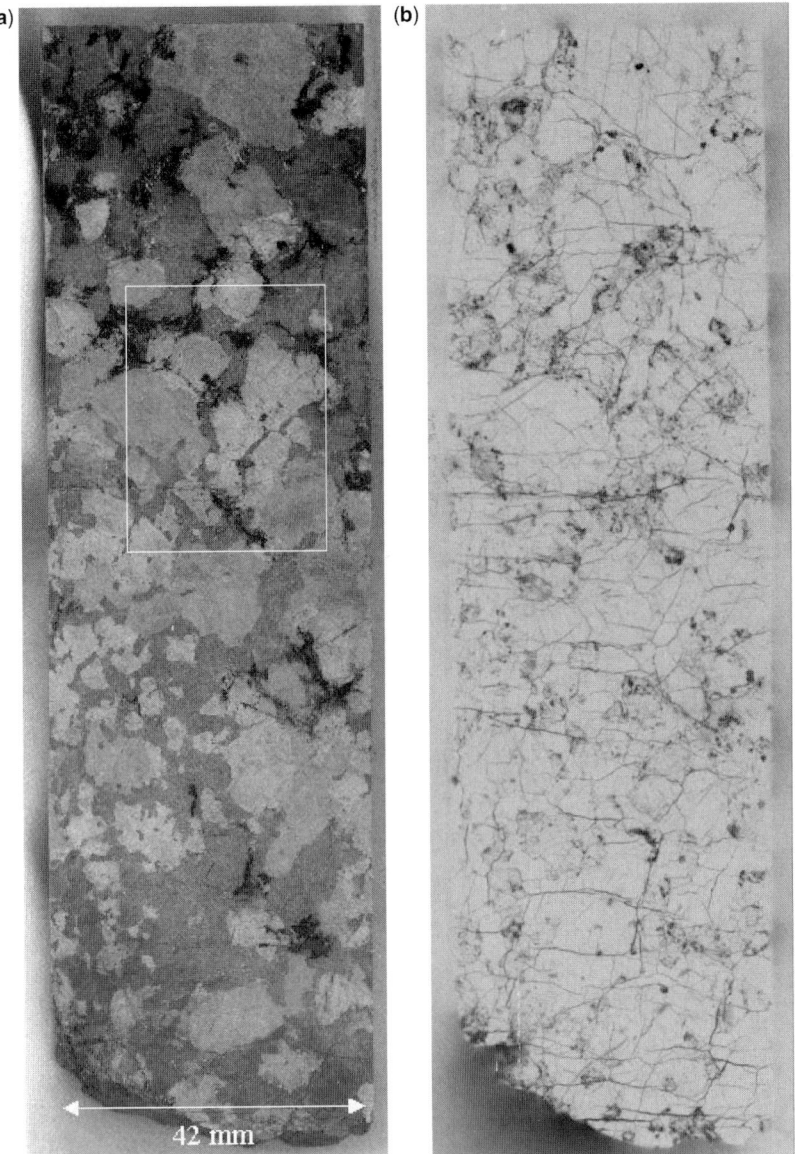

42 mm

Fig. 1. Photo-image of impregnated and stained R384 74 m rock sample (**a**) and the corresponding autoradiograph (**b**). The area used for digital image analyses is indicated on the photo-image (white square). The ^{14}C-PMMA method permits the study of the spatial distribution of pore spaces and heterogeneities in rock matrices, from micrometre to decimetre scales.

Porosity calculation from autoradiography

The grey levels of pixels on the autoradiographic film were converted into optical densities then into radio activities with the help of the calibration curve from [14]C- PMMA standards, and finally into porosities. Since the linear response of the scanner has been verified, the digitized grey levels of the pixels can be treated as intensities. Optical densities defined by Lambert are derived from intensities (grey-level values):

$$OD = -\log_{10}(I/I_0) \qquad (1)$$

where OD is the optical density here, I_0 is the intensity of the background and I is the pixel intensity. A calibration function was needed to convert each measured optical density into radio-activity (Keller & Waser 1982). The calibration curve was constructed by equating the radio-activities of standards to their corresponding optical densities, using an inversion calculation based on the least-squares method and determining a set of coefficients (a,k,c). For a given exposure time, the mathematical function describing the non-linear behaviour of the local radioactivity versus optical density of the film has the form:

$$OD = a(1 - e^{-kA}) + c \qquad (2)$$

where OD is the optical density and A is the radioactivity of the source. The radioactivity for each pixel was calculated using the relation:

$$A = -k^{-1} \ln[1 - ((OD - c)/a)] \qquad (3)$$

If the radioactivity of one pixel of the sample is A, and the radioactivity of the tracer used to impregnate the sample is A_0, the porosity ε can be calculated by

$$\varepsilon(\%) = \beta(A/A_0) \times 100\% \qquad (4)$$

where β is the beta-absorption correction factor, which corrects the difference in β-emissions from the PMMA and the sample. The absorption of β-radiation in a matrix is roughly linearly dependent on the density of the matrix (Tingle 1987). Therefore β can be approximated by:

$$\beta = \frac{\rho_s}{\rho_0} \qquad (5)$$

where ρ_s is the density of the sample and ρ_0 is the density of pure PMMA (1.18 g cm^{-3}). Our sample was assumed to consist of rock material and pores (containing PMMA), and therefore ρ_s could be expressed as:

$$\rho_s = \varepsilon \rho_0 + (1 - \varepsilon)\rho_r \qquad (6)$$

where ρ_r is the density of mineral grains. In this work, the average density values used for each mineral were taken from Landholt-Börnstein (1982). The porosity as a function of radioactivity can be solved from equations 4–6 (Hellmuth *et al.* 1994; Sammartino *et al.* 2002):

$$\varepsilon = \frac{\rho_r/\rho_0}{1 + (\rho_r/\rho_0 - 1)A/A_0} \cdot \frac{A}{A_0} \qquad (7)$$

Plotting a histogram of porosities calculated using equation (7) gave the relative frequency for individual porosity regions. The porosity of the whole measured area was obtained from the porosity distribution by taking a weighted average:

$$\varepsilon_{tot} = \frac{\sum_n \text{Area}_n \varepsilon_n}{\sum_n \text{Area}_n} \qquad (8)$$

where Area_n was the area of pixel n, and ε_n the local porosity corresponding to pixel n.

The blackening of the autoradiographic film caused by the radiation emitted from the rock surface corresponded to the amount of ^{14}C-MMA tracer that intruded into the rock. The major fraction of the emitted β-radiation was attenuated by silicates. The tracer was considered as diluted by the silicate.

Mineral staining

After autoradiography, the primary minerals were identified from the rock surface, using hydrofluoric acid etching and two coloration agents, K-ferrocyanide and Na-cobaltnitrite (Müller 1967). The K-ferrocyanide staining in combination with hydrofluoric acid etching caused a clear contrast between quartz, feldspars, and dark minerals. The hydrofluoric acid etching with Na-cobaltnitrite staining caused a clear contrast between K-feldspar and plagioclase. Sardini *et al.* (1999) have shown that these staining techniques can be used to discriminate between the primary mineral species in crystalline rocks, except muscovite which has yet to be tested.

The sample was first immersed for eight hours in a solution of K-ferrocyanide, to stain the ferro-magnesian minerals blue. Biotite grains appeared dark blue, and hematitized plagioclase grains were stained a light-blue colour because of the high iron content in the grains. The coloured surface was etched for one minute in hydrofluoric acid. After the immersion with K-ferrocyanide solution, the sample surface was treated with a dilute solution of HCl, and then washed with water. The surface was then immersed for two minutes in hydrofluoric acid, washed with distilled water, and dried at room temperature for one hour. After drying, the sample was placed horizontally in a container. Some 40 ml of Na-cobaltnitrite solution were poured uniformly on the sample surface and left for three minutes. The surface was carefully washed with water and dried. After this step, K-feldspar mineral grains were bright yellow and plagioclase grains were white. The Na-cobaltnitrite staining did not affect quartz or ferromagnesian minerals that had already been stained, such as biotite. The stained rock surface was digitized in 24-bit RGB mode, with the same resolution and orientation as the autoradiograph.

Thresholding of minerals
and superimposition

The four colours representing different minerals were thresholded simultaneously. Existing superimposition and colour thresholding programs (Sardini *et al.* 1999, 2001) for an SGI workstation were converted to a PC-environment, using the MatLab®, Image Processing Toolbox, and the developed porosity calculation program (Mankeli 2.0). The scanned autoradiograph and the thresholded mineral map were then precisely superimposed, and a region of interest (3.5 × 4.5 cm²) of the sample was selected for mineral-specific porosity calculations.

Results and discussion

From the 2D autoradiograph of the sawn surface (Fig. 1b), a well-developed connective pore network consisting of intragranular microfissures and porous patches was observable. Figure 2 shows the final mineral map obtained by thresholding a 24-bit image from the stained rock surface. The connected porosity of a 15 × 4.2 cm² sample area was determined as 0.7%. The porosity values measured by water gravimetry for the same Palmottu granite, but containing a lower amount of altered plagioclase, were 0.4–0.6% (Siitari-Kauppi *et al.* 1999).

Figure 3 presents magnifications of porosity patterns observed from the autoradiographs; they were obtained by superimposing mineral maps on to a porosity map. Differences between the porosity patterns of the primary minerals were clearly evident, and so the spatial distribution of porosity in the rock matrix was dependent on mineralogy.

Figure 4 shows the porosity histograms of the main minerals represented on a log-linear scale. The average values of the mineral-specific porosities were 0.48% for potassium feldspar, 0.56% for quartz, 1.14% for plagioclase and 1.19% for dark minerals (= biotite). Porous mineral grains of plagioclase contained both solution porosity and intragranular fissures. The porosity of biotite grains consisted mainly of porous patches. In many geological contexts, biotite is often described as the most altered phase, because of:

(1) its high internal surface area developed along numerous (001) cleavage planes, and

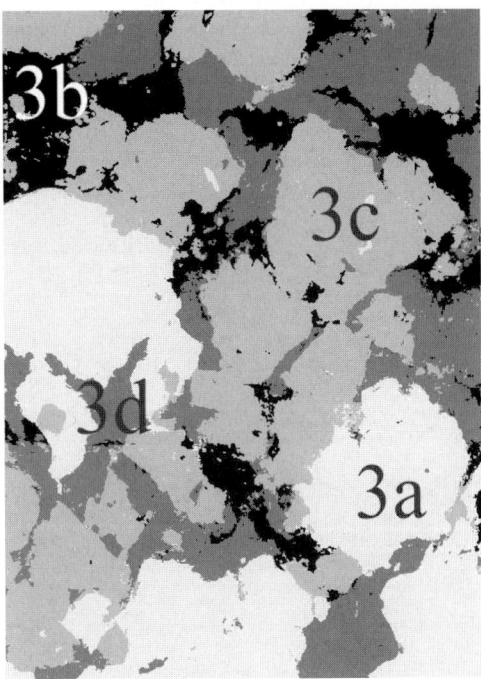

Fig. 2. Thresholded mineral map (3.5 × 4.5 cm²) after staining and thresholding. White corresponds to K-feldspar, light grey to plagioclase, dark grey to quartz, and black to dark minerals. 3a–d indicate locations of Figures 3a–d.

(2) its important geochemical reactivity against alteration processes (Parneix *et al.* 1985; White *et al.* 2001).

We can also note that the porosity histograms of the two most porous phases were asymmetrical toward the high porosity range. Potassium-feldspar and quartz grains showed mainly intragranular fissuring, and their porosity histograms present a symmetrical shape, thus revealing a log-normal distribution of porosity.

Autoradiographic resolution depends strongly on the range of ^{14}C β-radiation (Siitari-Kauppi *et al.* 1998). The features shown on the autoradiograph represent β-particles from the sample absorbed by the film. Figure 5 shows the backscattered electron image (BSE) of one region of the sample and the corresponding area on the autoradiograph, visualized through an optical microscope. Well-separated microfissures could be detected easily on the autoradiograph, but fissures less than 100 μm apart could not be differentiated. The mineral map obtained from thresholding the stained sample surface was compared to the mineral distribution seen

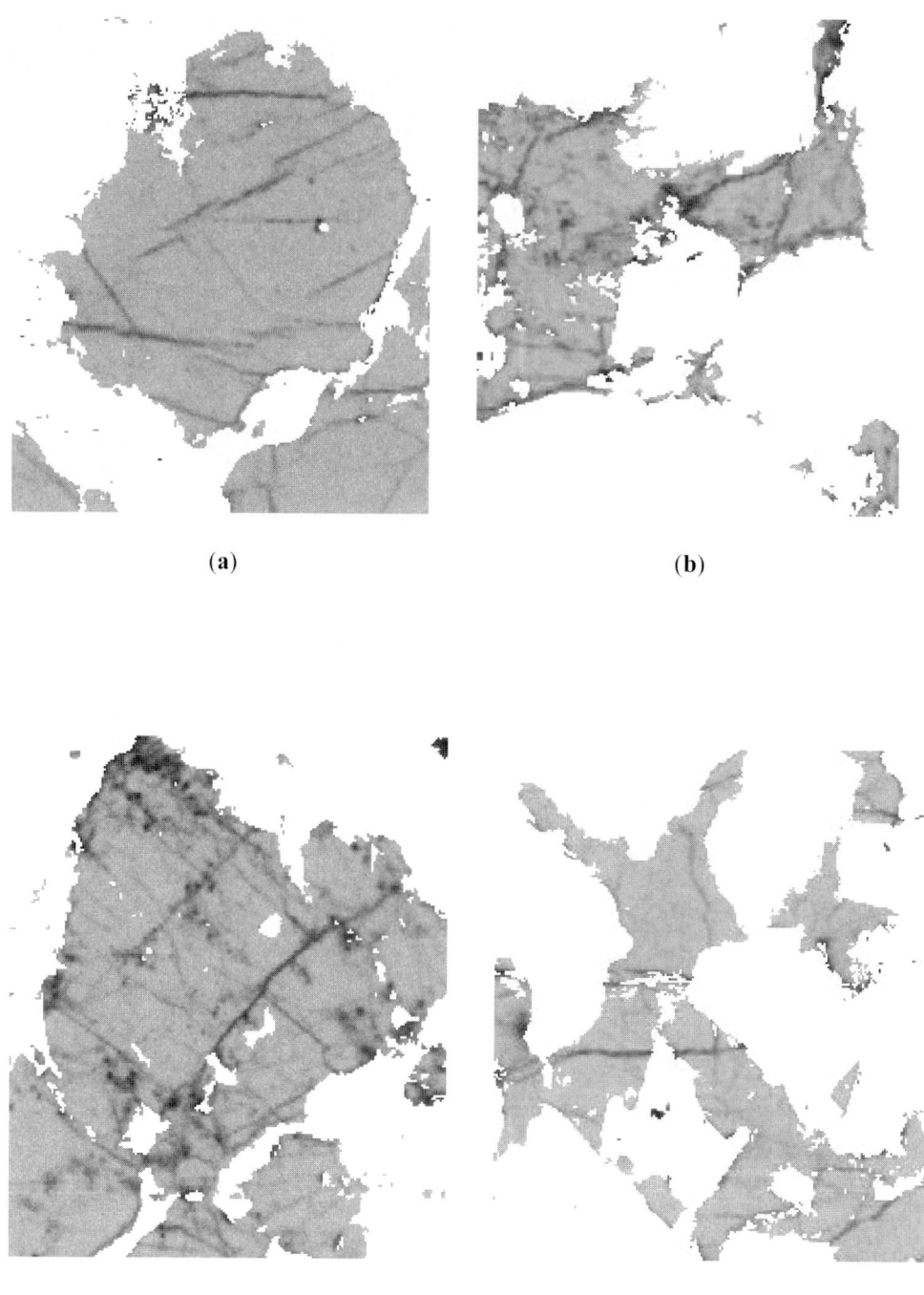

(a)   (b)

(c)   (d)

Fig. 3. Magnifications from an autoradiograph presenting individual mineral grains of K-feldspar (**a**), dark minerals (**b**), plagioclase (**c**) and quartz (**d**). The locations are numbered in Figure 2. Quartz and K-feldspar are clearly less porous than biotite and plagioclase.

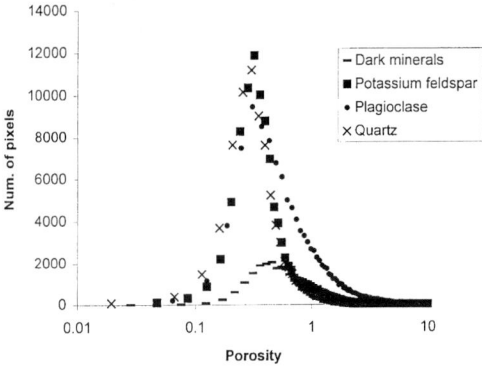

Fig. 4. Porosity histograms of primary minerals in the R384 74 m rock sample. Note the symmetrical shapes of the K-feldspar and quartz porosity histograms compared to the asymmetrical shapes of dark minerals and plagioclase porosity histograms.

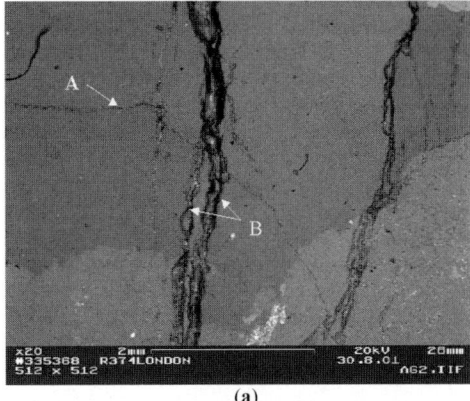

(a)

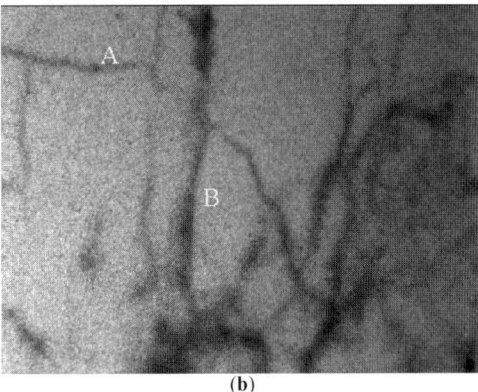

(b)

Fig. 5. The backscattered electron image (**a**) and the corresponding image of the autoradiograph viewed through an optical microscope (**b**) of the same area. Same scale in both images; codes A and B indicate a microfissure and fractures in the BSE image, as well as the corresponding features on the autoradiograph. Microfissures observed from BSE images are also detected on the autoradiograph, if they are sufficiently separated.

using scanning electron microscopy (SEM). The result of thresholding the mineral maps at the centimetre scale was comparable to SEM observations, but the accuracy of thresholding is improved when using SEM. The linking of PMMA autoradiographs to stained mineral maps functioned well in this type of rock, because small mineral grains were rare.

Conclusions

Full petrophysical evaluation of the host rock with regard to the construction of a deep geological nuclear-waste repository requires the use of a variety of complementary methods. Quantitative autoradiography employing ^{14}C-labelled polymethylmethacrylate supplemented the number of digital image analyses, and yielded valuable petrophysical information. Superimposition of porosity and mineral maps produced two main results:

(1) a realistic quantitative mineral-specific porosities, and
(2) a centimetre-scale view of the connective porous network that governs petrophysical properties in a low-permeability granite.

The present study focused on determining the porosity distribution relative to mineralogy, but the data analysis could also be used as a link between petrophysical properties such as diffusion and petrographic characterization (Sardini et al. 2003). Two necessary improvements in the method would be to quantify the porosity of boundaries between different mineral-species,

and also to analyse in detail the shapes of the mineral specific porosity histograms.

This study was financially supported by the Finnish Centre for Radiation and Nuclear Safety Authority (STUK).

References

BLOMQVIST, R., KAIJA, J. et al. 1998. The Palmottu Natural Analogue Project – Hydrogeological Evaluation of the Site. European Commission, Nuclear Science and Technology Series, Luxemburg, EUR 18202 EN.

BLOMQVIST, R., RUSKEENIEMI, T. et al. 2000. The Palmottu Natural Analogue Project – Transport of Radionuclides in Natural Flow System at Palmottu. European Commission, Nuclear

Science and Technology Series, Luxemburg, EUR 19611 EN.

CLENNEL, M. B. 1997. Tortuosity: a guide through the maze. *In*: LOVELL, M. A. & HARVEY, P. K. (eds) *Developments in Petrophysics*. Geological Society, London, Special Publications, **122**, 299–344.

GASCOYNE, M., STROES-GASCOYNE, S. & SARGENT, F. P. 1995. Geochemical influences on the design, construction and operation of a nuclear waste vault. *Applied Geochemistry*, **10**, 657–671.

HELLMUTH, K. H., LUKKARINEN, S. & SIITARI-KAUPPI, M. 1994. Rock matrix studies with carbon-14-polymethylmethacrylate (PMMA); been method development and applications *Isotopen-praxis Environmental Health Study*, **30**, 47–60.

HELLMUTH, K.-H., SIITARI-KAUPPI, M. & LINDBERG, A. 1993. Study of porosity and migration pathways in crystalline rocks by impregnation with ^{14}C-polymethylmethacrylate. *Journal of Contaminant Hydrology*, **13**, 403–418.

KATSUBE, T. I. & KAMINENI, D. C. 1983. Effect of alteration on pore structure of crystalline rocks: core samples from Atikokan, Ontario. *Canadian Mineralogist*, **21**, 637–646.

KATZ, A. J. & THOMPSON, A. H. 1987. Prediction of rock electrical conductivity from mercury injection measurements. *Journal of Geophysical Research*, **92**, 599–607.

KELLER, F. & WASER, P. G. 1982. Quantification in macroscopic autoradiography with carbon-14 – An evaluation of the method. *International Journal of Applied Radioactive Isotopes*, **33**, 1427–1432.

LANDOLT-BÖRNSTEIN, 1982. Numerical data and functional relationships in science and technology. *In*: AUGENHEISTER, A. (ed.) *Physical Properties of Rocks*, Vol. 1, Springer-Verlag, New York.

MÜLLER, G. 1967. Methods in sedimentary petrology *In*: ENGELHARDT, W. V., FUCHBAUER, H. & MÜLLER G. (eds) *Sedimentary Petrology*. Schweitzerbart, Stuttgart, 163–167.

NERETNIEKS, I. 1980. Diffusion in the rock matrix: an important factor in radionuclide migration? *Journal of Geophysics*, **85**, 4379–4397.

PARNEIX, J. C., BEAUFORT, P., DUDOIGNON, P. & MEUNIER, A. 1985. Biotite chloritization process in hydrothermally altered granite. *Chemical Geology*, **51**, 89–101.

RUSKEENIEMI, T., NISSINEN, P. & LINDBERG, A. 1998. *Mineralogical characterisation of major water-conducting fractures in boreholes R302,* R318, R332, R335, R373, R384, R388, R389 and R390. The Palmottu Natural Analogue Project, Technical Report, **98–07**. Geological Survey of Finland, Espoo.

SAMMARTINO, S., SIITARI-KAUPPI, M., MEUNIER, A., SARDINI, P., BOUCHET, A., & TEVISSEN, E. 2002. An imaging method for the porosity of sedimentary rocks: adjustment of the PMMA method: example of a characterization of a calcareous shale. *Journal of Sedimentary Research*, **72**, 937–943.

SARDINI, P., DELAY, F., HELLMUTH, K.-H., POREL, G. & OILA, E. 2003. Interpretation of out-diffusion experiments on crystalline rocks using random walk modelling. *Journal of Contaminant Hydrology*, **61**, 339–350.

SARDINI, P., MOREAU, E., SAMMARTINO, S. & TOUCHARD, G. 1999. Primary mineral connectivity of polyphasic igneous rocks by high-quality digitisation and 2D image analysis. *Computers & Geosciences*, **25**, 599–608.

SARDINI, P., MEUNIER, A. & SIITARI-KAUPPI, M. 2001. Porosity distribution of minerals forming crystalline rocks. *In*: CIDU, R. (ed.) *Water–Rock Interaction*, **10**, Balkema, Lisse, The Netherlands, 1375–1378.

SIITARI-KAUPPI, M., FLITSIYAN, E. S., KLOBES, P., MEYER, K. & HELLMUTH, K.-H. 1998. Progress in physical rock matrix characterization: structure of the pore space. Scientific basis for nuclear waste management XXI. *In*: MCKINLEY, I. G. & MCCOMBIE, C. (eds) *Material Research Society Symposium*, **506**, Warrendale, PA, 671–678.

SIITARI-KAUPPI, M., MARCOS, N., KLOBES, P., GOEBBELS, J., TIMONEN, J. & HELLMUTH, K.-H. 1999. Physical rock matrix characterization. *The Palmottu Natural Analogue Project, Technical Report*, **99–12**. Geological Survey of Finland, Espoo.

TINGLE, T. N. 1987. An evaluation of carbon-14 beta track technique: implications for solubilities and partition coefficients determined by beta track mapping. *Geochimica et Cosmochimica Acta*, **51**, 2479–2487.

WHITE, A. F., BULLEN, T. D., SCHULTZ, M. S., BLUM, A. E., HUNTINGTON, T. G. & PETERS, N. E. 2001. Differential rates of feldspar weathering in granitic regoliths. *Geochimica et Cosmochimica Acta*, **65**, 847–869.

Index

Page numbers in italic, e.g. *294*, refer to figures. Page numbers in bold, e.g. **291**, signify entries in tables.

From: HARVEY, P. K., BREWER, T. S., PEZARD, P. A. & PETROV, V. A. (eds) 2005. *Petrophysical Properties of Crystalline Rocks*. Geological Society, London, Special Publications, **240**, 343–351.
0305-8719/05/$15.00 © The Geological Society of London 2005.